Introduction
to
Plant   Biotechnology
Second Edition

Introduction
to
Plant Biotechnology
Second Edition

H.S. Chawla
Genetics & Plant Breeding Department
G.B. Pant University of Agriculture & Technology
Pantnagar, India

Science Publishers, Inc.
Enfield (NH), USA   Plymouth, UK

CIP data will be provided on request

SCIENCE PUBLISHERS, Inc.
Post Office Box 699
Enfield, New Hampshire 03748
United States of America

Internet site: *http://www.scipub.net*

sales@scipub.net (marketing department)
editor@scipub.net (editorial department)
info@scipub.net (for all other enquiries)

ISBN 1-57808-228-5

Published by Science Publishers, Inc., Enfield, NH, USA
Printed in India

Preface to the Second Edition

Plant biotechnology has emerged as an exciting area of plant sciences as it can create unprecedented opportunities for the manipulation of biological systems. No branch of biology has developed at such a fast pace as biotechnology has. The scope of this textbook has been broadened in this revised edition keeping in view the new developments in the field of biotechnology, but the basic philosophy remains the same. Protocols are given in those chapters for which laboratory exercises are essential at the undergraduate level. Emphasis is placed on application of a technique, its contribution and impact on crop improvement.

Courses on biotechnology are offered at various levels of undergrad and grad. In the sections relating to plant tissue culture and organization of genetic material the contents have been suitably amended, wherever needed, but the basic aspects remain the same except for few additions. Plant tissue culture courses usually include history, laboratory organization, nutrition media, micropropagation, organ culture, cell suspension culture, haploid culture, somatic fusion, secondary metabolite production, somaclonal variation and cryopreservation. For a good understanding of recombinant DNA technology, organization of DNA in the genome has been expanded and a chapter on basic techniques involved in recombinant DNA technology has been added. Since understanding of plant tissue culture at the undergraduate level has become very critical for realizing the full potential of biotechnology, in almost all the chapters protocols have been given.

In the revised edition recombinant DNA technology part has been broadened to a large extent. Gene cloning is an important aspect in biotechnology and this aspect has been split into three chapters. A chapter on isolating plant genes has been added to give an overview of different gene isolation procedures at an introductory level. Since transposons are used in isolating plant genes, and, is one of the major approaches for tagging of genes, a separate chapter on transposons and gene tagging has been added. A chapter on *in vitro* mutagenesis has been introduced. This will enable students to learn how changes can be introduced in DNA, selecting them in the absence of phenotypic expression and studying the effects of these deliberate alterations in the DNA either *in vitro* or after reintroudction *in vivo*. Genomics and bioinformatics have been introduced in this edition. In these chapters both functional and structural genomics, proteomics, sequencing status of different organisms and DNA chip technology have been discussed.

Chapters on gene transfer methods and PCR have been suitably amended wherever some lacunae were there. Plant breeders strive to meet the challenge of increasing production by developing plants with higher yield, resistance to biotic and abiotic stresses, and for quality characteristics. The chapter on transgenics in crop improvement has been updated and I have tried to give suitable examples of transgenics developed for various characteristics. Also an overview of the impact of biotechnology on crop improvement in context of transgenics has been given.

The molecular markers and marker-assisted selection chapter has been expanded to give more information on some of the new molecular markers and on genetic fingerprinting. The scope of chapter on intellectual property rights has been broadened by giving basic information on various forms of patenting, plant breeder's rights, biodiversity and some examples of patents, all being subjects which have generated debate.

The work was completed due to the grace of almighty God. I am highly indebted to God that He has given me strength and encouragement to complete this work and hope his gracious blessings will continue to be showered on me so that I can continue to improve this book.

I am highly indebted to reviewers of first edition of this book; that helped me immensely to improve this book. I am thankful to my well wishers, colleagues and innumerable students who have made great contributions directly or indirectly by giving suggestions for improvement in one way or the other for this revised edition. I would appreciate receiving your suggestions and criticism, covering any aspect relevant to this book.

I am thankful to my elders whose continuous support and inspirations led me to complete this work. I am fortunate in having a family which understands the preoccupation that goes with such projects. I am grateful to my wife and to my children, Komaljit and Jasmit, for their continued support and help.

March, 2002   H.S. Chawla

Preface to the First Edition

Plant biotechnology has emerged as an exciting area of plant sciences by creating unprecedented opportunities for the manipulation of biological systems. We are seeing the genes and genomes of a wide range of different organisms being manipulated by the use of new techniques for the benefit of man. One of the key techniques in genetic engineering is gene transfer, which encompasses a variety of methods for returning cloned genes to cells and to generate transgenic plants. Cell and tissue culture are the innovative breeding techniques applied to meet the increasing need for improved crop varieties. Tissue culture techniques can shorten the time and can lessen the labor and space requirements needed to produce a new variety. Cell and plant tissue culture and recombinant DNA technology constitutes an important aspect of plant biotechnology. Further, to understand gene technology it has become essential to understand the basic structure of gene and its organization in plant cell.

Courses on biotechnology are offered at various levels of undergraduate and graduate studies in various departments of Botany, Genetics, Plant Breeding, Horticulture, Plant Pathology, Entomology, Plant Science, Biotechnology and Bioscience. A good understanding of genetic engineering and plant tissue culture at the undergraduate level has become very critical for realizing the full potential of biotechnology. There are a number of books which deal with specialized aspects of plant tissue culture, cloning of genes and genome organization, but, I expect it will be useful to have a book describing the basic aspects of gene and genome organization in plant cells, basic tissue culture techniques and the fundamentals of cloning, gene transfer approaches and molecular markers. Basic tissue culture courses usually include history of the technique, laboratory organization, nutrition media, micropropagation, organ culture, cell suspension culture, anther culture, somatic fusion, secondary metabolite production and cryopreservation. I have given information on variability generated by tissue culture as somaclonal variation in one of the chapters. Gene cloning, gene transfer techniques, genome mapping and molecular markers have been described in relation to plants. A chapter on intellectual property rights has been included to give basic information on various aspects of patenting, copyright and plant breeders right.

Plant breeders are striving to meet the challenge of increased production by developing plants with higher yield, resistance to pests, diseases and weeds and tolerance to various abiotic stresses. I have tried to give suitable examples of transgenics developed for various characters in one of the chapters so that a student is aware of the impact of biotechnology on crop improvement. In most of the chapters protocols for conducting laboratory exercises have been given. A very important point is that in most chapters emphasis is placed on application of a technique and its contribution and impact on crop improvement.

I would very much appreciate receiving your suggestions, criticism and research contributions (as reprints) which relate to different aspects relevant to this book. It will be most helpful during the penetration of a revised edition. Please mention the errors you find with page numbers and describe mistakes. I'll highly appreciate your assistance in this regard.

I am thankful to Mr. Vijay Upadhaya and Mr. Fahim for their patience and agreeing to my suggestions for making diagrams. I am thankful to innumerable students who have made great contributions in one way or the other during the preparation of manuscript.

I am thankful to my elders whose continuous support and inspiration led me to complete this work. I am fortunate in having a family which understands the preoccupation that goes with such projects. I am grateful to my wife and to my children, Komaljit and Jasmit for their perseverance and help.

March, 2000

H.S. Chawla

Abbreviations

2,4-D	2,4-Dichlorophenoxyacetic acid
IAA	Indole-3-acetic acid
IBA	Indole-3-butyric acid
NAA	Naphthaleneacetic acid
PCPA	*p*-Chlorophenoxyacetic acid
BAP or BA	Benzylamino purine or benzyladenine
2iP	2-Isopentyladenine
Kin	Kinetin
Zea	Zeatin
ABA	Abscisic acid
A.t.	*Agrobacterium tumefaciens*
B_5	Gamborg's medium
DMSO	Dimethylsulfoxide
GA	Gibberellic acid
kb (p)	Kilobase (pairs)
LS	Linsmaier and Skoog medium
Mb	Megabase
Mdal	Magadalton
MS	Murashige and Skoog medium
ng	Nanogram
PAGE	Polyacrylamide gel electrophoresis
pg	Picogram
Ri	Root inducing plasmid of *Agrobacterium rhizogenes*
SH	Schenk and Hildebrandt medium
Ti	Tumor inducing plasmid of *Agrobacterium tumefaciens*

Contents

PART III: RECOMBINANT DNA TECHNOLOGY

Part I

Plant Tissue Culture

1

Introduction

Plants are the key to life on earth as they directly supply 90% of human calorie intake, and 80% of the protein intake, the remainder being derived from animal products, although these animals have also derived their nutrition from plants. Of the three thousand plant species which have been used as food by man, the world now depends mainly on around twenty crop species for the majority of its calories, with 50% being contributed by eight species of cereals. Minerals and vitamins are supplied by a further thirty species of fruits and vegetables. Most important of the staple foods are the cereals, particularly wheat and rice, with more than one-third of all cultivated land used to produce these two crops.

As the population continues to expand, concern has been expressed over the finite number of people world agriculture can support. An estimate of how many people could be supported by world food production can be obtained from consideration of the average daily calorie requirements and the net amount of grain yield from well-managed areas. It has been calculated that the earth can support about 15 billion people on a strictly vegetarian diet, or five billion on a mixed diet. The population is expected to be eight billion by 2001 and double by 2025.

The farming practises and crops cultivated today have developed over a relatively short span of time. Crop plants of today have changed in a number of ways so that they now bear very little resemblance to their wild type ancestors. These changes have come about through selection, either conscious or unconscious, for traits which are advantageous to the people growing the crops. Thus modern wheats do not disperse their seeds or legumes do not have pods which burst open. Today, varieties are the result of generation of plants cultivated under ideal conditions from the man's point of view.

From the beginning of crop cultivation to the late nineteenth century, all improvements in the species used were brought about by those who were directly involved, i.e. farmers themselves. In the following 100 years, the laws of genetic inheritance and rules governing species variation were laid down by Mendel, Darwin, and others, which redefined the breeding techniques by making them predictable and therefore quicker, more precise and more productive. Furthermore, despite the implementation of breeding techniques, the time taken to produce and test varieties is an important limiting consideration.

NEW TECHNOLOGIES

Over the past few years a number of methodologies have come that would seem to have much more to offer in terms of advancing current research in the plant sciences, and exploiting the knowledge gained to develop new crops. The first of these areas is concerned with manipulation and subsequent growth of cells, tissues, organs and naked plant cells (protoplasts) in tissue culture. The second field is genetic engineering or recombinant DNA (rDNA) technology which has grown out very fast from the work initially carried out on microorganisms. With genetic engineering, scientists have more exact methods for breeding better livestock and crop varieties. This technology allows for the detailed manipulation of genes. These two areas of research have in recent years become associated with the general field of biotechnology and are potentially applicable to a wide variety of plant species, as well as offer a precision in manipulating genetic material. Recombinant DNA techniques in particular have already contributed much to the elucidation of basic mechanisms in plants at the molecular level.

Our knowledge of the structure and expression of the plant genome has come largely through the use of recombinant DNA or gene cloning techniques. This technology allows the isolation and characterization of specific pieces of DNA and by cloning the DNA sequences into bacterial cells, they can be replicated to yield large quantities for analysis. In addition to supplying much basic information concerning gene structure and expression, recombinant-DNA technology provides the opportunity for specifically manipulating genetic material, and moving such material among different organisms.

ORIGIN OF BIOTECHNOLOGY

The term Biotechnology was coined by Karl Ereky, a Hungarian engineer, in 1919. The origin of biotechnology can be traced back to prehistoric times when microorganisms were already used for processes like fermentation, formation of yoghurt and cheese from milk, vinegar from molasses, production of butanol and acetone from starch by *Clostridium acetobutylicum* or the production of antibiotics like penicillin from *Penicillium notatum*. However, biotechnology got

a boost in the 1970's with the discovery of restriction enzymes which led to the development of a variety of gene technologies and is thus considered to be the greatest scientific revolution of this century. Biotechnology thus consists of a variety of techniques, designed to genetically improve and/or exploit living systems or their components for the benefit of man. In fact, biotechnology is the product of interaction between sciences of biology and technology. It is the technological exploitation and control of biological systems. Attempts have been made to define biotechnology and it has been interpreted in different ways by different groups of workers. However, the following definitions seems to be most appropriate.

i. The application of science and engineering in the direct or indirect use of living organisms, or parts or products of living organisms, in their natural or modified form.

ii. Biotechnology means any technological application that uses biological system, living organism or derivatives thereof to make or modify products or processes for specific use (Convention on Biodiversity).

iii. The controlled use of biological agents such as microorganisms or cellular components for beneficial use (US National Science Foundation).

The term genetic engineering refers to a number of new techniques involving the transfer of specific genetic information from one organism to another. These techniques do not rely on sexual methods, but instead involve genetic manipulation at the cellular and molecular levels. These are the nonsexual methods for gene transfer.

An important aspect of all plant biotechnology processes is the culture of either the microorganisms or plant cells or tissues and organs in artificial media. One of the problems in conventional plant breeding is that the range of organisms among which genes can be transferred is severely limited by species barriers. The new technologies provide a better approach for defining and manipulation of targets, and species-specific barriers are broken. They do not replace plant breeding but provide methods capable of achieving objectives not possible by other means.

HISTORY

The last three decades have seen a very rapid rise in the number of plant scientists using the techniques of organ, tissue and cell cultures for plant improvement. The term "plant tissue culture" broadly refers to the *in vitro* cultivation of plants, seeds, plant parts (tissues, organs, embryos, single cells, protoplasts, etc.) on nutrient media under aseptic conditions.

During the 1800s, the cell theory, which states that the cell is the basic structural unit of all living creatures, was very quick to gain acceptance. However, the second portion of the cell theory states that these structural units are distinct and potentially totipotent physiological and developmental units, failed to gain universal acceptance. The skepticism associated with the latter part was because of the inability of scientists such as Schleiden and Schwann to demonstrate totipotency in their laboratories. It was in 1902 that the well-known German plant physiologist, Gottlieb Haberlandt (1854–1945), attempted to cultivate plant tissue culture cells *in vitro*. He is regarded as the father of plant tissue culture. He clearly stated the desirability of culturing the isolated vegetative cells of higher plants. He stated: "To my knowledge, no systematically organized attempts to culture isolated vegetative cells from higher plants in simple nutrient solutions have been made. Yet the results of such culture experiments should give some interesting insights into the properties and potentialities which the cell as an elementary organism possesses. Moreover, it would provide information about the interrelationships and complementary influences to which cells within the multicellular whole organism are exposed." Haberlandt started his experiments in 1898 using single cells isolated from the palisade tissue of leaves, pith parenchyma,

epidermis and epidermal hair of various plants. He grew them on Knop's salt solution with sucrose and observed obvious growth in the palisade cells. However, unfortunately none of the cells divided but this did not deter others to perform experiments. In the early 1920s workers again attempted to grow plant tissues and organs *in vitro*. Molliard in 1921 demonstrated limited success with the cultivation of plant embryos and subsequently Kotte (1922a, b), a student of Haberlandt in Germany and, independently Robbins (1922a, b) were successful in the establishment of excised plant root tips *in vitro*. However, in 1934, the pioneering work of growing excised roots of tomato *in vitro* for periods of time without theoretical limits was demonstrated by White (1934). Initially, White used a medium containing inorganic salts, yeast extract and sucrose, but later yeast extract was replaced by three B-vitamins viz. pyridoxine, thiamine and nicotinic acid. In the same year, Gautheret was engaged in experimentation with excised root tips and the cultivation of cambial tissue removed under aseptic conditions from *Salix capraea, Populus nigra* and other trees on Knop's solution containing glucose and cysteine hydrochloride and recorded that they proliferated for a few months. These studies continued, and in 1939, Gautheret, Nobecourt and White, published independently studies on the successful cultivation for prolonged periods of cambial tissues of carrot root (Gautheret, 1939), tobacco (White, 1939) and carrot (Nobecourt, 1939). These were the first true plant tissue cultures in the strict sense of prolonged cultures of unorganized materials. The methods and media now used are in principle modifications of those established by these three pioneers. All the three workers used meristematic cells to generate continuously growing cultures. Philip R. White (1943, 1954, 1963) and Roger J. Gautheret (1959) can be credited with providing a significant impetus to the field with the publication of their authoritative handbooks.

From 1939 to 1950, experimental work with root cultures drew attention to the role of vitamins in plant growth and advanced the knowledge of the shoot – root relationship. During early 1950s, a number of lines of enquiry were initiated. The work of Miller and Skoog (1953) on bud formation from cultured pith explants of tobacco led to the discovery of kinetin. In 1952, Steward initiated work on cultured carrot explants and used coconut milk as a nutrient (Steward *et al.*, 1952) that ultimately led to the discovery of embryogenesis (Steward *et al.*, 1958). In 1953, Muir reported that if fragments of callus of *Tagetes erecta* and *Nicotiana tabacum* are transferred to liquid culture medium and the medium is agitated on a reciprocal shaker, then the callus fragments break up to give a suspension of single cells and cell aggregates (Muir, 1953). He further developed a nurse culture technique. An important technique of cloning large number of single cells of higher plants was developed by Bergmann (1960).

The differentiation of whole plants in tissue cultures may occur via shoot and root differentiation, or alternatively the cells may undergo embryogenic development to give rise to somatic embryos. Differentiation of plants from callus cultures has often been suggested as a potential method for rapid propagation. Ball in 1946 successfully raised transplantable whole plants of *Lupinus* and *Tropaeolum* by culturing their shoot tips with a couple of leaf primordia. The practical usefulness of this technique is credited to Morel and Martin (1952), who for the first time recovered virus-free Dahlia plants. Morel (1960) also realized the potential of this method for rapid propagation of virus-free orchids.

The release of protoplasts from root tip cells using a fungal cellulase in 0.6M sucrose was reported by Cocking in 1960. Protoplasts released by cell wall degrading enzymes have now been prepared from many plant tissues. The most universally used high-salt medium was developed by Skoog and his students (Murashige and Skoog, 1962). In addition to mineral salts, media contain an energy source, vitamins and growth regulators.

In 1970s, restriction enzymes were discovered, and the technique was developed by Lobban and Kaiser for joining two restriction fragments by ligase. Paul Berg, working at Stanford University, was the first to make a recombinant DNA molecule combining DNA from SV40 virus with that of lambda virus in 1972. However, realizing the dangers of his experiments, he terminated it immediately. He proposed, in what is known as "Berg Letter," a one-year moratorium on such research in order that safety concerns be worked out. In 1973 Stanley Cohen of Stanford University and Herbert Boyer of University of California produced world's first recombinat DNA organism. This set the foundation of modern biotechnology and genetic engineering. Genetic engineering became a reality when a man-made insulin gene was used to manufacture a human protein in bacteria. The 1980s saw the development of various genetic transformation techniques which revolutionized the rDNA technology and led to the development of transgenic plants for various crops.

It was in mid-1970s that techniques to sequence DNA were developed. Sanger *et al.* (1977) and Maxam and Gilbert (1977) were the first to report gene sequencing techniques. Sanger *et al.* (1977) sequenced virus ϕx 174 comprising 5375 nucleotides that code for 10 proteins. This was followed by report on complete DNA sequence of simian virus 40 (SV40). In recent years, entire genomes of a number of prokaryotes and eukaryotes have been sequenced. In 1990, the Human Genome Project (HGP) was launched to sequence the genome and identify all the genes. This laid the foundation for genomics.

Molecular breeding work started in 1980s with the development of the restriction fragment length polymorphism technique. The technique of DNA fingerprinting, based on the sequence of bases in segments of DNA, was introduced in 1986 by Jeffreys to identify individuals and it was first used as an evidence in the court room in USA in 1987. A large number of molecular markers have been reported that led to the development of methodologies for mapping of genes on the chromosomes and in molecular-marker-assisted selection.

Some of the contributions are mentioned below.

1902 First attempt of plant tissue culture (Haberlandt).

1904 Embryo culture of selected crucifers attempted (Hannig).

1922 Asymbiotic germination of orchid seeds *in vitro* (Knudson).

1922 *In vitro* culture of root tips (Robbins)

1925 Use of embryo culture technique in interspecific crosses of *Linum* (Laibach).

1934 *In vitro* culture of the cambial tissue of a few trees and shrubs, although failed to sustain cell division (Gautheret).

1934 Successful culture of tomato roots (White).

1939 Successful establishment of continuously growing callus cultures (Gautheret, Nobecourt and White).

1940 *In vitro* culture of cambial tissues of *Ulmus* to study adventitious shoot formation (Gautheret).

1941 Use of coconut milk containing a cell division factor for the first time in *Datura* (van Overbeek).

1941 *In vitro* culture of crown gall tissues (Braun).

1944 *In vitro* adventitious shoot formation in tobacco (Skoog).

1946 Raising of whole plants of *Lupinus* and *Tropaeolum* by shoot tip culture (Ball).

1950 Regeneration of organs from callus tissue of *Sequoia sempervirens* (Ball).

1952 Use of meristem culture to obtain virus-free dahlias (Morel and Martin).

1952 First application of micrografting (Morel and Martin).

1953 Production of haploid callus of the gymnosperm *Ginkgo biloba* from pollen (Tulecke).

1954 First plant from a single cell (Muir *et al.*).

1955 Discovery of kinetin a cell division hormone (Miller *et al.*).

1957 Discovery of the regulation of organ formation by changing the ratio of auxin: cytokinin (Skoog and Miller).

1958 Regeneration of somatic embryos *in vitro* from the nucellus of *Citrus* ovules (Maheshwari and Rangaswamy).

1959 Regeneration of embryos from callus clumps and cell suspensions of *Daucus carota* (Reinert, Steward).

1959 Publication of first handbook on plant tissue culture (Gautheret).

1960 First successful test tube fertilization in *Papaver rhoeas* (Kanta).

1960 Use of the microculture method for growing single cells in hanging drops in a conditioned medium (Jones *et al.*).

1960 Enzymatic degradation of cell walls to obtain large number of protoplasts (Cocking).

1960 Filtration of cell suspensions and isolation of single cells by plating (Bergmann).

1962 Development of Murashige and Skoog nutrition medium (Murashige and Skoog).

1964 Production of first haploid plants from pollen grains of *Datura* (Guha and Maheshwari).

1968 Restriction endonuclease term coined to a class of enzymes involved in cleaving DNA (Meselson and Yuan).

1970 Selection of biochemical mutants *in vitro* by the use of tissue culture derived variation (Carlson).

1970 First achievement of protoplast fusion (Power *et al.*).

1970 Discovery of first restriction endonuclease from *Haemophillus influenzae* Rd. It was later purified and named *Hind*II (Smith).

1971 Preparation of first restriction map using *Hind*II enzyme to cut circular DNA of SV 40 into 11 specific fragments was prepared (Nathans).

1971 Regeneration of first plants from protoplasts (Takebe *et al.*).

1972 First report of interspecific hybridization through protoplast fusion in two species of *Nicotiana* (Carlson *et al.*).

1972 First recombinant DNA molecule produced

using restriction enzymes (Berg *et al.*).

1972 Joining of two restriction fragments regardless of their origin produced by the same restriction enzyme by the action of DNA ligase (Mertz and Davis).

1972 Development of a procedure in which the appropriate enzyme can be added to fill in any single-stranded gap and the use of DNA ligase to join the two fragments thus giving rise to recombinant DNA (Lobban and Kaiser).

1972 Discovery of reverse transcriptase: In certain cancer-causing animal virus, genetic information flows in the reverse form (Temin).

1973 Use of the Lobban and Kaiser technique to develop hybrid plasmid – insertion of *Eco*RI fragment of DNA molecule into circular plasmid DNA of bacteria using DNA ligase. Gene from African clawed toad inserted into plasmid DNA of bacteria (Herbert Boyer and Stanley Cohen).

1973 Cytokinin found capable of breaking dormancy in excised capitulum explants of *Gerbera* (Pierik *et al.*).

1974 Regeneration of haploid *Petunia hybrida* plants from protoplasts (Binding).

1974 Biotransformation in plant tissue cultures (Reinhard).

1974 Discovery that the Ti plasmid is the tumor inducing principle of *Agrobacterium* (Zaenen *et al.*; Larebeke *et al.*).

1975 Positive *in vitro* selection of maize callus cultures resistant to T toxin of *Helminthosporium maydis* (Gengenbach and Green).

1975 Development of the high resolution two-dimensional gel electrophoresis procedure, which led to the development of proteomics (O'Farrel).

1976 Shoot initiation from cryopreserved shoot apices of carnation (Seibert).

1976 Octopine and nopaline synthesis and breakdown found to be genetically controlled by the Ti plasmid of *A. tumefaciens* (Bomhoff *et al.*).

1977 Successful integration of the Ti plasmid DNA from *A. tumefaciens* in plants (Chilton *et al.*).

1977 A method of gene sequencing based on degradation of DNA chain (Maxam and Gilbert).

1977 Discovery of split genes (Sharp and Roberts).

1978 Somatic hybridization of tomato and potato resulting in pomato (Melchers *et al.*).

1979 Cocultivation procedure developed for transformation of plant protoplasts with *Agrobacterium* (Marton *et al.*).

1980 Use of immobilized whole cells for biotransformation of digitoxin into digoxin (Alfermann *et al.*).

1980 Commercial production of human insulin through genetic engineering in bacterial cells (Eli Lilly and Co.).

1980 Restriction fragment length polymorphism (RFLP) technique developed.

1980 Studies on the structure of T-DNA by cloning the complete *Eco*RI digest of T37 tobacco crown gall DNA into a phage vector, thus allowing the isolation and detailed study of T-DNA border sequences (Zambryski *et al.*).

1981 Introduction of the term *somaclonal variation* (Larkin and Scowcroft).

1982 Incorporation of naked DNA by protoplasts resulting in the transformation with isolated DNA (Krens *et al.*).

1983 Polymerase chain reaction (PCR), a chemical DNA amplification process idea conceived (Kary Mullis).

1984 Transformation of tobacco with *Agrobacterium;* transgenic plants developed (De Block *et al.*; Horsch *et al.*).

1984 Development of the genetic fingerprinting technique for identifying individuals by analyzing polymorphism at DNA sequence level (Jeffreys).

1986 TMV virus-resistant tobacco and tomato transgenic plants developed using cDNA of coat protein gene of TMV (Powell-Abel *et al.*).

1987 Development of biolistic gene transfer method for plant transformation (Sanford *et al.*; Klein *et al.*).

1987 Isolation of *Bt* gene from bacterium (*Bacillus thuringiensis*) (Barton *et al.*).

1990 Formal launch of the Human Genome Project.

1990 Development of the random amplified polymorphic DNA (RAPD) technique (Williams *et al.*; Welsh and McClelland).

1991 Development of DNA microarray system using light directed chemical synthesis system (Fodor).

1995 Reporting by the Institute for Genomic Research of the complete DNA sequence of *Haemophilus influenzae* (Fleischmann *et al.*).

1995 Development of DNA fingerprinting by amplified fragment length polymorphism (AFLP) technique (Vos *et al.*).

1997 Sequencing of *E. coli* genome (Blattner *et al.*).

1998 Sequencing of the genome of a multicellular organism (*Caenorhabditis elegans*) (*C. elegans* sequencing consortium).

2001 Sequencing of human genome successfully completed (Human Genome Project Consortium and Venter *et al.*).

2

Laboratory Organization

It is generally accepted that the term "plant tissue culture", broadly refers to the *in vitro* cultivation of all plant parts, whether a single cell, a tissue or an organ, under aseptic conditions. The underlying principles involved in plant tissue culture are very simple. Firstly, a plant part from the intact plant and its inter-organ/inter-tissue is isolated. Secondly, it is necessary to provide the plant part with an appropriate environment in which it can express its intrinsic or induced potential. This means that a suitable culture medium and proper culture conditions must be provided. Finally, the above procedures must be carried out aseptically. A facility for plant tissue culture operations involving any type of *in vitro* procedures must include certain essential elements. The overall design in any laboratory organization must focus on maintaining aseptic conditions. The specific design can vary, but an effective laboratory organization for plant tissue culture must include certain elements irrespective of whether used by a scientist or group of workers. The size and proportion of different parts of the facility will be dependent upon the function, purpose and size of the operations. The basic rule essential in the planning of an efficient and functional laboratory facility is the order and continuous maintenance of cleanliness because all operations require aseptic conditions.

A general guideline for setting up a facility is to focus on a design that different units and activities need to be arranged to make opera-tional steps possible with the least amount of cross-traffic. Thus a tissue culture facility should have the following features.

Washing facility: An area with large sinks (some lead lined to resist acids and alkalis) and draining area is necessary. The conventional method for cleaning laboratory glassware involves chromic acid – sulphuric acid soak followed by thorough washing with tap water and subsequent rinsing with distilled water. However, due to the corrosive nature of acids, it is recommended that for routine procedures glassware should be soaked in a 2% detergent cleaner for 16 h followed by washing with 60 – 70°C hot tap water, and finally distilled water. The cleaned glassware should then be dried at 150 – 200°C in a convection-drying oven for 1-4 h. Only highly contaminated glassware should be first treated in a chromic acid bath for 16 h. With the availability of a wide range of reusable plasticware, it is recommended that these should be washed with mild detergents followed by a rinse with tap water and then distilled water.

The following precautions should be followed:

1. Contents of containers should be discarded immediately after completion of an experiment.

2. Contaminated containers should be autoclaved, contents discarded in a waste bottle or container and then washed. The waste

bottle containing autoclaved contaminated media should not be emptied in the laboratory sink.

3. Flasks or beakers used for agar-based media should be rinsed immediately after dispensing the media into culture vessels to prevent drying of the residual agar in the beaker prior to washing.

General laboratory and media preparation area: This part is a central section of the laboratory where most of the activities are performed. The general laboratory section includes the area for media preparation and autoclaving of media. It also includes many of the activities that relate to the handling of tissue culture materials. The area to be set aside for media preparation should have ample storage and bench space for chemicals, glassware, culture vessels, closures and other items needed to prepare media. The following equipment items are essential:

Laminar air flow hood
Autoclave (available as horizontal or upright) or pressure cookers
Refrigerator
Freezer
Balances, preferably electronic
pH meter
Water distillation unit
Magnetic stirrer (with hot plate)
Ovens
Water bath with temperature control
Hot plate/gas plate
Microwave oven for heating agar media
Centrifuge tabletop
Vortex
Shaker, gyratory with platform and clips for different size flasks
Dissecting microscope
Lab carts

A laboratory should have an inventory and a complete up-to-date record of all the equipment and their operating manuals. It is very essential to keep a detailed and up-to-date record of all the chemicals including the name of manufacturer and the grade. All chemicals should be assigned to specific areas preferably by arranging alphabetically. Special handling or storage procedures should be posted in the records so that retrieving chemicals is easy because chemicals need storage at different temperatures. Strong acids and bases should be stored separately. Solvents such as chloroform, alcohol, phenol which are volatile or toxic in nature must be stored in a fume hood. Several chemicals are hygroscopic and must be stored in desiccators to avoid caking. Chemicals kept in refrigerators or freezers are most conveniently arranged either alphabetically or in smaller baskets.

Transfer area: Tissue culture techniques can be successfully carried out in a very clean laboratory, dry atmosphere with some protection against air-borne microorganisms. But it is advisable that a sterile dust-free room should be available for routine transfer and manipulation work. The laminar airflow cabinet is the most common accessory used for aseptic manipulations. It is cheaper and easier to install than a transfer room. The cabinet should be designed with horizontal airflow from the back to the front, and should be equipped with gas cocks if gas burners are to be used. Vertical flow sterile cabinet is recommended if pathogens or *Agrobacterium*-mediated transformation work is to be carried out. In the airflow cabinets, air is forced into a cabinet through a bacterial HEPA (high efficiency particulate air) filter, it flows outwards (forward) over the working bench at a uniform rate. The constant airflow prevents particles from settling on the bench. One or more laminar airflow cabinets can be housed in such a transfer area. The roof of some of the cabinet houses an ultraviolet (UV) germicidal light often used to sterilize the interior of chamber. These are turned on about 30 min prior to using the chamber but must be turned off during operations. Even so, before starting work, it is desirable to swab down the inside of cabinet and especially the table top with 70% alcohol.

Culturing facilities: Plant tissue cultures should be incubated under conditions of well-controlled temperature, illumination, photoperiod, and humidity and air circulation. Incubation culture rooms and commercially available incubator cabinets, large plant growth chambers, and walk-in environmental rooms satisfy this requirement. These facilities can also be constructed by developing a room with proper air-conditioning, perforated shelves to support culture vessels, fluorescent tubes and a timing device to set light dark regimes (photoperiods) which are the standard accessories. In addition, a dark area simply closed off from the rest of room with thick black curtains is necessary if some cultures require continuous darkness. This type of culture room can also be used to house a variety of liquid culture apparatus including gyratory and reciprocating shakers or a variety of batch and continuous bioreactors etc. For most culture conditions the temperature control should be adequate to stay within ± 1°C in a range from 10 to 32°C. The photoperiod may be set according to the specific types of culture.

Light units: Lux or foot candles terms have been used but more acceptable terms are:

Joule per square meter (j/m²) is an expression of radiant energy per unit area.

Watts per square meter (W/m²) is used to express irradiance.

The preferred term for photosynthetically active radiation units is to use photon flux density which is µmol photons/m²/s.

A common term for photon flux is based on 1 mol of photons being equal to 1 Einstein (E); Thus, a unit of photon flux density may then be expressed as µE/m²/s.

Greenhouses: Greenhouses are required to grow regenerated plants for further propagation and for growing plants to maturity. This facility is required as a transitional step of taking plant materials from culture containers present in the controlled room to the field. Thus in the greenhouse, plants are acclimatized and hardened before being transferred to the field conditions. The plants are grown in the greenhouse to develop adequate root systems and leaf structures to withstand the field environment. The greenhouse should be equipped with cooling and heating systems to control temperature. For maximum use, the houses should have an artificial lighting system. The lights should include a mixture of fluorescent and incandescent lights or contain lamps designed to provide balanced wavelengths of light for plant growth and photosynthesis. The installation of a misting system is recommended on benches in the greenhouse that will be used for acclimatizing the plants.

LABORATORY AND PERSONAL SAFETY

Every new worker should be given some orientation before starting work. It should include instructions on how to operate various equipment such as pH meter, balance, laminar airflow, microscope, gas burners etc. Special instructions should be given on potential for fire, broken glass, chemical spills or accidents with sharp-edged instruments. The following regulations/instructions should be followed:

1. Toxic chemicals should be handled with appropriate precautions and should be discarded into separately labeled containers:
    Organic solvents (e.g. alcohol, chloroform, etc.)
    Mercury compounds (e.g. mercuric chloride)
    Halogens and mutagenic chemicals

2. Broken glass and scalpel blades must be disposed into separate marked containers.

3. Pipettes, tips, Pasteur pipettes and other items used for genetically modified organisms and various materials used in the pathogen work. It should be first collected in autoclavable bags, autoclaved and then disposed of.

4. A pipettor should always be used for pipetting any solution.

5. First-aid kits should be placed in every laboratory and the staff should know its location and how to use its contents.

6. Fire extinguishers should be provided in each laboratory. Some portable extinguishers should also be placed at convenient sites for easy handling by amateurs.

3

Nutrition Medium

The greatest progress towards the development of nutrition media for plant cells grown in culture took place in 1960s and 1970s. The basic nutritional requirements of cultured plant cells are very similar to those utilized by plants. However, the nutritional composition varies according to the cells, tissues, organs and protoplasts and also with respect to particular plant species. Thus in a discussion on nutrition media it is important to recognize what type of culture, i.e. callus, cell, organ, or protoplast is to be studied and what are the objectives and purposes of the investigation.

Different types of culture have unique requirements of one or more nutritional components. A wide variety of salt mixtures have been reported in various media. A nutrient medium is defined by its mineral salt composition, carbon source, vitamins, phytohormones and other organic supplements. When reference is made to a particular medium, the intention is to identify only the salt composition unless otherwise specified. Any number and concentration of amino acids, vitamins, phytohormones or organic supplements can be added in an infinite variety of compositions to a given salt composition to achieve the desired results.

FACILITIES AND EQUIPMENT

Autoclave and/or pressure cookers
Balances, preferably electronic
Centrifuge tabletop
Cotton plugs, aluminium foil, brown sheets, etc.
Culture vessels (flasks, tubes, petri dishes, jars, etc.)
Desiccators
Dissecting microscope
Filter units
Flasks, beakers, pipettes, etc.
Freezer
Hot plate/gas plate
Lab carts/trolley, trays, etc.
Laminar airflow hoods
Magnetic stirrer (with hot plate)
Microwave oven
Ovens
pH meter
Refrigerator
Shaker
Vortex
Water bath with temperature control
Water distillation unit

UNITS FOR SOLUTION PREPARATION

The concentration of a particular substance in the media can be expressed in various units that are as follows:

Units in weight

It is represented as milligram per liter (mg/l)

10^{-6} = 1.0 mg/l or 1 part per million (ppm)
10^{-7} = 0.1 mg/l
10^{-9} = 0.001 mg/l or 1 µg/l

Molar concentration

A molar solution (M) contains the same number of grams of substance as is given by its molecular weight.

1 molar (M) =	the molecular weight in g/l
1 mM =	the molecular weight in mg/l or 10^{-3} M
1 µM =	the molecular weight in µg/l or 10^{-6} M or 10^{-3} mM.

Conversion from milli molar (mM) to mg/l

For example, the molecular weight of auxin 2,4-D = 221.0
1 M 2,4-D solution consists of 221.0 g/l
1 mM 2,4-D solution consists of 0.221 g/l = 221.0 mg/l
1 mM 2,4-D solution consists of 0.000221 g/l = 0.221 mg/l

Conversion from mg/l to mM

The molecular weight of $CaCl_2 \cdot 2H_2O$
= 40.08 + 2 × 35.453 + 4 × 1.008 + 2 × 16 = 147.018
(the atomic weights of Ca, Cl, H and O being 40.08, 35.453, 1.008 and 16.0 respectively).
If 440 mg/l of $CaCl_2 \cdot 2H_2O$ is to be converted into mM, then

$$\text{The number of mM } CaCl_2 \cdot 2H_2O = \frac{\text{No. of mg } CaCl_2 \cdot 2H_2O}{\text{Molecular weight of } CaCl_2 \cdot 2H_2O}$$

$$= \frac{440}{147.018} = 2.99 \text{ mM}$$

Thus, 440 mg/l $CaCl_2 \cdot 2H_2O$ = 2.99 mM

MEDIA COMPOSITION

A number of basic media are listed in Table 3.1. The MS medium of Murashige and Skoog (1962) salt composition is very widely used in different culture systems. In the development of this medium and also other media, it was demonstrated that not only the presence of necessary nutrients but also the actual and relative concentrations of various inorganic nutrients are of crucial significance. The B_5 medium of Gamborg *et al.* (1968) or its derivatives have been useful for cell and protoplast culture. This medium was originally designed for callus and suspension cultures but it has also been effectively used for plant regeneration. It differs from MS medium in having much lower amounts of nitrate in the form of ammonium. The N_6 medium was developed for cereal anther culture and is used successfully with other types of cereal tissue culture (Chu, 1978). Any success with a medium is in all probability due to the fact that the ratios as well as concentrations most nearly match the optimum requirements for the cells or tissues for growth and/or differentiation. The nutritional milieu generally consists of inorganic nutrients, carbon and energy sources, vitamins, phytohormones (growth regulators), and organic supplements which include organic nitrogen, acids and complex substances.

Inorganic nutrients

Mineral elements are very important in the life of a plant. For example, calcium is a component of the cell wall, nitrogen is an important part of amino acids, proteins, nucleic acids, and vitamins, and magnesium is a part of chlorophyll molecules. Similarly iron, zinc and molybdenum are parts of certain enzymes. Besides, C, H, N, and O, 12 other elements are known to be essential for plant growth. According to the recommendations of International Association for Plant Physiology, the elements required by plants in concentration greater than 0.5 mmol/l are referred to as macroelements and those required in concentration less than that are microelements. A variety of salts supply the needed macro- and micronutrients that are the same as those required by the normal plant. The following are required in macro- or millimole quantities: N, K, P, Ca, S, and Mg. The essential

Table 3.1 Composition of plant tissue culture media (values expressed in mg/l).

Constituent	Nitsch & Nitsch (1956)	White (1963)	Murashige and Skoog (1962)– MS	Gautheret (1942)	Gamborg's (1968) – B_5	Chu (1978) – N_6
KCl	1500	65	-	-	-	-
$MgSO_4 \cdot 7H_2O$	250	720	370	125	250	185
$NaH_2PO_4 \cdot H_2O$	250	16.5	-	-	150	-
$CaCl_2 \cdot 2H_2O$	-	-	440	-	150	166
KNO_3	2000	80	1900	125	2500	2830
$CaCl_2$	25	-	-	-	-	-
Na_2SO_4	-	200	-	-	-	-
NH_4NO_3	-	-	1650	-	-	-
KH_2PO_4	-	-	170	125	-	400
$Ca(NO_3)_2 \cdot 4H_2O$	-	300	-	500	-	-
$(NH_4)_2SO_4$	-	-	-	-	134	463
$NiSO_4$	-	-	-	0.05	-	-
$FeSO_4 \cdot 7H_2O$	-	-	27.8	0.05	-	27.8
$MnSO_4 \cdot 4H_2O$	3	7	22.3	3	-	-
$MnSO_4 \cdot H_2O$	-	-	-	-	10	3.3
KI	-	0.75	0.83	0.5	0.75	0.8
$CoCl_2 \cdot 6H_2O$	-	-	0.025	-	0.025	-
$Ti(SO_4)_3$	-	-	-	0.2	-	-
$ZnSO_4 \cdot 7H_2O$	0.5	3	8.6	0.18	2	1.5
$CuSO_4 \cdot 5H_2O$	0.025	-	0.025	0.5	0.025	-
$BeSO_4$	-	-	-	0.1	-	-
H_3BO_3	0.5	1.5	6.2	0.05	3	1.6
H_2SO_4	-	-	-	1	-	-
$Na_2MoO_4 \cdot 2H_2O$	0.025	-	0.25	-	0.25	-
$Fe_2(SO_4)_3$	-	2.5	-	-	-	-
EDTA disodium salt	-	-	37.3	-	-	37.3
EDTA-Na ferric salt	-	-	-	-	43	-
m-inositol	-	-	100	-	100	-
Thiamine	-	0.1	0.1	0.1	1.0	1.0
Pyridoxine	-	0.1	0.5	0.1	1.0	0.5
Nicotinic acid	-	0.5	0.5	0.5	1.0	0.5
Glycine	-	3	2	3	-	-
Cysteine	-	1.0	-	-	10	-
Sucrose	34,000	20,000	30,000	30,000	20,000	30,000

micronutrients which are required in micromolar concentrations include iron (Fe), manganese (Mn), boron (B), copper (Cu), zinc (Zn), iodine (I), molybdenum (Mo) and cobalt (Co). For achieving maximum growth rate, the optimum concentration of each nutrient can vary considerably. When mineral salts are dissolved in water, they undergo dissociation and ionization. The active factor in the medium is the ions of different types rather than the compounds. Therefore, a useful comparison between the two media can be made by looking into the total concentrations of different types of ions in them (Table 3.2). For most

purposes a nutrient medium should contain from 25 to 60 mM inorganic nitrogen. The cells may grow on nitrate alone, but often there is a distinct beneficial effect and requirement for ammonium or another source of reduced nitrogen. Besides, nitrate alone in the medium drifts the pH towards alkalinity. Adding a small amount of an ammonium compound together with nitrate checks this drift. Nitrate is used in the range of 25–40 mM and ammonium in the range of 2–20 mM. The response to ammonium varies from inhibitory to essential depending upon the tissue and the purpose of culture. In case of amounts in excess

Table 3.2 Concentration of different inorganic salts (M) in different plant tissue culture media

Salt	MS	B$_5$	N$_6$
Major salts (mM)			
Ca	3	1.0	1.1
Cl	6	2.2	–
K	20.1	25	31
NH$_4$	20.6	2	6.6
NO$_3$	39.4	25	28
PO$_4$	1.25	1.1	2.94
SO$_4$	1.8	–	0.2
Mg	1.5	1.0	0.75
Micronutrients (mM)			
B	100	50	26
Co	0.11	0.1	0.11
Cu	0.1	0.1	–
Fe	1	1	1
I	5	4.5	4.8
Mn	92.5	60	19.5
Mo	1	1	–
Zn	30	7	5.2

of 8 mM of ammonium or if grown solely on this source of nitrogen, then citrate, malate, succinate or another TCA cycle acid should be present. Most plants prefer nitrate to ammonium, although the opposite is also true in some cases. Potassium is required at concentrations of 2 to 26 mM. Potassium is generally supplied as the nitrate or as the chloride form and sodium cannot be substituted for potassium. A concentration of 1 – 3 mM calcium, sulphate, phosphate and magnesium is usually adequate. Iron is generally added as a chelate with ethylene diamine tetra acetic acid (EDTA). In this form iron remains available up to a pH of 8.0.

Carbon and energy source

Without exception, the standard carbon source is sucrose or glucose. Fructose can also be used but is less efficient. The sucrose in the medium is rapidly converted into glucose and fructose. Glucose is then utilized first, followed by fructose. Sucrose is generally used at a concentration of 2 - 5%. Other carbohydrates which have been tested include lactose, maltose, galactose and starch, but these compounds are generally much inferior to sucrose or glucose. Most media contain myo-inositol at a concentration of

ca. 100 mg/l, which improves cell growth.

Vitamins

Normal plants synthesize the vitamins required for growth and development. But plant cells in culture have a requirement for vitamins. There is an absolute requirement for vitamin B$_1$ (thiamine). Growth is also improved by the addition of nicotinic acid and vitamin B$_6$ (pyridoxine). Some media contain pantothenic acid, biotin, folic acid, p-amino benzoic acid, choline chloride, riboflavine and ascorbic acid.

Growth regulators

Hormones are organic compounds naturally synthesized in higher plants, which influence growth and development. They are usually active at a site different from where they are produced and are only present and active in very small quantities. Apart from natural compounds, synthetic compounds have been developed which correspond to the natural ones. These are collectively called growth regulators. There are two main classes of growth regulators that are of special importance in plant tissue culture. These are the auxins and cytokinins, while the others viz. gibberellins, abscisic acid, ethylene, etc. are of minor importance. Some of the naturally occurring hormones are the auxin indole acetic acid (IAA) and the cytokinin zeatin, while the others are synthetic growth regulators. The characteristics of growth regulators are listed in Table 3.3.

Auxins

A common feature of auxins is their property to induce cell division and formation of callus. Auxin causes cell division, cell elongation and swelling of tissues, and the formation of adventitious roots. It often inhibits adventitious and axillary shoot formation. At low auxin concentrations, adventitious root formation predominates, whereas at high auxin concentrations, root formation fails to occur and callus formation takes place. The compound most frequently used and highly effective is 2,4-dichlorophenoxy ace-

Table 3.3 Characteristics of growth regulators and vitamins

Name	Chemical formula	Molecular weight	Solubility
p-Chlorophenoxy acetic acid	$C_8H_7O_3Cl$	186.6	Alcohol
2,4-Dichlorophenoxy acetic acid	$C_8H_6O_3Cl_2$	221.0	Alcohol
Indole-3 acetic acid	$C_{10}H_9NO_2$	175.2	1N NaOH
Indole-3 butyric acid	$C_{12}H_{13}NO_2$	203.2	1N NaOH
α-Naphthalene acetic acid	$C_{12}H_{10}O_2$	186.2	1N NaOH
β-Naphthoxy acetic acid	$C_{12}H_{10}O_3$	202.3	1N NaOH
Adenine	$C_5H_5N_5 \cdot 3H_2O$	189.1	H_2O
Adenine sulphate	$(C_5H_5N_5)_2 \cdot H_2SO_4 \cdot 2H_2O$	404.4	H_2O
Benzyl adenine	$C_{12}H_{11}N_5$	225.2	1N NaOH
N-isopentenylamino purine	$C_{10}H_{13}N_5$	203.3	1N NaOH
Kinetin	$C_{10}H_9N_5O$	215.2	1N NaOH
Zeatin	$C_{10}H_{13}N_5O$	219.2	1N NaOH
Gibberellic acid	$C_{19}H_{22}O_6$	346.4	Alcohol
Abscisic acid	$C_{15}H_{20}O_4$	264.3	1N NaOH
Folic acid	$C_{19}H_{19}N_7O_6$	441.4	1N NaOH
Colchicine	$C_{22}H_{25}NO_6$	399.4	H_2O

tic acid (2,4-D). Other auxins in use include naphthalene acetic acid (NAA), indole acetic acid (IAA), indole butyric acid (IBA), 2,4,5-trichlorophenoxy acetic acid (2,4,5-T), p-chlorophenoxy acetic acid (pCPA) and picloram (4-amino-3,5,6-trichloropicolinic acid).

Cytokinins

Cytokinins are derivatives of adenine and have an important role in shoot induction. The compounds that are most frequently used are kinetin, benzyl adenine (BA) or 6-benzyl amino purine (BAP), zeatin, and isopentenyladenine (2 iP). These are often used to stimulate growth and development. They usually promote cell division if added together with an auxin. At higher concentrations (1 to 10 mg/l), adventitious shoot formation is induced but root formation is generally inhibited. They promote axillary shoot formation by decreasing apical dominance.

Stock solutions of IAA and kinetin are stored in amber bottles or bottles covered with a black paper and kept in dark since they are unstable in light.

Other hormones

Gibberellins are normally used in plant regeneration. GA_3 is essential for meristem culture of some species. In general, gibberellins induce elongation of internodes and the growth of meristems or buds *in vitro*. Gibberellins usually inhibit adventitious root as well as shoot formation.

Abscisic acid is an important growth regulator for induction of embryogenesis. Ethylene is produced by cultured cells, but its role in cell and organ culture is not known.

Organic supplements

Organic nitrogen

Cultured cells are normally capable of synthesizing all the required amino acids, but it is often beneficial to include organic nitrogen in the form of amino acids such as glutamine and aspargine and nucleotides such as adenine. For cell culture, it is good to add 0.2 to 1.0 g/l of casein hydrolysate or vitamin-free casamino acids. In some cases L-glutamine (up to 8 mM or 150 mg/l) may replace the casein hydrolysate. The amino acids when added should be used with caution, since they can be inhibitory. The amino acids included in the media and amount in mg/l are: glycine (2), asparagine (100), tyrosine (100), arginine (10) and cysteine (10). Sometimes adenine sulphate (2–120 mg/l) is added to the agar media for morphogenesis.

Organic acids

Plant cells are not able to utilize organic acids as a sole carbon source. Addition of TCA cycle acids such as citrate, malate, succinate or fumarate permits growth of plant cells on ammonium as the sole nitrogen source. The cells can tolerate a concentration of upto 10 mM of the acid. Pyruvate may also enhance growth of cells cultured at low density.

Complex substances

A variety of extracts *viz.* protein hydrolysate, yeast extract, malt extract, coconut milk, orange and tomato juice have been tested. With the exception of protein hydrolysate and coconut milk, most others are used as the last resort. If used at higher concentration, the complex substances may adversely affect cell growth. It is advisable to make a preliminary test in a range of 0.1 to 1.0 g/l to assess its effect on growth. Coconut milk is commonly used at 2 – 15% (v/v). The present trend is, however, towards fully defined media and the use of complex mixtures is losing favour.

Activated charcoal at concentrations of 0.2 to 3.0% (w/v) is used where phenol-like compounds are a problem for *in vitro* growth of cultures. It can adsorb toxic brown/black pigments and stabilizes pH. Besides activated charcoal, polyvinylpyrrolidone (250–1000 mg/l), citric acid and ascorbic acid (100 mg/l each), thiourea or L-cysteine are also used to prevent oxidation of phenols.

Phloroglucinol, a phenolic compound, is sometimes added to inhibit the enzyme IAA oxidase responsible for the breakdown of IAA.

Gelling agents

Agar, a seaweed derivative, is the most popular solidifying agent. It is a polysaccharide with a high molecular mass and has the capability of gelling media. Solublized agar forms a gel that can bind water and adsorb compounds. The higher the agar concentration, the stronger the binding of water. Plant tissue culturists often use Difco Bacto agar at a concentration of 0.6 to 1.0% (w/v), although other forms of agar (agarose, phytagar, flow agar, etc.) are also becoming popular. *In vitro* growth may be adversely affected if the agar concentration is too high. With higher concentrations, the medium becomes hard and does not allow the diffusion of nutrients into the tissues. Besides agar, the following alternatives are also available.

i. Alginate can be used for plant protoplast culture.

ii. Gelrite at 0.2% can be used for solidification of media. Gelrite gels are remarkably clear in comparison to those formed by agar. Gelrite requires both a heating cycle and the presence of divalent cations (Mg^{++} or Ca^{++}) for gelation to take place.

iii. Synthetic polymer biogel P200 (polyacrylamide pellets) or a starch polymer can be used.

Liquid media with a support can also be used instead of solid media. These include

i. liquid medium without agar using clean foam plastic, glasswool or rockwool as support;

ii. filter paper-bridge which is hung in a liquid medium;

iii. growth on a liquid medium containing glass beads;

iv. viscose sponge underneath filter paper as a carrier for a liquid medium instead of agar.

pH

pH determines many important aspects of the structure and activity of biological macromolecules. Nutrient medium pH ranges from 5.0 to 6.0 for suitable *in vitro* growth of explant. pH higher than 7.0 and lower than 4.5 generally stops growth and development. The pH before and after autoclaving is different. It generally falls by 0.3 to 0.5 units after autoclaving. If the pH falls appreciably during plant tissue culture (the medium becomes liquid), then a fresh medium should be prepared. It should be known that a starting pH of 6.0 could often fall to 5.5 or even lower during growth. pH higher than 6.0 give a fairly hard medium and a pH below 5.0 does not allow satisfactory gelling of the agar.

PROTOCOLS

3.1: General methodology for medium preparation

Preparation of macro- and micronutrient stock solutions

Stock solutions of macro- and micronutrients, vitamins and growth regulators are prepared in distilled or high purity demineralized water. Chemicals should be of the highest grade. Usually the stock solutions of macro- and micronutrients are prepared as 10x and 100x concentrations as shown in Table 3.4. These nutrient solutions can be dispensed in plastic bags with zipper seals and stored frozen (e.g. 10x macronutrient solution dispensed into a bag containing 100 ml of solution for preparation of 1 l medium). While dissolving the nutrients in water, one compound is added at a time to avoid precipitation. For macronutrient stock solution,

calcium chloride is dissolved separately in water and then added to the rest of the solution to avoid precipitation.

Preparation of vitamin and growth regulator stock solutions

All the growth regulators are not soluble in water. Solubility of different growth regulators is given in Table 3.3. The compound should be dissolved in few ml of solvent and then water slowly added to make up to the requisite volume. Concentrations of compounds can be taken as mg/l or in molarity.

Concentration in mg/l: It is preferable to dissolve 50 mg/100 ml to give a concentration of 0.5 mg/ml.

Concentration in mM: The growth regulator solutions can be prepared as 1 mM for 100 ml as shown in Table 3.5. Concentration of growth regulators in mg/l converted to μM are also given

Table 3.4 Preparation of stock solutions of Murashige and Skoog (MS) medium

Constituent	Concentration in MS medium (mg/l)	Concentration in the stock solution (mg/l)	Volume to be taken/ liter of medium (ml)
Macronutrients (10x)			
NH_4NO_3	1650	16500	100
KNO_3	1900	19000	
$MgSO_4 \cdot 7H_2O$	370	3700	
$CaCl_2 \cdot 2H_2O$	440	4400	
KH_2PO_4	170	1700	
Micronutrients (100x)			
H_3BO_3	6.2	620	10
$MnSO_4 \cdot 4H_2O$	22.3	2230	
$ZnSO_4 \cdot 7H_2O$	8.6	860	
KI	0.83	83	
$Na_2MoO_4 \cdot 2H_2O$	0.25	25	
$CuSO_4 \cdot 5H_2O$	0.025	2.5	
$CoCl_2 \cdot 6H_2O$	0.025	2.5	
Iron source			
Fe·EDTA–Na salt	40	Added fresh	
Vitamins			
Nicotinic acid	0.5	50 mg/100 ml	1
Thiamine HCl	0.1	50 mg/100 ml	0.2
Pyridoxine HCl	0.5	50 mg/100 ml	1
Others			
Myo-inositol	100	Added fresh	
Glycine	2	50 mg/100 ml	4
Sucrose	30,000	Added fresh	
Agar	8000	Added fresh	
pH 5.8			

Table 3.5 Preparation of stock solutions of growth regulators

Compound	Abbreviation	mg/100 ml (1mM)
Auxins		
2,4-Dichlorophenoxy acetic acid	2,4-D	22.1
Indole-3 acetic acid	IAA	17.5
Indole-3 butyric acid	IBA	20.32
a-Naphthalene acetic acid	NAA	18.62
b-Naphthoxy acetic acid	NOA	20.23
2,4,5-Trichlorophenoxy acetic acid	2,4,5-T	25.56
p-Chlorophenoxy acetic acid	pCPA	18.66
Picloram	PIC	24.12
Cytokinins		
Adenine	Ade	18.91
Benzyl adenine or benzyl amino purine	BA or BAP	22.52
N-isopentenylamino purine	2-iP	20.33
Kinetin	KIN	21.52
Zeatin	ZEA	21.92
Others		
Gibberellic acid	GA_3	34.64
Abscisic acid	ABA	26.43
Colchicine	Col	39.94

in Table 3.6. If a culture medium is to contain 10 µM of the growth regulator (e.g. 2,4-D), then

1 mM = 221 mg/l or 0.221 mg/ml

The amount in 100 ml stock solution = 0.221 × 100 ml = 22.1 mg

10 µM = 2210 mg or 2.210 mg

The required volume of stock solution to be added is 10 ml (22.1/10 = 2.210 mg).

Add each component according to the list, including growth regulators, by using correctly sized graduated cylinders or pipettes or balance.

Add water to just below the final volume.

Adjust the pH of the medium to the required value (e.g. pH 5.8 for MS) by drop-wise addition of 1N KOH or 1N HCl with constant stirring.

Preparation of 1N HCl: Commercially available HCl has minimum assay of 35.4%, specific gravity = 1.18, M.W. = 36.46

Conversion of liquid into solid =

minimum assay × specific gravity

35.4 × 1.18 = 41.7 g

Thus 100 ml of HCl contains 41.77 g

Table 3.6: Molarity concentrations of some commonly used plant growth regulators

Concentration used	BA	Kinetin	Zeatin	2,4-D	IAA	IBA
Mol. Weight	225.3	215.2	219.2	221.0	175.2	203.2
mg/l	µM	µM	µM	µM	µM	µM
0.1	0.44	0.46	0.46	0.45	0.57	0.49
0.2	0.89	0.93	0.91	0.90	1.14	0.98
0.3	1.33	1.39	1.36	1.36	1.71	1.48
0.4	1.78	1.86	1.82	1.81	2.28	1.97
0.5	2.22	2.32	2.28	2.26	2.85	2.46
0.6	2.66	2.79	2.73	2.71	3.42	2.95
0.7	3.11	3.25	3.19	3.17	3.99	3.44
0.8	3.55	3.72	3.65	3.62	4.57	3.94
0.9	3.99	4.18	4.11	4.07	5.14	4.43
1.0	4.40	4.60	4.56	4.52	5.71	4.90
2.0	8.90	9.30	9.12	9.05	11.42	9.80
5.0	22.20	23.20	22.81	22.62	28.54	24.60
10.0	44.0	46.0	45.62	45.25	57.08	49.00

$$\text{Normality} = \frac{\text{Grams in one liter}}{\text{Equivalent weight}}$$

E.wt. = M.W./acidity or basicity

Therefore, E.wt of HCl = 36.46/1 = 36.46

1N HCl = 417.7/36.46 = 11.5 ml of HCl to make volume of one liter

Make up the solution to the final volume.

Add the needed amount of agar (e.g. 6–8 g/l) or any other gelling agent. Heat the solution while stirring until the agar is dissolved. The solution becomes transparent when the agar is completely dissolved. Distribute the medium in glass or polypropylene vessels and plug it with cotton or cover it with aluminium foil.

The culture medium is sterilized in an autoclave for 20 min at 121°C at 15 p.s.i (105 kPa). If filter-sterilized hormones/compounds are to be added, then these are added to the autoclaved medium. Medium is cooled in laminar airflow to 50 – 60°C, the hormones are added by Millipore or any other filter assembly using 0.22 mm filter, and then dispensed into sterile vessels. After cooling, the media are preferably stored at 4 – 10°C.

It is advisable to mark each vessel to show the precise medium and date of preparation. The culture medium should be used 3 – 4 days after preparation, so that if it is not properly sterilized, contamination will start appearing and this medium can be discarded.

Further reading

Evans, D.A., Sharp, W.R., Ammirato, P.V. and Yamada, Y. 1983. *Handbook of Plant Cell Culture.* Macmillan Publishing Company, New York.

Pierik, R.L.M. 1987. *In vitro Culture of Higher Plants.* Martinus Nijhoff Publishers, Dordrecht.

Reinert, J. and Bajaj, P.P.S. (eds.). 1977. *Applied and Fundamental Aspects of Plant Cell, Tissue and Organ Culture.* Springer-Verlag, Berlin.

Street, H.S. (ed.) 1973. *Plant Tissue and Cell Culture.* Blackwell Scientific Publications, Oxford.

Thorpe, T.A. 1981. *Plant Tissue Culture: Methods and Applications in Agriculture.* Academic Press Inc., New York.

4

Sterilization Techniques

Sterilization is a procedure used for elimination of microorganisms. The maintenance of aseptic (free from all microorganisms) or sterile conditions is essential for successful tissue culture procedures. The need for asepsis requires that all culture vessels, media, and instruments used in handling tissues, as well as explant itself be sterilized. The importance is being laid on keeping the air, surface and floors free of dust. All operations are carried out in laminar airflow sterile cabinets. An obvious precaution is not to share these cabinets and areas of tissue culture work being carried out with other microbiologists and pathologists. If tissue culture work itself requires screening against pathogens or isolation of toxins etc., then the pathological work area must be separated. The cabinets for plant tissue culture work should have a horizontal airflow from back towards the front, while vertical airflow cabinets are preferable for working with pathogens or *Agrobacterium*-mediated transformation.

The objective in this chapter is to explain specific procedures/guidelines for preparing sterile media, containers/vessels, explant materials and for maintaining sterile conditions.

In general different sterilization procedures can be grouped under three categories.
1. Preparation of sterile media, containers and small instruments
2. Maintenance of aseptic conditions
3. Preparation of sterilized explant material

PREPARATION OF STERILE MEDIA, CONTAINERS AND SMALL INSTRUMENTS

The most popular method of sterilizing culture media and apparatus is by autoclaving the material using steam/dry sterilization.

Steam sterilization

Most nutrient media are sterilized by using an autoclave. The standard conditions for autoclaving media are 121°C with a pressure of 15 psi for 20 min. These conditions are followed for test tubes or other containers containing between 20 – 50 ml of nutrient media. An autoclave has a temperature range of 115–135°C. Good sterilization relies on time, pressure, temperature and volume of the object to be sterilized. The following guidelines must be followed:

Test tubes/flasks containing between 20 – 50 ml nutrient media – 20 min. at 121°C, 15 psi.

Flasks containing between 50–500 ml nutrient media – 25 min. at 121°C, 15 psi.

Flasks containing between 500–5000 ml nutrient media – 35 min. at 121°C, 15 psi.

It must be realized that heat penetration is very important in an autoclave and large volumes must be sterilized for longer periods, as the heat will take longer to penetrate than with smaller volumes.

Bottles when autoclaving should not be tightly packed and their tops should be loose.

After autoclaving, these are kept in the laminar airflow and tops of bottles are tightened when they are cool.

The advantages of an autoclave are speed, simplicity and destruction of viruses, while disadvantages are change in pH by 0.3 – 0.5 units, components can separate out and chemical reactions can occur resulting in a loss of activity of media constituents.

Units of energy and pressure

Pounds per square inch (psi) is very commonly used but it can also be expressed as Joules per kilogram (J/kg) which is an expression of energy per unit of mass and should be used.

Pascal units are used to express pressure. 1 J/kg = 1 kPa (kilo Pascal)

Thus, 1 psi = 6.9 kPa or 15 psi = ~ 105 kPa During and after autoclaving, the following should be considered:

i. The pH of the media is lowered by 0.3 – 0.5 units.

ii. Autoclaving at too high temperature can caramelize sugars, which may be toxic.

iii. Volatile substances can be destroyed by the use of an autoclave.

iv. Autoclaving for too long periods can precipitate salts, and depolymerize the agar.

v. Care should be taken to use the correct duration and temperature.

Dry sterilization

The sterilization of glassware and metallic instruments can be carried out in dry heat for 3 h at 160–180°C. An exposure of 160°C dry heat for 2 h is regarded as equivalent to moist heat (steam) sterilization at 121°C for 15 min. Dry goods (instruments, empty glassware) can either be wrapped in aluminium foil, brown paper or sealed metal containers to maintain sterility.

Present day autoclaves come with both dry and steam sterilization programs. If material to be autoclaved is dry *viz.* instruments, glassware, pipettes, tips, etc. then dry cycle in an autoclave should be used. For sterilization of liquid material (media, water) wet or steam sterilization employing either autoclave or domestic pressure cooker can be used but with a precaution of slow exhaust. Domestic pressure cookers are very useful when small amount of media is to be sterilized.

Filter sterilization

Some growth factors, amino acids, vitamins, and toxins are heat labile and get destroyed during autoclaving with the rest of nutrient medium. Such chemicals are therefore sterilized by filtration through a sieve or filtration assembly using filter membranes of 0.45 to 0.22 μm. A range of bacteria proof membrane filters (Millipore - Millipore Co., Nalgene - Nalgene Co., Sartorius - Sartorius Filters, Inc., etc.) and associated equipment are available for sterilization of different volumes of liquid in the range of 1 – 200 ml. Most filters are of cellulose acetate and/or cellulose nitrate, and are available in pre-sterilized, plastic disposable units. During filter sterilization, all the particles, microorganisms and viruses that are bigger than the pore diameter of the filter used are removed. Filter sterilization is often used if a thermolabile substance (e.g. X) is needed in the medium (e.g. A). The basic medium without X substance is first autoclaved in a flask, while the medium is still liquid (i.e. ~50°C), the substance X is injected with the hypodermic syringe fitted to a membrane filter. These operations are carried out in the laminar airflow hood. The medium containing X is then stirred and dispensed into pre-sterilized containers. For larger volumes, filter sterilization can be carried out using a vacuum filtering setup marketed by various companies.

Ultra violet sterilization

Sterilization of nutrient media by irradiation (via gamma rays) is hardly ever used for tissue culture media preparation because it is extremely expensive as compared to the usual method of autoclaving. However, the pre-sterilized disposable plastic wares are generally UV- radiation sterilized. The sterilized autoclaved media are then dispensed into these sterilized containers in the laminar airflow cabinet. Radiation treatment as such is generally not used in research laboratories.

MAINTENANCE OF ASEPTIC CONDITIONS

The following procedures/techniques are employed during aseptic transfer:

Alcohol sterilization

It is essential that worker hands be relatively aseptic during manipulation work. A wash with an antibacterial detergent followed by spraying of 70% ethanol on hands is quite effective. The laminar airflow cabinets should also be sprayed with 70% ethanol before use.

Flame sterilization

This method is used for sterilization of instruments that are continuously used during manipulation work. Instruments are soaked in 70–80% alcohol followed by flaming on a burner in the laminar airflow hood, which should be carried out repeatedly while aseptic manipulation work is in progress. This is essential even if instruments have been subjected to prior dry or steam sterilization.

Culture containers should be covered as quickly as possible after an operational step is completed.

Talking in the hood should be avoided. If necessary, turn your face either left or right, but the hands should remain inside the cabinet.

Set up all materials including containers, media, etc. in one side of the cabinet. It should not disturb the airflow pattern. Material should not come in the path of transfer work to be carried out in the cabinet.

STERILIZATION OF EXPLANT

Chemical sterilization

Plant material can be surface sterilized by a variety of chemicals. It is the eradication of microorganisms with the aid of chemicals. The type and concentration of chemical sterilant to be used and exposure time must be decided upon empirically. Some of the commonly used chemical sterilants and their effectiveness is shown in a comparative form in Table 4.1. It is frequently seen that over-zealous sterilization leads not only to the complete removal of all microorganisms but is also lethal to the plant tissue. It is therefore important to determine the optimal conditions for each tissue.

1. 1% Solution of sodium hypochlorite – NaClO, commercial bleach: It is generally available with 5% active chlorine content, so 20% can be used for normal sterilization.
2. Calcium hypochlorite $Ca(ClO)_2$: This comes in a powder form. It is mixed well with water, allowed to settle and the clarified filtered supernatant solution is used for sterilization. Generally, 4–10 g/100 ml of $Ca(ClO)_2$ is used. Calcium hypochlorite enters the plant tissue slowly as compared to sodium hypochlorite. It can be stored only for a limited period of time because of deliquescent (takes up water) nature.
3. 1% solution of bromine water.
4. 0.01 to 1% solution of mercuric chloride: It is dissolved in water to make the solution. Mercuric chloride is an extremely toxic substance for plants, so rinsing must be very

Table 4.1 A comparison of the effectiveness and properties of common surface sterilants for explant

Sterling agent	Concentration used	Ease of removal	Treatment time (min)	Remarks
Sodium hypochlorite	1–1.4%[a]	+++	5 – 30	Very effective
Calcium hypochlorite	9–10%	+++	5 – 30	Very effective
Hydrogen peroxide	10–12%	+++++	5 – 15	Effective
Bromine water	1–2%	+++	2 – 10	Very effective
Silver nitrate	1%	+	5 – 30	Effective
Mercuric chloride	0.01–1%	+	2 – 10	Satisfactory
Antibiotics	4–50 mg/l	++	30 – 60	Effective

[a]Common usage rate is 20% v/v of commercial solution

thorough. It should be disposed of in a separately marked container.

5. Alcohol: 70% alcohol is used for sterilization of plant material by dipping them for 30 sec to 2 min. Generally, alcohol alone is not sufficient to kill all the microorganisms and the plant material after alcohol treatment is treated with another chemical sterilant.
6. 10% hydrogen peroxide.
7. 1% silver nitrate solution.
8. 4–50 mg/l of antibiotic solution.

It is important that the plant surface be properly wetted by the sterilizing solution. A 30 sec immersion in 70% (v/v) ethanol or the addition of a few drops of liquid detergent (e.g. Teepol, Tween 20) to the other chemical sterilant solutions during use will ensure effectiveness of the sterilizing agent. Explants after treatment with sterilants must be thoroughly rinsed with several changes of sterile distilled water because retention of such noxious chemicals will seriously affect the establishment of culture.

PROTOCOLS

4.1. Sterilization of seeds

1. Take seeds in a tea strainer/nylon net pouch/muslin cloth and wash in running tap water. If seeds are highly infected then first wash them in a beaker with few drops of liquid detergent.
2. After washing, treat seeds with 70% alcohol for 30 sec to 2 min in a beaker. Decant the alcohol.
3. Then treat seeds with 20–40% solution of sodium hypochlorite solution (preferably 40%) containing 2–3 drops of Tween 20 for 10–20 min with stirring. Alternatively, treat the seeds with 0.1 to 1% solution of mercuric chloride for 2–10 min or saturated solution of calcium hypochlorite and use a 10–100% dilution of this solution for 10–20 min.
4. In a sterile laminar airflow hood, decant the chemical sterilant and wash the seeds thoroughly 5x with sterile distilled water.

4.2. Sterilization of buds, leaf, stem, roots, tubers, scales, etc.

1. Treat the tissue for 30 sec to 2 min in 70% alcohol by submerging in it. Tubers, roots and bulbs should be thoroughly washed in detergent and running tap water before alcohol treatment.
2. Tissue is then treated with 20% sodium hypochlorite solution (1% active chlorine content) for 5–10 minutes. Concentration and/or time can be varied depending upon the tissue and degree of contamination.
3. In a sterile laminar airflow hood, decant the chemical sterilant and wash the tissue thoroughly 3–5x with sterile distilled water.

4.3. Sterilization of tissue for immature embryos, ovules, and flower buds for anther culture

1. The whole enclosing tissue is surface sterilized, e.g. ovary, developing seeds of wheat/barley/maize/rice etc. for immature embryos, flower buds for anther culture.
2. Tissue is submerged in 20% sodium hypochlorite solution containing 2–3 drops of Tween 20 for 5–10 minutes with stirring.
3. In a sterile laminar airflow hood, decant the chemical sterilant and wash the tissue thoroughly 3–5x with sterile distilled water.
4. Dissect out/excise the required explant.

Further reading

Evans, D.A., Sharp, W.R., Ammirato, P.V. and Yamada, Y. 1983. *Handbook of Plant Cell Culture*. Macmillan Publishing Company, New York.

Pierik, R.L.M. 1987. *In vitro Culture of Higher Plants*. Martinus Nijhoff Publishers, Dordrecht.

Reinert, J. and Bajaj, P.P.S. (eds.). 1977. *Applied and Fundamental Aspects of Plant Cell, Tissue and Organ Culture*. Springer-Verlag, Berlin.

Street, H.S. (ed.) 1973. *Plant Tissue and Cell Culture*. Blackwell Scientific Publications, Oxford.

Thorpe, T.A. 1981. *Plant Tissue Culture: Methods and Applications in Agriculture*. Academic Press Inc., New York.

Vasil, I.K. 1984. *Cell Culture and Somatic Cell Genetics of Plants: Laboratory Procedures and Their Applications*. Academic Press, Inc., New York.

5

Types of Culture

A plant consists of different organs, and each organ is composed of different tissues, which in turn are made up of different individual cells. Plant tissue culture refers to the *in vitro* cultivation of plants, seeds, plant parts (tissues, organs, embryos, single cells, protoplasts) on nutrient media under aseptic conditions. The procedures of plant tissue culture have developed to such a level that any plant species can be regenerated *in vitro* through several methodologies. Unlike animal cells, plant cells, even highly mature and differentiated, retain the ability to change to a meristematic state and differentiate in to a whole plant if it has retained an intact membrane system and a viable nucleus.

In plant tissue culture, more often an explant (an excised piece of differentiated tissue or organ) is used to initiate their growth in culture. The non-dividing, differentiated, quiescent cells of the explant when grown on a nutrient medium first undergo changes to achieve the meristematic state. The phenomenon of mature cells reverting to a meristematic state and forming undifferentiated callus tissue is termed as *dedifferentiation*. Since the multicellular explant comprises cells of diverse types, the callus derived will be heterogeneous. The ability of the component cells of the callus to differentiate into a whole plant or a plant organ is termed as *redifferentiation*. These two phenomena of dedifferentiation and redifferentiation are inherent in the capacity of a plant cell, and thus giving rise

to a whole plant is described as *cellular totipotency*. Generally, a callus phase is involved before the cells can undergo redifferentiation leading to the regeneration of a whole plant. The dedifferentiated cells can rarely give rise to whole plants directly without an intermediate callus phase.

CYTODIFFERENTIATION

The cells in a callus are parenchymatous in nature; the differentiation of these cells into a variety of cells is required during redifferentiation of cells into whole plants. This redifferentiation of cells is known as cytodifferentiation. *In vitro* and *in vivo,* the main emphasis in plant cytodifferentiation has been laid on vascular tissue differentiation (xylem, phloem), particularly the xylem elements. In an intact plant, tissue differentiation goes on in a fixed manner that is characteristic of the species and the organ, while callus cultures which lack vascular elements offer a valuable system for the study of the effect of various chemicals and physical factors on vascular tissue differentiation. Two substances, auxins and sucrose, have a major effect on vascular tissue differentiation, while cytokinins and gibberellins promote differentiation into xylem tissue (xylogenesis). Generally, auxin at low concentration (0.05 mg/l) stimulates xylogenesis and there is an inverse relationship between the auxin concentration and degree of xylem differentiation. Further, it has been observed that the effect of auxin on vascular tissue differentiation seems to be closely dependent on the presence of sugars. Varying the sucrose concentration in the presence of a low concentration of auxin could change the relative amounts of xylem and phloem formed in callus cultures. In studies on *Syringa*, keeping the concentration of IAA auxin at 0.05 mg/l constant and sucrose level at 1% induced little xylem formation; 2% sucrose favoured better xylem formation with little or no phloem; 2.5–3.5% sucrose resulted in both xylem and phloem differentiation and with 4% sucrose, phloem was formed with little or no xylem. Besides sucrose, other sugars such as glucose, fructose, maltose and trehalose have also been used to stimulate vascular differentiation. The role of cytokinins and gibberellins in xylem differentiation has also been reported. Besides light and temperature, physical factors have a pronounced effect on vascular differentiation.

ORGANOGENIC DIFFERENTIATION

In nature, totipotentiality of somatic cells has been observed in several taxa where stem, leaf and root pieces are able to differentiate into shoots and roots. *In vitro* studies have indicated that totipotentiality is not restricted to few species; most plant species if provided with appropriate conditions would show differentiation. For the regeneration of a whole plant from a cell or callus tissue, cytodifferentiation is not enough and there should be differentiation leading to shoot bud or embryo formation. This may occur either through organogenesis or somatic embryogenesis. In organogenesis, a monopolar structure that has a connection with the pre-existing vascular tissue within the callus, while in embryogenesis a bipolar structure with no vascular connection with the maternal callus tissue or explant is formed. Regeneration of plants via organogenesis and embryogenesis has been discussed in micropropagation (see Chapter 6).

TYPES OF CULTURE

Plant tissue culture, which covers all types of aseptic plant culture should be used in a restricted sense and it is possible to distinguish it into various types of cultures (Fig. 5.1).

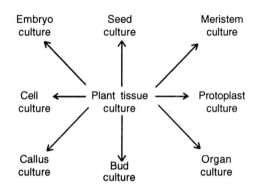

Fig. 5.1   Types of culture.

Seed culture	Culture of seeds *in vitro* to generate seedlings/plants
Embryo culture	Culture of isolated mature or immature embryos
Organ culture	Culture of isolated plant organs. Different types can be distinguished, e.g. meristem, shoot tip, bud culture root culture, anther culture
Callus culture	Culture of a differentiated tissue from explant allowed to dedifferentiate *in vitro* and a so-called callus tissue is produced
Cell culture	Culture of isolated cells or very small cell aggregates remaining dispersed in liquid medium
Protoplast culture	Culture of plant protoplasts, i.e. cells devoid of their cell walls

Seed culture

Seed culture is an important technique when explants are taken from *in vitro*-derived plants and in propagation of orchids. Sterilizing procedures are needed for plant materials that are to be used directly, as explant source can cause damage to the tissues and affect regeneration. In that case, culture of seeds to raise sterile seedlings is the best method. Besides, in several species during *in vivo* conditions, seeds do not germinate well, e.g. orchids. It has been reported that orchids live in symbiotic relationship with the fungus from the moment of germination. Symbiosis is the association of two organisms to their mutual advantage. The research carried out by Knudson around 1922 showed that seeds of *Cattleya, Laelia, Epidendrum,* and many other orchids were able to germinate asymbiotically *in vitro.* In nature germination of orchid seedlings is dependent on a symbiotic relationship with a fungus. However, *in vitro* it is possible to be independent of the fungus by substituting its action with a nutrient medium, and this is known as asymbiotic germination. Orchid seeds are propagated vegetatively as well as generatively. With vegetative propagation, the progeny is identical to the parent plants. However, with generative propagation (by seed), identical progeny is rarely obtained. If seeds from a cultivated orchid are used (mainly obtained from a cross and strongly heterozygous), the progeny will be extremely heterogeneous and seldom identical to the starting material. In principle, cultivated orchids can only be propagated vegetatively. Orchid cloning *in vivo* is a very slow process. Thus seeds can be germinated *in vitro* and mericloning (vegetative propagation by meristem culture) is then carried out on a large scale.

Most of the orchids are sown *in vitro* because

i. Orchid seeds are very small and contain very little or no food reserves. Their small size (1.0–2.0 mm long and 0.5–1.0 mm wide) makes it very likely that they can be lost if sown *in vivo*, and the limited food reserves also make survival *in vivo* unlikely. The seed consists of a thickened testa, enclosing an embryo of about 100 cells. The embryo has a round or spherical form. Most orchid seeds are not differentiated: there are no cotyledons, roots and no endosperm. The cells of the embryo have a simple structure and are poorly differentiated;

ii. Sowing *in vitro* makes it possible to germinate immature orchid embryos, thus shortening the breeding cycle; and

iii. Germination and development take place much quicker *in vitro* since there is a conditioned environment and no competition with fungi or bacteria.

Orchid seeds imbibe water via the testa and become swollen. After cell division has begun, the embryo cracks out of the seed coat. A protocorm-like structure is formed from the clump of cells and on this a shoot meristem can be distinguished. Protocorm has a morphological state that lies between an undifferentiated embryo and a shoot. Protocorms obtained by seed germination have many close similarities with those produced from isolated shoot tips; the term *protocorm-like bodies* has been introduced while cloning orchids by meristem culture. The vegetative propagation of orchids follows culture of seeds, transformation of meristem into protocorm-like bodies, the propagation of protocorms by cutting them into pieces and the development of these protocorms to rooted shoots.

A large number of factors influence the germination and growth of orchids. The mineral requirement of orchids is generally not high and a salt poor medium of Knudson (1946) and Vacin and Went (1949) are good. Some of the orchids require darkness (*Paphiopedilum ciliolare*) for germination while others require low irradiance. Sugar is extremely important as an energy source, especially for those that germinate in darkness. Regulators are usually not necessary for seed germination, and their addition often leads to unwanted effects like callus formation, adventitious shoot formation, etc.

Embryo culture

Embryo culture is the sterile isolation and growth of an immature or mature embryo *in vitro*, with the goal of obtaining a viable plant. The first attempt to grow the embryos of angiosperms was made by Hannig in 1904, who obtained viable plants from *in vitro* isolated embryos of two crucifers *Cochleria* and *Raphanus* (Hannig, 1904). In 1924, Dietrich grew embryos of different plant species and established that mature embryos grew normally but those excised from immature seeds failed to achieve the organization of a mature embryo (Dietrich, 1924). They grew directly into seedlings, skipping the stages of normal embryogenesis and without the completion of dormancy period. Laibach (1925, 1929) demonstrated the practical application of this technique by isolating and growing the embryos of interspecific cross *Linum perenne* and *L. austriacum* that aborted *in vivo*. This led Laibach to suggest that in all crosses where viable seeds are not formed, it may be appropriate to excise their embryos and grow them in an artificial nutrient medium. Embryo culture is now a well-established branch of plant tissue culture.

There are two types of embryo culture:

Mature embryo culture: It is the culture of mature embryos derived from ripe seeds. This type of culture is done when embryos do not survive *in vivo* or become dormant for long periods of time or is done to eliminate the inhibition of seed germination. Seed dormancy of many species is due to chemical inhibitors or mechanical resistance present in the structures covering the embryo, rather than dormancy of the embryonic tissue. Excision of embryos from the testa and culturing them in the nutrient media may bypass such seed dormancy. Some species produce sterile seeds, which may be due to incomplete embryo development. Embryo culture procedures may yield viable seedlings. Embryos excised from the developing seed at or near the mature stage are autotrophic and grow on a simple inorganic medium with a supplemental energy source.

Immature embryo culture/embryo rescue: It is the culture of immature embryos to rescue the embryos of wide crosses. This is mainly used to avoid embryo abortion with the purpose of producing a viable plant. Wide hybridization, where individuals from two different species of the same genus or different genera are crossed, often leads to failure. There are several barriers, which operate at pre- and post-fertilization levels to prevent the successful gene transfer from wild into cultivated species. The pre-fertilization barriers include all factors that hinder effective fertilization, which is usually due to inhibition of pollen tube growth by the stigma or upper style. Post-fertilization barriers hinder or retard the development of the zygote after fertilization and normal development of the seed. This frequently results from the failure of the hybrid endosperm to develop properly, leading to starvation and abortion of the hybrid embryo or results from embryo–endosperm incompatibility where the endosperm produces toxins that kills the embryo.

Raghavan (1976) has discussed evidence which suggests that embryos of inviable hybrids possess the potential for initiating development, but are inhibited from reaching adult size with normal differentiation. Endosperm development precedes and supports embryo development nutritionally, and endosperm failure has been implicated in numerous cases of embryo abortion. Endosperm failure generally results in abnormal embryo development and eventual starvation. Thus, isolation and culture of hybrid embryos prior to abortion may circumvent these strong post-zygotic barriers to interspecific and intergeneric hybridization. The production of interspecific and intergeneric hybrids is the most conspicuous and impressive application of embryo rescue and culture technique, particularly for subsequent valuable gene transfer from wild species.

The underlying principle of embryo rescue technique is the aseptic isolation of embryo and its transfer to a suitable medium for development under optimum culture conditions. With embryo culture there are normally no problems with disinfecting. Florets are removed at the proper time and either florets or ovaries are sterilized. Ovules can then be removed from the ovaries. The tissue within the ovule, in which the embryo is embedded, is already sterile. For mature embryo culture either single mature seeds are disinfected or if the seeds are still unripe then the still closed fruit is disinfected. The embryos can then be aseptically removed from the ovules. Utilization of embryo culture to overcome seed dormancy requires a different procedure. Seeds that have hard coats are sterilized and soaked in water for few hours to few days. Sterile seeds are then split and the embryos excised.

The most important aspect of embryo culture work is the selection of medium necessary to sustain continued growth of the embryo. In most cases a standard basal plant growth medium with major salts and trace elements may be utilized. Mature embryos can be grown in a basal salt medium with a carbon energy source such as sucrose. But young embryos in addition require different vitamins, amino acids, and growth regulators and in some cases natural endosperm extracts. Young embryos should be transferred to a medium with high sucrose concentration (8–12%) which approximate the high osmotic potential of the intracellular environment of the young embryo-sac, and a combination of hormones which supports the growth of heart-stage embryos (a moderate level of auxin and a low level of cytokinin). Reduced organic nitrogen as aspargine, glutamine or casein hydrolysate is always beneficial for embryo culture. Malic acid is often added to the embryo culture medium. After one or two weeks when embryo ceases to grow it must be transferred to a second medium with a normal sucrose concentration, low level of auxin and a moderate level of cytokinin which allows for renewed embryo growth with direct shoot germination in many cases. In some cases where embryo does not show shoot formation directly, it can be transferred to a medium for callus induction followed by shoot induction. After the embryos have grown into plantlets *in vitro*, they are generally transferred to sterile soil and grown to maturity.

Applications of embryo culture

1. *Prevention of embryo abortion in wide crosses*: Successful interspecific hybrids have been seen in cotton, barley, tomato rice, legume, flax and well known intergeneric hybrids include wheat × barley, wheat × rye, barley × rye, maize × *Tripsacum*, *Raphanus sativus* × *Brassica napus*. Distant hybrids have also been obtained via embryo rescue in *Carica* and *Citrus* species. Embryo rescue technique has been successfully used for raising hybrid embryos between *Actidinia deliciosa* × *A. eriantha* and *A. deliciosa* × *A. arguata*. Embryo rescue from a cross between an early Japanese cv. Sunago Wase and a Chinese cv. Yuhualu obtained "Zaoxialu" an extra early maturing peach cultivar. It matures in 55 days in Hanzhou, China. Some of the hybrid plants raised by embryo culture have recombined desirable genes such as earliness, disease and pest resistance are listed in Table 5.1. Embryo culture is also used in crosses between diploids and tetraploids.

2. *Production of haploids*: Embryo culture can be utilized in the production of haploids or monoploids. Kasha and Kao (1970) have

Table 5.1 Resistance traits transferred to cultivated species through embryo rescue technique

Crossing species	Resistance trait(s)
Lycopersicon esculentum × *L. peruvianum*	Virus, fungi and nematodes
Solanum melongena × x *S. khasianum*	Brinjal shoot and fruit borer (*Leucinodes arbonalis*)
Solanum tuberosum × *S. etuberosum*	Potato leaf roll virus
Brassica oleracea × *B. napus*	Triazine resistance
Brassica napus × *B. oleracea* (Kale)	Cabbage aphid
Triticum aestivum × *Thinopyrum scirpeum*	Salt tolerance
Hordeum sativum × *H. vulgare*	Powdery mildew and spot blotch
Hordeum vulgare × *H. bulbosum*	Powdery mildew
Oryza sativa × *O. minuta*	Blast (*Pyricularia grisea*) and Bacterial blight (*Xanthomonas oryzae*)

developed a technique to produce barley monoploids. Interspecific crosses are made with *Hordeum bulbosum* as the pollen parent, and the resulting hybrid embryos are cultured, but they exhibit *H. bulbosum* chromosome elimination resulting in the monoploids of the female parent *H. vulgare*.

3. *Overcoming seed dormancy:* Embryo culture technique is applied to break seed dormancy. Seed dormancy can be caused by numerous factors including endogenous inhibitors, specific light requirements, low temperature, dry storage requirements and embryo immaturity. These factors can be circumvented by embryo excision and culture. Seed dormancy in *Iris* is due to the presence of stable chemical inhibitor in the endosperm. American basswood (*Tilia americana*) seed is borne within a tough indehiscent pericarp where the resistance is mechanical. Thus, there is difficulty in germinating seeds of this species. By excising embryos, germination occurred without delay at all stages of seed maturity. A potential use of the technique is in the production of seedlings from seed of naturally vegetatively propagated plants such as bananas (*Musa balbisiana*) and *Colocasia* whose seeds do not germinate in nature, probably due to recalcitrant dormancy. Embryo culture techniques may thus be capable of producing viable seedlings in these species.

4. *Shortening of breeding cycle:* There are many species that exhibit seed dormancy that is often localized in the seed coat and/or in the endosperm. By removing these inhibitions, seeds germinate immediately. Seeds sometimes take up water and O_2 very slowly or not at all through the seed coat, and so germinate very slowly if at all, e.g. Brussels sprouts, rose, apple, oil palm and iris. Hollies (*Ilex*) are important plants for Christmas decorations. *Ilex* embryos remain in the immature heart-shaped stage though the fruits have reached maturity. Under proper germination conditions, three years are required for seeds of *I. opaca* from mature berries to complete their embryonic development and to begin germination. The excised embryo goes through all the developmental sequences *in vitro* and plants can be obtained in 2–3 weeks time. *Rosa* normally takes a whole year to come to flowering and through embryo culture it has been possible to produce two generations in a year.

5. *Prevention of embryo abortion with early ripening stone fruits:* Some species produce sterile seeds that will not germinate under appropriate conditions and will eventually decay in soil, e.g. early ripening varieties of peach, cherry, apricot, plum. Seed sterility may be due to incomplete embryo development, which results in the death of the germinating embryo. In crosses of early ripening stone fruits, the transport of water and nutrients to the yet immature embryo is sometimes cut off too soon resulting in abortion of the embryo. Macapuno coconuts are priced for their characteristic soft endosperm, which fills the whole nut. These nuts always fail to germinate because the endosperm invariably rots before germinating embryo comes out of the shell. Embryo culture has been practised as a general method in horticultural crops including avocado, peach, nectarine and plum. Two cultivars "Goldcrest peach" and "Mayfire nectarine" have resulted from embryo culture and are now commercially grown.

6. Embryos are excellent materials for *in vitro* clonal propagation. This is especially true for conifers and members of Gramineae family.

7. Germination of seeds of obligatory parasites without the host is impossible *in vivo*, but is achievable with embryo culture.

Callus culture

Callus is basically a more or less non-organized tumor tissue, which usually arises on wounds of differentiated tissues and organs. Thus, in principle, it is a non-organized and little-differentiated tissue. The cells in callus are of a parechymatous nature. When critically examined, callus culture is not homogeneous mass of cells, because it is usually made up of two types of tissue: differentiated and non-differentiated. Explant tissue is a differentiated tissue (roots, stem, leaves, flowers, etc.), which is used as a starting material for callus induction. These explant tissues generally show distinct planes of cell division, cell proliferation and organization into specialized structures such as vascular systems. If there are only differentiated cells present in an isolated explant, then dedifferentiation must take place before cell division can occur. Parenchyma cells present in the explant usually undergo this differentiation. If the explant already contains meristematic tissue when isolated, then this can divide immediately without dedifferentiation taking place. Dedifferentiation plays a very important role, enabling mature cells in an explant isolated from an adult plant to be redetermined. In this process, adult cells are temporarily able to revert from the adult to the juvenile state. The rejuvenated cells have a greater growth and division potential and under special circumstances are able to regenerate into organs and/or embryos.

Callus formation takes place under the influence of exogenously supplied growth regulators present in the nutrient medium. The type of growth regulator requirement and its concentration in the medium depends strongly on the genotype and endogenous hormone content of an explant. These requirements can be put into three categories:

1. Auxin alone (especially in monocotyledons)
2. Cytokinin alone
3. Both auxin and cytokinin (carrot)

If the callus is difficult to induce, or if juvenile callus is needed, then immature embryos or seedlings or parts of these are used. It should be taken into account that the type of starting material (juvenile or adult) and the original position of the explant in the plant reflects the endogenous hormone level which have an important influence on processes such as cell division and organ and embryo formation.

Many other factors are important for callus formation: genotype, composition of the nutrient medium, physical growth factors (light, temperature, etc.). The Murashige and Skoog (1962) mineral medium, or modifications of this are often used. Sucrose or glucose (2–4%) is usually employed as the sugar source. The effect of light on callus formation is dependent on the plant species; light may be required in some cases and darkness in other cases. A temperature of 22–28°C is normally advantageous for callus formation.

Callus tissue from different plant species may be different in structure and growth habit: white or coloured, soft (watery) or hard, friable (easy to separate into cells) or compact. The callus growth within a plant species may also depend on factors such as the original position of the explant within the plant, and the growth conditions.

After callus induction, the callus is grown further on a new medium, which is referred to as subculturing. When subcultured regularly on agar medium, callus cultures will exhibit an S-shaped or sigmoid pattern of growth during each passage. There are five phases of callus growth:

1. Lag phase, where cells prepare to divide.
2. Exponential phase, where the rate of cell division is highest.
3. Linear phase, where cell division slows but the rate of cells expansion increases.
4. Deceleration phase, where the rates of cell division and elongation decreases.
5. Stationary phase, where the number and size of cells remain constant.

Callus growth can be monitored by fresh weight measurements, which are convenient for observing the growth of cultures over time in a non-destructive manner. Dry weight measurements are more accurate than fresh weight, but this method requires sacrifice of the samples. Mitotic index measurement of cell division rates require extensive sampling to reduce sample error and are not easy to perform.

Organ culture

It is an isolated organ grown *in vitro*. It can be given different names depending upon the organ used as an explant. For example, meristem or shoot tip culture (see Chapter 6), root culture, nucellus, endosperm, ovule culture, anther culture for production of androgenic haploids while ovule and ovary culture for *in vitro* production of gynogenic haploids (see Chapter 8). The culture of plant organs results in three types of *in vitro* culture (De Fossard *et al.*, 1977).

Organized: The culture of whole plants (embryos, seeds) and organ has been termed as organized culture. In this, characteristic organized structure of a plant or the individual organ is maintained. If the organized structure is not broken down, then progeny arise which are identical to the original plant material (e.g. meristem culture).

Non-organized: If cells and/or tissues are isolated from an organized part of a plant, dedifferentiate and are then cultured, a non-organized growth in the form of a callus tissue results. If the callus disperses into clumps of cells (aggregates) and/or single cells results, it is referred to as suspension culture. Non-organized cultures have very low genetic stability.

Non-organized/organized: This type of culture is intermediate between the above two types. Cells in an isolated organ or tissue, first dedifferentiate and then form callus tissue which then redifferentiate to form organs (roots or shoots) or embryos. Thus organized structures can develop from non-organized cultures either through special techniques or spontaneously. In this the progeny are often not completely identical to the original plant material.

Two types of organ culture are discussed below:

NUCELLUS CULTURE

Nucellus culture has been utilized to study factors responsible for the formation of adventive embryos. Normally the embryo originates from the zygote, but sometimes in plants such as citrus and mango, the embryos arise adventitiously from cells of the nucellus or integument. Rangaswamy (1958) studied for the first time nucellar embryony *in vitro*. He excised the nucellus from unpollinated ovules of *Citrus microcarpa*, a natural polyembryonic species and cultured on a modified White medium supplemented with casein hydrolysate. The nucellus proliferated into a callus mass, and differentiated into embryo-like structures termed pseudobulbils, which eventually developed into plants. Rangan *et al.* (1968) demonstrated that it is possible to induce nucellar embryogenesis in monoembryonic citrus species as well. Adventive embryos were successfully initiated in nucellus cultures of *C. grandis, C. limon* and *C. reticulata* on MS medium supplemented with either malt extract (500 mg/l) or a combination of adenine sulphate (25 mg/l) + NAA (0.5 mg/l) + orange juice (5%). In these species nucellar explants did not produce callus or pseudobulbils, but gave rise directly to highly organized multiple embryos.

Pollination and fertilization are generally considered essential prerequisites for the induction of nucellar embryogenesis. However, in later studies, embryogenesis could be induced *in vitro* from nucelli of unfertilized ovules of *C. sinensis* and *C. aurantifolia*. MS medium is generally adequate for both poly- and mono-embryonic varieties. A high sucrose concentration of 5% is generally recommended. Medium should be supplemented with complex substances rich in amino acids, such as casein hydrolysate, malt extract, or yeast extract. In some systems, malt extract can be replaced by a combination of adenine sulphate and kinetin. In general, it appears that auxins are not essential for citrus proliferation particularly from immature fruit tissues. But auxin together with cytokinin may stimulate embryogenesis.

In a procedure of nucellus culture for direct embryogenesis in citrus, immature green fruits, 2.5–3.0 cm in diameter, are harvested from trees 100–120 days after pollination. After sterilization, a shallow incision around the circumference of the fruit is given and ovules are excised. With the aid of a microscope, the zygotic embryos along with nucellus are rolled away from the integument. At the micropylar end, it is pierced and embryo is pushed and the entire nucellus can be placed on the media.

Reports for induction of nucellar embryogenesis in other species have been very few. Mullins and Srinivasan (1976) succeeded in inducing nucellar embryony in ovule cultures of *Vitis vinifera*, a normally monoembryonic species. Unfertilized ovules grown on a media supplemented with BA and B-naphthoxy acetic acid formed a nucellar callus, which subsequently produced embryoids. These embryoids when transferred to a semi-solid media containing GA and 2iP produced plantlets. Eichholtz *et al.* (1979) reported the formation of adventive embryos in nucellus cultures of apple (*Malus domestica*). Micropylar halves of nucellus formed embryo-like structures 50 days after culture in darkness. Subsequent reculture of these embryos in fresh medium resulted in the formation of secondary embryo-like structures from the cotyledons. Litz and Conover (1982) reported embryogenesis and plant regeneration from nucellar callus in *Carica papaya*.

Application

The adventive embryos are of considerable importance to the horticulturists in that they are genetically uniform and reproduce the characters of the maternal parent without inheriting the variations brought about by gametic fusion. Pseudobulbils of citrus represent tissue banks capable of initiating clones of adventive embryos. In monoembryonic *Citrus* taxa, a serious problem is that of obtaining disease-free clones. It is generally believed that most of the viruses of citrus are not transmitted through seedlings, whether of zygotic or nucellar origin. Thus, in contrast to clones established from cuttings or by conventional methods of propagation, plants derived from nucellar embryos are free from most viruses. Moreover, many of the desirable plant vigor and fruiting characteristics associated with juvenility are restored in trees established from nucellar seedlings. The formation of nucellar embryos from unpollinated ovules is also of considerable interest in that it can be employed to import seedless citrus varieties without the risk of introducing new viral diseases, thereby avoiding the necessity of subjecting vegetative material to long periods of quarantine.

ENDOSPERM CULTURE

In flowering plants, nutrition of the embryo is an important and vital aspect of the life cycle. In angiosperms, it is accomplished by the formation of a new structure, the endosperm, produced by a unique event of double fertilization. Endosperm is formed in most cases by fusion of two polar nuclei and one of the male gametes, resulting in a triploid number of chromosomes. It is the main source of reserve food for the developing embryo and influences its differentiation. Failure of proper development of endosperm resulted in the abortion of embryo. Generally, the endosperm is a short-lived structure and is consumed during the development of embryo (exalbuminous seed). In some plants (e.g. castor), it persists as a massive tissue even in the mature seed (albuminous). The endosperm is an excellent experimental system for morphogenic studies since it lacks any differentiation and consists mostly of parenchymatous cells.

Attempts to grow endosperm tissue in culture began in the 1930s and immature and mature endosperms of various angiosperm taxa (autotrophic as well as parasitic) have been successfully cultured. Lampe and Mills (1933) were the first to report the proliferation of immature endosperm tissue of maize, grown on a medium containing extract of potato or young kernels. La Rue (1947) observed that in nature, in some maize kernels, the pericarp ruptured and the endosperm exhibited a white tissue mass. Sehgal (1974) cultured the immature endosperm of *Hordeum* and *Triticum* and obtained only proliferation; all efforts to induce organogenesis failed. However, in *Oryza sativa*, organogenesis was successfully induced in cultures of immature endosperm by Nakano *et al.* (1975).

Organogenesis has been observed in apple too (Mu *et al.*, 1977). The reports on immature endosperm culture indicate that the tissue requires substances such as yeast extract or casein hydrolysate in addition to basal medium. The age of endosperm at culture is critical for its growth *in vitro*. For example, in maize, wheat, and barley, endosperm tissue younger than 8 or older than 12 days after pollination did not grow in cultures. The endosperm of *Lolium* and *Cucumis* can be grown only when excised 7–10 days after pollination. Nakano *et al.* (1975) pointed out that in rice, only cellular endosperm cultured 4–7 days after pollination showed proliferation.

The mature endosperm of several taxa has now been grown successfully. Mostly, the mature endosperm of parasitic taxa shows optimal growth on a medium containing either only a cytokinin or a cytokinin and an auxin. However, in autotrophic members, in addition to an auxin and cytokinin, casein hydrolysate or yeast extract is also essential. The endosperms of *Santalum album, Croton bonplandianum,* and *Ricinus communis* could be grown on White's medium containing 2,4-D, Kin, and yeast extract. Organ formation has been demonstrated mainly in parasitic species (viz. *Dendrophytoe, Exocarpus, Taxillus*) and in some autotrophic species (viz. *Croton, Citrus, Santalum, Petroselinum,* etc.). In almost all parasitic members, differentiation of shoot buds from the mature endosperm occurred without callusing, whereas in autotrophic members the endosperm usually formed a callus mass followed by the differentiation of shoot buds or roots.

Organogenesis from immature or mature endosperm has been observed only in few species such as rice, apple, *Citrus, Santalum, Croton,* etc. In *Ricinus communis, Zea mays,* and *Cucumis* mature endosperm proliferation resulted in a callus tissue of unlimited growth but organogenesis did not occur, even though endosperm callus showed groups of meristematic cells. The induction of organogenesis in endosperm cultures has always been a challenging problem.

Application

Investigations have provided ample evidence that triploid plants can be raised through endosperm culture. Lakshami Sita *et al.* (1980) developed triploid plants of *Santalum album* and Wang and Chang (1978) produced triploid plantlets from *Citrus*. The conventional method for raising triploids is through a cross between a tetraploid and diploid. The crosses may not be successful due to many problems associated in such crosses. In these cases the culture of endosperm may prove to be advantageous. Presently there are a number of crop species (viz. banana, apple, beet, tea, mulberry) in which triploids are in commercial use and nucellus culture will be helpful.

Endosperm can be used as a nurse tissue for raising hybrid embryos. Kruse (1974) was able to culture 9- to 12-day-old hybrid embryos of *Hordeum × Triticum, Hordeum × Agropyrum* and *Hordeum × Secale* on a medium that had endosperm of *Hordeum*. Using *Hordeum* endosperm as a nurse tissue, young embryos of these hybrids can be induced to germinate and form normal hybrid plants. It is desirable that plant breeders apply these techniques to economically important plants where conventional breeding methods prove futile.

Cell culture

The growing of individual cells that have been obtained from an explant tissue or callus is referred to as cell suspension culture. These are initiated by transferring pieces of explant/callus into flasks with a liquid medium and are then placed on a gyratory shaker to provide aeration and dispersion of cells. It has been discussed in Chapter 7.

Protoplast culture

Protoplasts are the cells without cell wall. Culture of protoplasts and fusion of protoplasts from different strains has been discussed in detail (see Chapter 9).

PROTOCOLS

5.1: Protocol for seed germination (*Nicotiana*)

Plant material: Nicotiana tabacum

1. Wash seeds by submerging in water with a few drops of detergent in a beaker and shake by hand, or wrap seeds in two layers of cheese cloth/muslin cloth/nylon pouch and then wash with water.
2. Submerge the seeds in 70% alcohol for 30–60 sec. Decant the alcohol.
3. Transfer the seeds to a flask or beaker containing 20–40% commercial sodium hypochlorite for 15–20 min. Rinse 5x with sterile distilled water.
4. Place 2–3 seeds per culture vessel on the surface of MS agar medium without growth regulators.
5. Incubate the cultures at 25°C under 16 h photoperiod with ~1000 lux light intensity for 1–2 weeks.
6. Observe weekly for root initiation. If need be, transfer the individual plantlets to half MS medium.

Plant establishment

7. Gently remove well rooted plantlets from the culture vessel, keeping the roots intact. Gently rinse the roots with warm water to remove agar attached to it.
8. Plant the regenerants in small plastic pots with sterile soil mix. Make sure the soil is moist with water. After transfer of plant to the soil, either cover it with inverted beaker or put a plastic wrap and use a plastic stake to allow the plastic wrap to form a tent over the plant. Plants do not need to be watered for first few days.
9. Place the pots in diffused light. Open the tents or remove the inverted container to allow air exchange briefly everyday. After 1 week, let some air in the tent for 1 h each day. After another week, increase gradually to several hours per day. After 2–3 weeks remove the wrap or inverted container from the plants to adjust to natural environmental conditions.

5.2: Protocol for embryo culture (cereals–wheat, maize, barley, rice, etc.)

There are numerous examples in which hybrid embryo rescue operations have been carried out. Because hybrid embryos are difficult to obtain for practical classes, the immature embryos of any cereal or legume can be isolated. The same procedure can be applied to hybrid embryo rescue when such crosses are available.

1. Harvest ears of maize/panicle of rice/spikes of wheat or barley after 12–14 days of crossing or pollination. Developing seeds should be in the milky stage. At this stage embryos are less than 1.5 mm in diameter. Check the size of embryo under binocular microscope. With the help of forceps and/or needle take out the seeds from the ear/panicle/spike and remove the lemna and palea wherever it is present.
2. Place few such seeds in a beaker and sterilize them with 20% commercial sodium hypochlorite solution for 10 min. These operations are to be carried out in the laminar airflow hood. Rinse 3–5x with sterile distilled water.
3. Place a sterilized grain on a sterile slide/petri dish and make cuts with needles above the embryo in the endosperm while holding embryo stationary with the second needle.
4. Dissect out immature embryos (1–1.5 mm diameter) from the seeds under a dissection microscope in the hood of laminar flow.
5. Place the embryos on MS or B_5 medium without growth regulators for direct plant formation.
6. Incubate the cultures in light at 25°C.
7. Transfer the plantlets after 2 weeks to half strength MS or B_5 media with reduced sucrose (1%) concentration.
8. Transfer the plants to plastic pots and follow the conditions of plant establishment as mentioned in Protocol 5.1.

5.3: Protocol for embryo culture (legumes–green gram, black gram, french bean, soybean, etc.)

1. Collect developing pods in which the embryos are at the heart stage and the beginning of cotyledon stage. Depending upon the species this stage will approximately be after 9–12 days of pollination.
2. Dip the pods in 70% alcohol for 1 min and then sterilize with 20% commercial sodium hypochlorite solution for 10 min. These operations are to be carried out in the laminar flow hood. Rinse 3–5x with sterile distilled water.
3. Place the pods on a sterile slide/petri dish and slit open the pods and place the developing seeds on a slide or petri dish.
4. Remove the seed coat pieces directly above the embryo and carefully dissect out the embryos with forceps and needle.
5. Inoculate the embryos on the following medium:
   MS + casein hydrolysate (200 mg/l) + L-glutamine (100–200 mg/l) + NAA (0.01 µM) + BAP (1.0 µM) + sucrose (3–8%).
6. Incubate the cultures in light at 25°C.
7. Transfer the plantlets after 3–4 weeks to half strength MS medium without growth regulators but with reduced sucrose (1%) concentration.
8. Transfer the plants to plastic pots and follow the conditions of plant establishment as mentioned in Protocol 5.1.

5.4: Protocol for callus induction (Nicotiana tabacum)

Plant material: Nicotiana tabacum

1. Follow the Steps 1–5 for *in vitro* germination of seeds.
2. Collect the aseptically germinated seedlings when the cotyledons are fully expanded and the epicotyl is beginning to emerge. Usually this will occur when the seedlings are 2 weeks old. Place each seedling on a sterile slide/petri dish and prepare the explant.
3. Various explants can be used. Excise the shoot apex with and without cotyledons and insert the stem base into the medium, cotyledons, and hypocotyl section from the decapitated seedling. Place the explants on the following medium:

   MS + 1–2 mg/l 2,4-D

4. Incubate the cultures in dark at 25°C. Callus will be produced in 3–4 weeks.
5. Compare the callus induction growth from various explants.
6. Cut small pieces of callus ~ 0.5 g fresh weight and subculture on the same fresh medium for proliferation.

5.5: Protocol for callus induction (cereals–wheat, rice, maize, barley, etc.)

1. Follow the Steps 1–4 of Protocol 5.2 for isolation of immature embryos.
2. Place 5–6 immature embryos per vessel with scutellum side up on the following medium for callus induction:

   MS + 2 mg/l 2,4-D

3. Incubate the cultures in dark at 25°C. Initially the embryos will show swelling followed by callus proliferation from the scutellum in 3–4 weeks. (Embryos bigger than 2 mm diameter will show precocious plant formation.)
4. Scutellar callus can be multiplied at least once on the same callus initiation medium by breaking into 2–3 pieces after 4–6 weeks of initiation of callus.
5. Place hard and compact callus pieces of approximately 0.5 cm^2 on the same callus induction medium for proliferation.

Further reading

Johri, B.M. 1982. *Experimental Embryology of Vascular Plants*. Springer-Verlag, Berlin.

Pierik, R.L.M. 1987. *In vitro Culture of Higher Plants*. Martinus Nijhoff Publishers, Dordrecht.

Reinert, J. and Bajaj, P.P.S. (eds.). 1977. *Applied and Fundamental Aspects of Plant Cell, Tissue and Organ Culture*. Springer-Verlag, Berlin.

Thorpe, T.A. 1981. *Plant Tissue Culture: Methods and Applications in Agriculture*. Academic Press Inc., New York.

Vasil, I.K. 1984. *Cell Culture and Somatic Cell Genetics of Plants: Laboratory Procedures and Their Applications*. Academic Press, Inc., New York.

6

Micropropagation

Tissue culture techniques are becoming increasingly popular as alternative means of plant vegetative propagation. Plant tissue culture involves asexual methods of propagation and its primary goal is crop improvement. The success of many *in vitro* selection and genetic manipulation techniques in higher plants depends on the success of *in vitro* plant regeneration. Crop breeding and rDNA technology require the widespread use of reliable true to type propagation and better regeneration methods. In this general context, the types of propagation required will vary immensely.

Clonal propagation *in vitro* is called micropropagation. The word 'clone' was first used by Webber for cultivated plants that were propagated vegetatively. The word is derived from Greek (clon = twig, spray or a slip, like those broken off as propagules for multiplication). It signifies that plants grown from such vegetative parts are not individuals in the ordinary sense, but are simply transplanted parts of the same individual and such plants are identical. Thus, clonal propagation is the multiplication of genetically identical individuals by asexual reproduction while clone is a plant population derived from a single individual by asexual reproduction. With the current techniques of single cell and protoplast culture that enable many thousands of plants to be ultimately derived from a single cell in a comparatively short span of time, the products of this

rapid vegetative propagation should by definition be considered a single clone.

The significant advantage offered by the aseptic methods of clonal propagation (micropropagation) over the conventional methods is that in a relatively short span of time and space, a large number of plants can be produced starting from a single individual. It has been estimated that axillary bud proliferation approach typically results in an average 10-fold increase in shoot number per monthly culture passage. In a period of 6 months, it is feasible to obtain as many as 1,000,000 propagules or plants starting from a single explant.

Suitable explants from vascular plants including angiosperms, gymnosperms and pteridophytes can be cultured *in vitro* and induced to form adventitious buds, shoots, embryoids or whole plants. Murashige (1974) outlined three major stages involved in micropropagation (Fig. 6.1).

Stage I: Selection of suitable explants, their sterilization and transfer to nutrient media for establishment, i.e. initiation of a sterile culture of the explant.

Stage II: Proliferation or multiplication of shoots from the explant on medium.

Stage III: Transfer of shoots to a rooting medium followed later by planting into soil.

Sometimes a Stage IV is also included in cases where establishment of plantlets in soil is

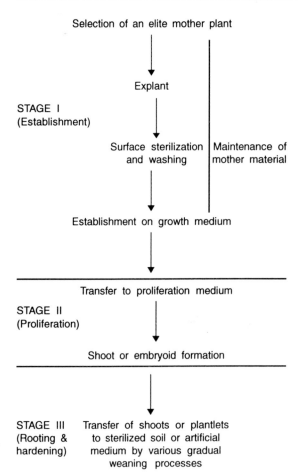

Fig. 6.1 Major stages of micropropagation.

upon the presence of organized explant tissue. *De novo* plant regeneration occurs from callus and cell cultures in the absence of organized tissue. Thus adventitious or *de novo* plant regeneration can occur by organogenesis and embryogenesis. Micropropagation and plant regeneration can be grouped into the following categories:

a) Enhanced release of axillary bud proliferation, i.e. by multiplication through growth and proliferation of existing meristems. It can be through apical shoots excised from the parent plant (meristem and shoot tip culture) and by multiplication of existing meristems within axillary shoots, which proliferate on explants after removal from the parent plant (single node and axillary bud culture).

b) Organogenesis is the formation of individual organs such as shoots and roots either directly on the explant where a preformed meristem is lacking or *de novo* origin from callus and cell culture induced from the explant.

c) Somatic embryogenesis is the formation of a bipolar structure containing both shoot and root meristems either directly from the explant (adventitive origin) or *de novo* origin from callus and cell culture induced from the explant.

AXILLARY BUD PROLIFERATION APPROACH

It is the stimulation of axillary buds, which are usually present in the axil of each leaf, to develop into a shoot. This technique exploits the normal ontogenetic route for branch development by lateral (axillary) meristems. In nature these buds remain dormant for various periods depending upon the growth pattern of the plant. In some species, removal of terminal bud is necessary to break the apical dominance and stimulate the axillary bud to grow into a shoot. Due to continuous application of cytokinin in a culture medium the shoot formed by the bud, which is present on the explant (nodal segment or shoot

particularly elaborate. Further, not all crop species need to be propagated *in vitro* by means of all three stages. These stages have been designed simply to describe micropropagation processes and to facilitate comparison between two or more systems. Plant regeneration can be achieved by culturing tissue sections either lacking a preformed meristem (adventitious origin) or from callus and cell cultures (*de novo* origin). Axillary bud proliferation approach of plant regeneration is based on preformed meri-stems. In contrast, adventitious regeneration occurs at unusual sites of a cultured tissue (e.g. leaf blade, cotyledon, internode, petiole, etc.) where mer-istems do not naturally occur, but it is dependent tip cutting), develops axillary buds. The shoots are then separated and rooted to produce plants or shoots are used as propagules for further propagation. The merit of using axillary bud proliferation from meristem, shoot tip, or bud culture as a means of regeneration is that the incipient shoot has already differentiated *in vivo*. Thus, only elongation and root differentiation are required to establish a complete plant. The induction of axillary bud proliferation seems to be applicable as a means of micropropagation in many cases. This approach can be subdivided into meristem and shoot tip culture, where the explant is taken from the tip of growing shoots while bud cultures are initiated from either terminal or axillary buds, usually with the stem segment attached.

Meristem and shoot tip culture

Morel and Martin (1952) developed the technique of meristem culture for *in vivo* virus eradication of *Dahlia*. Georges Morel (1965) was the pioneer in applying shoot tip culture for micropropagation of orchid *Cymbidium*. *In vitro* clonal propagation gained momentum in the 1970s when Murashige (1974) gave the concept of developmental stages by defining establishment, proliferation, and rooting and hardening stages. This concept stimulated awareness that a single medium is not sufficient for *in vitro* multiplication, but the propagule needs a series of transfers in specially designed physical and chemical environments to reach regeneration. This method has been more successful in herbaceous plants because of weak apical dominance and strong root regenerating capacities as compared to woody species. Many crops are being currently propagated commercially using these *in vitro* procedures.

Shoots of all angiosperms and gymnosperms grow by virtue of their apical meristems. The apical meristem is usually a dome of tissue located at the extreme tip of a shoot and measures ~0.1 mm in diameter and ~0.25 to 0.3 mm in length. The apical meristem together with one to three young leaf primordia measuring 0.1

to 0.5 mm constitutes the shoot apex. The explant of meristem tip culture may either be the apical meristematic dome or the apical dome with few subjacent leaf primordia. Meristem or shoot tip is cut or isolated from stem by applying a V-shaped cut with a sterilized knife. For meristem tip culture, the cut is applied 0.3 to 0.5 mm below the tip of the dome and the excised tissue (explant) is removed along with portions of procambial tissue and is immediately planted on the media. Size of the tip explant is the decisive factor which governs the success of culture. When very small explants are used, the presence of leaf primordia appears to determine the capability of an explant to develop. In general, the larger the explant, the better are the chances for survival. It is generally agreed that large explants such as shoot tips measuring up to 2 cm in length should be selected for *in vitro* micropropagation. However, when disease-free plants are the objective, meristem tips should be cultured. Meristem tips between 0.2 and 0.5 mm most frequently produce virus-free plants and this method is referred to as meristem tip culture. Like all asexual propagation methods, success of meristem and shoot tip cultures is affected by the season during which the explants are obtained. For plant species with a dormant period, best results may be expected when the explants are dissected at the end of their dormancy period. Further, actively growing shoot tips are recommended for their culture because of their strong growth potential and low virus concentration. A general protocol of shoot tip culture has been given in Fig. 6.2.

MS medium salts have been very satisfactory for such cultures though White's medium and the Gautheret's (1959) were the most widely used media during the early days of meristem culture. The growth additives vary from species to species and in some woody species culture one needs to add certain polyphenol oxidative compounds. Some general conclusions from the published reports have been given.

Murashige reported that there are three stages of culture (Fig. 6.1). Stage I is the culture

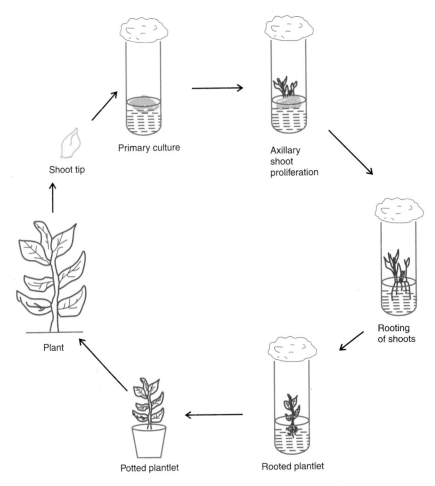

Fig. 6.2 Schematic representation of shoot tip culture.

establishment stage when explants may develop either into single shoots or multiple shoots. It is unlikely that meristem, shoot tip or bud explants have sufficient endogenous cytokinin to support growth and development. Thus at this stage media are supplemented with cytokinin like BA, kinetin, and 2iP. Auxin is another hormone required for shoot growth. Since the young shoot apex is an active site for auxin biosynthesis, exogenous auxin is not always needed in Stage I medium, especially when relatively large shoot tip explants from actively growing plants are used. If an auxin is to be added, it should be a weak one like NAA or IBA rather than 2,4-D. The usual range of concentration is between 0.45 and 10.0 µM.

Many plants are rich in polyphenolic compounds. After tissue injury during dissection, such compounds will be oxidized by polyphenol oxidases and the tissue will turn brown or black. The oxidation products are known to inhibit enzyme activity, kill the explants, and darken the tissues and culture media which severely affects the establishment of explant. Some of the procedures used by various workers to combat this problem are: (i) Adding antioxidants to culture medium viz. ascorbic acid, citric acid, polyvinylpyrrolidone (PVP), dithiothreitol, bovine serum albumin, etc.; (ii) Presoaking explants in antioxidant before inoculating into the culture medium; (iii) Incubating the initial period of primary cultures in reduced light or darkness because it is known that phenolic oxidation products are formed under illumination; and (iv) Frequently transferring explants into fresh medium whenever browning of the medium is observed.

In Stage II the objective is to multiply the propagule and for this axillary shoot proliferation is followed as it maintains higher genetic stability and is more easily achievable by most plant species than organogenic or embryogenic path. In axillary shoot proliferation, high levels of cytokinin are utilized to overcome the apical dominance of shoots and to enhance the branching of lateral buds from leaf axils. The cytokinin concentration varies but, 4.5 to 25.0 µM or higher has been used. In general BA is the most effective cytokinin followed by kinetin and 2iP. Exogenous auxin does not promote axillary shoot proliferation, however, if it is used in low concentration it is to nullify the suppressive effect of high cytokinin concentrations on axillary shoot elongation and to restore normal shoot growth.

The purpose of Stage III purpose is *de novo* regeneration of adventitious roots from the shoots obtained at Stage II or in some cases at Stage I itself. Usually *in vitro* produced shoots of sufficient length are cut or separated and transferred to a medium containing auxin. It is well known that *de novo* root initiation depends on a low cytokinin to a high auxin ratio. Since the hormonal requirements for Stage II culture medium are opposite to this particular balance, Stage II medium is rarely used in Stage III. Numerous studies have indicated that NAA followed by IBA, IAA, 2,4-D and other auxins are used for induction of root regeneration. In many species addition of auxin is unnecessary because of its production in young shoots and in such cases it should be avoided because high level of auxin will induce callus formation. Sometimes if roots are unable to initiate in such

high salt media irrespective of the hormone present, then the salt concentration in the medium should be lowered to one-half, one-third, or one-fourth of the standard strength.

Optimum inoculation conditions vary but generally the cultures are kept in the range of 20–28°C with the majority in the middle range of 24–26°C. Light intensity and duration tends to vary with species but commonly 16 h day vs. 8 h night is used. The light intensity is usually 1–10 Klux. In tissues where browning is the problem, lower light intensity is better.

Bud culture

Buds contain quiescent or active meristems depending upon the physiological state of the plant. Most vascular plants have an indeterminate mode of growth in which the leaf axils contain subsidiary meristems, each of which is capable of growing into a shoot that is identical to the main axis. Depending on the degree of branching displayed by a particular species, only a limited number of axillary meristems develop, the majority being inhibited by apical dominance. The various types used in bud culture are described below.

Single node culture

Here a bud is isolated along with a piece of stem, with the purpose of forming a shoot by allowing the bud to develop. This method is the most natural method for vegetatively propagating plants *in vitro*, since it is also applicable *in vivo*. The bud found in the axil of a leaf is similar to the one in the stem tip. Bud is isolated and cultured on a nutrient medium for *in vitro* development. The buds in the axil of the newly formed leaves can then also be subcultured and allowed to develop into plantlet. When enough shoots have been obtained, these must be rooted and then transferred to soil. In this technique where buds and shoot tips are isolated and cultured, in principle no cytokinin is added (Fig. 6.3).

The first research into isolation of buds and subsequent rooting of the shoots formed was

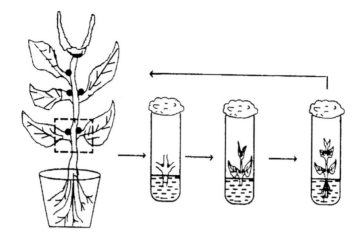

Fig. 6.3 Schematic representation of vegetative propagation by single node technique.

carried out on asparagus. The single node method has had a good success with many different plants such as potato, *Vitis,* peas, rose, tomato, cucumber, aubergine, etc. Recently, this method has also been applied to *Eucalyptus grandis* and *Araucaria cunninghammii.*

The following points should be kept in mind when applying this method:

i. Isolation by this method is practically impossible when dealing with rosette plants such as gerbera, and when the chances of infection are high.

ii. To reduce the chances of infection it is best to isolate closed buds.

iii. The rate of propagation is dependent on the number of buds available. The more leaves that are initiated with time, the more buds became available, and more rapid is the propagation.

iv. Dormancy, especially of woody plants from temperate climates often presents problems.

v. Starting material, position, age, season of the year at the time of isolation, and physical factors are of great importance as is the composition of nutrient medium.

Axillary bud method

In this method a shoot tip is isolated, from which the axillary buds in the axils of leaves develop under the influence of a relatively high cytokinin concentration. This high cytokinin concentration stops the apical dominance and allows axillary buds to develop. If a shoot tip has formed number of axillary or side shoots, then they can be inoculated on to fresh medium containing cytokinin. A schematic diagram is shown in Fig. 6.4. In practice, the single node method is often used in combination with the axillary bud method.

The axillary shoot method was first used for carnation. However, in recent years this method is rarely used for carnation, while it is used on a large scale for strawberry and gerbera. A few of the most important points concerning axillary shoot formation must be kept in mind:

i. The cytokinin requirement is extremely variable and should be adjusted according to the plant species and particular cultivar in use. Most plants react very well to BA and often less to kinetin or 2iP.

ii. Cytokinin concentration correlates with the developmental stage of the material. Juvenile material requires less cytokinin than adult material.

iii. Cytokinin/auxin ratio is usually 10 : 1.

iv. It is sometimes recommended that the shoot tip should be allowed to develop before the

a

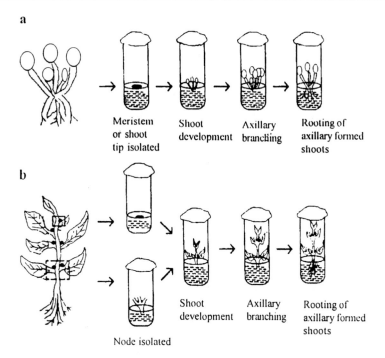

| Meristem or shoot tip isolated | Shoot development | Axillary branching | Rooting of axillary formed shoots |

b

Node isolated

Shoot development   Axillary branching   Rooting of axillary formed shoots

Fig. 6.4 Schematic representation of axillary bud method of vegetatively propagating plants. (a) rosette plants; (b) elongate plants.

cytokinin concentration is increased to induce axillary shoot formation.

v. If axillary shoot development is not satisfactory then remove or kill the apical meristem.

vi. Liquid medium sometimes promotes axillary shoot formation.

vii. It has been demonstrated in wide range of species that thidiazuron and several substituted pyridylphenylurea compounds rather then classical cytokinins stimulate more axillary branching.

ORGANOGENESIS

In plant tissue culture, organogenesis is a process of differentiation by which plant organs viz. roots, shoots, bud flowers, stem, etc. are formed while adventitious refers to the development of organs (roots, buds, shoots, flowers, etc.) or embryos (embryo-like structures) from unusual points of origin of an organized explant where a preformed meristem is lacking. Adventitious shoots or roots are induced on tissues that normally do not produce these organs. Adventitious shoots are stem and leaf structures that arise naturally on plant tissues located in sites other than at the normal leaf axil regions. Plant development through organogenesis is the formation of organs either *de novo* or adventitious in origin. Whole plant regeneration via organogenesis is a monopolar structure. It develops procambial strands which establish a connection with the pre-existing vascular tissue dispersed within the callus or cultured explant. This organogenesis process is much more common than somatic embryogenesis and has far more potential for mass clonal propagation of plants.

Plant production through organogenesis can be achieved by two modes: (i) Organogenesis through callus formation with *de novo* origin and (ii) Emergence of adventitious organs directly from the explant. The schematic representation has been shown in Fig. 6.5.

Rooted plantlet

Initial explant

Plantlet

Organogenesis via
callus formation

Direct adventitious
shoot formation

Callus
3 wks.

Callus formation

Shoot formation

Fig. 6.5   Protocol for plant regeneration via organogenesis.

Organogenesis via callus formation

Plant regeneration from cultured explant involves the initiation of basal callus and then shoot bud differentiation. Establishment of callus growth with subsequent organogenesis has been obtained from many species of plants and from numerous explants viz. cotyledons, hypocotyl, stem, leaf, shoot apex, root, young inflorescence, flower petals, petioles, embryos, etc. cultured *in vitro*. For any given species or variety, a particular explant may be necessary for successful plant regeneration. Explants from both mature and immature organs can be induced to form callus and then plant regeneration. However, explants with mitotically active cells are generally good for callus initiation. Immature tissues and organs are invariably more morphogenetically plastic *in vitro* than mature tissues and organs. The size and shape of the explant is also crucial. The increased cell number present in bigger explants increases the probability of obtaining a viable culture. It has been seen that only a small percentage of the cells in a given explant contribute to the formation of callus. The site of initiation of callus is generally at the peripheral surfaces of the inoculum or at the excised surface. Callus is produced on explants *in vitro* as a result of wounding and in response to hormones, either endogenous or exogenously supplied in the medium. Meristematic tissues or organs like shoot tip, lateral bud, inflorescence, rachis, leaf, petiole, root should be selected in preference to other tissues because of their clonal properties, culture survival, growth rates and totipotency *in vitro*. Meristems, shoot tips, axillary buds, immature leaf, and immature embryos in cereals are particularly suited as good explants. The explants like mature leaves, roots, stems, petioles, and flower parts from herbaceous species can often be successfully cultured to initiate plantlets through organogenesis.

The season of the year, donor conditions of the plant, the age and physiological state of the parent plant contribute to the success of organogenesis in cell cultures. Various culture media are used for organogenesis which includes MS, B_5, White's medium (White, 1963) and SH (Schenk and Hildebrandt, 1972). MS medium contains a high concentration of nitrogen as ammonium unlike other media. It may be difficult to transfer cells from White's, B_5 or SH medium to MS medium. General requirements of vitamins, carbon source etc. have already been discussed in Chapter 3 on nutrition media.

Two modes of cell culture are generally used for organogenic path: (i) the cultivation of cell clusters on a solid medium and (ii) the cultivation of cell suspensions in liquid medium. A suspension cell culture is usually initiated by placing friable callus into liquid culture medium. The suspension usually consists of free cells and aggregates of 2–100 cells. Suspension cultures should be subcultured at least once a week while callus culture should be subcultured after 3–4 weeks.

Growth regulator concentration in the culture medium is critical for morphogenesis. Skoog and Miller (1957) first indicated its role in morphogenesis. Auxin at a moderate to high concentration is the primary hormone used to produce callus. Often 2,4-D, a very potent auxin, is used alone to initiate callus. In some species a high concentration of auxin and a low concentration of cytokinin in the medium promotes abundant cell proliferation with the formation of callus. Cytokinins, if supplied are kinetin or benzyladenine. Callus tissue comprises a wide range of cell types and characteristically consists of irregularly differentiated, vacuolated cells interspersed with smaller more meristematic cells. The nature of any callus will depend on the explant tissue or tissues from which it has arisen and also on the composition of medium used to induce and maintain it. Callus may be serially subcultured and grown for extended periods, but its composition and structure may change with time as certain cells are favored by the medium and come to dominate the culture. Under conditions favoring unorganized growth, the meristems in a callus are random and scattered. Transfer of callus pieces to conditions supporting organized growth leads to the formation of meristemoids, also termed as growth centers. The meristemoids are localized clusters of cambium-like cells which may become vascularized due to the appearance of tracheidal cells in the center. These are the sites of organ formation in the callus and can form shoots or roots.

By varying the growth regulator levels and types, one can determine the route of morphogenesis *in vitro*. As mentioned earlier, medium containing high auxin levels will induce callus formation. Lowering the auxin and increasing the cytokinin concentration is traditionally performed to induce shoot organogenesis from callus. In tobacco, the presence of adenine or KIN in the medium leads to the promotion of bud differentiation and development. Auxins inhibit bud formation. The next phase involves the induction of roots from the shoots developed. IAA or IBA auxins alone or in combination with a low concentration of cytokinin are important in the induction of root primordia. Organogenesis can be induced from either cell suspension or callus cultures. One needs to transfer only cells from callus medium to regeneration medium and then continued subculturing on regeneration medium. Thus organ formation is determined by quantitative interaction, i.e. ratios rather than absolute concentrations of substances participating in growth and development.

Direct adventitious organ formation

The somatic tissues of higher plants are capable, under certain conditions, of regenerating adventitious plants. Adventitious buds are those which arise directly from a plant organ or a piece thereof without an intervening callus phase. Induction of adventitious shoots directly on roots, leaves, bulb scales and other organs of intact plants is a common method of propagation. In culture, this method is particularly suitable to herbaceous species. The literature records numerous examples throughout the plant kingdom of shoots arising adventitiously on many different organs. In *Begonia,* for example, buds normally originate along the leaf. But in a medium containing BAP, the bud formation from the cut end of a leaf is so profuse that the entire surface of the cutting becomes covered with shoot buds. Promotion of bud formation by cytokinin occurs in several plant species. However, the requirement for exogenous auxin and cytokinin in the process varies with the tissue system, apparently depending on the endogenous levels of the hormones in the tissue. These observations led to the concept of totipotency, i.e. the capacity of all cells to regenerate a complete new plant even after differentiation within the somatic tissues of the plant. Every cell in the plant is derived from the original zygote through mitotic divisions and should contain the complete genome. Thus, the formation of adventitious organs will depend on the reactivation of genes concerned with the embryonic phase of development.

In conventional propagation, the main stimulus for adventitious shoot formation arises from the physical separation of cutting from the parent plant, causing changes in the production and distribution of endogenous hormones. The same applies to explants used for *in vitro* procedures, and in some species shoot formation may occur spontaneously on a medium lacking any growth regulators. But in most species the addition of growth regulators to the medium is required to initiate shoot formation. Two main types of growth regulating substances, auxins and cytokinins, are employed at different concentrations with respect to one another and depending upon the explant taken, age of the plant and growing conditions. Adventitious *in vitro* regeneration may give a much higher rate of shoot production than is possible by proliferating axillary shoots. Adventitious shoot proliferation is the most frequently used multiplication technique in micropropagation system.

EMBRYOGENESIS

In somatic embryogenesis the embryos regenerate from somatic cells, tissues or organs either *de novo* or directly from the tissues (adventive origin), which is the opposite of zygotic or sexual embryogenesis. Various terms for nonzygotic embryos have been reported in literature such as adventive embryos (somatic embryos arising directly from other organs or embryos), parthenogenetic embryos (those formed by the unfertilized egg), androgenetic embryos (formed by the male gametophyte). However, in general context, somatic embryos are those which are formed from the somatic tissue in culture, i.e. *in vitro* conditions. In sexual embryogenesis, the act of fertilization triggers the egg cell to develop into an embryo. However, it is not the monopoly of the egg to form an embryo. Any cell of the gametophytic (embryo-sac) or sporophytic tissue around the embryo-sac may give rise to an embryo. Cells of the nucellus or inner integument of members of Rutaceae family (e.g. *Citrus*) may develop into embryos. There are examples of embryos arising from endospermal cells also. However, occurrence of asexual embryogenesis is generally restricted to intraovular tissues. What is particularly striking about embryogenesis in plant cultures is the development of embryos from somatic cells (epidermis, parenchymatous cells of petioles or secondary root phloem) in addition to their formation from unfertilized gametic cells and tissues typically associated with *in vivo* sexual embryogenesis (e.g. nucellus). Somatic embryogenesis differs from organogenesis in the embryo being a bipolar structure with a closed radicular end rather than a monopolar structure. The embryo arises from a single cell and has no vascular connection with the maternal callus tissue or the cultured explant. Further, induction of somatic embryogenesis requires a single hormonal signal to induce a bipolar structure capable of forming a complete plant, while in organogenesis it requires two different hormonal signals to induce first a shoot organ, then a root organ.

The initiation and development of embryos from somatic tissues in plant culture was first recognized by Steward *et al.* (1958) and Reinert (1958, 1959) in cultures of *Daucus carota*. In addition to the development of somatic embryos from sporophytic cells, embryos have been obtained from generative cells, such as the classic work of Guha and Maheshwari (1964) with *Datura innoxia* microspores. Although the list of species from which somatic embryogenesis has been reported is so long, clear-cut examples are far less. In addition to the members of Umbellifereae and Solanaceae, a range of dicotyledonous families have produced somatic embryos. The Leguminoseae and many monocots of Gramineae family, which are so important

agronomically, have proven difficult to grow in culture and regenerate somatic embryos. Though there are reports of successes in these species but manipulation is not so easy as with solanaceous crops. Somatic embryos should closely resemble their bipolar nature as in the case of zygotic embryos. There should be appropriate root, shoot and cotyledonary development. There should be no vascular connection with the mother tissue. In this chapter embryogenesis will be restricted to sporophytic tissue, the discusssion on androgenesis will be dealt separately (see Chapter 8).

Sharp *et al.* (1980) described two routes to somatic embryogenesis.

1. *Direct embryogenesis*: The embryos initiate directly from the explant tissue in the absence of callus proliferation. This occurs through 'pre-embryogenic determined cells' (PEDC) where the cells are committed to embryonic development and need only to be released. Such cells are found in embryonic tissues (e.g. scutellum of cereals), certain tissues of young *in vitro* grown plantlets (e.g. hypocotyl in *Daucus carota*, *Ranunculus scleratus*, *Linum usitatissimum*, *Brassica napus*), nucellus and embryo-sac (within ovules of mature plants).

2. *Indirect embryogenesis*: Cell proliferation, i.e. callus from explant takes place from which embryos are developed. The cells from which embryos arise are called embryogenically determined cells and forms embryos which are induced to do so, also called as 'induced embryogenic determined cells' (IEDC), e.g. secondary phloem of carrot, leaf tissues of coffee, *Petunia*, *Asparagus,* etc. In majority of cases embryogenesis is through indirect method. Here, specific growth regulator concentrations and/or cultural conditions are required for initiation of callus and then redetermination of these cells into the embryogenic pattern of development.

Views expressed favor the idea that the two processes are actually the same and that somatic embryos arise from the continuation of special cells in the original explant. In either case, (embryogenic) callus, once developed, continues to produce embryos. A diagrammatic representation of somatic embryogenesis in carrot has been depicted in Fig. 6.6. Somatic embryos arise from single cells located within clusters of meristematic cells either in the callus mass or in suspension. Such cells develop into proembryos, with polarity following a pattern that tends to mimic the general pattern associated with the development of embryos in the ovule. Proembryo initials may be single cells or multicellular groups. When the conditions are suitable these embryos germinate to produce plantlets.

For some species any part of the plant body serves as an explant for embryogenesis (e.g. carrot) whereas in some species only certain regions of the plant body may respond in culture e.g. cereals. Floral or reproductive tissue in general has proven to be an excellent source of embryogenic material. The physiological state of the plant from which the explant is taken is also extremely important, as is the season during which the material is removed.

Somatic embryogenesis encompasses various stages from callus initiation to embryo development and maturation and subsequently plantlet formation. Equally important is the sequence of media and especially the growth regulators. For many species, one medium is used for initial callusing and for the maintenance of callus, a second medium is used for somatic embryo maturation, and a third to allow their growth into plants. An elaborate sequence of media is essential where somatic embryogenesis is lacking or difficult.

The presence of auxin in the medium is generally essential for embryo initiation. Tissue or calluses maintained continuously in an auxin-free medium generally do not form embryos. The callus is initiated and multiplied on a medium rich in auxin which induces differentiation of localized group of meristematic cells called embryogenic clumps. Somatic embryo

Fig. 6.6 Protocol for somatic embryogenesis in carrot.

development generally follows the transfer of cells or callus to media lacking auxin, or with reduced levels of the same auxin, or with similar or reduced levels of a weaker auxin. When transferred to a medium with low auxin or no auxin, the embryogenic clumps develop into mature embryos. In a number of species, somatic embryo initiation and maturation occurs on the primary medium. Transfer to a secondary medium is needed for their growth into plants. Some variables that affect somatic embryogenesis and maturation of embryos in culture are discussed.

Somatic embryos have been grown on a range of media from the dilute White's medium to very high salt MS medium, but the latter has been extensively employed. The addition of reduced nitrogen in the medium helps in both embryo initiation and maturation. The sources of reduced nitrogen have already been explained. Of all the amino acids, L-glutamine seems to play a special role. Another factor is the chelated form of iron in the media. In the absence of iron, embryo development fails to pass from the globular to the heart-shaped stage.

Growth regulators in the medium, especially auxin or auxin in combination with cytokinin appear essential for the onset of growth and the induction of embryogenesis. Of all the auxins, 2,4-D followed by NAA have proven to be extremely useful. Effective concentration ranges are 0.5–27.6 µM for 2,4-D and 0.5–10.7 µM for NAA. The auxin for the primary (callus initiation) and secondary media (embryo development) may be same or different. One auxin or several may be used in the same medium. Cytokinins have been important in a number of species. Cytokinins have been used in the primary medium invariably during embryogenesis of crop plants. The effective concentration range for kinetin is 0.5–5.0 µM. Cytokinins are important in fostering somatic embryo maturation and especially cotyledon development. Cytokinins are sometimes required for growth of embryos into plantlets. Gibberellins are rarely incorporated

in primary culture media. However, they have proven useful in fostering embryo maturation or in stimulation of rooting and subsequent growth of plants in a number of cases. The role of growth inhibitor ABA in somatic embryogenesis has emerged when added at non-inhibitory levels (0.1–1.0 µM). ABA promotes somatic embryo development and maturation and at the same time inhibits abnormal proliferation and initiation of accessory/adventitious embryos. ABA, anti-auxins (5-hydroxynitrobenzyl bromide, 7-azaindole, 2,4,6-trichlorophenoxy acetic acid, p-chlorophenoisobutyric acid) and other growth inhibitors may serve to promote somatic embryo maturation by countering the effects of growth promoters. Their addition to culture media may permit somatic maturation to proceed under conditions when it normally would not occur.

The addition of activated charcoal to the medium has proved to be useful for somatic embryo development. Charcoal media shows lower levels of phenylacetic acid and p-OH benzoic acid compounds, which inhibit somatic embryogenesis. Also, it absorbs 5-hydroxymethyl furfural, an inhibitor formed by sucrose degradation during autoclaving.

Environmental conditions of light, temperature, and density of embryogenic cells in medium are important. Regarding culture vessel, the position of embryos (floating or submerged) and the physical state of the medium (semisolid or liquid) have little effect.

Somatic embryogenesis as a means of propagation is seldom used because:

i. There is a high probability of mutations arising.

ii. The method is usually rather difficult.

iii. The chances of losing regenerative capacity become greater with repeated subcultures.

iv. Induction of embryogenesis is often very difficult or impossible with many plant species.

v. A deep dormancy often occurs with somatic embryogenesis that may be extremely difficult to break.

ADVANTAGES OF MICROPROPAGATION

In vitro micropropagation techniques are now often preferred to conventional practices of asexual propagation because of the following advantages not only from commercial point of view but also with regard to crop improvement programs:

1. A small amount of plant tissue is needed as the initial explant for regeneration of millions of clonal plants in one year. In comparison it would take years to propagate an equal number of plants by conventional methods.

2. Many plant species are highly resistant to conventional bulk propagation practices. Micopropagation techniques provide a possible alternative for these species. Micropropagation helps in bulking up rapidly new cultivars of important trees that would otherwise take many years to bulk up by conventional methods.

3. The *in vitro* techniques provide a method for speedy international exchange of plant materials. If handled properly, the sterile status of the culture eliminates the danger of disease introduction. Thus, the period of quarantine is reduced or unnecessary.

4. The *in vitro* stocks can be quickly proliferated at any time of the year. Also, it provides year round nursery for ornamental, fruit and tree species.

5. Production of disease free plants: Meristem tip culture is generally employed in cases where the aim is to produce disease-free plants. It has been demonstrated that the shoot apices of virus-infected plants are frequently devoid of viral particles or contain very low viral concentration. Though chemotherapeutic and physical agents have been used for production of virus-free plants but with limited success. *In vitro* culture has become the only effective technique to obtain virus-free plants in potato, *Dianthus*, chrysanthemum, gladiolus, *Pelargonium*, sweet potato (*Ipomea batatus*), yam (*Dioscorea rotundata*), cassava (*Manihot esculenta*), etc. from stocks systemically infected not only with virus but with various other pathogens.

6. Seed production: For seed production in some of the crops, a major limiting factor is the high degree of genetic conservation required. In such cases micropropagation (axillary bud proliferation method) can be used. For example, production of F_1 hybrid seed lines in crops like cauliflower where individual parent clones can be bulked for the production of more uniform seed, the production of male sterile lines in crops like onion to provide an alternative to difficult backcrossing methods, and the production of asparagus for producing high quality supermale and female homozygous lines from which desirable all male hybrids can be produced.

7. Germplasm storage: Plant breeding programs rely heavily on the germplasm. Preservation of germplasm is a means to assure the availability of genetic materials as the need arises. Most seeds and vegetative organs have a limited storage life, research on germplasm preservation has concentrated on the development of procedures to extend usable life spans. Since meristem cells are highly cytoplasmic and non-vacuolated, a high percentage of cells can be expected to survive cryopreservation procedures (for details read Chapter 11 on germplasm storage and cryopreservation). Meristems are genetically stable and can be regenerated into pathogen-free plants. Meristems have been identified as excellent material for germplasm preservation of crop species with seed borne viruses also.

8. Artificial seeds: The concept of artificial or synthetic seeds (i.e. encapsulated embryos) produced by somatic embryogenesis has become popular. The aim of somatic embryo (SE) encapsulation is to produce an analog to true seeds, which is based on the similarity of SE with zygotic embryo with

respect to gross morphology, physiology and biochemistry. The procedure is to mix somatic embryos with an encapsulation matrix so that embryos are well protected in that matrix and it is rigid enough to allow for rough handling. Encapsulation matrix consists of hydrogels such as agar, carrageenan, alginate or plant exudates of arabic (*Acacia senegal*), karaya (*Sterculia* species) or seed gums of guar (*Cyamosis tetragonolobia*) tamarind (*Tamarindus indica*) or microbial products like dextran, xantham gum and divalent salts which forms a complex with ionic bonds of hydrogel matrix. Various divalents salts which can be used are calcium chloride, calcium hydrooxide, ferrous chloride, cobaltous chloride and lanthanum chloride. Sodium alginate has been used successfully with calcium chloride for encapsulation of somatic embryos. Besides somatic embryos, efforts have also been made to use buds, leaves, and meristems.

The production of seeds by coating somatic embryos and obtaining plants from these embryos have some problems: (i) artificial seeds that are stable for several months requires the procedures for making the embryos quiescent, (ii) artificial seeds need to be protected against desiccation, (iii) recovery of plants from encapsulated embryos is often very low due to incomplete embryo formation or difficulties in creating an artificial endosperm and (iv) the embryo must be protected against micro-organisms.

Disadvantages with artificial seeds:
1. Facilities required are costly and economic considerations may not justify their use in commercial propagation of many kinds of plants.
2. Special skills are required to carry out the work.
3. Errors in maintenance of identity, introduction of an unknown pathogen, or appearance of an unobserved mutant may be multiplied to very high levels in a short time.

4. Specific kinds of genetic and epigenetic modifications of the plant can potentially develop with some cultivars and some systems of culture.

PROBLEMS ASSOCIATED WITH MICROPROPAGATION

1. Expensive requirements and sophisticated facilities with trained manpower are needed.
2. Though precautions of high order are normally taken during culture, but there are chances of contamination by various pathogens. Contamination could cause very high losses in a short time.
3. The problem of genetic variability may be pronounced in some forms of culture. Shoot tip culture tends to remain stable while those systems that utilize adventitious shoots or multiplication of callus, genetic variability is pronounced. Source of variability may be due to chimeral breakdown, aberrant cell division in callus or cell suspension, epigenetic effects and pre-existing genetic variability.
4. It has been experienced that during repeated cycles of *in vitro* shoot multiplication, cultures show water-soaked, almost translucent leaves. Such shoots exhibit a decline in the rate of growth multiplication and may eventually die. This phenomenon is referred as vitrification.

PROTOCOLS

6.1. Meristem and node culture of potato (*Solanum tuberosum*)

Plant material: Potato tuber
1. Soak potato tubers in 0.03 M gibberellic acid (GA$_3$) for 1 h or warm potato tubers to room temperature for 2 days before using to break dormancy.
2. Rinse potato tubers with tap water. Surface sterilize the tubers by soaking in 10% commercial sodium hypochlorite for 10–20 min.

3. Place whole tubers or cut into 2 – 3 cm² of 20 – 30 g sections on the surface of sterile moist vermiculite and keep it in a growth chamber.
4. Harvest sprouts when they reach 2 – 5 cm in length.

Establishment of *in vitro* plantlets from plant material:

1. Excise nodal sections 1 – 2 cm from third and fourth nodes from the stem apex with a scalpel and detach leaves.
2. Surface sterilize the nodal tissue by soaking in 10% commercial sodium hypochlorite for 10 min. Rinse 3x with sterile distilled water. Recut the base of tissue with a scalpel.
3. Place the nodal sections or sprouts on MS medium.
4. Allow growth to 4 – 6 nodes/stem stage.

Propagation by nodal cutting

1. In the laminar flow hood remove a plantlet from the culture vessel with the help of large forceps and place it on a sterile filter paper or sterile petri dish.
2. Cut the main stem above and below each node into sections of 1–2 cm length. Place one node in each culture vessel containing MS medium.

Propagation by meristem

1. Take nodal cuttings from third and fourth nodes of *in vitro* raised plants obtained from plants or sprouts.
2. In the laminar flow hood, peel away protecting leaves on the buds under a dissection microscope at 10–25x.
3. Dissect out meristem domes with one subjacent leaf primordium ~0.5 to 0.75 mm.
4. Culture the meristem domes on the surface of MS agar medium supplemented with KIN (0.1 mg/l = 0.046 µM) and GA_3 (0.1 mg/l = 0.29 µM).

6.2. Proliferation of axillary buds (strawberry-*Fragaria chiloensis*)

Plant material: Runner shoot tips of strawberry

1. Collect runners from strawberry plants

2. Surface sterilize the runners in 10% commercial sodium hypochlorite for 10 min and then rinse 4x with sterile distilled water. It is preferable to sterilize further with 0.02% $HgCl_2$ for 3–4 min followed by washing with 1% KCl for 1 min to neutralize the mercury ions.
3. Dry the shoots in a sterile petri dish.
4. Cut the runners into 1–2 cm or peel the shoot tips and then cut the tip portion of 0.1–0.5 mm size. Inoculate on the establishment medium
   MS + 0.5 – 1.0 mg/l BAP + 0.1 – 0.2 mg/l IBA
5. Incubate the cultures at 25°C under 16 h photoperiod.
6. After 3–4 weeks on this medium, axillary buds will appear.
7. The buds can be proliferated or multiplied if cultured on the following medium:
   MS + 0.8 – 1 mg/l BAP + 0.2 mg/l IBA + 0.2 mg/l GA_3
8. The developed shoots alongwith buds can be further multiplied on the same proliferation medium after 3–4 weeks of culture. Or shoots are separated and transferred to MS medium with auxin (0.5–1 mg/l) IBA. The development of axillary buds ceases and the young plantlets with roots are developed. Profuse rooting should occur within 4–5 weeks.
9. Rooted plantlets can be transferred to soil and follow the procedure of plant establishment and hardening (Protocol 5.1).

6.3. Organogenesis – Adventitious shoot formation

Plant material: Leaf explants of tobacco or African violet

1. Take leaf explants from *in vitro* grown plants or potted plants.
2. If leaves are taken from potted plants, gently wash with a mild detergent and then rinse under tap water. Dip each leaf into 70% alcohol. Leaves are then surface sterilized with 10% commercial sodium hypochlorite

for 5–10 min followed by rinsing 3x in sterile distilled water.

3. Place the leaves in sterile petri dish. Cut leaves into approximately 1 cm² sections using sterile forceps and scalpel, without vein in the explant.

4. Place the leaf sections on the following media:
   African violet: MS+1.0 mg/l KIN+0.1 mg/l IAA
   Nicotiana: MS + 1–2 mg/l BAP + 0.1–0.2 mg/l NAA

5. Seal the cultures. Incubate at 25°C under 16 h photoperiod at 1000-lux light intensity.

6. Observe the shoot formation after 3 weeks and record the data at regular intervals.

7. Cut tissue masses into small pieces for shoot multiplication, each piece should include at least one tip or rosette, and place them on their respective media.

8. For rooting, transfer the well developed shoots to MS medium without growth regulators or little concentration of auxin (0.5 mg/l) to enhance root production.

9. Transfer the rooted plantlets to soil and follow the procedure of plant establishment and hardening (Protocol 5.1).

6.4. Organogenesis via callus formation (*Nicotiana*)

Plant material: *Nicotiana*

1. Follow Steps 1–4 of Protocol 6.3. Place the leaf sections on the following medium:
   MS + 1–2 mg/l 2,4-D

2. Incubate the cultures in dark at 25°C. Callus will be produced in 3–4 weeks.

3. For shoot induction, subculture pieces of callus approximately 0.5 cm² to the medium:
   MS + 1–2 mg/l BAP + 0.1 – 0.2 mg/l NAA

4. Follow Steps 5 to 9 of Protocol 6.3.

6.5. Organogenesis via callus formation (cereals–wheat, barley, maize, rice, etc.)

Plant material: Immature embryos of cereals (wheat, barley, maize, rice)

1. Collect spikes (for example wheat/barley) 14–15 days after anthesis from plants.

Separate each developing seed and remove the lemna and palea.

2. Surface sterilize the seeds in 10% commercial sodium hypochlorite containing 1–2 drops of Tween 20 for 10 min. Rinse 3–5x with sterile distilled water.

3. Dissect out immature embryos (1 – 1.5 mm diameter) from the seeds under dissection microscope in the hood of laminar flow.

4. Place the immature embryos with scutellum side up on the following medium for callus induction: MS + 2 mg/l 2,4-D

5. Incubate the cultures in dark at 25°C. Initially the embryos will show swelling followed by callus proliferation from the scutellum in 3–4 weeks. (Embryos bigger than 2 mm diam. will show precocious plant formation.)

6. Scutellar callus can be multiplied at least once on the same callus initiation medium by breaking into 2–3 pieces after 4–6 weeks of initiation of callus.

7. Place hard and compact calli of approximately 0.5 cm² on the following shoot induction medium: MS + 2 mg/l BAP + 0.2 mg/l NAA or MS without growth regulators.

8. Incubate the cultures at 25°C under 16 h photoperiod with ~2 klux light intensity. Shoots would emerge in 3–5 weeks.

9. Transfer shoots to MS medium with 1 mg/l of auxin (IAA/IBA/NAA) for root induction under the same cultural conditions.

10. Transfer rooted plantlets to soil and follow the procedure of plant establishment and hardening (Protocol 5.1).

6.6. Embryogenesis (carrot)

Plant material: Hypocotyl of carrot seedling

1. Wash seeds by submerging in water with a few drops of detergent in a beaker and shake by hand, or wrap seeds in two layers of cheese cloth/muslin cloth/nylon pouch and then wash with water.

2. Submerge the seeds in 70% alcohol for 30–60 s. Decant the alcohol.

3. Transfer the seeds to a flask or beaker containing 20–40% commercial sodium

hypochlorite for 15–20 min. Rinse 4× with sterile distilled water.

4. Place 2–3 seeds per culture vessel on the surface of MS agar medium.

5. Incubate the cultures at 25°C under 16 h photoperiod with ~1000 lux light intensity for 1–2 weeks.

6. Collect the germinated seedlings when the cotyledons are fully expanded. Place each seedling on a sterile petri dish and excise the hypocotyl from each seedling and cut them transversely into two parts.

7. Place the hypocotyl sections on the following medium: MS + 1–2 mg/l 2,4-D

8. Incubate the cultures in dark at 25°C for 4–8 weeks.

9. Maintain the callus by subculturing small pieces on fresh medium every 3 – 4 weeks. Callus will contain pro-embryo initial cells as well as minute microscopic embryos in the early stages of development.

10. Place 0.5 to 1 cm² callus pieces on MS agar medium without growth regulators and incubate the cultures at 25°C under 16 h photoperiod with ~1000 lux light intensity. Within 2–3 weeks cultures will exhibit embryos and green plantlets.

11. Tease out individual or group of plantlets from the callus mass and transfer on half strength MS medium under 16 h photoperiod with high light intensity of ~5 klux. Within 4–5 weeks the cultures will resemble seedling carrots.

12. Transfer the plantlets to small pots containing sterile peat moss and vermiculite in a 1 : 1 ratio. Enclose the plantlets with plastic containers to maintain high humidity.

13. Transfer the plants to soil and follow the procedure of plant establishment and hardening (Protocol 5.1).

Further reading

Bhojwani, S.S. and Razdan, M.K. 1990. *Plant Tissue Culture: Applications and Limitations.* Elsevier Science Publishers, Amsterdam.

Chawla, H.S. 1998. *Biotechnology in Crop Improvement.* International Book Distributing Co., Lucknow.

Evans, D.A., Sharp, W.R., Ammirato, P.V. and Yamada, Y. 1983. *Handbook of Plant Cell Culture.* Macmillan Publishing Company, New York.

Ignacimuthu, S. 1998. *Plant Biotechnology.* Oxford & IBH Publishing Co. Pvt. Ltd., New Delhi.

Mantell, S.H., Mathew, J.A. and McKee, R.A. 1985. *Principles of Plant Biotechnology: An Introduction to Genetic Engineering in Plants.* Blackwell Scientific Publications, Oxford.

Pierik, R.L.M. 1987. *In vitro Culture of Higher Plants.* Martinus Nijhoff Publishers, Dordrecht.

Razdan, M.K. 1983. *An Introduction to Plant Tissue Culture.* Oxford and IBH Publishing Co. Pvt Ltd., New Delhi.

Street, H.S. (ed.) 1973. *Plant Tissue and Cell Culture.* Blackwell Scientific Publications, Oxford.

Thorpe, T.A. 1981. *Plant Tissue Culture: Methods and Applications in Agriculture.* Academic Press Inc., New York.

Vasil, I.K. 1984. *Cell Culture and Somatic Cell Genetics of Plants: Laboratory Procedures and Their Applications.* Academic Press, Inc., New York.

7

Cell Suspension and Secondary Metabolites

Haberlandt (1902) made the first attempt to isolate and culture single cells from leaves of flowering plants. Although he failed to achieve the division of free cells, his work stimulated workers to pursue investigation on these lines. To date, the progress in this field is spectacular that it is possible not only to culture-free cells but also to induce divisions in a cell and to raise whole plant from it.

A cell suspension culture consists of cell aggregates dispersed and growing in moving liquid media. It is normally initiated by transferring pieces of undifferentiated and friable calli to a liquid medium, which is continuously agitated by a suitable device. Suspension cultures have also been started from sterile seedlings or imbibed embryos or leaves by mechanical method. Leaves or the other tissue can be gently grinded or soft tissues can be broken up in a hand operated glass homogenizer. This homogenate, containing intact living cells, dead cells and cell debris is cleared by filtration and centrifugation and then transferred to moving liquid medium. An enzymatic method for isolation of single cells by the use of pectinases, which digest the pectin wall, has also been employed. But the single cell system used in basic and applied research is isolated from cultured tissues. Cultures are initiated by placing freshly cut sections of plant organs on a solidified nutrient medium containing suitable growth regulators. On such a medium the explant exhibits callusing, which is separated

from the parent explant and transferred to a fresh medium to build up reasonable amount of callus tissue. This callus is transferred to a liquid medium and agitated to raise fine suspension of cells. Successful establishment of the suspension culture depends upon the initial callus. A wide variety of explants and media compositions have been used with success for callus induction viz. MS, B_5, LS (Linsmaier and Skoog, 1965), Blaydes (Blaydes, 1966) etc. Variations and modifications of these media are widely used. To these media are added vitamins, inositol, sucrose and growth regulators, especially auxin (2,4-D at ~1 – 5 µM concentration), for cells to divide. The addition of 1 µM kinetin may be beneficial. The growth rate and friability of callus produced can vary widely due to explants. The culture medium for suspension culture should be such that it maintains good growth of callus.

Platform (orbital) shakers are widely used for the initiation and serial propagation of plant cell suspension culture. They should have a variable speed control (30–150 rpm) and the stroke should be in the range of 4–8 cm orbital motion. Agitation of medium on a shaker serves two purposes. First it exerts a mild pressure on cell aggregates, breaking them in to smaller clumps and single cells and secondly agitation maintains uniform distribution of cell and cell clumps in the medium. Movement of the medium also provides good gaseous exchange between the culture medium and air. The shaker should be kept in the air-conditioned room with good temperature control. Wide-mouthed Erlenmeyer flasks are widely used as culture vessels. The volume of the culture should be appropriate to the culture vessel e.g. 20 ml per 100 ml flask and 70 ml per 250 ml flask. The flasks are normally sealed with aluminium foil. Flask closure must maintain sterility, allow gas exchange and reduce evaporation. Cotton wool plugs may be used for sealing flasks during autoclaving but not for culturing cells. They are a common source of contamination in flasks that are sitting on a shaker for several weeks. For the first passage, suspensions are selectively subcultured by (i) allowing the cultures to settle briefly and pipetting only suspended material, (ii) gradually reducing the bore size of the pipette, or (iii) sieving before subculture. A reduction in mean aggregate size can be brought by addition of enzymes such as cellulases and pectinases in the medium.

TYPES OF SUSPENSION CULTURES

Batch culture

A batch culture is a cell suspension culture grown in a fixed volume of nutrient culture medium. Cell suspension increases in biomass by cell division and cell growth until a factor in the culture environment (nutrient or oxygen availability) becomes limiting and the growth ceases. The cells in culture exhibit the following five phases of a growth cycle (Fig. 7.1).

 i. Lag phase, where cells prepare to divide.
 ii. Exponential phase, where the rate of cell division is highest.
 iii. Linear phase, where cell division slows but the rate of cells expansion increases.
 iv. Deceleration phase, where the rates of cell division and elongation decreases.
 v. Stationary phase, where the number and size of cells remain constant.

When cells are subcultured into fresh medium there is a lag phase. It is the initial period of a batch culture when no cell division is

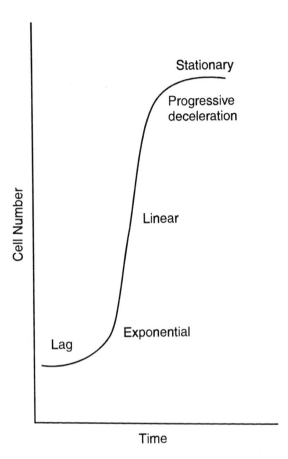

Fig. 7.1 A model curve showing cell number per unit volume of culture to time in a batch grown cell suspension culture.

apparent. It may also be used with reference to the synthesis of a specific metabolite or the rate of a physiological activity. Then follows a period of cell division (exponential phase). It is a finite period of time early in a batch culture during which the rate of increase of biomass per unit of biomass concentration (specific growth rate) is constant and measurable. Biomass is usually referred to in terms of the number of cells per ml of culture. After 3 to 4 cell generations the growth declines. Finally, the cell population reaches a stationary phase during which cell dry weight declines. It is the terminal phase of batch culture growth cycle where no net synthesis of biomass or increase in cell number is apparent.

Cultures can be maintained continuously in exponential phase by frequent (every 2–3 days) subculture of the suspensions. Cells may remain viable for many days by utilizing intracellular reserves or recycling metabolites released by lysed cells. The stability of cells in stationary phase depends upon the species and the nature of the growth limiting factor (cells brought to stationary phase by nitrogen limitation retain viability longer than when carbohydrate starved). When such a stationary phase cell suspension is subcultured, the cells in succession pass through a lag phase, a short period of exponential growth, a period of declining relative growth rate and then again enter stationary phase. Prior to the last phase, the cells expand in size. Traditionally such cultures at each subculture are initiated from a relatively high cell density and the cells therefore accomplish only a limited number of divisions (cell number doublings) before entering stationary phase again. For example, a suspension culture of *Acer pseudoplatanus* initiated at 2×10^5 cells/ml will reach a final cell density of ca. 3×10^6 cells/ml corresponding to four successive doublings of the initial population. The cell generation time (doubling time) in suspension cultures varies from 24 to 48 h in well established cell cultures. Doubling time (td) is the time required for the concentration of biomass of a population of cells to double.

For general purposes, the objective with cell suspension cultures is to achieve rapid growth rates, and uniform cells, with all cells being viable. Frequent subculturing in a suitable medium ensures these qualities. Cells should be subcultured at weekly intervals or less if they are to be used for experimental purposes. The size of the inoculum (dilution rate) will determine the lag period. Optimally suspension cultures passage length is 1–2 weeks, but the exact time and the dilution required must be determined for each cell line. Dilutions of 1 : 4 after 1 week or 1 : 10 after 2 weeks are commonly used. There is generally a minimum inoculum size below which a culture will not recover.

Continuous culture

A culture is continuously supplied with nutrients by the inflow of fresh medium but the culture volume is normally constant. It is of two types:

i) *Open continuous culture:* A continuous culture is one in which inflow of fresh medium is balanced by outflow of corresponding volumes of culture including harvest of cells. Cells are constantly washed out with the outflowing liquid. The rate of inflow of medium and culture harvest are adjusted so that the cultures are maintained indefinitely at a constant sub-maximal growth rate. In a steady state, the rate of cells washout equals the rate of formation of new cells in the system. A situation of balanced growth is achieved, i.e. majority of cells in the culture are in a similar metabolic state. The growth rate and cell density are held constant by a fixed rate of input of growth limiting nutrients and removal of cells and spent medium. These open continuous systems may be: (a) *chemostats* in which growth rate and cell density are held constant by a fixed rate of input of a growth limiting nutrient medium (nitrogen, phosphorus or glucose). In such a medium, all the constituents other than growth limiting nutrients are present at concentrations higher than that required to maintain the desired rate of cell growth. The growth limiting substance is so adjusted that its increase or decrease is reflected by a corresponding increase or decrease in the growth rate of cells; (b) *turbidostats* in which fresh medium flows in response to increase in the turbidity so as to maintain the culture at a fixed optical density of suspension. A pre-selected biomass density is maintained by the washout of cells.

ii) *Closed continuous culture:* A closed culture is one in which cells are retained and inflow of fresh medium is balanced by outflow of corresponding volumes of spent medium only. The cells from the out-flowing medium are separated mechanically and added back to the culture. Thus in closed continuous culture, cell biomass continues to increase as the growth proceeds. It has potential value in studies on cytodifferentiation, where it may be important to grow cells under a particular regulated environment and then maintain them for a considerable period in a non-dividing but viable state. It can also be used in cases where secondary products e.g. coumarins, lignins, polysaccharides, monoterpene derivatives produced by cell suspension cultures have been shown to be released in significant amounts into their culture medium. In such cases, a maintenance culture in a closed continuous system should enable the chemical product to be continuously harvested from a fixed culture biomass.

Semi-continuous culture

In this type of culture the inflow of fresh medium is manually controlled at infrequent intervals by a "drain and refill" process, such that the volume of culture removed is always replaced by an equivalent volume of fresh medium. In these cultures, although the number of cells increases exponentially, the cell density is maintained within fixed limits by the periodic replacement of harvested culture by fresh medium. Such a culture is termed open. However, during the intervals between additions of new medium, both the cell density and the nutrient environment are continuously changing.

GROWTH MEASUREMENTS

1. *Cell number*: This is an indispensable growth parameter for suspension cultures. To count the number of cells in aggregates of cells, these are treated with chromium trioxide. One volume of suspension cell culture is added to 4 volumes of 12% (w/v) aqueous chromium trioxide and heated at 70°C until the cells are stained and plasmolysed. Aggregates of cells are partially macerated by vigorously pumping the

culture with a syringe. The incubation time and maceration required will vary with the culture age. Shrinkage of the protoplasts makes counting of cells in the aggregates easier. The number of cells can be counted under a simple microscope.

2. *Fresh weight*: This can be determined by collecting cells on pre-weighed (in wet condition) nylon fabric filters which can be supported in a funnel. Cells are washed with water to remove the medium, drained under vacuum and weighed. A large sample is always needed to get a reasonably accurate fresh weight estimation.

3. *Dry weight/packed cell volume (PCV)*: An appropriate volume of suspension culture is centrifuged in a graduated conical centrifuge tube at 2000× g for 5 min and pellet volume or packed cell volume is noted. PCV is expressed as ml pellet/ml culture. The supernatant is discarded, the pellet washed onto a preweighed Whatman filter paper and dried overnight at 80°C, cooled in a desiccator and reweighed to get dry weight.

SYNCHRONIZATION OF SUSPENSION CULTURE CELLS

Cultured plant cells vary greatly in size and shape, nuclear volume and DNA content and cell cycle time. This variation makes the culture asynchronous which complicates the studies on biochemical, genetic, and physiological aspects of cell division, metabolism and other cellular properties. A synchronous culture is defined as a culture in which the cell cycles (or a specific phase of the cycles) of a proportion of cells (often a majority) are synchronous, i.e. majority of cells proceed through each cell cycle phase simultaneously. In a suspension culture, low percentage of actively dividing cells and the tendency of cells to form aggregates are the major obstacles. To make the cell synchronous a pattern of cell division cycle must be understood. The cell cycle can be divided into interphase and mitosis (M). Interphase is divided into a DNA synthesis phase (S) and preceded by a pre-synthetic phase (G1) and followed by a post-synthetic phase (G2). Thus, mitosis represents a small proportion of the total cell cycle, which is bordered by G1 and G2. Synchronization would require that all cells proceed simultaneously through specific stages of the cell cycle. The various methods employed to bring this synchronization have been listed below.

1. ***Cold treatment***: Temperature shocks at 4°C for few days can be used.

2. ***Starvation method***: Deprivation of an essential growth compound from suspension cultures leads to a stationary growth phase. Resupplying the missing compound or subculturing to a fresh complete medium will allow growth to resume and may result in synchronization of cell growth. It has been reported that sycamore cells are arrested in G1 and G2 by deprivation of phosphorus and carbohydrate respectively. Nitrogen-starved cells accumulate in G1.

3. ***Use of inhibitors***: Synchronization is achieved by temporarily blocking the progression of events in the cell cycle by addition of DNA synthesis inhibitors and accumulating cells in a specific stage. Upon release of block, cells will synchronously enter the next stage. Generally, 5-fluorodeoxyuridine (FudR), excess thymidine (TdR), and hydroxyurea (HU) are used which accumulate cells at G1/S interface stage. In this method, the cell synchrony is limited to only one cell cycle.

4. ***Colchicine method***: Colchicine is one of the most effective spindle fiber inhibitors, which arrests the cells at metaphase stage of cell division. Suspension cultures in exponential growth are supplemented with 0.02% (w/v) colchicine. This synchronization is measured in terms of mitotic index, i.e. the percentage of cells in the mitotic phase of the cell cycle.

Procedures effective for synchronization with one species may not be useful for another. Thus, combinations of these methods are often used.

TECHNIQUE FOR SINGLE CELL CULTURE – BERGMANN CELL PLATING TECHNIQUE

Bergmann (1960) first used the most popular technique of plating out cell suspensions on agar plates. This technique is particularly useful where attempts are being made to obtain single-cell clones. The objective is to establish evenly distributed suspension in a thin layer of culture medium with as low a cellular density as is compatible with the growth of colonies from a high proportion of the cellular units of single cells and cell aggregates (Fig. 7.2). The basic technique of plating is to first count the cell number without maceration stage (see growth measurements). This will enable a known number of cell units to be established per unit volume of plating media. The counted suspension is adjusted by dilution or concentrated by low speed centrifugation. Both the suspension and nutrient medium containing agar are prepared in double the concentration separately. The equal volumes of suspension and the agar medium cooled at 35°C are mixed and then dispensed rapidly into petri dishes in such a manner that cells are evenly distributed in a thin layer (~1 mm thick). The suspension culture, which carries cell aggregates should be filtered through a sieve and only the fine suspension should be plated. The number of cellular units in per square mm are directly counted *in situ* under inverted microscope. The dishes are then sealed and incubated at 25°C for 21 days. After incubation, the number of colonies per plate are counted by making shadowgraph prints of each plate by resting the plate on a sheet of photographic document paper under a photographic enlarger as the light source. This photographic method clearly records colonies not easily observed on the standard type of

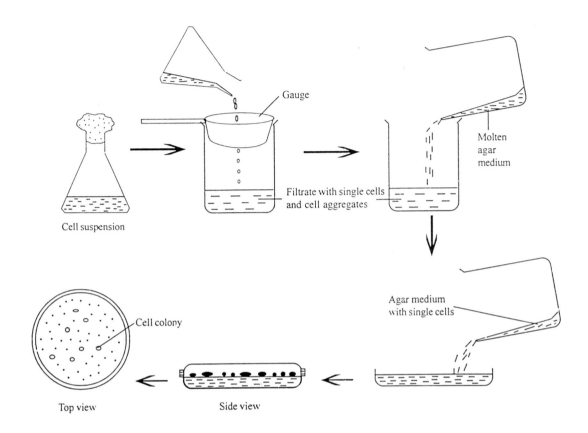

Fig. 7.2 Diagrammatic summary of steps involved in Bergmann cell planting technique.

bacterial colony counter. Plating efficiency (PE) is then calculated as:

$$PE = \frac{\text{No. of colonies / plate at the end of experiment} \times 100}{\text{No. of cellular units initially/plate}}$$

Applications

1. Embryogenic cell suspension offers the possibility for large-scale clonal propagation.
2. Embryos are natural organs of perennation, many of which typically become dormant. Because of their innate properties, somatic embryos from cell suspensions can prove useful for long-term storage in germplasm banks.
3. Embryogenic suspension cultures present an excellent tool for both theoretical studies and practical applications. Selection schemes with embryogenic cultures give rise to variants for various abiotic and biotic stresses which has been discussed in Chapter 10 on somaclonal variation.
4. Embryogenic cell suspension cultures and population of somatic embryos are a source for the production of important chemicals. Somatic embryos of celery produce the same flavor compounds present in the mature plant but absent in celery callus cultures. Somatic embryos of cacao produce the same lipids, including cacao butter as their zygotic counterparts.

PRODUCTION OF SECONDARY METABOLITES

Plants are a valuable source of a vast array of chemical compounds including fragrances, flavors, natural sweeteners, industrial feedstocks, anti-microbials and pharmaceuticals. In most instances these compounds belong to a rather broad metabolic group, collectively referred to as secondary products or metabolites. If plant cells produce metabolites *in vivo* or *in vitro,* which are not directly needed by the plant itself, then these are termed secondary metabolites. These secondary metabolites do not perform vital physiological functions as primary compounds such as amino acids or nucleic acids, but these are produced to ward off potential predators, attract pollinators, or combat infectious diseases.

Plant cell suspension culture is potentially valuable for studying the biosynthesis of secondary metabolites. If plant cells are cultured under conditions that allow the expression of a certain degree of differentiation, they have the potential to produce either by *de novo* or by biotransformation of specific precursors, a wide range of secondary products.

Despite limitations, there are some major advantages expected from cell culture systems over the conventional cultivation of whole plants for production of secondary metabolites. These are:

a) Independence from various environmental factors, including climate, pests and microbial diseases, geographical and seasonal constraints.

b) Useful compounds could be produced under controlled conditions and geared more accurately to market demands.

c) Any cell of a plant could be multiplied to yield specific metabolites.

d) A consistent product quality could be assured with the use of characterized cell lines.

e) There would be greater control over production levels since they would not be so much at the mercy of political interference.

f) Cell growth could be automatically controlled and metabolic processes could be regulated rationally, all contributing to the improvement of productivity and the reduction of labor and costs.

g) Production of substances in chemically controlled environment facilitates later processing and product recovery steps.

h) Culture of cells may prove suitable in cases where plants are difficult or expensive to grow in the field due to their long life cycles, e.g. *Papaver bracteatum*, the source of thebaine takes two to three seasons to reach maturity.

i) New routes of synthesis can be recovered from mutant cell lines. These routes of synthesis can lead to novel products not previously found in whole plants.

j) Some cell cultures have the capacity for biotransformation of specific substrates to more valuable products by means of single or multiple step enzyme activity.

As in microorganisms, attempts have been made to obtain substances from cell suspension cultures of higher plants, either through accumulation in the cells (biomass) or sometimes by the release into the nutrient medium. With respect to production of secondary metabolites two main approaches have been followed: (i) The rapid growth of suspension cultures in large volumes, which are subsequently manipulated to produce secondary metabolites. (ii) The growth and subsequent immobilization of cells, which are used for the production of compounds over a prolonged period.

Rapidly growing plant cell suspension cultures initiated and maintained by repeated subculturing on the same medium are the result of an unconscious but systematic selection for those cells in a heterogeneous population with the best growth rate. These cultures often consist mainly of small meristematic type cells. Further, there are evidences to indicate that in many cases, secondary product synthesis typical of intact plant cannot occur in rapidly growing undifferentiated cell cultures, but requires some degree of morphological or biochemical differentiation and slow growth. It is therefore not surprising that the production and yield of metabolite in cell culture will be dependent on a number of factors which have been discussed below.

Morphological and chemical differentiation

Some secondary metabolites accumulate entirely in specific structures, showing so-called morphological differentiation. Essential oils are found in glandular scales (*Mentha piperita*) and secretory ducts, latex in laticifiers (*Papaver somniferum*) and tobacco alkaloids are primarily synthesized in roots. Thus, accumulation of such products occurs only if specific structures are present in the cultures. The yields in most callus and liquid cell suspension cultures is low because of lack of tissue and organ differentiation. Thus, root differentiating calluses of *Atropa belladona* are capable of producing tropane alkaloids like atropine, while non-differentiating calluses are not. Some secondary metabolites accumulate in the vacuoles of cells, while others are formed elsewhere in the cell get accumulated and is then excreted from cells via the vacuole system or by other means. Failure to excrete a metabolite from cells might lead to cessation of biosynthesis, in some cases by feedback regulation mechanisms. As a general routine cell suspensions are established from friable, fast growing types which are suitable for growth and not necessarily for their ability for secondary metabolite production. This fast growth ability is linked in most cases with little tissue and/or biochemical differentiation so that it was believed that cell suspensions capable of good growth in bioreactors were not suitable for product formation. However, lately it has come to notice that under some culture conditions, various degrees of cell differentiation are expressed in cell suspension cultures.

Medium composition for secondary product formation

Secondary product formation can be obtained by one-stage or two-stage culture systems. In a two-stage culture system, the first stage involves growing the cells on a standard growth medium and the second stage involves transferring these cells into a production medium suitable for secondary product synthesis. In the one-stage culture system both growth and production steps are combined together.

There are dramatic effects on secondary product synthesis by altering the nutrients in a production medium. In cell cultures of *Morinda citrifolia* a 50% increase in anthraquinone accumulation occurred when phosphate was increased to a concentration of 5 mg/l. In suspension cultures of *Catharanthus roseus*, the overall accumulation of secondary metabolites like tryptamine and indole alkaloids have been shown to occur rapidly when cells are shifted to a medium devoid of phosphate.

An insight into the importance of specific media components or production of secondary product shikonin by *Lithospermum erythrorhizon* has been obtained. It had been shown that callus cultures of *L. erythrorhizon* could be induced to produce shikonin on Linsmaier–Skoog (LS) medium supplemented with 1 µm IAA and 10 µm KIN (Tabata *et al.*, 1974). No shikonin derivatives were produced in cell suspension cultures grown in LS medium, which is suitable for cell growth, while White medium favored the production of shikonin derivatives but it could not support good cell growth. Optimal concentrations of different nutrients were identified, e.g. 6.7 mM for nitrate, 1.2 µM for copper and 13.5 mM for sulfate as well as optimal levels of key organic components. The resultant medium showed 13 times higher shikonin production.

Sucrose is frequently used as a carbon and energy source and leads to optimal growth rate of plant cell cultures. Mannose, galactose, glucose and raffinose have also been used for this purpose. Both the nature and concentration of the sugar generally affects the yield of secondary product.

The type and concentration of growth regulators in cell suspension is probably one of the most important factors influencing their potential for secondary product synthesis. For example 2,4-D can stimulate both cell division and cell expansion, but it can also bring about a dramatic suppression of secondary metabolite synthesis. A range of hormone combinations and concentrations should be examined for optimum effect on secondary product formation. For example nicotine production by tobacco cell cultures is best achieved by a two-stage approach. The growth phase is supported on MS medium with 10^{-5} M NAA and 10^{-6} M KIN, while the production phase is optimal in MS with reduced levels of growth regulators (10^{-6} M NAA and 10^{-7} M KIN) and by omitting casein hydrolysate from the growth medium.

The buffering capacity of plant tissue culture media is often very low which causes drift in the pH of medium. Thus, biological buffers such as N-2 hydroxyethyl piperazine-N'-2-ethanesulphonic acid (HEPES) or 2-(N-morpholino) ethane sulphonic acid (MES) are added at a concentration of approximately 100 mM. Undefined medium additives such as casein hydrolysate can also act as buffers.

An exogenous supplement of the biosynthetic precursor added to the culture medium may increase the yield of the end product. However, administration of a direct precursor does not always produce the desired effect.

Growth production patterns

Secondary plant metabolites may be obtained from cell cultures produced by batch or less commonly by continuous processes. Batch culture consists of a heterogeneous mixture of cells at different stages of biochemical and morphological differentiation. Thus during initial and middle phases of culture, varying proportions of cells are in rapid division. As substrates in the medium become more limiting, the proportion of cells undergoing cell divisions decreases and leads to stationary phase, where the majority of cells become quiescent and ageing processes occur in more and more cells. It is during these later stages of batch culture that most of the secondary metabolites are produced by cells. The accumulation of metabolites at any one particular instant in the culture cycle is the result of a dynamic balance between the rate of its biosynthesis and its biodegradation. Therefore, culture conditions that promote biosynthesis of the desired compound while at the same time limiting those parts of the metabolism which are involved in biodegradation are sought. There is no hard and fast rule as to the best time in the culture cycle to harvest cells in order to obtain maximum yield of the metabolites. Thus, it is best to produce large amounts of biomass under rapid growth conditions and to then transfer the accumulated biomass to a second stage of culture that promotes secondary metabolite production.

Continuous cultures may also be operated in closed and open systems for metabolite production. In closed system the cells are retained and cell density progressively increases as growth continues. Nutrients in excess of requirements are supplied by the continuous inflow of fresh medium, balanced by the continuous harvesting of spent medium. The open system involves the input of new medium and a balancing harvest of an equal volume of culture and cells. This system allows the steady state of growth and metabolism. Regulation involves the use of either chemostats or turbidostats. In the former case, continuous input of fresh medium is set at a fixed rate and determines the nature of the resulting equilibrium. In the latter case, cell density is set at levels monitored by reading the absorbance of the culture.

For products which accumulate most actively when the growth rate is declining, it is possible to maintain cells at a growth rate consistent with the highest cellular content of the compound by chemostat culture. A considerably higher level of the product will be built up in such steady state cells than can be achieved in batch culture. Variation in the medium during this steady state may further enhance biosynthesis of the secondary metabolite. Once the highest level of product is achieved consistent with the maintenance of cell viability, the product can be harvested either through continuous run off of surplus culture or through a single harvest of the whole culture. The latter alternative would probably be used where a relatively slow growth rate is needed for optimal product accumulation. The turbidostat system is used in cases where maximum levels of secondary product develop early during the growth cycle, either during or at the beginning of the decay of exponential growth. This enables studies on the influence of growth regulators, physical factors and precursors.

Semi-continuous culture systems in which large volumes of culture are removed from bioreactors and then replaced with an equivalent volume of fresh medium can be successfully operated as chemostats to either generate biomass or for producing certain metabolites, e.g. different levels of anthocyanins have been produced in carrot cell lines. However, despite clear-cut advantages, continuous cultures are not widely used by plant tissue culture workers, as these cultures probably require a lot of attention.

Environmental factors

For a number of plant tissue cultures, it has been shown that environmental conditions, especially light, stimulate the production of secondary compounds. For example, compounds such as carotenoids, flavonoids, polyphenols and plastoquinones are stimulated by light. On the other hand, biosynthesis of certain metabolites is not significantly affected by light and may even be inhibited by it. Thus, irradiation with white or blue light has been seen to result in an almost complete inhibition of alkaloid biosynthesis. Lipid compositions of tobacco cell suspensions and alkaloid accumulation in *Papaver* spp. have shown significant fluctuations in response to temperature changes.

Selection of cell lines producing high amounts of a useful metabolite

It is important that cultured cells should produce high amounts of secondary metabolites from the induced callus. Recently cell strains containing amounts greater than those found in intact plants have been isolated by clonal selection. The successful selection of these cells producing high amounts of secondary metabolites has been made possible due to the heterogeneity of cultured plant cells.

In selecting specific cell lines producing high amounts of a useful metabolite, two methods have been generally used: (i) Single cell cloning and (ii) cell aggregate cloning. There are difficulties with the isolation and culture of single cells that limit its applicability. Cell aggregate cloning appears to be time consuming, but it is far easier and has been more successful.

Cell aggregate cloning is simply a tissue culture method that permits the identification of cells with increased biosynthetic capacity for a particular secondary product. It exploits the genetic variation for chemical production of a compound present in the cell population. The variation may be spontaneous, i.e. already present in the original explant material, or that induced during the process of tissue culture initiation and subculture, i.e. somaclonal variation.

A typical scheme for the selection of high secondary metabolite producing cell lines by cell aggregate cloning was given by Yamada and Fujita (1973), which has been shown in Fig. 7.3. The first step in obtaining highly productive cell cultures is to initiate the growth of callus cultures on solid media. Cell suspension cultures are initiated from calli which promote generation of cell aggregates. In this technique, cell aggregates are plated on solid media and cultured. The resulting calluses are then divided; one half for continued growth and the other half for analysis of secondary compounds. In this way, Yamada and Fujita selected high anthocyanin producing cell lines from cultured cells of *Euphorbia millii*. The highest pigmented calluses

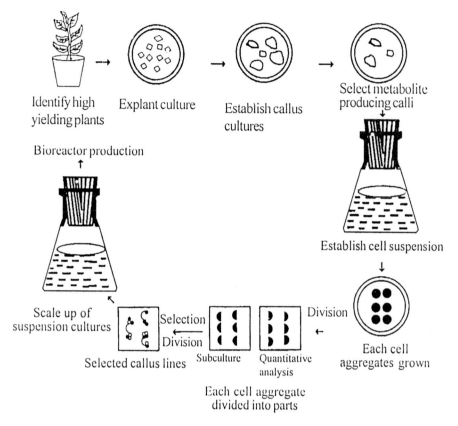

Identify high yielding plants → Explant culture → Establish callus cultures → Select metabolite producing calli

Establish cell suspension

Each cell aggregates grown

Each cell aggregate divided into parts

Quantitative analysis — Subculture — Selected callus lines — Division ← Selection

Division ←

Scale up of suspension cultures

Bioreactor production

Fig. 7.3 Induction and selection of high producing cell lines by cell aggregate cloning method.

were continually subcultured in this manner for 24 selections. The amount of pigment found in the highest producing lines after this period of time was seven times greater than in the original cells.

Product analysis

Procedures for the separation and purification of desired metabolic compounds produced by cultured cells are basically the same as those used for plants. Separation–purification processes, however, are expensive and a reduction in their costs is very important. To make these processes cheaper, the content of the target compound should be as high as possible. Also, the production of side products that are difficult to remove must be prevented. When a high level of side product(s) are produced, a cell line that gives only the desired compound must be the object of selection. When the target compound is released from the cultured cells into the medium, the separation–purification process is greatly simplified if the compound can be recovered easily, e.g. by extraction.

Application

Cell cultures have been shown to produce a number of valuable secondary metabolites, which includes alkaloids, glycosides (steroids and phenolics) terpenoids and a variety of flavors, perfumes, agrochemicals, commerical insecticides of plant origin, etc. The most successful example of production of valuable chemical plant cell culture is shikonin, a naphthaquinone used both as dye and a pharmaceutical. According to WHO, nearly 70–80%

of the global population depends on herbal drugs. The market value for herbal products is estimated to be over $20 million. *Catharanthus roseus,* a tropical perennial plant, is important as a rich source of terpenoid indole alkaloids. The *Catharanthus* alkaloids represent a market of considerable economic size, viz. vincristine ($3,500,000/kg), vinblastine ($ 1,000,000/kg). The most economically important natural plant insecticides are the pyrethrins. These compounds are of great interest because of their lethal activity against insects, low toxicity to mammals and low persistence after use. Pyrethrins are found in some members of Compositeae, like *Chrysanthemum, Tagetus,* etc. The rotenoids are found in *Derris, Lonochocarpus* and *Tephrosia.* Some of the antineoplastic compounds and other useful compounds produced by plants are listed in Table 7.1.

Problems associated with secondary metabolite production

1. Plant cells have a very long doubling time (16–24 h) when compared to microorganisms (approximately 1 h). This means that plant cells produce relatively low biomass and therefore only small amounts of secondary metabolites.
2. If the production of secondary metabolites in the intact plant is restricted to a particular organ, for instance to a differentiated tissue (e.g. laticifiers or glands), then little production takes place *in vitro* because the cells *in vitro* are often non-differentiated.
3. The production of secondary metabolites *in vitro* is generally very low, certainly when compared to an intact plant.
4. Cell and suspension cultures are genetically unstable; the production of compounds often falls to a level lower than that at the start of the *in vitro* culture through mutation.
5. Plant cells form aggregates *in vitro,* thus vigorous stirring is needed to prevent this clumping, but this may result in damaging the cells.

6. The use of high sugar concentration in the nutrient medium (up to 5% w/v) increases the cost considerably.
7. Infections can considerably limit the process. When a bacterial infection occurs, it quickly spreads over the culture as a result of its shorter doubling time.
8. Much more intensive research for bioreactors and the development of good methods of isolation is needed.
9. Lack of knowledge about the growth of plant cells in bioreactors and fermenters is severely hindering their further development.
10. It is important to consider the cost of the final product per kilogram so that it should not be cost inhibitive.

Immobilized cell systems

There are an increasing number of reports of immobilization of plant cells producing secondary metabolites on a variety of supports. Presently, different processes based on immobilized biocatalysts are in commercial operation, which include production of high fructose syrups, 6-aminopencillanic acid and various amino acids. The technique of immobilization is based on confinement of biocatalyst on or within a matrix by entrapment, adsorption or covalent linkage. Entrapment within a polymeric network is the most appropriate method used for the immobilization of large sensitive plant cells and protoplasts. The cells are physically entrapped in a gel structure and substrate(s)/product(s) can diffuse to/from the immobilized cells. The cells are commonly immobilized by entrapment in calcium alginate, potassium carrageenan or in agarose beads. It is possible to immobilize large clumps of aggregate cells or even calluses, but the most suitable cells for immobilization are fine homogeneous suspensions.

The stages in development of an immobilization system for plant cells are as follows:

i) Obtain a cell culture with defined and desirable properties, e.g. a cell line with high product yield and a low growth rate.
ii) Determine a suitable immobilization method

Table 7.1 Secondary metabolites produced by plant species used for increased production through tissue culture techniques

Compound	Species	Activity
Pyrethrins	*Chrysanthemum cincerariefolium*	Insecticidal
	Tagetus erecta	,,
Nicotine	*Nicotiana tabacum*	,,
	Nicotiana rustica	,,
Rotenoids	*Derris elliptica*	,,
	Lonchocarpus utilis	,,
	Tephrosia vogaeli	,,
	Tephrosia purpurea	,,
Azadirachtin	*Azadirachta indica*	,,
Phytoecdysones	*Trianthema portulacastrum*	,,
Baccharine	*Baccharis megapotamica*	Antineoplastic
Bruceantine	*Brucea antidysenterica*	,,
Cesaline	*Caesalpinia gillisesii*	,,
3-Deoxycolchicine	*Colchicum speciosum*	,,
Ellipticine,		
9-methoxyellipticine	*Ochrosia moorei*	,,
Fagaronine	*Fagara zanthoxyloides*	,,
Harringtonine	*Cephalotaxus harringtonia*	,,
Indicine N-oxide	*Heiotropium indicum*	,,
Maytansine	*Maytenus bucchananii*	,,
Podophyllotoxin	*Podophyllum peltatum*	,,
Taxol	*Taxus brevifolia*	,,
Thalicarpine	*Thalictrum dasycarpum*	,,
Tripdiolide, triptolide	*Tripterygium wilfordii*	,,
Vinblastine, ajmalicine, vincristine	*Catharanthus roseus*	,,
Quinine	*Cinchona officinalis*	Antimalarial
Digoxin, reserpine	*Digitalis lanata*	Cardiac tonic
Diosgenin	*Dioscorea deltoidea*	Antifertility
Morphine	*Papaver somniferum*	Analgesic
Thebaine	*Papaver bracteatum*	Source of codeine
Scopolamine	*Datura stramonium*	Antihypertension
Atropine	*Atropa belladonna*	Muscle relaxant
Codeine	*Papaver* spp.	Analgesic
Shikonine	*Lithospermum erythrorhizon*	Dye, pharmaceutical
Anthraquinones	*Morinda citrifolia*	Dye, laxative
Rosamarinic acid	*Coleus blumei*	Spice, antioxidant
Jasmine	*Jasmium* spp.	Perfume
Stevioside	*Stevia rebaudiana*	Sweetner
Crocin, picrocrocin	*Crocus sativus*	Saffron
Capsaicin	*Capsicum frutescens*	Chilli
Vanillin	*Vanilla* spp.	Vanilla

which maintains cell viability and product biosynthesis.

iii) Obtain product release from the cells into the medium while maintaining culture survival.

iv) Design a suitable vessel for culture of the immobilized cells and operating conditions which maintain cellular characteristics.

During the last 30 years, immobilized enzymes and microbial cells have been extensively studied. Some of the advantages of an immobilized system are as follows:

i) Reuse of expensive biomass.

ii) It results in continuous operation.

iii) Automatic physical separation of product from biocatalyst (cells).

iv) Stabilization of biocatalyst.

v) Efficient utilization of expensive fermenters by using high biomass/low volume vessels.

When immobilized whole cells are used they offer additional advantages as: (i) more stable enzyme(s) in natural environment; (ii) possibility of using whole biosynthetic pathways; (iii) regeneration of coenzymes and (iv) *in situ* activation when viable cells are used.

The immobilization technique applies only to those plant cell suspensions that show non-growth associated type of products. That means after generation of biomass (cell growth in suspension), the cells can be immobilized and used for various purposes.

Polymers for immobilization

A number of polymers are available for entrapment of plant cells. These include synthetic-polyacrylamide, epoxy resin, polyurethane; carbohydrates-cellulose, agar, agarose, kappa-carrageenan, alginate, chitosan; and proteins-collagen, gelatin and fibrin. The list is rapidly growing and it appears possible to immobilize any plant cell line with preserved viability. Agar and agarose, alginate and kappa-carrageenan have been used quite often to entrap the cells. A different approach is to allow cells to become entrapped within the spaces of a pre-formed matrix such as nylon, polyurethane or metal mesh. When cells in suspension are mixed with these materials (usually in the form of spheres or cubes up to 1 cm diam.), they are rapidly incorporated into the network and subsequently grow into the cavities of the mesh and become entrapped either by physical restriction or by attachment to the matrix material. Some success with this simple technique has been obtained. Modifications to the methods of using agarose, alginate is to use the mixtures of alginate and gelatin, agarose and gelatin, polyacrylamide and xanthan gum or polyacrylamide and alginate gels.

After the immobilization, tests may be employed to study the viability of immobilized plant cells. The methods used for viability are same as used for cell suspension, e.g. staining with FDA, calcoflour white, phenosafranine, cell growth, mitotic index, etc. (refer to chapters 8 and 9). In general, all methods employed to study the viability of immobilized plant cells have shown that the cells are not affected to a great extent by the immobilization process.

Cells raised in suspension cultures and immobilized on a suitable matrix are then placed in aerated columns. Most biosynthetic experiments have been carried out in simple batch reactors but there are examples of continuous operation of packed bed reactors. The continuous reactor types are used where the product is extracellular. From the studies made so far, it can be concluded that the immobilized cells can carry out the same biosynthetic reactions as the corresponding freely suspended cells. In fact, in certain cases, a higher productivity (up to 100 times) has been observed for plant cells. Immobilized plant cells may be employed for biosynthetic purposes over an extended period of time. The same preparation of cells has in one case been used for 220 days. The total productivity per biomass unit is in such cases considerably increased.

Product release

As mentioned earlier, continuous reactors can only be applied to extracellular products. In order to realize the products of intracellular origin, the product is sometimes spontaneously released. It has been observed that products normally stored within the cells leak out into the medium. If it is not so, the induced release is performed. Secondary products are often stored within the vacuoles of cultivated plant cells. In order to reach such products, two membranes i.e. the plasma membrane and the tonoplast have to be penetrated. For this purpose, a number of permeabilizing agents such as DMSO (dimethyl sulfoxide) have been used. The concentration of DMSO required for quantitative permeabilization varies from 5 to 25% with cell species. For cells of *Catharanthus roseus*, 5% DMSO appears to be sufficient for the release of substances stored in the vacuoles.

The immobilization of plant cells is a relatively new technique. This has not been used on an industrial scale for biosynthesis of plant metabolites. However, its potential is considerable and at least in one instance for the pilot industrial scale biotransformation of β-methyl digitoxin to valuable β-methyl digoxin has been conducted.

Biotransformation

An area of biotechnology that has gained considerable attention in recent years is the ability of plant cell cultures to catalyze the transformation of a readily available or inexpensive precursor into a more valuable final product. The conversion of a small part of molecule in the structure and composition to industrially important chemicals by means of biological systems is termed as biotransformation. A partial listing of biotransformations with plant cells that can be implemented in a diverse array of enzyme reactions has been shown in Table 7.2.

The use of plant cell cultures for biotransformation requires the selection of cell types that express the enzymatic capabilities to catalyze the specific reaction of interest. Another factor that weighs heavily in the cell selection process is the specificity of enzyme reaction. Plant cells have been shown to perform more than one biotransformation with a given substrate. Therefore, the selection process needs to focus on the detection of cell lines that specifically catalyze the desired reaction with little or no contamination from other reaction products. Undoubtedly, a complete knowledge of the enzymology of the

Table 7.2 Biotransformations which can be achieved by plant cell cultures

Reaction	Substrate	Product
Reduction	$C = C$ $C = C - CO$	$CH_2 - CH_2$ $CH_2 - CH_2 - CO$ or $CH_2 - CH_2 - CHOH$
	CO CHO	CHOH CH_2OH
Oxidation	CH_3 CH_2OH CHOH $= S$	CH_2OH or COOH CHO CO $S = O$
Hydroxylation	CH CH_2 NH2	C – OH CH – OH NH – OH
Glycosylation	OH COOH CH N	O – glucose COO – glucose C – glucose N^+–arabinose
Esterification	OH COOH NH	O–palmitate COO–malate N–acetate
Methylation and demethylation	OH N N^+-CH_3	O-CH_3 N^+-CH_3 $= NH$
Isomerization	*trans* dextra β–OH	*cis* laevo rotation α–OH
Epoxidation	$CH = CH$	$- HC - CH -$ $\diagdown$ O $\diagup$

desired reaction needs to be incorporated into the cell culture selection process.

Two highly desirable features of plant cell biotransformation have been described. First, more than one reaction can be accomplished by a particular cell culture. In other words, multistep transformation can be accomplished using cell cultures that express a series of enzyme activities. The biotransformation of valencene into nootkatone by cell suspensions of *Citrus* is produced by two steps through the 2-hydroxy-derivative nootkatol. Up to 66% conversion rate was achieved with specific cell cultures (Table 7.3). The second unique feature of biotransformation is that in some instances, non-producing cell cultures can be used to synthesize the desired end-product. The most celebrated example of this application is the biotransformation of digitoxin into digoxin. To date, plant tissue cultures have been used in achieving a specific biotransformation in the production of a cardiovascular steroid digoxin or β-methyl digoxin from digitoxin or β-methyl digoxin obtained from *Digitalis lanata* (foxglove). Although undifferentiated cell cultures of *D. lanata* do not produce cardiac glycosides, they are nevertheless able to perform special biotransformations on cardenolide glycosides added to the medium. *Datura* cell cultures possess ability to convert hydroquinone into arbutin through glycosylation. Cell cultures of *Stevia rebaudiana* and *Digitalis purpurea* can convert steviol into steviobiocide and stevioside which are 100 times sweeter than cane sugar.

PROTOCOLS

7.1. Protocol for cell suspension culture (*Nicotiana tabacum*)

Plant material: *Nicotiana tabacum*

1. Follow Steps 1–5 of Protocol 5.1 for *in vitro* germination of seeds.
2. Collect the aseptically germinated seedlings when the cotyledons are fully expanded and the epicotyl is beginning to emerge. Usually this will occur when the seedlings are 2-week old. Place each seedling on a sterile slide/petri dish and prepare the explant.
3. Various explants can be used. Excise the shoot apex with and without cotyledons and insert the stem base into the medium, cotyledons, and hypocotyl section from the decapitated seedling, etc. Place the explants on the following medium (MSN):
   MS + 1–2 mg/l 2,4-D
4. Incubate the cultures in dark at 25°C. Callus will be produced in 3–4 weeks.
5. Cut small pieces of callus of ~0.5 g fresh weight and subculture on the same fresh medium for proliferation.
6. Prepare MSN liquid medium without agar. For experiment purposes 15 ml medium in a 125 ml Erlenmeyer flask is put. Sterilize the opening of a flask with the flame of burner in the hood.
7. Transfer a piece of callus to a sterile petri dish. Gently break the callus of ~2 cm diameter with forceps into 20–30 small pieces.

Table 7.3 Biotransformations performed by plant cells

Plant cell culture	Substrate	Product
Papaver somniferum	Codeinone	Codeine
Nicotiana tabacum	Carvoxiome	Cavoxone
Citrus spp.	Valencene	Nootkatone
Solanum tuberosum	Solavetivone	Hydroxylated derivatives
Choisya ternate	Ellipticine	5-formyl ellipticine
Galium mollugo	2-succinyl benzoate	Anthraquinones
Digitalis lanata	Digitoxin	Digoxin
Datura spp.	Hydroquinone	Arbutin
Stevia rebaudiana / *Digitalis purpurea*	Steviol	Steviocide, steviobiocide

8. Transfer the small pieces of callus to the liquid media using forceps.

9. Heat sterilize the opening of a flask and place the sterile cap on the flask. Prepare replicated samples.

10. Incubate on a gyratory shaker set at 125 rpm which should be placed in a temperature controlled room.

11. Subculture every week. For the first few subcultures, remove a portion of the spent medium and replace with fresh medium with the help of a large bore sterilized pipette. When the cell mass has about doubled, carefully split the culture into two flasks with an equal volume of fresh medium. Repeat the incubation cycle.

12. Once the suspension culture becomes established and consists of finely dispersed cell clusters and aggregates, a dilution ratio of 1 : 4 to 1 : 10 old culture to fresh medium should be possible on a 7–10-day basis to maintain the cell line. Cell suspension can be seen under a microscope by taking an aliquot from the flask on a glass slide under sterile conditions.

Cell suspension growth curve

1. Use an established cell suspension line to follow the growth curve over time. Carefully combine several cultures into a larger single batch of suspension, in order to provide a uniform inoculum for each replicate.

2. Pipette 5 ml of inoculum into 25 ml of fresh MSN liquid medium in a 125 ml Erlenmeyer flask. Prepare at least four replicates.

3. Aseptically transfer 10 ml of each culture into separate sterile, conical, calibrated centrifuge tubes. Spin at 2000 g for 5 min. Measure the packed cell volume (PCV) in each tube. This is the zero time value.

4. Resuspend the cells in the centrifuge tube with 10 ml of medium and return to their respective culture flasks. Incubate the suspensions on the gyratory shaker. Excellent sterile techniques and handling is required as this is the most likely source of contamination.

5. Repeat Steps 3–4 every 2–3 days for a total of 3 weeks. Calculate the mean PCV and standard deviation for each sampling time and plot the growth curve.

7.2. Protocol for cell suspension culture (cereals–wheat, rice, maize, barley, etc.)

1. Follow Steps 1–5 of Protocol 5.4 for induction of callus from embryos of cereals.

2. Follow Steps 6–12 of Protocol 7.1 for initiation of cell suspension culture and then Steps 1–5 for measurement of cell suspension growth. Same liquid medium should be used for cell suspension which was employed for initiation of callus i.e. MS + 2 mg/l 2,4-D.

Further reading

Chawla, H.S. 1998. *Biotechnology in Crop Improvement.* International Book Distributing Co., Lucknow.

Evans, D.A., Sharp, W.R., Ammirato, P.V. and Yamada, Y. 1983. *Handbook of Plant Cell Culture.* Macmillan Publishing Company, New York.

Reinert, J. and Bajaj, P.P.S. (eds.). 1977. *Applied and Fundamental Aspects of Plant Cell, Tissue and Organ Culture.* Springer-Verlag, Berlin.

Street, H.S. (ed.) 1973. *Plant Tissue and Cell Culture.* Blackwell Scientific Publications, Oxford.

Thorpe, T.A. 1981. *Plant Tissue Culture: Methods and Applications in Agriculture.* Academic Press Inc., New York.

Vasil, I.K. 1984. *Cell Culture and Somatic Cell Genetics of Plants: Laboratory Procedures and Their Applications.* Academic Press, Inc., New York.

Yamada, Y. and Fujita, Y. 1983. Production of useful compounds in culture. In: *Handbook of Plant Tissue Culture* (eds. Evans, D.A., Sharp, W.R., Ammirato, P.V. and Yamada, Y.), Macmillan Publishing Company, New York, pp. 717-728.

8

In vitro Production of Haploids

The term haploid refers to those plants which possess a gametophytic number of chromosomes (single set) in their sporophytes. The interest in haploids stems largely from their considerable potential in plant breeding, especially for the production of homozygous plants and in their studies on the detection of mutations.

The significance of haploids in the fields of genetics and plant breeding has been realized for a long time. However, their exploitation remained restricted because of low frequency (.001–.01%) with which they occur in nature. Spontaneous production of haploids usually occurs through the process of apomixis or parthenogenesis (embryo development from an unfertilized egg). Artificial production of haploids was attempted through distant hybridization, delayed pollination, application of irradiated pollen, hormone treatment and temperature shock. However, none of these methods was dependable and repeatable. It was in 1964 that Guha and Maheshwari reported the direct development of embryos and plantlets from microspores of *Datura innoxia* by the culture of excised anthers. Later, Bourgin and Nitsch (1967) obtained complete haploid plants of *Nicotiana tabacum*. Since then, anthers containing immature pollen have been successfully cultured for a wide range of economically important species.

Haploids may be grouped into two broad categories: (a) monoploids (monohaploids), which possess half the number of chromosomes from a diploid species, e.g. maize, barley; and (b) polyhaploids, which possess half the number of chromosomes (gametophytic set) from a polyploid species, e.g. potato, wheat. Here, the general term haploid is applied to any plant originating from a sporophyte (2n) and containing (n) number of chromosomes.

Haploid production through anther culture has been referred to as androgenesis while gynogenesis is the production of haploid plants from ovary or ovule culture where the female gamete or gametophyte is triggered to sporophytic development.

ANDROGENIC METHODS

The androgenic method of haploid production is from the male gametophyte of an angiosperm plant, i.e. microspore (immature pollen). The underlying principle is to stop the development of pollen cell whose fate is normally to become a gamete, i.e. a sexual cell, and to force its development directly into a plant. Haploids can be obtained by the culture of excised anthers and culture of isolated pollen (microspore). Here it is worthwhile to explain the structure of anther and development of pollen *in vivo* (Fig. 8.1). An angiosperm stamen consists of a filament, connective tissue and anther. A typical anther shows two anther lobes and each lobe possesses

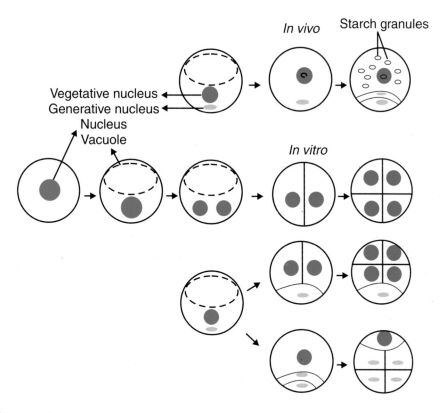

Fig. 8.1 Diagrammatic representation showing various modes of division of the microspores under *in vivo* and *in vitro* conditions.

two microsporangia or pollen sacs. During microsporogenesis in a young anther, there are four patches of primary sporogenous tissue, which either directly function as pollen mother cells (PMCs) or undergo several divisions. The PMCs form pollen tetrads by meiosis and when the callose wall of tetrad dissolves, the four pollen grains or microspores are liberated. The newly released microspore is uninucleate, densely cytoplasmic with the nucleus centrally located. As vacuolation occurs, the nucleus is pushed towards the periphery. At the first division or first pollen mitosis, the microspore nucleus produces a large vegetative and a small generative nucleus. The second pollen mitosis is restricted to generative nucleus and forms two sperms and takes place in either the pollen or pollen tube.

Anther culture

The technique of anther culture is rather simple, quick and efficient. Young flower buds with immature anthers in which the microspores are confined within the anther sac at the appropriate stage of pollen development are surface sterilized and rinsed with sterile water. The calyx from the flower buds is removed by flamed forceps. The corolla is slit open and stamens are removed and placed in a sterile petri dish. One of the anthers is crushed in acetocarmine to test the stage of pollen development. If it is found to be correct stage, each anther is gently separated from the filament and the intact uninjured anthers are inoculated horizontally on nutrient media. Injured anthers may be discarded because wounding often stimulates callusing of the anther wall tissue. When dealing with plants having minute flowers such as *Brassica* and *Trifolium*, it may be necessary to use a stereo microscope for dissecting the anthers. In case of cereals, spikes are harvested at the uninuclear stage of microspore development and surface sterilized. The inflorescence of most cereals at this stage is covered by the flag leaf. Anthers can be plated on solid agar media in petri dishes. Normally 10–20 anthers are plated in a 6 cm petri dish. Anthers can be cultured on a liquid media and 50 anthers can be cultured in 10 ml of liquid medium (Fig. 8.2). In responsive anthers, the wall tissue gradually turn brown and within 3–8 weeks they burst open due to the pressure exerted by the growing pollen callus or pollen plants. After they have attained a height of about 3–5 cm, the individual plantlets or shoots emerging from the callus are separated and transferred to a medium that would support further development. The rooted plants are transferred to sterile soil mix in the pots.

Anther culture has an advantage over microspore culture in being very quick for practical purposes, and also sometimes the anther wall has an influence on the development of microspores in it, acting as a conditioning factor.

Fig. 8.2 Diagrammatic illustration of anther and microspore culture for production of haploid plants and diploidization

Microspore Culture

Haploid plants can be produced through *in vitro* culture of male gametophytic cells, i.e. microspores or immature pollen. In a general procedure for microspore culture, anthers are collected from sterilized flower buds in a small beaker containing basal media (e.g. 50 anthers of *Nicotiana* in 10 ml media). The microspores are then squeezed out of the anthers by pressing them against the side of beaker with a glass rod (Fig. 8.2). Anther tissue debris is removed by filtering the suspension through a nylon sieve having a pore diameter which is slightly wider than the diameter of pollen (e.g. 40 μ for *Nicotiana*, 100 μ for maize, etc.). It has been observed that smaller microspores do not regenerate, thus larger, good and viable microspores can be concentrated by filtering the microspore suspension through nylon sieves. This pollen suspension is then centrifuged at low speed ca. 150× g for 5 min. The supernatant containing fine debris is discarded and the pellet of pollen is resuspended in fresh media and it is washed atleast twice. The microspores obtained are then mixed with an appropriate culture medium at a density of 10^3 to 10^4 microspores/ml. The final suspension is then pipetted into small petri dishes. To ensure good aeration, the layer of liquid in the dish should be as thin as possible. Each dish is then sealed with parafilm to avoid dehydration and is incubated. The responsive microspores form embryos or calli and its subsequent development to plant formation can be achieved by transferring to suitable media.

Culture of anthers has proved to be an efficient technique for haploid induction. But it has one main disadvantage that plants not only originate from pollen but also from various other parts of the anther (especially in dicots) with the result that a population of plants with various ploidy levels are obtained. This difficulty can be removed by culture of isolated microspores which offer the following advantages:

1. Uncontrolled effects of the anther wall and other associated tissue are eliminated and various factors governing androgenesis can be better regulated. But disadvantageous where anther wall has a stimulatory effect.
2. The sequence of androgenesis can be observed starting from a single cell.
3. Microspores are ideal for uptake, transformation and mutagenic studies, as microspores may be evenly exposed to chemicals or physical mutagens.
4. Higher yields of plants per anther could be obtained.

The various factors governing the success of androgenesis have been discussed below.

Genotype: Success in anther culture is predominantly dependent on the genotype of the anther donor material. It has been repeatedly observed that various species and cultivars exhibit different growth responses in culture. It is now known that anther culture ability is genetically controlled. Thus, a general survey for tissue culture response of various cultivars must be undertaken with simple media, as complex media rich in growth regulators tend to favor proliferation of somatic anther tissue giving rise to callus of various ploidy levels.

Physiological status of the donor plants: The physiological status of the plants at the time of anther excision influences the sporophytic efficiency of microspores. In order to allow for the *in vitro* development of pollen into an adult plant it is very important to start with healthy pollen cells. Therefore, it is good to culture anthers from plants grown under the best environmental conditions. The donor plant should be taken care of from the time of flower induction to the sampling of pollen. The use of any kind of pesticide, whether externally applied or systemic should be avoided for 3–4 weeks preceding sampling. The response in culture is predominantly influenced by the different growth conditions during various seasons. The variation in response of anthers from plants grown under different environmental conditions may be due to the differences in the endogenous level of growth regulators. Critical environmental factors are light intensity, photoperiod, temperature, nutrition and concentration of carbon dioxide. It has been generally observed that plants grown outdoors during natural growing seasons are more responsive than greenhouse-grown material. Flowers from relatively young plants at the beginning of the flowering season are more responsive. It is therefore of prime importance that plants be grown under optimal growth conditions, watered with minimal salt solutions periodically, and that relatively young plants be used.

Stage of pollen: It has been established that selection of anthers at an appropriate stage of pollen development is most critical. Anthers with microspores ranging from tetrad to the binucleate stage are responsive. But as soon as starch deposition has begun in the microspore, no sporophytic development and subsequently no macroscopic structure formation occurs. Data has established that uninucleate microspores are more prone to experimental treatment for culture just before or during first mitosis. There is an optimum stage for each species. For example, pollen at or just after the pollen mitosis in *Datura innoxia, Nicotiana tabacum*, early bicellular stage in *Atropa balladonna* and *Nicotiana sylvestris* and mid- to late uninucleate stage of microspore development in cereals has been found to give the best results.

Pretreatment of anthers: The underlying principle of androgenesis is to stop the development of the pollen cell whose fate is normally to become a gamete, and to force its development directly into a plant. This abnormal pathway is possible if the pollen cell is taken away from its normal environment in the living plant and placed in other specific conditions. This induction of androgenesis is enhanced by giving certain treatments to the whole spike, or flower bud or to the anthers.

i) *Cold pretreatment:* In general, cold treatment between 3 and 6°C for 3 to 15 days gives good response. Maize responds better to a temperature of 14°C. The degree of cold that should be given is dependent on the species. As a result of cold treatment, weak or non-viable anthers and microspores are killed and the material gets enriched in vigorous anthers. It is possible that cold pretreatment retards aging of the anther wall, allowing a higher proportion of microspores to change their developmental pattern from gametophytic to sporophytic.

ii) *Hot treatment:* Floral buds or entire plants in some species when subjected to 30°C for 24 h or 40°C for 1 h stimulates embryogenesis (e.g. *Brassica*). The temperature shock appears to cause dissolution of microtubules and dislodging of the spindle which causes abnormal division of the microspore nucleus.

iii) *Chemical treatment:* Various chemicals are known to induce parthenogenesis. 2- Chloro ethylphosphonic acid (ethrel) has a pronounced effect in increasing the haploid production in various species. Plants are sprayed with an ethrel solution (e.g. 4000 ppm in wheat) just before meiosis in PMCs which results in multinucleated (4–6) pollen with fewer starch grains. It is possible that multinucleated pollen might be induced to form embryos when cultured.

Culture media: The composition of medium is one of the most important factors determining not only the success of anther culture but also the mode of development. Normally only two mitotic divisions occur in a microspore, but androgenesis involves repeated divisions. It is difficult to draw a conclusion as to which medium is most suitable as species or even genotypes may demand different nutritional conditions. Basal medium of MS, White (1963), Nitsch and Nitsch (1969) and N_6 for solanaceous crops; B_5 and its modifications for *Brassica*; and B_5, N_6, LS (Linsmaier and Skoog, 1965) and Potato2 medium (Chuang *et al.*, 1978) for cereals have been used. Thus, no general recommendations can be given but some general observations have been mentioned.

Sucrose is essential for androgenesis. Sugars are indispensable in the basal medium as they are not only the source of carbon but are also involved in osmo-regulation. The usual level of sucrose is 2–4%; however, higher concentration (6–12%) favors androgenesis in cereals. Chelated iron has been shown to play an important role in the differentiation of globular embryos into heart-shaped embryos and further into complete plants.

Nitrogen metabolism is quite an important feature. The presence of nitrate, ammonium salts as well as amino acids appear to play a very special role at different stages of the developmental process. However, glutamine is probably beneficial for most plant species as an aid to achieving the *in vitro* differentiation of a cell to a complete plant.

Pollen embryogenesis can be induced on a simple mineral-sucrose medium in plants like tobacco, yet for androgenesis to be completed, addition of certain growth regulators is required. For example, cereal anthers require both auxins and cytokinins and optimal growth response depends on the endogenous level of these growth regulators. However to promote direct embryogenesis simple media with low levels of auxins are advisable. When the response of cereal cells and tissues to different phytohormones concentrations and combinations are compared with dicots, generally growth hormones are needed in the former whereas high amount of phytohormone autotropy is present in the dicots.

Activated charcoal in the medium enhances the percentage of androgenic anthers in some species presumably by removing the inhibitors from the medium.

Process of androgenesis

The anthers normally start undergoing pollen embryogenesis within 2 weeks and it takes about 3–5 weeks before the embryos are visible bursting out of the anthers. In rice this may take up to 8 weeks. Haploid plantlets are formed in two ways: (i) direct androgenesis: embryos originating directly from the microspores of anthers without callusing or (ii) indirect androgenesis i.e. organogenic pathway: microspores undergo proliferation to form callus which can be induced to differentiate into plants.

The process of androgenesis has shown that microspores undergo divisions, which continues until a 40–50-celled proembryo is formed. The embryos, mostly at globular stage, burst out of the exine and are released. The embryos undergo various stages of development, simulating those of normal zygotic embryo formation. However, when the microspore takes organogenetic pathway, it looks to be larger than embryonal type of micropspore after 2–3 weeks and contain only a few cells. These cells increase in size, exerting pressure on the exine which bursts open and the contents are released in the form of a callus. These calluses then differentiate into plantlets. The plants with well developed shoots and roots are then transferred to pots. The plantlets originating from the callus generally exhibit various levels of ploidy.

The physical environmental conditions in which the cultures are to be placed can enhance the differentiation. The cultures are incubated at 24–28°C. In the initial stages of induction of morphogenesis, darkness is normally more effective or cultures should be kept in low light intensity (500 lux). After induction, macroscopic structures are transferred to a regeneration medium (in cereals with reduced sucrose and auxin concentration) and kept at 14 h day–light regime at 2000–4000 lux.

The ploidy level and chromosome doubling

The ploidy of plants derived from anther or microspore culture is highly variable. The wide range of ploidy levels seen in androgenetic plants has been attributed to endomitosis and/or fusion of various nuclei. Variations in ploidy observed in anther cultures appears to be a function of the developmental stage of anthers at the time of excision and culture, with higher ploidy levels more prevalent in anthers cultured following microspore mitosis. Moreover, haploid tissues are quite susceptible to changes in ploidy level during cell proliferation and growth *in vitro*. For obtaining homozygous lines, the plants derived through anther culture must be analysed for their ploidy status.

1. **Counting of plastids in the stomata:** The ploidy level of a plant may be determined by counting the number of plastids in the stomata of a leaf. For example, in potato monohaploids have 5–8, dihaploids 10–15, and tetraploids 18–24 chloroplasts per guard cell.

2. **Chromosome number:** It can be counted from pollen mother cells of buds which can be collected from the regenerated plants and fixed in Carnoy's solution. Acetocarmine or propionocarmine can stain cells. Root tips are also utilized for chromosome counting. Fixed root tips are normally hydrolyzed in 1 N HCl at 60°C for 10 min followed by staining with acetocarmine.

3. **Number of nucleoli:** Haploid plants contain one nucleolus while diploids contain 2 nucleoli. Number of nucleoli is directly related to the ploidy status of a plant. Leaves can be incubated overnight with orcein and number of nucleoli can be counted.

4. **Flow cytometric analysis:** Leaves of potential haploid plants are finely chopped and intact interphase nuclei are freed from the cells. At this stage nuclear DNA content reflects the ploidy state of the donor which is determined by flow cytometry. This method is very quick.

Diploidization

Haploids can be diploidized to produce homozygous plants by following methods:

1. **Colchicine treatment:** Colchicine has been extensively used as a spindle inhibitor to induce chromosome duplication. It can be applied in the following ways:
   i) The plantlets when still attached to the anther are treated for 24–48 h with 0.5%

colchicine solution, washed thoroughly and replanted.

ii) Anthers can be plated directly on a colchicine-supplemented medium for a week and when the first division has taken place, these are transferred to colchicine-free medium for androgenesis process to take place. This method can be followed in maize where male and female flowers are borne separately and diploidization is a problem.

iii) Colchicine–lanolin paste (0.4%) may be applied to the axils of leaves when the plants are mature. The main axis is decapitated to stimulate the axillary buds to grow into diploid and fertile branches.

iv) Repeated colchicine treatment to axillary buds with cotton wool plugs over a period of time (e.g. 14 days in potato).

v) In cereals, vigorous plants at 3–4 tiller stage are collected, soil is washed from the roots and are cut back to 3 cm below the crown. The plants are placed in glass jars or vials containing colchicine solution (2.5 g colchicine dissolved in 20 ml dimethyl sulfoxide and made up to a liter with water). The crowns are covered with colchicine solution. The plants are kept at room temperature in light for 5 h, the roots are washed thoroughly with water and potted into light soil.

Plants should be handled with extra care after colchicine treatment for few days and should be maintained under high humidity.

2. **Endomitosis:** Haploid cells are in general unstable in culture and have a tendency to undergo endomitosis (chromosome duplication without nuclear division) to form diploid cells. This property of cell culture has been exploited in some species for obtaining homozygous plants or for diploidization. A small segment of stem is grown on an auxin-cytokinin medium to induce callus formation. During callus growth and differentiation there is a doubling of chromosomes by endomitosis to form diploid homozygous cells and ultimately plants.

Significance and uses of haploids

1. *Development of pure homozygous lines*: In the breeding context, haploids are most useful as source of homozygous lines. The main advantage is the reduction in time to develop new varieties. A conventional plant breeding program takes about 6–8 years to develop a pure homozygous line, whereas by the use of anther/microspore culture, the period can be reduced to few months or a year. Thus, homozygosity is achieved in the quickest possible way making genetic and breeding research much easier. Homozygosity is still more important for those plants which have a very long juvenile phase (period from seed to flowering) such as fruit trees, bulbous plants and forestry trees. Even if repeated self-pollination is possible, achievement of homozygosity in this group of plants is an extremely long process.

2. *Hybrid development*: As a result of complete homozygosity obtained from diploidization of haploids, one can rapidly fix traits in the homozygous condition. Pure homozygous lines can be used for the production of pure F_1 hybrids.

3. *Induction of mutations*: Haploid cell cultures are useful material for studying somatic cell genetics, especially for mutation and cell modification. Majority of mutations induced are recessive and therefore is not expressed in the diploid cells because of the presence of dominant allele. Single cells and isolated pollen have the advantage over the entire plant in that they can be plated and screened in large numbers, in a manner similar to microbiological technique. Mutants which are resistant to antibiotics, herbicides, toxins, etc. have been isolated in a number of plant species. By subjecting haploid *Nicotiana tabacum* cells to methionine sulfoximine, Carlson (1973) regenerated mutant plants which showed a considerably lower level of infection to *Pseudomonas tabaci*. Wenzel and Uhrig (1981) developed mutants in potato through anther culture which were resistant to potato cyst nematode. Mutants have also been isolated for various temperatures, radiosensitivity to ultraviolet light and gamma radiation, amino acid, e.g. valine, and various antibiotics and drugs.

4. *Induction of genetic variability*: By anther culture not only haploids but also plants of various ploidy levels and mutants are obtained and can be incorporated into the breeding programs.

5. *Generation of exclusively male plants*: By haploid induction followed by chromosome doubling it is possible to obtain exclusively male plants. For example, in *Asparagus officinalis* male plants have a higher productivity and yield earlier in the season than female plants. If haploids are produced from anthers of male *Asparagus* plants (XY) these are either X or Y, chromosome doubling of Y results in super male plants YY which can subsequently be vegetatively propagated.

6. *Cytogenetic research*: Haploids have been used in the production of aneuploids. Monosomics in wheat, trisomics with 2n = 25 in potato and in tobacco nullisomics were derived from haploids obtained from monosomics which could not produce nullisomics on selfing. Haploids also give evidence for the origin of basic chromosome number in a species or a genus. For example in pearl millet (*Pennisetum americanum*), occurrence of pairing (upto two bivalents in some cells) suggested that the basic chromosome number may be x = 5.

7. *Significance in the early release of varieties*: Based on anther culture many varieties have been released which are listed in Table 8.1. In Japan, a tobacco variety F 211 resistant to bacterial wilt has been obtained through anther culture. In *Brassica napus*, anther-derived doubled haploid lines had low erucic acid and glucosinolate content. Similarly, in sugarcane, selection among anther culture derived haploids led to the development of

Table 8.1 Varieties released through anther culture

Crop	Varieties
Wheat	Lunghua 1, Huapei 1, G. K. Delibab, Ambitus, Jingdan 2288, Florin, Zing Hua-1, Zing Hua-3
Rice	Hua Yu-1, Hua Yu- 2, Zhong Hua 8, Zhong Hua 9, Xin Xiu, Tanfong 1, Nonhua 5, Nonhua 11, Zhe Keng 66, Aya, Hirohikari, Hirohonami
Tobacco	Tan Yuh No.1, Tanyu 2, Tanyu 3, F 211, Hai Hua 19, Hai Hua 30

superior lines with tall stem and higher sugar content. In bell peppers, dihaploid lines exhibited all shades of color ranging from dark green to light green. These reports have encouraged many plant breeders to incorporate anther culture in breeding methods.

8. *Hybrid sorting in haploid breeding*: One of the essential steps in haploid breeding involves selection of superior plants among haploids derived from F_1 hybrids through anther culture. It is properly described as hybrid sorting and virtually means selections of recombinant superior gametes. The haploid method of breeding involving hybrid sorting is considered superior to pedigree and bulk methods, firstly, because the frequency of superior gametes is higher than the frequency of corresponding superior plants in F_2 generation, and secondly, because haploid breeding reduces significantly the time required for the development of a new variety. For instance if one assumes that the frequency of superior F_1 gametes is one in one hundred, then the frequency of homozygous F_2 plants derived from the fusion of two such superior similar gametes will be one in ten thousand. Therefore, under the haploid method a smaller population of doubled haploids derived from haploids will need be handled.

9. *Disease resistance*: Haploid production has been used for the introduction of disease resistance genes into cultivars. An established cultivar is crossed with a donor for disease resistance. Either F_1 or F_2 anthers are plated and haploids are developed. These haploids are screened for resistance and then diploidized. Resistance to barley yellow

mosaic virus has been introduced into susceptible breeding lines by haploid breeding (Foroughiwehr and Friedt, 1984). A barley accession Q 21681 was found to be resistant to various diseases. This line was crossed to susceptible breeding lines and anthers of F_1 plants were cultured to develop double haploid lines which were resistant to stem rust, leaf rust and powdery mildew (Steffenson *et al.*, 1995). Rice varieties ZhongHua No. 8 and No. 9 have been developed with blast resistance genes, high yield and good quality using haploids integrated in conventional breeding approaches. Hwasambye, a rice variety bred through anther culture showed resistance to leaf blast, bacterial leaf blight and rice stripe tenui virus (Byeong-Geun *et al.*, 1997). In tobacco, a variant that showed resistance to a highly necrotic strain of potato virus Y (PVY) from a population of doubled haploids was reported by Witherspoon *et al.* (1991).

10. *Insect resistance*: A medium-late maturing rice variety 'Hwacheongbyeo' derived from anther culture showed resistance to brown plant hopper. This variety was also resistant to blast, bacterial blight and rice tenui virus, and showed cold tolerance (Lee *et al.*, 1989). In rice, promising anther culture lines have been developed which show resistance to rice water weevil (N'Guessan *et al.*, 1994), pests (Zapata *et al.*, 1991).

11. *Salt tolerance*: Salt tolerant breeding lines have been developed in different crop species which have been integrated in conventional breeding. Miah *et al.* (1996) developed doubled haploid salt tolerant rice breeding line that showed tolerance at the level of 8–10 decisiemens/m (ds/m). Sathish

et al. (1997) established stable NaCl-resistant rice plant line (EC 16–18 ms) by *in vitro* selection of anther derived callus exposed to NaCl. Likewise *in vitro* screening of wheat anther culture derived callus to NaCl resulted in line Hua Bain 124-4 which showed salt tolerance as well as high yield and desiccation tolerance (Zao *et al.*, 1994; 1995).

12. *Doubled haploids in genome mapping.* A rather recent application of DH lines are their use in genome mapping. For molecular screening studies a much smaller sample of doubled haploids is required for desirable recombinants. In a population of DH lines, the identification of markers is much more secure, as most intermediate phenotypic expressions are excluded due to heterozygosity. A gene will segregate in a 1 : 1 ratio for both molecular marker and the phenotype at the plant level. DH is used for genome mapping for major genes and/or quantitative traits in barley, rice, oilseed rape, etc.

Problems

1. High level of management and expertise is required to operate the tissue culture production of haploids.
2. Diploids and tetraploids often regenerate at the same rate as the haploids.
3. Selective cell division must take place in the haploid microspores and not in other unwanted diploid tissues. This selective cell division is often impossible.
4. Callus formation whether it has arisen spontaneously or has been induced by regulators is usually detrimental.
5. The relatively high incidence of albinism in some types of anther and pollen culture.
6. The lack of selection of traits during the derivation of haploid material.
7. There is little chance of isolating a haploid from a mixture of haploids and higher ploidy levels since latter ones are easily outgrown.
8. The doubling of a haploid does not always result in the production of a homozygote.

GYNOGENIC HAPLOIDS

Recent advances in plant tissue culture have resulted in the successful induction of haploid plants from ovary and ovule culture. This means that megaspores or female gametophytes of angiosperms can be triggered *in vitro* to sporophytic development. These plants have been described as gynogenic as compared to androgenic plants derived from microspores. Figure 8.3 shows the various parts of a female reproductive organ. *In vitro* culture of unpollinated ovaries and ovules represents an alternative for the production of haploid plants in species for which anther culture has given unsatisfactory results (e.g. albino plants) or has proven insufficient, e.g. *Gerbera*. The first successful report on the induction of gynogenic haploids was in barley by San Noeum (1976). Subsequently haploid plants were obtained from ovary and ovule cultures of rice, wheat, maize, sunflower, sugar beet, tobacco, poplar, mulberry, etc.

Ovaries can be cultured as pollinated and unpollinated. For haploid production flower buds are excised 24–48 h prior to anthesis for

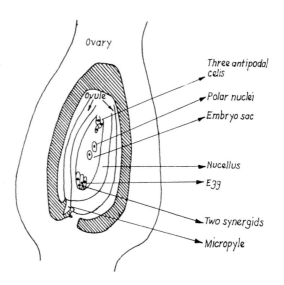

Fig. 8.3 A sectional view of an ovary.

unpollinated ovaries. The calyx, corolla, and stamens are removed and ovaries are then surface sterilized. Before culturing the tip of the distal part of the pedicel is cut off and the ovary is implanted with the cut end inserted in the nutrient medium. The normal Nitsch's (1951), or White's (1954) or MS or N_6 inorganic salt media supplemented with growth substances are used. When liquid medium is to be employed, the ovaries can be placed on a filter paper raft or float with the pedicel inserted through the filter paper and dipping into the medium. Sucrose as a carbon source is essential, although maltose and lactose have been shown to be equally favorable. The various species studied so far seem to have few requirements in common for growth *in vitro*. Some species require only the basal medium and sucrose for growth, although addition of an auxin brought about greater stimulation of growth.

Factors affecting gynogenesis

1. *Genotype*: Genotype of the donor plant is one of the most important factors since each genotype shows a different response. For each genotype a specific protocol must be followed for maximal efficiency.
2. *Growth condition of the donor plant*: It has been found that embryo induction from ovules harvested from lateral branches gives a higher response than ovules harvested from stem apex. Ovules harvested from first formed lateral branches (at the base of plant) gives a higher response than that from the sixth formed lateral branches in sugar beet.
3. *Stage of harvest of ovule*: Excision of ovules at an early developmental stage where no self-pollination can occur or use of highly self-incompatible donor plants help to minimize the problem of fertilization in sugar beet and other crops so that proper stage of unfertilized ovule can be harvested.
4. *Embryo-sac stage*: It has been reported that complete maturation of ovule is not necessary for induction of gynogenesis but an appropriate stage of embryo-sac is more important.
5. *Culture conditions*: Although solid media have been used more frequently for gynogenic culture, few investigators have used liquid medium for the induction of gynogenic calluses followed by dissection and transfer of embryogenic structures to solid medium for differentiation. MS, B5, Miller's basal medium have been most commonly used. The amount of sucrose has been reported to be important for embryogenesis. In rice 3–6% sucrose was effective, whereas in onion 10% sucrose gives maximum response.
6. *Seasonal effects*: Seasonal variation is an important factor. In sugar beet the highest embryo yield was obtained from the summer grown plants. Callus induction from ovules of Gerbera occurred at a higher frequency during autumn as compared to spring.
7. *Physical factors*: Certain physical factors (treatments) given to explants or plant parts from which explants are taken prior to culture may have a strong influence on embryo induction. For *Beta vulgaris* cold pretreatment of flower buds at 4°C for 4–5 days increases embryo yield from cultured ovules. Similarly, cold pretreatment of inflorescence prior to ovule isolation increased haploid callus frequency in *Salvia selarea*. Other physical factors influencing ovary and ovule culture are light and temperature of incubation.

In vitro ovary culture has been used to develop haploid plants and has the potential to overcome the problem of albino plant formation in anther culture. The technique has also got superiority over anther culture due to its potential use in the male sterile genotypes.

The origin of gynogenic haploids differs in species as reported to be from synergids in rice, whereas egg or antipodal cell develops into embryos in barley. *In vitro* culture of unfertilized ovules has been the most efficient and reliable technique for the production of haploid and doubled haploid plants of sugar beet. In tree species, gynogenic plants were reported in mulberry (*Morus indica*), an important tree for nourishing silkworms for use in the silk industry.

Ovaries were cultured on MS medium supplemented with 1 mg/l BA and 1 mg/l kinetin which resulted in direct regeneration of plants without a callus phase.

Though haploid plants have been produced by ovary and ovule culture, they have not been used as much as androgenic haploids in crop improvement. Production of gynogenic haploids through female gametophyte still needs more refinement and also there are problems in dissection of unfertilized ovules and ovaries. It has an inherent disadvantage of one ovary per flower to large number of microspores in one anther. A positive beginning has already been made in sugar beet and gymnosperms and the future is promising.

Chromosome elimination technique for production of haploids in cereals (barley and wheat)

Kasha and Kao (1970) reported a novel technique for the production of haploids in barley. The method involves crossing of *Hordeum vulgare* (2n = 14) with *H. bulbosum* (2n = 14). In nature, the seeds produced from such a cross develops for about 10 days and then begins to abort. However, if the immature embryos are dissected two weeks after pollination and cultured on B_5 medium without 2,4-D they continue to grow. The plants obtained from such embryos are monoploids with *H. vulgare* chromosomes (n = 7) (Fig. 8.4). Evidence indicates that these monoploids are not caused by parthenogenesis, but by the elimination of *H. bulbosum* chromosomes. The evidences were: (i) The occurrence of cells in the embryo with more than seven chromosomes; (ii) the percentage of seed set is higher than observed by induced parthenogenesis; (iii) when diploid *H. vulgare* is pollinated with tetraploid *H. bulbosum*, the percentage of seed set is similar to the diploid cross. Monoploid wheat has also been obtained by crossing *Triticum aestivum* (2n = 6x = 42) with diploid *H. bulbosum* (2n = 2x = 14) where the chromosomes of bulbosum are eliminated and haploids of *T. aestivum* are obtained (Barclay, 1975).

This technique represents a considerable advancement in the production of barley haploids and it has a number of advantages over anther culture. The frequency of haploid formation is quite high by this method. In addition, no aneuploids are obtained by this method. But this technique has a disadvantage of being highly laborious because it involves crossing, embryo excision and then regeneration. In barley varieties namely Mingo, Rodeo, Craig and Gwylan have been released through *bulbosum* technique of chromosome elimination. The varieties were more tolerant to powdery mildew and barley yellow mosaic virus diseases and superior in yield contributing characters as compared to parent.

PROTOCOL

8.1. Protocol for anther culture of cereals (rice, barley, wheat, etc.)

1. Harvest panicles or spikes at the uninucleate stage of microspore development. Stem containing panicle (rice) with pollen at this stage can be identified by the relative position of the flag leaf. Stems are cut with a scalpel below the node from which the panicle arises.

2. Cold pretreatment for a week at 4°C can be given by keeping the stems which have been cut in a beaker containing water. This treatment increases the percentage of cultured anthers that give rise to callus.

3. Remove the flag leaf and surface sterilize the panicles/spikes in a 20% commercial sodium hypochlorite solution for 10 min. Rinse them 4x in sterile distilled water.

4. Place the excised panicle/spike on a sterilized glass slide or petri dish. Anthers are squeezed out with the help of forceps. Each rice floret has 6 anthers.

5. Anthers are placed horizontally on the following N_6 medium for callus induction: About 60 anthers in 60 × 15 mm petri dish can be inoculated.

   N_6 basic salts + 6% sucrose + 2 mg/l 2,4-D (for rice)

Fig. 8.4 Scheme of monoploid and doubled monoploid barley productin via the interspecific hybridization method followed by somatic chromosome elimination.

Constitution of N_6 medium is given in Table 3.1.

6. Seal the petri dishes and incubate the cultures in dark at 27°C for rice or 20–22°C for barley.

7. Remove anthers 3–4 weeks after colonies or calluses start coming out and transfer to the following medium: MS + 0.05 mg/l NAA + 2 mg/l KIN.

8. Place the cultures under moderate light intensity with 16 h light : 8 h dark photoperiod.

9. Separate clumps of shoots that form individual shoot when about 1 cm in length and place in individual containers containing the following medium for root induction: MS + 1 mg/l (NAA or IBA).

10. Transfer plants to pots when roots and shoots have developed. Follow the procedure of hardening of plants.

11. Identify the haploid plants.

12. Carry out diploidization of chromosomes by colchicine treatment. Collect vigorous plants at 3–4 tiller stage. Wash soil from the roots and cut back to 3 cm below the crown. Place the plants in glass jars or vials containing colchicine solution (2.5 g colchicine dissolved in 20 ml dimethyl sulphoxide and made up to a liter with water). Cover crowns with the colchicine solution. Keep the plants at room temperature in light for 5 h, wash the roots thoroughly with water and pot into a light soil.

Further reading

Bhojwani, S.S. and Razdan, M.K. 1990. *Plant Tissue Culture: Applications and Limitations*. Elsevier Science Publishers, Amsterdam.

Chawla, H.S. 1998. *Biotechnology in Crop Improvement*. International Book Distributing Co., Lucknow.

Evans, D.A., Sharp, W.R., Ammirato, P.V. and Yamada, Y. 1983. *Handbook of Plant Cell Culture*. Macmillan Publishing Company, New York.

Mohan, M., Nair, S., Bhagwat, A., Krishna, T.G., Yano, M., Bhatia, C.R. and Sasaki, T. 1997. Genome mapping, molecular markers and marker-assisted selection in crop plants. *Mol. Breed.*, **3**: 87-103.

Pierik, R.L.M. 1987. *In vitro Culture of Higher Plants*. Martinus Nijhoff Publishers, Dordrecht.

Reinert, J. and Bajaj, P.P.S. (eds.). 1977. *Applied and Fundamental Aspects of Plant Cell, Tissue and Organ Culture*. Springer-Verlag, Berlin.

Street, H.S. (ed.) 1973. *Plant Tissue and Cell Culture*. Blackwell Scientific Publications, Oxford.

Thorpe, T.A. 1981. *Plant Tissue Culture: Methods and Applications in Agriculture*. Academic Press Inc., New York.

Vasil, I.K. 1984. *Cell Culture and Somatic Cell Genetics of Plants: Laboratory Procedures and Their Applications*. Academic Press, Inc., New York.

9

Protoplast Isolation and Fusion

The protoplast, also known as naked plant cell refers to all the components of a plant cell excluding the cell wall. Hanstein introduced the term protoplast in 1880 to designate the living matter enclosed by plant cell membrane. The isolated protoplast is unusual because the outer plasma membrane is fully exposed and is the only barrier between the external environment and the interior of living cell. The isolation of protoplasts from plant cells was first achieved by microsurgery on plasmolyzed cells by mechanical method (Klercker, 1892). However, the yields were extremely low and this method is not useful. Cocking (1960) first used enzymes to release protoplasts. He isolated the enzyme cellulase from the culture of fungus *Myrothecium verrucaria* to degrade the cell walls. He applied an extract of cellulase hydrolytic enzyme to isolate proto-plasts from tomato root tips. Since that time many enzyme formulations have been used to isolate protoplasts, and the most frequently used enzymes are now commercially available. The use of cell wall degrading enzymes was soon recognized as the preferred method to release large numbers of uniform protoplasts. Under appropriate conditions, in a number of plant species, these protoplasts have been success-fully cultured to synthesize cell walls and now genetic manipulations have also been made. Protoplasts are not only useful for cell fusion studies but, these can also take up, through their naked plasma membrane foreign DNA, cell

organelles, bacteria or virus particles.

PROTOPLAST ISOLATION

Protoplasts are isolated by (i) mechanical and (ii) enzymatic methods.

Mechanical method

In this method, large and highly vacuolated cells of storage tissues such as onion bulb scales, radish root and beet root tissue could be used for isolation. The cells are plasmolysed in an iso-osmotic solution resulting in the withdrawal of contents in the center of cell. Subsequently, the tissue is dissected and deplasmolyzed to release the preformed protoplasts. Klercker (1892) isolated protoplasts from *Stratiotes aloides* by this method. Mechanically, protoplasts have been isolated by gently teasing the new callus tissue initiated from expanded leaves of *Saintpaulia ionantha* which was grown in a specific auxin environment to produce thin cell walls. However, this method is generally not followed because of certain disadvantages: (i) It is restricted to certain tissues which have large vacuolated cells. (ii) Yield of protoplasts is generally very low. Protoplasts from less vacu-olated and highly meristematic cells do not show good yield. (iii) The method is tedious and laborious. (iv) Viability of protoplasts is low because of the presence of substances released by damaged cells. The mechanical method is useful when there are side effects of cell wall degrading enzymes.

Enzymatic method

Cocking in 1960 demonstrated the possibility of enzymatic isolation of protoplasts from higher plants. He used concentrated solution of cellulase to degrade the cell walls. However, there was little work for the next 10 years until the commercial enzyme preparations became available. Takabe *et al.* (1968) for the first time employed commercial enzyme preparation for isolation of protoplasts and subsequently regenerated plants in 1971. Protoplasts are now routinely isolated by treating tissues with a mixture of cell wall degrading enzymes in solution, which contain osmotic stabilizers. A general procedure for protoplast isolation has been shown in Fig. 9.1. The relative ease with which protoplast isolation can be achieved depends upon a variety of factors which have been discussed below.

Physiological state of tissue and cell material

Protoplasts have been isolated from a variety of tissues and organs including leaves, petioles, shoot apices, roots, fruits, coleoptiles, hypocotyls, stem, embryos, microspores, callus and cell suspension cultures of a large number of plant species. A convenient and most suitable source of protoplasts is mesophyll tissue from fully expanded leaves of young plants or new shoots. Leaf tissue is popular because it allows the isolation of a large number of relatively uniform cells without the necessity of killing the plants. Moreover the mesophyll cells are loosely arranged and enzymes have an easy access to the cell wall. Leaves are taken, sterilized and lower epidermis from the excised leaves is peeled off. These are cut into small pieces and then proceeded for isolation of protoplasts. Since the physiological condition of the source tissue markedly influences the yield and viability of isolated protoplasts, the plants or tissues must be grown under controlled conditions. Callus

tissue and cell suspension cultures used for protoplast isolation should be in the early log or exponential phase of growth. Friable tissue with low starch content generally gives better results.

Enzymes

The release of protoplasts is very much dependent on the nature and composition of enzymes used to digest the cell wall. There are three primary components of the cell wall which have been identified as cellulose, hemicellulose and pectin substances. Cellulose and hemicellulose are the components of primary and secondary structure of cell wall, while pectin is a component of middle lamella that joins the cells. Pectinase mainly degrades the middle lamella, while cellulase and hemicellulase are required to digest the cellulosic and hemicellulosic components of the cell wall respectively. Cellulase (Onozuka) R10 generally used for wall degradation has been partially purified from the molds of *Trichoderma reesei* and *T. viride*. Sometimes additional hemicellulase maybe necessary for recalcitrant tissues and for this Rhozyme HP 150 has been used. The most frequently used pectinase is macerozyme (macerase) which was derived from the fungus *Rhizopus*. Driselase, another enzyme, has both cellulolytic and pecteolytic activities and has been successfully used alone for isolation of protoplasts from cultured cells. Besides, there are several enzymes, e.g. helicase, colonase, cellulysin, glusulase, zymolyase, meicelase, pectolyase, etc. which are used to treat a tissue that does not release protoplasts easily (Table 9.1). Aleurone cells of barley treated with cellulase results in protoplasts with the cellulase resistant cell wall. These cells called sphaeroplasts are to be treated with glusulase to digest the remaining cell wall.

The activity of the enzyme is pH dependent and is generally indicated by the manufacturer. However, in practice the pH of enzyme solution is adjusted between 4.7 and 6.0. Generally the temperature of 25–30°C is adequate for isolation of protoplasts. Duration of enzyme treatment is

Leaf sterilization, epidermis removed to expose mesophyll

Peeled leaf segment

Plasmolyzed cells in an enzyme mixture

Protoplasts harvested by density centrifugation

Hanging droplet   Liquid droplet   Feeder layer

Liquid culture   Agar culture   Co-culturing

Callus culture

Regenerated plant

Fig. 9.1 Schematic illustration of protoplast isolation and culture procedure.

to be determined after trials. However, it may be short for 30-min duration to long duration for 20 h. The enzymatic isolation of protoplasts can be performed in two different ways.

a) *Two-step or sequential method*: The tissue is first treated with a macerozyme or pectinase enzyme which separates the cells by degrading the middle lamella. These free cells are then treated with cellulase which releases the protoplasts. In general, the cells are exposed to different enzymes for shorter periods.

b) *One step or simultaneous method*: The tissue is subjected to a mixture of enzymes in a one step reaction which includes both

Table 9.1 Some commercially available enzymes used for protoplast isolation (after M.K. Razdan, 1993)

Enzyme	Source
Cellulase Onozuka R-10	*Trichoderma viride*
Cellulase Onozuka RS	*Trichoderma viride*
Cellulysin	*Trichoderma viride*
Driselase	*Irpex lacteus*
Meicelase P-1	*Trichoderma viride*
Hemicellulase	*Aspergillus niger*
Hemicellulase H-2125	*Rhizopus* spp.
Rhozyme HP 150	*Aspergillus niger*
Helicase	*Helix pomatia*
Macerase	*Rhizopus arrhizus*
Macerozyme R-10	*Rhizopus arrhizus*
Pectolyase	*Aspergillus japonicus*
Zymolyase	*Arthrobacter luteus*

macerozyme and cellulase. The one step method is generally used because it is less labor-intensive.

During the enzyme treatment, the protoplasts obtained need to be stabilized because the mechanical barrier of cell wall which offered support has been broken. For this reason an osmoticum is added which prevents the protoplasts from bursting.

Osmoticum

Protoplasts released directly into standard cell culture medium will burst. Hence, in isolating protoplasts, the wall pressure that is mechanically supported by cell wall must be replaced with an appropriate osmotic pressure in the protoplast isolation mixture and also later in the culture medium. Osmotic stress has harmful effects on cell metabolism and growth. A condensation of DNA in cell nuclei and decreased protein synthesis are the two common effects of osmotic stress, although both can be reversed.

Lower (more negative) osmotic potentials are usually generated by the addition of various ionic or non-ionic solutes. Non-ionic substances include soluble carbohydrates such as mannitol, sorbitol, glucose, fructose, galactose or sucrose. Ionic substances are potassium chloride, calcium chloride and magnesium sulphate. Mannitol and sorbitol are the most frequently used with mannitol preferred for isolation of leaf

mesophyll protoplasts. Mannitol is considered to be relatively inert metabolically and infuses slowly into the protoplasts. Mineral salts particularly potassium chloride and calcium chloride have also been used, but there has never been good evidence that they are preferable over mannitol or sorbitol. In general, 0.3 to 0.7 M sugar solution can generate a suitable osmotic potential. Upon transfer of protoplasts to culture medium, it may be appropriate to use metabolically active osmotic stabilizers (e.g. glucose and sucrose) along with metabolically inert osmotic stabilizers, such as mannitol. Active substances will be gradually metabolized by the protoplasts during early growth and cell wall regeneration, resulting in a gradual reduction of the osmoticum. This eliminates sudden changes in osmotic potential when regenerated cells are transferred to a nutrient medium for plant regeneration. In a solution of proper osmolarity, freshly isolated protoplasts appear completely spherical.

Protoplast purification

The enzyme digested mixture obtained at this stage would contain sub-cellular debris, undigested cells, broken protoplasts and healthy protoplasts. This mixture is purified by a combination of filtration, centrifugation and washing. The enzyme solution containing the protoplasts is filtered through a stainless steel or nylon mesh (50–100 μm) to remove larger portions of undigested tissues and cell clumps. The filtered protoplast - enzyme solution is mixed with a suitable volume of sucrose in protoplast suspension medium to give a final concentration of about 20% and then centrifugation is done at about 100×g for 7–10 min. The debris moves down to the bottom of tube and a band of protoplasts appear at the junction of sucrose and protoplast suspension medium. The protoplast band is easily sucked off with a Pasteur pipette into another centrifuge tube. Following the repeated centrifugation and suspension, the protoplasts are washed thrice and finally suspended in the culture medium at an appropriate density. Numerous other gradients have been suggested to aid in protoplast isolation. (i) Two step Ficoll (polysucrose) gradients with 6% on top and 9% below, dissolved in MS medium with 7% sorbitol for purification using centrifugation at 150× g for 5 minutes has been suggested. Following centrifugation, debris settle at the bottom while protoplasts float at the top. (ii) Protoplasts can be separated on to 20% Percoll with 0.25 M mannitol and 0.1 M calcium chloride. After centrifugation at 200× g, protoplasts were recovered above the Percoll. As these high molecular weight substances such as Ficoll and Percoll are osmotically inert, their use may be preferable over sucrose flotation.

Protoplast viability and density

The most frequently used staining methods for assessing protoplast viability are:

i. *Fluorescein diacetate (FDA) staining method:* FDA, a dye that accumulates inside the plasmalemma of viable protoplasts can be detected by fluorescence microscopy. As FDA accumulates in the cell membrane, viable intact protoplasts fluoresce yellow green within 5 min. FDA is dissolved in acetone and used at a concentration of 0.01%. Protoplasts treated with FDA must be examined within 5-15 min after staining as FDA disassociates from the membranes after 15 min.

ii. *Phenosafranine staining:* Phenosafranine is also used at a concentration of 0.01% but it is specific for dead protoplasts that turn red. Viable cells remain unstained by phenosafranine.

iii. *Calcofluor White (CFW) staining:* Calcofluor White can ascertain the viability of protoplasts by detecting the onset of cell wall formation. CFW binds to the beta- linked glucosides in the newly synthesized cell wall which is observed as a ring of fluorescence around the plasma membrane. Optimum staining is achieved when 0.1 ml of protoplasts is mixed with 5.0 µl of a 0.1% w/v solution of CFW.

iv. Exclusion of Evans blue dye by intact membranes.

v. Observations on cyclosis or protoplast streaming as a measure of active metabolism.

vi. Variation of protoplast size with osmotic changes.

vii. Oxygen uptake measured by an oxygen electrode which indicates respiratory metabolism.

viii. Photosynthetic studies.

However, the true test of protoplast viability is the ability of protoplasts to undergo continued mitotic divisions and regenerate plants.

Protoplasts have both maximum as well as minimum plating densities for growth. Published procedures suggest that protoplasts should be cultured at a density of 5×10^3 to 10^6 cells/ml with an optimum of about 5×10^4 protoplasts/ ml. The concentration of protoplasts in ‧a given preparation can be determined by the use of hemocytometer.

Culture techniques

Isolated protoplasts are usually cultured in either liquid or semisolid agar media plates. Protoplasts are sometimes allowed to regenerate cell wall in liquid culture before they are transferred to agar media.

Agar culture

Agar of different qualities are available but agarose is most frequently used to solidify the culture media. Bergmann cell plating technique as explained earlier (Chapter 7) is followed for plating of protoplasts. Protoplast suspension at double the required final density is gently mixed with an equal volume of double the agar concentration in the medium kept molten at 45°C. The concentration of agar should be chosen to give a soft agar gel when mixed with the protoplast suspension. The petri dishes are sealed with parafilm and incubated upside down. By agar culture method, protoplasts remain in a fixed position so that protoplast clumping is avoided. Protoplasts immobilized in semisolid media give rise to cell clones and allow accurate determination of plating efficiency. Once immobilized, however, hand manipulations are required for transfer to other culture media.

Liquid culture

It has been generally preferred in earlier stages of culture because (i) it allows easy dilution and transfer; (ii) protoplasts of some species do not divide in agarified media; (iii) osmotic pressure of the medium can be effectively reduced; (iv) density of cells can be reduced after few days of culture. But it has the disadvantage of not permitting isolation of single colonies derived from one parent cell.

Various modifications to these culture methods have been developed.

Liquid droplet method

It involves suspending protoplasts in culture media and pipetting 100–200 µl droplets into 60 × 15 ml plastic petri dishes. Five to seven drops can be cultured per plate. The plates are then sealed and incubated. This method is convenient for microscopic examination, and fresh media can be added directly to the developing suspensions at 5–7 days interval. A problem encountered in this method is that sometimes the cultured protoplasts clump together at the center of the droplet.

Hanging droplet method

Small drops (40–100 µl) of protoplast suspension are placed on the inner side of the lid of a petri dish. When the lid is applied to the dish, the culture drops are hanging or suspended from the lid. This method allows culture of fewer protoplasts per droplet than the conventional droplet technique.

Feeder layer

In many cases it is desirable to reduce the plating density to a minimum for a given protoplast preparation. A feeder layer consisting of X-irradiated non-dividing but living protoplasts are plated in agar medium in petri dishes. Protoplasts are plated on this feeder layer at low density in a thin layer of agar medium. This is especially important when particular mutant or hybrid cells are to be selected on agar plates.

Co-culturing

It is the culturing of two types of protoplasts viz. slow growing and fast growing. A reliable fast growing protoplast preparation is mixed in varying ratios with protoplasts of a slow growing recalcitrant species. The fast growing protoplasts presumably provide the other species with growth factors and undefined diffusable chemicals, which aid the regeneration of a cell wall and cell division.

Culture medium

Protoplasts have nutritional requirements similar to those of cultured plant cells. The mineral salt compositions established for plant cell cultures have been modified to meet particular requirements by protoplasts.

Protoplast culture medium should be devoid of ammonium as it is detrimental to its survival and quantity of iron and zinc should be reduced. On the other hand calcium concentration should be increased 2–4 times over the concentration normally used for cell cultures. Increased calcium concentration may be important for membrane stability.

Mannitol and sorbitol are the most frequently used compounds for maintaining the osmolarity, which has already been discussed.

Glucose is perhaps the preferred and most reliable carbon source. Plant cells grow about equally well on a combination of glucose and sucrose, but sucrose alone may not always be satisfactory for plant protoplasts. The vitamins used for protoplast culture include those present in standard tissue culture media. If protoplasts are to be cultured at very low density in defined media there may be a requirement for additional vitamins. Protoplast media frequently contain one or more amino acids. A convenient approach is to add 0.01 to 0.25% vitamin-free casamino acids or casein hydrolysate. Such mixtures appear to meet the needs of protoplasts if inorganic nitrogen is inadequate. The addition of 1–5 mM L-glutamine can improve growth.

The majority of protoplast culture media contain one or more auxins plus one or two cytokinins to stimulate protoplast division and growth. Only a few protoplast systems grow without added growth regulators in the culture media. In general, protoplast culturing starts with a relatively high concentration 1–3 mg/l of NAA or 2,4-D alongwith a lower concentration (0.1 to 1.0 mg/l) of BAP or zeatin. When protoplast division has started, it is often recommended to change the exogenous hormone supply. A change in the auxin to cytokinin balance is often used for stimulating morphogenesis.

Environmental factors

Generally high light intensity inhibits protoplast growth when applied from the beginning of culture. It is better to initiate protoplast culturing in darkness or dim light for few days and later transfer the cultures to a light of about 2000 – 5000 lux. There are reports of better protoplast growth when the cultures are kept in continuous darkness. In contrast, it has been shown for legume species that light is necessary for initiating protoplast division. Protoplast cultures are generally cultured at temperatures ranging between 20–28°C. A pH in the range of 5.5 to 5.9 is recommended for protoplast culture media and it seems to be satisfactory.

PROTOPLAST DEVELOPMENT

Cell wall formation

Protoplasts in culture generally start to regenerate a cell wall within a few hours after isolation and may take two to several days to complete the process under suitable conditions. The protoplasts lose their characteristic spherical shape, which is an indication of new wall regeneration. However, newly synthesized cell wall by protoplasts can be demonstrated by staining it with 0.1% calcofluor white fluorescent stain.

In general, protoplasts from actively growing suspension cultures exhibit more rapid deposition of microfibrils than those from differentiated mesophyll cells. A freshly formed cell wall is composed of loosely arranged microfibrils which subsequently become organized to form a typical cell wall.

Growth, division and plant regeneration

Regeneration of a cell wall is not necessarily a prerequisite for the initiation of nuclear division in protoplast cultures but cell wall formation is required before cytokinesis occurs. Once a cell wall is formed, the reconstituted cells show a considerable increase in cell size. The first cell division generally occurs within 2–7 days. Protoplasts from actively dividing cell suspension enter the first division faster than those from highly differentiated cells of the leaf. The second division occurs within a week, and by the end of the second week in culture, small aggregates of cells are present. After 3 weeks, small cell colonies are visible and colonies of *ca.* 1 mm diameter are present after approximately 6 weeks in culture. Once small colonies have formed, their further growth is inhibited if they are allowed to remain on the original high osmotic medium. The colonies should, therefore, be transferred to a mannitol or sorbitol-free medium. Macroscopic colonies are transferred to an osmotic free medium to develop callus. The callus may then be induced to undergo organogenic or embryogenic differentiation leading to the formation of plants.

The first report of plant regeneration was in *Nicotiana tabacum* (Takabe *et al.*, 1971). Since then the list of species exhibiting the totipotency has steadily increased. Several reports have suggested that frequency of plant regeneration reported from calluses derived from plant organs differ from the calluses raised from protoplasts. The calluses from intact plant organs often carry preformed buds or organized structures, while such structures are absent in the callus from protoplast origin.

SOMATIC HYBRIDIZATION

Sexual hybridization between closely related species has been used for years to improve cultivated plants. Unfortunately, sexual hybridization is limited in most cases to cultivars within a species or at best to a few wild species closely related to a cultivated crop. Species barriers thereby limit the usefulness of sexual hybridization for crop improvement. Somatic cell fusion leading to the formation of viable cell hybrids has been suggested as a method to overcome the species barriers to sexual hybridization. Plant protoplasts offer exciting possibilities in the fields of somatic cell genetics and crop improvement. The technique of hybrid production through the fusion of isolated somatic (body) protoplasts under *in vitro* conditions and subsequent development of their product (heterokaryon) to a hybrid plant is known as somatic hybridization. This procedure eliminates sex altogether in hybridization. In somatic hybridization the nucleus and cytoplasm of both parents are fused in the hybrid cell. Sometimes, nuclear genome of only one parent but cytoplasmic genes (plastome) from both the parents are present in the fused hybrid, which is known as *cybrid* or cytoplasmic hybrid. Thus, protoplast fusion technique can be used to overcome the barriers of incompatibility and acts as a method for the genetic manipulation of plant cells. It provides us with an opportunity to construct hybrids between taxonomically distant plant species beyond the limits of sexual crossability. It also creates cells with new genetic, nuclear as well as cytoplasmic constitutions that otherwise cannot be obtained. Somatic hybridization involves the fusion of protoplasts, selection of hybrid cells and identification of hybrid plants.

Protoplast fusion

It involves mixing of protoplasts of two different genomes and can be achieved by either spontaneous or induced fusion methods.

Spontaneous fusion method

Cell fusion is a process integral to plant development. The most prominent process is egg fertilization. The breakdown of cell wall during protoplast isolation led people to believe that there would be spontaneous fusion leading to the formation of homokaryons because multinucleate cells were detected as soon as enzymatic protoplast isolation techniques were applied. The argument that cell wall degradation would permit dilation of plasmodesmata, fusion and complete mixing of cells was supported by electron micrographic studies. Spontaneous fusion of protoplasts is observed when protoplasts are isolated from callus cultures. However, spontaneous fusion products do not regenerate in to whole plants except for undergoing a few divisions. Later studies revealed that isolated protoplasts are usually characterized by smooth surfaces and fusion has to be induced by one of a variety of treatments.

Induced fusion method

Spontaneous fusion is of no value as fusion of protoplasts of different origins is required in somatic hybridization. To achieve this, a suitable agent (fusogen) is added to fuse the plant protoplasts of different origins. Some of the methods that have been employed are explained below. Protoplast fusion method has been outlined in Fig. 9.2.

Treatment with sodium nitrate

Induced fusion by $NaNO_3$ was first reported by Power *et al.* (1970). Isolated protoplasts are suspended in a mixture of 5.5% sodium nitrate in a 10% sucrose solution. The solution containing the protoplasts is incubated in a water bath at 35°C for 5 min and then centrifuged for 5 min at 200× g. Following centrifugation, most of the supernatant is decanted and the protoplast pellet is transferred to a water bath at 30°C for 30 min. During this period, most of the protoplasts undergo cell fusion. The remaining aggregation mixture is gently decanted and replaced with the culture medium containing 0.1% $NaNO_3$. The protoplasts are left undisturbed for sometime after which they are washed twice with the culture medium and plated. This technique results in low frequency of heterokaryon formation, especially when mesophyll protoplasts are involved.

Calcium ions at high pH

Keller and Melchers (1973) studied the effect of high pH and calcium ions on the fusion of tobacco protoplasts. In their method, isolated protoplasts are centrifuged for 3 min at 50× g in a fusion-inducing solution of 0.5 M mannitol containing 0.05 M $CaCl_2 \cdot 2H_2O$ at a pH of 10.5. The centrifuge tubes containing the protoplasts are then incubated in a water bath at 37°C for 40–50 min. After this treatment, 20–50% of the protoplasts were involved in fusion.

Polyethylene glycol method

Kao and Michayluk (1974) and Wallin *et al.* (1974) developed PEG method of fusion of protoplasts. This is one of the most successful techniques for fusing protoplasts. The protoplasts are suspended in a solution containing high molecular weight PEG, which enhances agglutination and fusion of protoplasts in several species. When sufficient quantities of protoplasts are available, 1 ml of the protoplasts suspended in a culture medium are mixed with 1 ml of 28–56% PEG (1500–6000 MW) solution. The tube is then shaken for 5 sec and allowed to settle for 10 min. The protoplasts are then washed several times by the addition of protoplast culture medium to remove PEG. The protoplast preparation is then resuspended in the culture medium. The PEG

method has been widely accepted for protoplast fusion because it results in reproducible high-frequency heterokaryon formation, low cytotoxicity to most cell types and the formation of binucleate heterokaryons. PEG-induced fusion is non-specific and is therefore useful for interspecific, intergeneric or interkingdom fusions.

Electrofusion

Protoplasts are placed in to a small culture cell containing electrodes, and a potential difference is applied due to which protoplasts line up between the electrodes. If now an extremely short wave electric shock is applied, protoplasts can be induced to fuse. In this fusion method, two-step procedure is followed beginning with application of an alternating current (AC) of low intensity to protoplast suspension. Dielectrophoretic collectors adjusted to 1.5V and 1 MHz and an electrical conductivity of the suspension medium less than 10^{-5} sec/cm generate an electrophoresis effect that make the cells attach to each other along the field lines. The second step of injection of an electric direct current (DC) field pulse of high intensity (750–1000 V/cm) for a short duration of 20–50 μsec leads to breakdown of membranes in contact areas of adjacent cells resulting in fusion and consequent membrane reorganization. This electrofusion technique has been found to be simple, quick and efficient. Cells after electrofusion do not show cytotoxic response. However, this method did not receive much acceptance because specialized equipment is required.

Mechanism of fusion

Protoplast fusion consists of three main phases

i. *Agglutination or adhesion*: Two or more protoplasts are brought into close proximity. The adhesion can be induced by a variety of treatments, e.g. concanavalin A, PEG, high pH and high Ca^{++} ions.

ii. *Plasma membrane fusion at localized sites*: Membranes of protoplasts agglutinated by fusogen get fused at the point of adhesion.

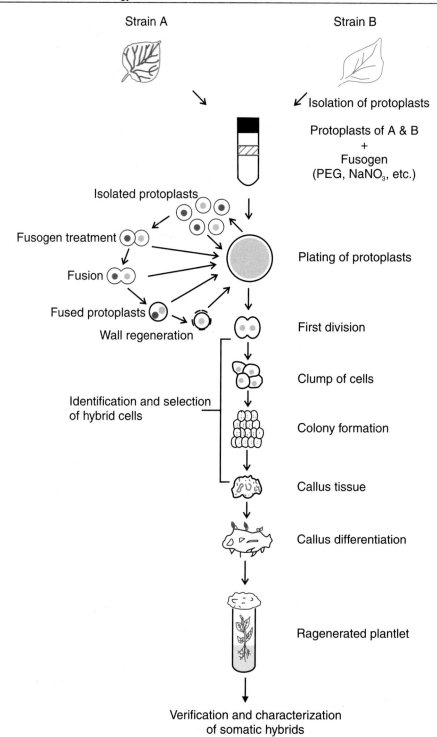

Fig. 9.2. Schematic illustration of protoplast fusion and regeneration of plants.

It results in the formation of cytoplasmic bridges between the protoplasts. Plant protoplasts carry a negative charge from -10 to -30 mV. Due to common charge, the plasma membranes of two agglutinated protoplasts do not come close enough to fuse. Fusion requires that membranes must be first brought close together at a distance of 10 Å or less. The high pH–high Ca^{++} ions treatment has shown to neutralize the normal surface charge so that agglutinated protoplasts can come in intimate contact. High temperature promotes membrane fusion due to perturbance of lipid molecules in plasma membrane and fusion occurs due to intermingling of lipid molecules in membranes of agglutinated protoplasts. PEG agglutinates to form clumps of protoplasts. Tight adhesion may occur over a large or small localized area. Localized fusion of closely attached plasma membranes occurs in the regions of tight adhesion and results in the formation of cytoplasmic bridges. It has been further suggested that PEG, which is slightly negative in polarity can form hydrogen bonds with water, protein, carbohydrates, etc. which are positive in polarity. When the PEG molecule chain is large enough it acts as a molecular bridge between the surface of adjacent protoplasts and adhesion occurs.

iii. *Formation of heterokaryon*: Rounding off of the fused protoplasts due to the expansion of cytoplasmic bridges forming spherical heterokaryon or homokaryon.

Identification and selection of hybrid cells

Following fusion treatment, the protoplast population consists of parental types, homokaryotic fusion products of parental cells, heterokaryotic fusion products or hybrids and a variety of other nuclear cytoplasmic combinations. Despite efforts to increase the efficiency of protoplast fusion, usually not more than 1–10% of the protoplasts in a treated population have actually undergone fusion. The proportion of viable heterospecific binucleate A + B type fusion products may be even lower. In the absence of systems that would fuse different protoplasts specifically to produce heterokaryocytes at a very high rate, there is an obvious need to select the products of fusion amongst the unfused and homofused parental cells. Identification and recovery of protoplast fusion products have been based on observation of visual characters, or hybrid cells display genetic complementation for recessive mutations and physiological complementation for *in vitro* growth requirements. In complementation, fusion of two protoplasts each carrying a different recessive marker, will generate a fusion product which is functionally restored since each parent contributes a functional allele that corrects the respective deficiency of the other parent.

Chlorophyll deficiency complementation

Melchers and Labib (1974) first used genetic complementation to isolate green somatic hybrids following fusion of two distinct homozygous recessive albino mutants of *Nicotiana tabacum*. They used chlorophyll deficient light sensitive varieties "sublethal" and "virescent". After two months of incubation under high light conditions, green colonies develop. This is the most frequently used method to isolate somatic hybrids (Fig. 9.3). Chlorophyll deficient mutants have also been used to raise somatic hybrids of *N. tabacum* + *N. sylvestris*.

In most cases, it has not been necessary to use two albino mutants to recover somatic hybrid shoots. A single recessive albino mutation could be very useful in combination with a morphological trait or a growth response in the isolation of somatic hybrid plants. Single recessive albino mutation marker in combination with morphological marker(s) to distinguish putative somatic hybrids in *Daucus* has been used (Fig. 9.4). Dudits *et al.* (1977) fused albino *D. carota* protoplasts with wild type *D. capillifolius*. Thus, both hybrid protoplasts and *D. capillifolius* will regenerate into green shoots. However, the morphology of leaves of fused product cnx⁻ and nia⁻ type mutants regularly complement each other upon fusion.

Fig. 9.3 A selection scheme involving complementation of chlorophyll deficient mutations.

Fig. 9.5 A summary of somatic hybrid selection based on auxotroph complementation.

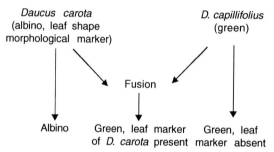

Fig. 9.4 A scheme illustrating selection of interspecific hybrid plants in *Daucus* utilizing morphological marker with albino mutant.

resembles those of *D. carota* leaves and can be isolated.

Auxotroph complementation

This method of selection has been applied to higher plants on a limited scale. The limitation is due to the paucity of higher plant auxotrophs. There is a report by Glimelius *et al.* (1978) where two parents were nitrate reductase deficient mutants of *N. tabacum* and could not be grown with nitrate as sole nitrogen source, while hybrids could regenerate shoots in the nitrate medium (Fig. 9.5). The lack of nitrate reductase (NR) activity causes an absolute requirement for reduced nitrogen and is caused by a deficiency either in the NR apoenzyme (nia–type mutant) or in the molybdenum cofactor (cnx⁻ type mutant).

Complementation of resistance markers

Dominant characters such as traits conferring resistance to antibiotics, herbicides, amino acid analogues or other toxic compounds have been recognized as potent selectable markers. When protoplasts from two lines are being fused together, the sensitivity trait of each parent will be dominated by the respective resistance trait from the other parent and will grow on a medium containing both the metabolites because of double resistance as compared to single resistance of the parents. This is also referred to as biochemical selection. Harms *et al.* (1981) selected double resistant somatic hybrids in *Daucus carota* following fusion of protoplasts from S 2-aminoethyl cysteine (2AEC) and 5-methyl tryptophan (5MT) resistant parental lines but were sensitive to other compound (Fig. 9.6). Maliga *et al.* (1977) used a kanamycin resistant variant of *N. sylvestris*, KR103, isolated from the cultured cells, as a genetic marker to recover fusion products between *N. sylvestris* and *N. knightiana*. Complementation system involving antibiotics (e.g. methotrexate, streptomycin, cycloheximide, and chloramphenicol) resistance has also been used.

Use of metabolic inhibitors

In this method, parental cells are treated with an irreversible biochemical inhibitor such as

Daucus carota
(Strain A–2AEC⁵, 5MTʳ)

D. carota
(Strain B–2AECʳ, 5MT⁵)

Fusion
Culture medium supplemented
with 2AEC & 5MT

No colony

Colony formation
(Hybrid resistant to
2AEC & 5MT)

No colony

Fig. 9.6 Scheme illustrating selection of somatic hybrids utilizing complementation of resistance markers.

iodoacetate or diethylpyrocarbamate and following treatment only hybrid cells are capable of division. Iodoacetate pretreatment has been used to aid recovery of somatic hybrids between *N. sylvestris* and *N. tabacum* and *N. plumbaginifolia* and *N. tabacum*. In each case the parent protoplasts treated with iodoacetate were unable to reproduce, while the newly formed hybrid protoplasts continued to develop and yield hybrid plants.

Use of visual characteristics

The most efficient but the most tedious method to select products of protoplast fusion is to visually identify hybrid cells and mechanically isolate individual cells. The approaches used by various workers have been summarized:

i. Use of morphologically distinct cells: Fusion of mesophyll protoplasts containing green chloroplasts with colorless cell culture protoplasts containing distinct starch granules due to growth on sucrose supplemented medium. Fused products can be seen immediately after PEG treatment with one half containing chloroplast and other half with distinct starch granules.

ii. Use of petal protoplasts with leaf protoplasts, and petal + cell culture protoplasts can be readily distinguished. The usually vacuolar petal pigment is originally separated within the fused cell but eventually becomes evenly distributed throughout the fused cell.

iii. Fluorescent labeling: Parental protoplasts that are isolated from the same type of cells and not sufficiently different for visual distinction may be loaded with different fluorescent dyes prior to fusion. Fluorescein isothiocyanate (fluorescing in the green) and rhodamine isothiocyanate fluorescing in the red have been used to label two separate batches of mesophyll protoplasts of *Nicotiana*. The labeling is achieved by adding 0.5 mg/l of the compound into the enzyme mixture during incubation period (Galbraith and Mauch, 1980). The double fluorescent label of fusion products can be recognized in a fluorescence microscope and fusion products then cultured after micro-isolation. The double fluorescence provides a possibility to separate fusion products from parental protoplasts when fusion mixtures are run through a fluorescence activated cell sorter (FACS).

iv. The growth pattern of hybrid callus is different from either parental line. Hybrid callus is often more vigorous than parental callus.

v. Differences in the morphology of callus: Parental type protoplasts as well as hybrid cells can develop to form different callus types (Fig. 9.7). No selection pressure is applied at the protoplast or cell level. There are differences in the morphology of callus in hybrid of *Datura innoxia* and *Atropa belladonna* (Krumbiegel and Schieder, 1979).

Compound selection system

Most of the somatic hybrid plants that are known today have been selected using compound selection strategies rather than straightforward marker complementation. Compound selection systems have used differential growth characteristics, regenerability, morphological features and complementation marker systems to distinguish somatic hybrids from parental cells/plants. Whatever the system employed, the utility of a selective system lies in the fact that it provides enrichment for fused cells by reducing or

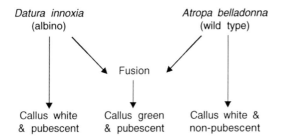

Fig. 9.7. Selection system based on morphology of callus employed in intergeneric somatic hybridization.

eliminating the non-desired parental type cells. This enrichment thus greatly reduces the number of colonies or plants among which hybrids must be sought.

Verification and characterization of somatic hybrids

Protoplasts from any two species can be fused together. However, there are a number of limitations to widespread utilization of somatic hybridization in higher plants, including aneuploidy, species barriers to hybridization, and the inability to regenerate plants from protoplasts. Besides these limitations, hybrids have been developed and these must be verified as products of somatic fusion of two different protoplasts. Successful passage through a selection system provides the first circumstantial evidence for the somatic hybrid nature of selected plant materials. Further evidence must be added based on the traits not involved in the selection. Proof of hybridity requires a clear demonstration of genetic contribution from both fusion partners and hybridity must be sought only from euploid and not from aneuploid hybrids.

1. *Morphology:* When plant regeneration has been accomplished from protoplast fusion, products a wide range of morphological features can be drawn upon for hybrid verification. In most cases morphological characteristics of either somatic or sexual hybrids are intermediate between the two parents. Both vegetative and floral charac-

ters such as leaf shape, leaf area, petiole size, root morphology, trichome (hairs on the leaf surface) length and density, flower shape, color size and structure, corolla morphology are considered, intensity of flower pigment and seed capsule morphology. The genetic basis for most of these morphological traits has not yet been elucidated, but the intermediate behavior in hybrids suggests control of traits by multiple genes. Sometimes intermediate morphology is not observed in the somatic hybrids because these traits behave as dominant single gene traits as they are present only in one parent, but are also expressed in the somatic hybrids. Such traits include stem anthocyanin pigment, flower pigment, heterochromatic knobs in interphase cells and leaf size. Pollen viability character in general shows a decrease in viability in the hybrids. Whenever possible, additional genetic data should be presented to support hybridity.

2. *Isoenzyme analysis:* Electrophoretic banding patterns of isoenzymes have been extensively used to verify hybridity. Isoenzymes are the different molecular forms of the enzyme that catalyze the same reaction. For isoenzymic studies, both starch and polyacrylamide gels are employed where electrical properties of proteins (enzymes) are used for separation into different bands. Somatic hybrids may display isoenzyme bands of certain enzymes specific to one or the other parent or to both parents simultaneously. Enzymes that have unique banding patterns versus either parental species and that have been used for identification of somatic hybrids include esterase, aspartate aminotransferase, amylase, isoperoxidase phosphatase, alcohol, malate, and lactate dehyrogenase and phosphodiesterase. If the enzyme is dimeric, the somatic hybrids contain a unique hybrid band intermediate in mobility in addition to the sum of parental bands. This probably

represents formation of a hybrid dimeric enzyme unique to the somatic hybrids. Since isoenzymes are extremely variable within plant tissues, it is important to use the same enzyme from each plant and to use plants/tissue at identical developmental ages when comparing parental plants with somatic hybrids. For comparison, zymograms (a diagrammatic representation of isoenzyme banding pattern) should therefore be prepared and interpreted with caution.

3. *Chromosomal constitution:* Counting chromosomes in presumed hybrid cells can be an easy and reliable method to verify hybrid cells and it also provides information on the ploidy state of the cells. Cytologically the chromosome number of the somatic hybrids should be the sum of chromosome number of two parental protoplasts used for fusion. Variation in chromosome number (aneuploidy) is generally observed in hybrids but frequently chromosome number is more than the total number of both the parental protoplasts. Besides numerical number, structural and size differences of their chromosomes should be studied. The presence of marker chromosomes can greatly facilitate the analysis of genetic events in hybrid cells. Extensive use of banding techniques can be made to identify specific chromosomes and investigate rearrangements in somatic hybrid cells.

4. *Molecular techniques:* Recent progress in the development of molecular biological techniques has greatly improved our capabilities to analyze the genetic constitution of somatic plant hybrids. Specific restriction patterns of chloroplast and mitochondrial DNA has been used to great advantage to characterize the nature of plastoms and chondrioms of somatic hybrids and cybrids. Species-specific restriction fragments of nuclear DNA coding for ribosomal RNA have been shown to verify somatic hybrids of *Nicotiana glauca* + *N. langsdorffii* (Uchimiya *et al.*, 1983). With the availability of numerous molecular markers such as RFLP, AFLP, RAPD, microsatellites etc., the hybrid identification can be done using these techniques. These are more accurate methods of analysis based on DNA, so no environmental effect is there. RFLP markers have been used for the identification of somatic hybrids between potato and tomato. PCR technology has been utilized for hybrid identification (Baird *et al.*, 1992). RAPD markers established the hybridity between *Solanum ochranthum* and *Lycopersicon esculentum* (Kobayashi *et al.*, 1996).

5. *Genetic characterization:* Somatic hybrid plants that produce fertile flowers can be analyzed by conventional genetic methods. Results obtained from segregation data of selfed F_2 generations are generally in good agreement with the expected values and have provided a sound confirmation of the somatic nature of the plants analyzed.

CHROMOSOME NUMBER IN SOMATIC HYBRIDS

Variation in chromosome number (aneuploidy) is generally observed in hybrids but frequently chromosome number is more than the total number of both the parental protoplasts.

Interspecific and intergeneric somatic hybrids are mostly polyploids. The chromosome number of these hybrids indicates that only few hybrids have the sum total chromosome number of both the parents as expected in an amphiploid. There is an indication that closely related species would yield true amphidiploids through somatic hybridization. Even sexually compatible parents show deviation in the number of chromosomes in their somatic hybrids. The chromosome number of parent species and somatic hybrids obtained has been listed in Table 9.2. The variability in chromosome number of hybrids could be due to following reasons:

i) Multiple fusions give a higher chromosome number.

ii) Fusion of more than two protoplasts with subsequent mitotic irregularities.

iii) In PEG and electro-induced fusions about one-third of fusion products result from fusions among more than two protoplasts.

iv) Asymmetric hybrids result from fusion of protoplasts isolated from actively dividing tissue of one parent and quiescent tissue of the other parent.

v) Unequal rates of DNA replication in two fusing partners may also give asymmetric hybrids.

vi) Somaclonal variation in cultured cells used for protoplast isolation may also lead to variation in chromosome number.

CYBRIDS

Sexual hybridization is a precise mixture of parental nuclear genes but the cytoplasm is or interkingdohe maternal parent only, while in somatic hybrids the cytoplasm is derived from both the parents. However, somatic hybrids can be obtained where nucleus is derived from one parent and cytoplasm is derived from both, thus producing cytoplasmic hybrids, also called as cybrids (Fig. 9.8). Early segregation of nuclei in a fused product can be stimulated and directed so that one protoplast contributes the cytoplasm while the other contributes the nucleus alone or both nucleus and cytoplasm. There are different ways of inactivating the nucleus of one protoplast. Thus, there will be fusion between protoplasts containing the full complement of nucleus, mitochondria and chloroplasts with functional cytoplasmic component of second parent. The various approaches to achieve this type of fusion are:

1. By application of lethal dosages of X or gamma irradiation to one parental protoplast population: This treatment renders the protoplasts inactivate and non-dividing but they serve as an efficient donor of cytoplasmic genophores when fused with recipient protoplasts. *Nicotiana* protoplasts can be inactivated by 5-kr dose of X-rays. Other protoplasts may require different doses.

2. By treatment with iodoacetate to metabolically inactivate the protoplasts: Pretreatment with iodoacetate will cause the degeneration of non-fused and autofused protoplasts while fusion of iodoacetate pretreated protoplasts with non-treated protoplasts will cause metabolic complementation and result in viable hybrids. In an experiment, iodoacetate-treated *Nicotiana plumbaginifolia* cell suspension was fused with X-irradiated *N. tabacum* mesophyll protoplasts. All regenerated cybrid plants had *N. plumbaginifolia* morphology but most of them contained *N. tabacum* chloroplasts. The iodoacetate treatment does not impair the nucleus of the treated protoplasts. Thus the latter can complement an X-irradiated protoplast.

3. Fusion of normal protoplasts with enucleated protoplasts: Enucleated protoplasts can be obtained by high-speed centrifugation

Table 9.2 Chromosome number in some of the interspecific hybrids produced through protoplast fusion

Plant species with their chromosome number			Chromosome number in hybrid
Brassica oleracea (2n = 18)	+	*B. campestris* (2n = 18)	Wide variation
Brassica napus (2n = 38)	+	*B. juncea* (2n = 36)	Wide variation
Datura innoxia (2n = 24)	+	*D. stramonium* (2n = 24)	46, 48, 72
Nicotiana tabacum (2n = 48)	+	*N. glutinosa* (2n = 24)	50 – 58
Nicotiana tabacum (2n = 48)	+	*N. nesophila* (2n = 48)	96
Nicotiana tabacum (2n = 48)	+	*N. sylvestris* (2n = 24)	72
Lycopersicon esculentum (2n = 24)	+	*L. peruvianum* (2n = 24)	72
Petunia parodii (2n = 48)	+	*P. hybrida* (2n = 14)	44 – 48
Solanum tuberosum (2n = 24, 48)	+	*S. chacoense* (2n = 14)	60

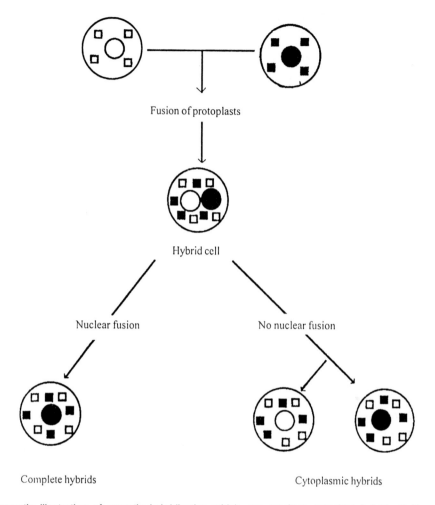

Fig. 9.8 Schematic illustration of somatic hybridization which can produce complete hybrids (left) or cytoplasmic hybrids (right).

(20,000–40,000× g) for 45–90 minutes in an iso-osmotic density gradient with 5–50% percoll.

4. Fusion of a normal protoplast with another in which nuclear division is suppressed.

Cybridization thus opens an exciting avenue to achieve alloplasmic constitution in a single step without the need to perform a series of 8–12 time-consuming backcrosses. Alloplasmic lines contain nucleus of one parent genome with the other parent's cytoplasmic constituents. Application of cybrids would be the directed transfer of cytoplasmic male sterility or herbicide resistance from a donor to a recipient crop plant species. Transfer of cytoplasmic male sterility (CMS) from *N. tabacum* to *N. sylvestris* by protoplast fusion was first reported by Zelcer *et al.* (1978). Resistance of plants to herbicide atrazine has been transferred from *Brassica campestris* to *B. napus* via fractionated protoplast fusions. CMS has been successfully transferred in various crop species of *Oryza*, *Lycopersicon*, *Brassica*, *Nicotiana*, etc. (Table 9.3).

POTENTIAL OF SOMATIC HYBRIDIZATION

1. Production of novel interspecific and intergeneric crosses between plants that are difficult or impossible to hybridize conventionally. It overcomes sexual incompatibility barriers. For example, fusion between protoplasts of *Lycopersicon esculentum* (tomato) and *Solanum tuberosum* (potato) created the pomato first achieved by Melchers *et al.* (1978). Asymmetric hybrids also develop when there is partial hybridization. These asymmetric hybrids have abnormal or wide variation in chromosome number than the exact total of two species (Table 9.2). Efficient cell fusion can be achieved between sexually compatible and incompatible parents involving interspecific or intergeneric combination (Tables 9.2 and 9.4). Attempts to overcome conventional breeding barriers by interspecific fusion of rice with four different wild species including *Oryza brachyantha, O. eichingeri, O. officinalis* and *O. perrieri* were more successful. Mature plants with viable pollen could be obtained in all but the first *Oryza* combination. The production of fertile interspecific diploid rice hybrid plants as well as cybrids demonstrates that the fusion technology can be extended to graminaceous crops.

2. Somatic hybridization for gene transfer

i) Disease resistance: The families Solanaceae and Brassiceae contain the most commonly used species for somatic hybridization. Both interspecific and intergeneric hybrids have been obtained. Many disease resistance genes *viz.* potato leaf roll virus, leaf blight, *Verticillium, Phytophthora*, etc. have been transferred to *Solanum tuberosum* from other species where normal crossings would not be possible due to taxonomic or other barriers. Resistance to blackleg disease (*Phoma lingam*) has been found in *B. nigra, B. juncea* and *B. carinata* and after production of symmetric as well as asymmetric somatic hybrids between these gene donors and *B. napus,* resistant hybrids have been developed. Attempts were made to introduce tolerance in *Brassica napus* against *Alternaria brassiceae* from *Sinapis alba* (Primard *et al.,* 1988) and beet cyst nematode from *Raphanus sativus* (Lelivelt *et al.,* 1993). Resistance has been introduced in tomato against various diseases like TMV, spotted wilt virus, insect pests and also cold tolerance. Kobayashi *et al.* (1996) reported the incorporation of disease resistance genes from wild species *Solanum ochranthum,* a woody vine like tomato relative, to *L. esculentum.* Hansen and Earle (1995) reported that black rot disease caused by *Xanthomonas campestris* is a serious disease in cauliflower. The somatic hybrids were produced by protoplast fusion of *B. oleracea* with *B. napus.* Some of the examples of incorporation of resistance genes via protoplast fusion technique have been listed in Table 9.3.

ii) Abiotic stress resistance: Work related to somatic hybridization for abiotic stress has been mainly done on families Fabaceae, Brassicaceae, Poaceae, Solanaceae and relates to cold and frost resistance. Rokka et al. (1998) developed somatic hybrids between cultivated potato (*S. tuberosum*) and wild relative (*S. acaule*) possessing several disease and early frost resistance characters.

iii) Quality characters: Somatic hybrids produced between *B. napus* and *Eruca sativa* were fertile and had low concentration of erucic acid content (Fahleson *et al.,* 1993). This hybrid material has been introduced into breeding program. Likewise, nicotine content character has been transferred to *N. tabacum.*

iv) Cytoplasmic male sterility: Several agriculturally useful traits are cytoplasmically encoded, including some types of male sterility and certain antibiotic and herbicide resistance factors. Pelletier *et al.* (1988) reported *Brassica raphanus* cybrids that contain the nucleus of *B. napus,* chloroplasts

Table 9.3 Genetic traits transferred via protoplast fusion

Somatic hybrids		Traits (resistance)
Nicotiana tabacum	+ *N. nesophila*	Tobacco mosaic virus
Solanum tuberosum	+ *S. chacoense*	Potato virus X
N. tabacum	+ *N. nesophila*	Tobacco horn worm
S. tuberosum	+ *S. brevidens*	Potato leaf roll virus
		Late blight and PLRV Potato virus Y
S. circalifolium	+ *S. tuberosum*	Phytophthora
S. melongena	+ *S. sanitwongsei*	Pseudomonas solanacearum
S. tuberosum	+ *S. commersonii*	Frost tolerance
B. napus	+ *B. nigra*	Phoma lingam
B. oleracea	+ *Sinapis alba*	Alternaria brassiceae
B. napus	+ *Sinapis alba*	Alternaria brassiceae
Citrus sinensis	+ *Poncirus trifoliata*	Phytophthora
Lycopersicon esculentum	+ *L. peruvianum*	TMV, spotted wilt virus, cold tolerance
S. lycopersicoides	+ *L. esculentum*	Cold tolerance
Solanum ochranthum	+ *L. esculentum*	Tomato diseases and insect pests
Brassica oleracea	+ *B. napus*	Black rot (*Xanthomonas campestris*)
B. oleracea sp. *capitata*	+ *B. oleracea* (Ogura CMS line)	Cold tolerance
S. tuberosum	+ *S. cammersonii*	Frost tolerance
R. sativa (Japanese radish)	+ *B. oleracea* var. *botrytis*	Club rot disease
B. oleracea var. *botrytis*	+ *S. alba* + *B. carinata*	Alternaria brassicola and *Phoma lingam*
Citrullus lanatus	+ *Cucumis melo*	Club rot resistance
Hordeum vulgare	+ *Daucus carota*	Frost and salt tolerance
Solanum melongena	+ *S. sisymbrifolium*	Nematode
S. tuberosum	+ *S. bulbocastanum*	Nematode
S. tuberosum	+ *S. bulbocastanum*	Root knot nematode
Raphanus sativus	+ *Brassica napus*	Beet cyst nematode
Sinapis alba	+ *R. sativus* + *B. napus*	Beet cyst nematode
Quality characters		
N. rustica	+ *N. tabacum*	High nicotine content
B. napus	+ *Eruca sativa*	Low erucic acid
Agronomic characters (transferred via cybrid formation)		
N. tabacum	+ *N. sylvestris*	Streptomycin resistance
N. tabacum spp.		Triazine resistance
S. nigrum	+ *S. tuberosum*	Triazine resistance
B. nigra	+ *B. napus*	Hygromycin resistance
B. napus spp.		Triazine resistance
N. tabacum	+ *N. sylvestris*	CMS
B. campestris	+ *B. napus*	CMS
N. tabacum	+ *N. sylvestris*	CMS
B. napus + *B. campestris*	+ *Raphanus sativa*	CMS and Triazine resistance
Oryza spp.		CMS
L. esculentum	+ *Solanum acaule*	CMS
B. napus	+ *B. tournefortii*	CMS

of atrazine resistant *B. campestris* and mitochondria that confer male sterility from *Raphanus sativus*. Cybridization has been successfully used to transfer cytoplasmic male sterility in rice (Kyozuka *et al.*, 1989). Sigareva and Earle (1997) produced cold tolerant cytoplasmically male sterile (CMS)

cabbage (*Brassica oleracea* sp. *capitata*) by the fusion of cabbage protoplasts with cold tolerant ogura CMS broccoli lines. Rambaud *et al.* (1997) reported that CMS chicory (*Cichorium intybus*) cybrids have been obtained by fusion between chicory and CMS sunflower protoplasts. Resistance to

Table 9.4 Intergeneric hybrids produced though protoplast fusion

Plant species with chromosome number		New genus
Raphanus sativus (2n = 18)	+ B. oleracea (2n = 18)	Raphanobrassica
B. oleracea (2n = 18)	+ Moricandia arvensis (2n = 27,28)	Moricandiobrassica
Eruca sativa (2n = 22)	+ B. napus (2n = 38)	Erucobrassica
Diplotaxis muralis (2n = 42)	+ B. napus (2n = 38)	Diplotaxobrassica
Nicotiana tabacum (2n = 24)	+ Lycopersicon esculentum (2n = 24)	Nicotiopersicon
Solanum tuberosum (2n = 24)	+ L. esculentum (2n = 24)	Solanopersicon
Datura innoxia (2n = 48)	+ Atropa belladona (2n = 24)	Daturotropa
Oryza sativa (2n = 24)	+ Echinochloa oryzicola (2n = 24)	Oryzochloa
Arabidopsis thaliana (2n = 10)	+ B. campestris (2n = 20)	Arabidobrassica

antibiotics, herbicide as well as CMS has been introduced in so many cultivated species (Table 9.3).

3. Production of autotetraploids: Somatic hybridization can be used as an alternative to obtain tetraploids and, if this is unsuccessful, colchicine treatment can be used.
4. Protoplasts of sexually sterile (haploid, triploid, aneuploid, etc.) plants can be fused to produce fertile diploids and polyploids.
5. Hybridization becomes possible between plants that are still in the juvenile phase.
6. Production of heterozygous lines within a single species that normally could only be propagated by vegetative means, e.g. potato and other tuber and root crops.
7. Somatic cell fusion is useful in the study of cytoplasmic genes and their activities: This information can be applied in plant breeding experiments.
8. The production of unique nuclear-cytoplasmic combinations: Evidence from a number of somatic hybrids suggests that although two types of cytoplasm are initially mixed during protoplast fusion, resulting in heteroplasmons, eventually one parent type cytoplasm predominates, resulting in cytoplasmic segregation. Mitochondrial and chloroplast recombination has also been reported to result in unique nuclear-cytoplasmic combinations. These unique combinations using protoplasts will aid the development of novel germplasm not obtainable using conventional methods.

PROBLEMS AND LIMITATIONS OF SOMATIC HYBRIDIZATION

There are limitations, however, to the use of these types of somatic hybridization since plants regenerated from some of the combinations are not always fertile and do not produce viable seeds. The successful agricultural application of somatic hybridization is dependent on overcoming several limitations.

1. Application of protoplast methodology requires efficient plant regeneration from protoplasts. Protoplasts from any two species can be fused. However, production of somatic hybrid plants has been limited to a few species.
2. The lack of an efficient selection method for fused product is sometimes a major problem.
3. The end-products after somatic hybridization are often unbalanced (sterile, misformed, and unstable) and are therefore not viable, especially if the fusion partners are taxonomically far apart.
4. The development of chimaeric calluses in place of hybrids: This is usually due to the nuclei not fusing after cell fusion and dividing separately. Plants that are regenerated from chimeras usually lose their chimeric characteristics, since adventitious shoots or embryos usually develop from a single cell.
5. Somatic hybridization of two diploids leads to the formation of an amphidiploid which is generally unfavorable (except when tetraploids

are formed intentionally). For this reason, in most cases, the hybridization of two haploid protoplasts is normally recommended.

6. Regeneration products after somatic hybridization are often variable due to somaclonal variation, chromosome elimination, translocation, organelle segregation etc.
7. It is never certain that a particular characteristic will be expressed after somatic hybridization.
8. The genetic stability during protoplast culture is poor.
9. To achieve successful integration into a breeding program, somatic hybrids must be capable of sexual reproduction. In all cases reported, somatic hybrids containing a mixture of genes from two species must be backcrossed to the cultivated crop to develop new varieties. All diverse intergeneric somatic hybrids reported are sterile and therefore have limited value for new variety development. It may be necessary to use back-fusion or embryo culture to produce gene combinations that are sufficiently stable to permit incorporation into a breeding program.
10. To transfer useful genes from a wild species to a cultivated crop, it is necessary to achieve intergeneric recombination or chromosome substitution between parental genomes.

PROTOCOL

9.1. Protocol for protoplast isolation and fusion

Plant material: *Nicotiana tabacum*

1. Detach fully expanded leaves including petiole. Sterilize in 10% sodium hypochlorite solution containing two drops of Tween 20 for 10 min. Wash 3–4x with sterile distilled water and blot dry on sterile tissue paper.
2. Peel the lower epidermis from sterilized leaves with fine forceps and cut into pieces with a fine scalpel.
3. Transfer the leaf pieces (peeled areas only) with lower surface down into 30 ml of 13% mannitol-inorganic salts of cell and protoplast washing media (CPW) solution contained in a petri dish for 30 min–1 h for plasmolysis. Composition of CPW salt solution is as follows:

KH_2PO_4	27.2 mg/l
KNO_3	101.0 mg/l
$CaCl_2 \cdot 2H_2O$	1480.0 mg/l
$MgSO_4 \cdot 7H_2O$	246.0 mg/l
KI	0.16 mg/l
$CuSO_4 \cdot 5H_2O$	0.025 mg/l
pH	5.8

4. Remove the mannitol–CPW salt solution with a Pasteur pipette and replace it with filter sterilized enzyme mixture in 13% mannitol solution (~ 20 ml). Enzyme mixture contains macerozyme (0.1–0.5%) and cellulase (0.5 1%).
5. Gently agitate the leaf pieces with sterile Pasteur pipette to facilitate the release of protoplasts, pushing the larger pieces of leaf material to one side and keeping the petri dish at an angle of 15°C or remove the larger debris by filtering through a 60–80 μm mesh.
6. Transfer the filtrate to a screw cap centrifuge tube.
7. Centrifuge at 100x g for 5–10 min to sediment the protoplasts. Remove the supernatant and resuspend the protoplast pellet in CPW + 21% sucrose (prepared in CPW) and gently disperse the protoplasts in the solution.
8. Centrifuge again at 100x g for 10 min. The viable protoplasts will float to the surface of the sucrose solution while the remaining cells and debris will sink to the bottom of tube.
9. Remove the band of protoplasts from the top with a Pasteur pipette and transfer into another centrifuge tube. Resuspend in CPW + 10% mannitol. Centrifuge again at 100x g for 10 min to separate the contaminating debris.
10. Repeat the washing procedure at least 3 times.
11. After the final washing add enough MS

protoplast culture medium with 9% mannitol to achieve a protoplast density of ~ 5 × 10⁴/ ml. Kao and Michayluk (1975) media has been widely used for protoplast culture (Table 9.5). Plate the protoplasts as small droplets (100–150 µl) or a thin layer in small petri dishes for 24 h at 25°C.

12. Keep the cultures at low light intensity of 500 lux for the next 2 days and then for rest of the experiment at 2000 lux.

13. Protoplasts form cell walls and begin to divide after 3–5 days.

Table 9.5 Kao and Michyaluk 8p medium for protoplast culture

Component	Concentration	Component	Concentration
Mineral salts (mg/l)			
NH_4NO_3	600	KI	0.75
KNO_3	1900	H_3BO_3	3.00
$CaCl_2 \cdot 2H_2O$	600	$MnSO_4 \cdot H_2O$	10.00
$MgSO_4 \cdot 7H_2O$	300	$ZnSO_4 \cdot 7H_2O$	2.00
KH_2PO_4	170	$Na_2MoO_4 \cdot 2H_2O$	0.25
KCl	300	$CuSO_4 \cdot 5H_2O$	0.025
Sesquestrene®330Fe	28	$CoCl_2 \cdot 6H_2O$	0.025
Vitamins and carbohydrates (mg/l)			
Inositol	100	Biotin	0.01
Nicotinamide	1	Choline chloride	1.00
Pyridoxine HCl	1	Riboflavin	0.20
Thiamine HCl	1	Ascorbic acid	2.00
D-calcium pantothenate	1	Vitamin A	0.01
Folic acid	0.4	Vitamin D_3	0.01
p-Amino benzoic acid	0.02	Vitamin B_{12}	0.02
Sucrose and glucose (g/l)			
Sucrose	0.25		
Glucose	68.4		
Hormones (mg/l)			
2,4-D	0.2		
Zeatin	0.5		
NAA	1.0		
Organic acids (mg/l) adjusted to pH 5.5 with NH_4OH			
Sodium pyruvate	20	Malic acid	40
Citric acid	40	Fumaric acid	40
Other sugars and sugar alcohols (mg/l)			
Fructose	250	Rhamnose	250
Ribose	250	Cellobiose	250
Xylose	250	Sorbitol	250
Mannose	250	Mannitol	250
Amino acids (mg/l)			
Glycine	0.1	Glutamine	5.6
Alanine	0.6	Glutamic acid	0.6
Cysteine	0.2	Asparagine	0.1
Nucleotides (mg/l)			
Adenine	0.10	Uracil	0.03
Guanine	0.03	Hypoxanthine	0.03
Thymine	0.03	Xanthine	0.03
Others			
Casein hydrolysate	250 mg/l	Coconut water	20 ml/l
pH	5.6		

14. Protoplasts may also be cultured in an agarose medium by mixing 1.5 ml of protoplast suspension with an equal volume of 1.2% agarose medium at 45°C in 35 mm petri dish.
15. Cultures continue to grow in the media and the first colonies become visible after 3–4 weeks.
16. Transfer the colonies to new medium with a reduced mannitol level.
17. Transfer the colonies to a suitable medium for embryo differentiation and plantlet regeneration.

Protoplast fusion

Fusion treatment: PEG

Material: Green leaf (mesophyll) and cultured cells (albino) of *N. tabacum*

1. Perform the fusion experiment in continuation with protoplast isolation experiment.
2. Take 4–6 ml of freshly isolated protoplast suspension from two sources in CPW salt medium—13% mannitol (CPW13M) with a density of 5×10^5/ml.
3. Using a Pasteur pipette place a small drop (~50 µl) of sterile silicone onto the center of a plastic petri dish (35 mm) and gently lower a cover glass onto the drop of silicone fluid.
4. Pipette ~150 µl of protoplast suspension directly onto the middle of cover glass and allow it to settle for 10–15 min. (Protoplasts suspended in CPW13M for convenience are mixed in screw capped centrifuge tube in a 1 : 1 ratio.)
5. Using a Pasteur pipette add 450 µl of PEG fusion solution drop-wise to the edge of the protoplast culture, placing the last drop in the center. PEG fusion solution contains
   22.5% (w/v) PEG molecular weight 6000
   1.8% (w/v) sucrose
   154 mg/l $CaCl_2 \cdot 2H_2O$
   9.52 mg/l KH_2PO_4
   pH 5.8 (adjust with 1M KOH or 1N HCl, autoclaved and stored in dark at 4°C)
6. Leave the protoplasts undisturbed for 20–40 min at room temperature and then add one drop of washing solution very gradually whilst withdrawing some of the fusion solution from the coalesced drops. Repeat this procedure at 5 min intervals for the next 20 min. During this procedure it is important that the protoplasts are subjected to minimal disturbance. Washing solution: CPW13M medium to which 0.74g/l $CaCl_2 \cdot 2H_2O$ is added (sterile).
7. Five minutes after the last wash, carefully suck off the solution with a Pasteur pipette leaving the protoplasts with a thin film of medium and replace it with culture medium (MS protoplast medium with 9% mannitol). Wash at least 3 times with culture medium.
8. For the culture of protoplasts flood the cover glass with culture medium and scatter several drops of medium on the base of petri dish prior to sealing with parafilm.
9. Incubate the cultures initially in the dark for 24 h at 25°C and then transfer to white light (1000 lux). The fusion products can be easily observed using an inverted microscope.
10. Subsequently transfer the colonies for regeneration.
11. Test the hybridity of fusion product and somatic hybrid as per equipment available in the lab, which has been explained in the text.

Further reading

Bhojwani, S.S. and Razdan, M.K. 1990. *Plant Tissue Culture: Applications and Limitations.* Elsevier Science Publishers, Amsterdam.

Evans, D.A., Sharp, W.R., Ammirato, P.V. and Yamada, Y. 1983. *Handbook of Plant Cell Culture.* Macmillan Publishing Company, New York.

Fowke, L.C. and Constabel, F. 1985. *Plant Protoplasts.* CRC Press, Inc, USA.

Ignacimuthu, S. 1998. *Plant Biotechnology.* Oxford & IBH Publishing Co. Pvt. Ltd., New Delhi.

Pierik, R.L.M. 1987. *In vitro Culture of Higher Plants.* Martinus Nijhoff Publishers, Dordrecht.

Razdan, M.K. 1983. *An Introduction to Plant Tissue Culture.* Oxford and IBH Publishing Co. Pvt Ltd., New Delhi.

Street, H.S. (ed.) 1973. *Plant Tissue and Cell Culture.* Blackwell Scientific Publications, Oxford.

Vasil, I.K. 1984. *Cell Culture and Somatic Cell Genetics of Plants: Laboratory Procedures and Their Applications.* Academic Press, Inc., New York.

10

Somaclonal Variation

Genetic variability is an essential component of any breeding program designed to improve the characteristics of crop plants. In recent years plant cell culture has been hailed as one of the potential sources of useful genetic variation. The variability generated by the use of a tissue culture cycle has been termed somaclonal variation by Larkin and Scowcroft (1981). They defined a tissue culture cycle as a process that involves the establishment of a dedifferentiated cell or tissue culture under defined conditions, proliferation for a number of generations and the subsequent regeneration of plants. In other words, one imposes a period of callus proliferation between an explant and the regeneration of plant. The initiating explant for a tissue culture cycle may come virtually from any plant organ or cell type including embryos, microspores, roots, leaves and protoplasts.

Historically, a culture cycle was essentially seen as a method of cloning a particular genotype. Thus, it became an accepted dictum that all plants arising from tissue culture should be exact copies of the parental plant. However, phenotypic variants were frequently observed amongst regenerated plants. These were usually dismissed earlier as tissue culture artifacts due to the recent exposure of exogenous phytohormones, and sometimes they were labeled as epigenetic events. However, evidences have now shown that these variants are not artifacts but variation arising due to culture of cells, and this has been termed as somaclonal variation. The cause of variation is attributed to changes in the chromosome number and structure. Genetic heterogeneity in cultures arises mainly due to (i) the expression of chromosomal mosaicism or genetic disorders in cells of explant and (ii) new irregularities brought about by culture conditions through spontaneous mutations. Earlier, terms such as "calliclones" and "protoclones" were coined to indicate variation arising in regenerated plants from stem and protoplast derived callus respectively. However, a general term somaclones has been given for plants derived from any form of somatic cell culture and **somaclonal variation** is the variation displayed amongst such plants. Lately, the term **gametoclonal variation** has been introduced for variation observed among plants regenerated from cultured gametic cells.

NOMENCLATURE

Alhough different letters and symbols have been used, two symbols are generally used. Chaleff (1981) has labeled the plants regenerated from tissue culture as R or R_0 plants and the self-fertilized progeny of R_0 plants as R_1. Subsequent generations produced by self-fertilization are termed R_2, R_3, R_4, etc. Larkin and Scowcroft (1981) have referred regenerated plants as SC_1 (=R_0) and subsequent self-fertilized generations as SC_2, SC_3, SC_4, etc.

SCHEMES FOR OBTAINING SOMACLONAL VARIATION

Two schemes, with and without *in vitro* selection, have been generally followed for getting somaclonal variation in crop plants.

Without *in vitro* selection

An explant is cultured on a suitable medium, such as small shoot segments (1–2 cm) of sugarcane, cotyledons, hypocotyls, protoplasts, leaves, embryos, etc. The basal medium is supplemented with growth regulators, which support the dedifferentiation stage, i.e. callus. Normally these cultures are subcultured and then transferred to shoot induction medium for plant regeneration. The plants so regenerated are transferred to pots, grown to maturity and analyzed for variants (Fig. 10.1).

The approach is to find somaclonal variants among the regenerated plants for various characters. Here no directed approach is used and appearance of desirable variant is a chance

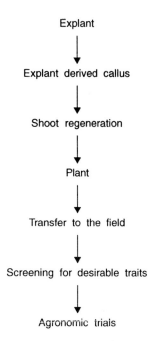

Explant

↓

Explant derived callus

↓

Shoot regeneration

↓

Plant

↓

Transfer to the field

↓

Screening for desirable traits

↓

Agronomic trials

Fig. 10.1 A flow diagram for generation of somaclonal variation without *in vitro* selection.

event. In such a method, both dominant and homozygous recessive traits can be directly selected. If the regenerants are heterozygous, then recessive traits can be selected in the progenies of regenerants. Epigenetic variation will also be avoided when progenies are used. Thus, in case of self-fertilizing crops it is recommended to screen the progenies of the regenerants. Unfortunately, a disadvantage of this approach is that it is time consuming due to the fertilization step and requires screening of many plants.

With *in vitro* selection

It is now well known that *in vitro* culture of higher plants can be used for selection of mutants. Protoplast, cell suspension and callus cultures are handled like microorganisms to search for biochemical mutants. Selection for resistance is the most straightforward method for mutant selection, whereby resistant cells in a large population can be selected by their ability to grow in the presence of an inhibitor while the sensitive cells do not. The protocol has been described in Fig. 10.2. Here the dedifferentiated culture (callus) is subjected to selection against inhibitors like antibiotics, amino acid analogs, pathotoxins, etc. These compounds are put in the medium at a concentration such that some cell population survives and can be further grown on a selective medium. Different selection cycles are performed to get tolerant cells/callus cultures that are subsequently regenerated into plants. These plants are then *in vivo* screened against the inhibitor. If the plants are resistant to the inhibitor, then stable transmission of that character is analyzed in subsequent generations. In this approach, variants for a particular character are selected rather than the general variation obtained in first case where selection is done at the plant level.

Factors influencing somaclonal variation

When attempting to produce somaclones for a new crop plant, some factors that are valid for both the schemes must be considered.

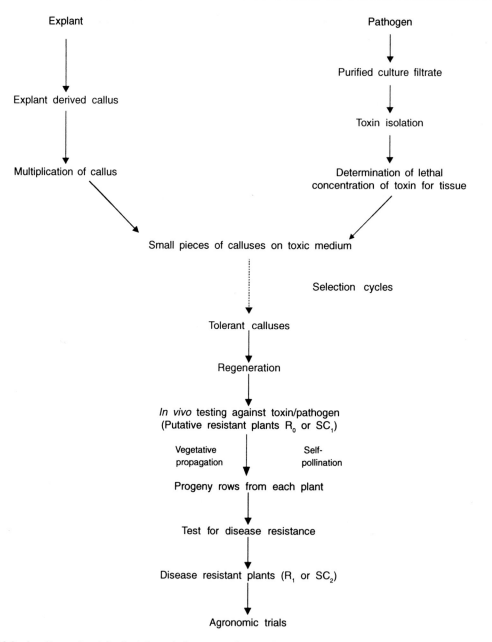

Fig. 10.2 *In vitro* protocol for isolation of disease resistant plants.

Genotype: The genotype of plants used for somaclonal variation is an important variable. Genotype can influence both frequency of regeneration and the frequency of somaclones.

Explant source: The source of explant has been considered a critical variable for somaclonal variation. For example, work with Geranium has demonstrated that variants were recovered from *in vivo* root and petiole cuttings, but not from stem cuttings. Similarly in sugarcane, which is a chromosome mosaic, propagation by stem cuttings can result in plants with variable chromosome number. In addition, since not all explants are equal in terms of regenerability, it is likely that different selective pressures would be exerted against different explants. This could result in different frequencies and spectrums of somaclonal variation among plants from different explants.

Duration of cell culture: Chromosome variation has been reported for many established plant cell cultures. It is widely accepted that most long-term cultures are chromosomally variable. Polyploid and aneuploid plants have been frequently recovered in large number of species e.g. Geranium, *Nicotiana*, tomato, alfalfa, *Saccharum,* etc. It has been stated that variation increased with increasing duration of culture.

Culture conditions: It has been known that growth regulator composition of the culture medium can influence frequency of karyotypic alterations in cultured cells. The growth regulators 2,4-D, NAA, BAP, etc. have been most frequently considered to be responsible for chromosome variability.

All the above variables are important in the recovery of somaclones among regenerated plants. In many species, these variables cannot be modified and still permit recovery of plants. However, these variables should be carefully examined.

The following factors are important during *in vitro* selection of somaclonal variants:

Selective propagule (cells, protoplasts or calli): *In vitro* selection can be applied to any tissue, e.g. cells, protoplasts or calli. There are some advantages as well as disadvantages associated with each type of tissue, but high frequency regeneration must be obtained from that tissue. A main advantage of the use of plant suspension cultures or protoplasts is the capacity to apply nearly standard microbiological techniques. Selection of resistant plant cell lines from protoplasts offer an advantage that these can be prepared not only from suspension cells and calli, but also directly from whole plant tissue. The clonal origin of cell lines arising from selections using protoplasts is more assured since the digestion of the cell wall frees and separates the individual plant cells. However, the advantages may be offset by the technical requirements necessary to produce viable protoplasts. Generally, protoplast preparation protocols must be developed independently for each plant species or tissue used.

However, selection from calli is probably one of the least complex methods. Callus culture is easy to perform and maintenance and handling can also be done without too much difficulty. The initiation and maintenance of suspension, protoplasts or callus cultures have been explained earlier.

Selective agent: Inhibitor used can be an amino acid analog, herbicide, a synthetic toxin(s) isolated from fungal liquid culture. Some of the inhibitors/toxins are commercially available and can be used directly for *in vitro* selection experiments. However, when pathotoxins produced by fungi in liquid culture are to be used, then first these toxins must be isolated. For abiotic stresses of salt and aluminium tolerance, specific compounds like sodium chloride and aluminium chloride can be used respectively. The inhibitor concentration used in the selection experiments will depend on the sensitivity of the plant cell line. Therefore, a dose response curve with respect to growth inhibition must be performed for each selection agent of interest. For amino acid analogs, typical gradients range logarithmically from 1 µM to 10mM. For sodium chloride tolerance, range may be from 0.1% to 2% of sodium chloride in the medium. The dilution steps are arbitrarily chosen to facilitate graphic representation of the final results.

Serial dilutions can be made as shown below:

1 mM———→0.1———→0.01———→0.001 mM

The inactivating effect of an inhibitor on cells can be assessed by determining increase in fresh or dry weight or using characters such as plating efficiency, dye uptake, cytoplasmic streaming and incorporation of labeled metabolites into macromolecules. A dose response curve should be prepared to assess the effect of an inhibitor.

For selection of disease resistant mutants, the toxicity of toxin/crude toxin/purified culture filtrate can be tested by leaf or root bioassays. Then, a particular concentration of toxin is used for selection of calluses/cells/protoplasts. For this, pieces of calluses (3–5) with 50 mg fresh weight can be inoculated on a medium with different concentrations of toxin and incubated for 3–4 weeks. On the basis of fresh weight increase, a particular toxin/inhibitor concentration can be selected for *in vitro* selection experiments.

Selection technique: *In vitro* selection of cell lines against an inhibitor is best carried out at a concentration that will kill 70 to 90% of wild type cells. Doses that are too low should be avoided because cells escaping for purely physiological reasons could outnumber true mutants. On the other hand, inhibitor concentrations that are too high may mean that potentially resistant variants are also unable to grow and are thus lost. Cells should be incubated in an inhibitor-containing medium for a sufficient length of time to ensure expression and growth of the resistant clones. The selection can be carried out at a very high toxin/inhibitor concentrations with many or just one cycle of selection, or many cycles with a pause on nontoxic media after 2nd or 3rd selection cycle, or on an increasingly higher inhibitor/toxin level. A selection cycle can be of 3 to 4 weeks duration.

The selection technique involves placing pieces of cultured tissue on a medium containing a near-lethal concentration of inhibitor and selecting out fast growing visible sectors of tissue from the remainder of the cultured material. Resistant tissue obtained in this way continues to proliferate on a medium containing high level of inhibitor concentration.

Regeneration of plants: The next step in the procedure is to regenerate plants from the cultured material. In many instances, the capacity of tissue cultures to regenerate plants is lost over a long period of continuous subculture. Regeneration is generally achieved by reducing the level of auxin in the media and incubating the cultures in light.

In vivo testing: It is important that the plants obtained from cultures selected for a particular inhibitor/toxin should be critically examined for expression and inheritance of resistance character. Only in this manner it can be determined whether the resistance observed *in vitro* is due to a mutational event or to the epigenetic adaptation by the cultured cells to the presence of inhibitor/toxin. Normally plants developed against toxin(s) of pathogen are *in vivo* tested against the pathogen. Conditions required for the development of disease symptoms in plants should be provided.

Agronomic analysis: The regenerated plants selected against inhibitor/toxin should be analyzed for their agronomic value in the first generation and also in the second generation (SC_2) for stability of resistance and other desirable features.

Resistance stability: A major difficulty often encountered with plant cell lines selected for resistance to inhibitor/toxin is the instability of the characteristic when removed from selective pressure. There are a variety of reasons for this instability. The genetic lesion responsible may have a high reversion rate or the alteration may simply be a transient genetic or epigenetic adaptation. The most common explanation, however, is the presence of wild type escapes in the cell population. Under nonselective conditions, these cells would rapidly proliferate eventually taking over the culture.

In order to test stability of the resistance trait, establish at least two cultures for each selected line—one subcultured routinely in the presence and one in the absence of selective conditions. At various intervals over a period of several months perform dose response test. Further, the R_0 plants that were found to be resistant should be analyzed for resistance to that character in R_1 and subsequent generations.

APPLICATION OF SOMACLONAL VARIATION

To have an impact on a plant breeding program, an *in vitro* scheme must either produce plants useful for breeding or useful directly.

1. *Novel variants:* An implication of somaclonal variation in breeding is that novel variants can arise and these can be agronomically used. A number of breeding lines have been developed by somaclonal variation. An improved scented Geranium variety named 'Velvet Rose' has been developed. An example of heritable somaclonal variation is the development of pure thornless blackberries Lincoln Logan (*Rubus*), Hasuyume, a protoplast-derived rice cultivar, and somaclone T-42 has been released. In India two varieties namely Pusa Jai Kisan in mustard *Brassica* and CIMAP/Bio13 in *Citronella* have been released for cultivation. Some of the variants developed as varieties in *Rubus, Torenia, Apium* and others have been shown in Table 10.1, while variation for other useful morphological characters observed in diverse species has been shown in Table 10.2.

2. *Somaclonal variation and disease resistance:* The greatest contribution of somaclonal variation towards plant improvement is in the development of disease resistant genotypes in various crop species. Resistance was first reported in sugarcane for eye spot disease (*Helminthosporium sacchari*), downy mildew (*Sclerospora sacchari*) and Fiji virus disease by regenerating plants from the callus of susceptible

Table 10.1 A sample of cultivars derived from tissue culture via somaclonal variation

Crop
Eustoma grandiflorum
Hemerocallis Yellow Tinkerbell
Paulownia tomentosa Snowstrom
Pelargonium Velvet Rose
Torenia fournieri U Conn White
Rubus laciniatus Lincoln Logan
Apium UC-T3 somaclone
Ipomea batatus Scarlet
Oryza sativa Hasuyume
Mustard *Brassica* Pusa Jai Kisan
Citronella CIMAP/Bio13

Table 10.2 A sample of somaclonal variants obtained in different crop species

Crop	Characters
Rice	Tiller no., panicle size, flowering date, plant height, heading date, culm length, leaf shape, leaf color, ear shape, sterility mutants
Maize	Glyphosate resistance
Wheat	Grain color, height, tiller no., seed storage protein
Sorghum	Improved acid soil stress tolerance
Oats	Avenin banding pattern variation, increased seed weight, seed number, grain yield
Barley	Grain yield, test weight, leaf shape, 1000 grain weight, height, lodging resistance, heading date, β-glucan content, ash
Soybean	Varied height, lodging, maturity, seed protein and oil content
Brassica napus	Growth habit, glucosinolate content
Sugarcane	Yield, sugar content, disease resistance
Tomato	Violet cotyledon color, dwarf habit, modification of first floral node, early flowering, orange fruit color
Carrot	Higher carotene content
Potato	Yield, vine quality, maturity and morphology
Fennel	Chlorotic foliage, altered stem height, reduced fruit production, varied oil composition
Datura	Sulfonylurea, imidazolinone and triazolopyrimidine herbicide resistance

clones and screening the somaclones. Variability from protoplast-derived plants was also examined in the tetraploid Russet Burbank variety of potato. Although many of the variants were aberrant, some showed resistance to *Phytophthora infestans*. Table 10.3 lists few successful examples of somaclonal variants obtained without *in vitro* selection at the plant level with an increased disease resistance. Somaclonal variation without *in vitro* selection system has the same approach as that for conventional or mutation breeding for screening of disease-resistant variants.

The most successful strategy employed so far to select *in vitro* for resistance is that in which either the pathotoxin or partially purified culture filtrate of the pathogen is used as a screening agent. Many facultative fungal and bacterial plant pathogens produce low molecular weight toxic metabolites *in vitro* and *in vivo*. The induction of chlorosis and/or necrosis by a pathogen in a susceptible host suggests that toxins are involved. Carlson (1973) first tested this possibility of *in vitro* selection for disease resistance for wild fire disease of tobacco caused by *Pseudomonas syringae* pv *tabaci*. Haploid cells

and protoplasts were first treated with ethylmethane sulfonate mutagen and then cultured in the presence of methyl sulphoximine (MSO) thought to be an analog of tabtoxin which gave chlorotic halos on tobacco leaves. Most of the calluses obtained from the surviving cells soon lost their ability to grow in the presence of MSO. However, three gave resistant plants that were also resistant to tabtoxin. Later, Gengenbach *et al.* (1977) used recurrent selection to select variants carrying resistance to T-toxin of *Helminthosporium maydis*. A very high level of resistance to T-toxin was selected in originally susceptible TmS corn lines by serial subculture of embryo-derived callus cultures on a medium containing a just sublethal dose of T-toxin. Most significantly, plants regenerated from resistant TmS cultures proved to be fully toxin resistant. The character was transmitted to the progeny of regenerated plants through the female parent, and gene for this is located on the mitochondrial genome.

The work with T-toxin had a highly significant enriching effect for *in vitro* resistance to a pathotoxin. Nevertheless, several studies have strongly suggested that purified or partially

Table 10.3 A list of disease resistant crop plants obtained by screening of somaclones at the plant level without *in vitro* selection

Crop	Pathogen
Food crops	
Barley	*Rhynchosporium secalis*
Maize	*Helminthosporium maydis*
Rice	*Helminthosporium oryzae*
Rape	*Phoma lingam, Alternaria brassicicola*
Sugar-cane	Fiji virus, *Sclerospora saccharii, Helminthosporium sacchari, Puccinia melanocephala*
Horticultural and other crops	
Potato	*Alternaria solani, Phytophthora infestans,* Potato virus X & Y, *Streptomyces scabie, Verticillium dahlie*
Tobacco	*Phytophthora parasitica*
Tomato	*Fusarium oxysporum, Pseudomonas solanacearum*
Alfalfa	*Verticillium albo-atrum, Fusarium solani*
Celery	*F. oxysporum* f. sp. *apii, Septoria apii*
Lettuce	Lettuce mosaic virus
Apple	*Phytophthora cactorum*
Banana	*F. oxysporum* f. sp. *cubense*
Peach	*Xanthomonas campestris* pv *pruni, Pseudomonas syringae* pv *syringae*
Poplar	*Septoria musiva, Melamspora medusae*

purified toxin preparations may be used to select for novel disease resistance *in vitro*. Behnke (1979) was the first to perform experiments on calli of dihaploid potato clones using a medium containing crude culture filtrate extract of the fungal media employed for *Phytophthora infestans*. However, subsequent workers used purified culture filtrate or crude toxin for *in vitro* selection experiments.

Chawla and Wenzel (1987a, 1987b) used immature embryo-derived calli of barley and wheat for screening against crude toxin isolated from culture filtrate of *Helminthosporium sativum* and also synthetic fusaric acid toxin of *Fusarium* species. Two selection methods were employed (a) continuous: four selection cycles were performed continuously on toxic media and (b) discontinuous: calluses were transferred to non-toxic media after the second or third selection cycle, but four cycles were performed on toxic media. Discontinuous approach for *in vitro* selection resulted in a larger number of resistant plants. Table 10.4 lists a few successful examples of *in vitro* selection for disease resistance in crop plants.

Selection of somaclonal variants with disease resistance has resulted in the release of only a few cultivars or genotypes, *viz.* sugarcane cv. with resistance to Fiji disease, tomato variety DNAP-17 a somaclonal variant with monogenic *Fusarium* wilt race 2 resistance, celery line UC-T3 somaclone with resistance to *F. oxysporum* f. sp. *apii*.

3. **Somaclonal variation and abiotic stress resistance:** Somaclonal variation has resulted in several interesting biochemical mutants, which are being successfully used in plant metabolic pathway studies, i.e. amino acid and secondary metabolic pathways. Investigations have shown that the level of free amino acids, especially

Table 10.4 Disease resistant crop plants obtained by *in vitro* selection

Crop	Pathogen	Selection agent
Food crops		
Rape seed	*Phoma lingam, Alternaria brassicicola*	CF
Rice	*Helminthosporium oryzae*	CT
	Xanthomonas oryzae	Bacterial cells
Barley	*Helminthosporium sativum*	CT
	Fusarium spp.	Fusaric acid
Maize	*Helminthosporium maydis*	Hm T toxin
Oats	*Helminthosporium victoriae*	Victorin
Wheat	*Helminthosporium sativum, Fusarium graminearum*	CT
	Pseudomonas syringae	Syringomycin
Sugar-cane	*Helminthosporium sacchari*	Toxin
Horticultural and other crops		
Tobacco	*Pseudomonas syringae* pv. *tabaci*	Methionine sulfoximine
	Alternaria alternata; *P. syringae* pv *tabaci*	Toxin
	Tobacco mosaic virus	Virus
	Fusarium oxysporum f. sp. *nicotianae*	CF
Potato	*Phytophthora infestans, Fusarium oxysporum*	CF
	Erwinia carotovora	Pathogen
Alfalfa	*F. oxysporum* f. sp. *medicagnis*	CF
Tomato	Tobacco mosaic virus X	Virus
	Pseudomonas solanacearum	CF
Eggplant	*Verticillium dahlie*	CF
	Little leaf disease	Pathogen
Peach	*Xanthomonas campestris* pv. *pruni*	CF
Hop	*Verticillium albo-atrum*	CF
Celery	*Septoria apiicola*	CF

CF: culture filtrate; CT: crude toxin

proline, increases during cold hardening. Lazar *et al.* (1988) developed somaclonal variants for freezing tolerance in Norstar winter wheat. A significant positive correlation between proline level and frost tolerance has been found in a broad spectrum of genotypes. *In vitro* selection and regeneration of hydroxyproline resistant lines of winter wheat with increased frost tolerance and increased proline content has been reported (Dorffling *et al.*, 1997). The results showed strong correlation of increased frost tolerance with increased proline content.

In vitro selection has also been used to obtain plants with increased acid soil, salt, aluminium, and herbicide resistance.

Salt tolerance: Plant tissue culture techniques have been successfully used to obtain salt tolerant cell lines or variants in several plant species, *viz.* tobacco, alfalfa, rice, maize, *Brassica juncea, Solanum nigrum,* sorghum, etc. In most cases, the development of cellular salt tolerance has been a barrier for successful plant regeneration, or if plants have been obtained they did not inherit the salt tolerance. Only in a few cases it was possible to regenerate salt-tolerant plants. Mandal *et al.* (1999) developed a salt-tolerant somaclone BTS24 from indigenous rice cultivar Pokkali. This somaclone yielded 36.3 q/ha under salt stress conditions as compared to 44 q/ha under normal soil. Some of the reports for successful production of healthy, fertile, and

genetically stable-salt tolerant regenerants from various explants, level of selection, and other details in various species have been summarized in Table 10.5.

Aluminium tolerance: It has long been established that plant species or cultivars greatly differ in their resistance to aluminium stress. In recent years, considerable research has been focused on the understanding of physiological, genetic and molecular processes that lead to aluminium tolerance. Despite the problems encountered in adapting culture media for *in vitro* selection for aluminium resistance, cell lines have been isolated in several species, e.g. alfalfa, carrot, sorghum, tomato, tobacco. Aluminium-tolerant somaclonal variants from cell cultures of carrot were selected by exposing the cells to excess ionic aluminium in the form of aluminium chloride (Arihara *et al.*, 1991). Jan *et al.* (1997) elicited aluminium toxicity during *in vitro* selection in rice by making several modifications in the media *viz.* low pH, low phosphate and calcium concentrations, and unchelated iron and aluminium along with aluminium sulphate. After selection on aluminium toxic media, callus lines were maintained on aluminium toxic-free media and 9 tolerant plants were obtained. Transmission of aluminium resistance character was identified till the fourth generation.

Drought tolerance: Wang *et al.* (1997) reported a sorghum somaclonal variant line

Table 10.5 *In vitro* selection and regeneration of salt tolerant plants

Plant	Explant	Salt and conc. (g/l)	Selection	Inheritance
Rice	Embryogenic callus	NaCl (10,20)	One step	R_0, R_1, R_2, R_3 and R_4
Rice	Mature seeds	NaCl (15)	One cycle	R_0, R_1
Wheat	Immature embryo	NaCl (2 - 7)	Two steps	R_0, R_1
Wheat	Mature embryo	NaCl (5)	Several cycles	R_0, R_1
Canola	Morphogenic cotyledons	NaCl (7.5 –10)	One step	R_0, R_1
Canola	Somatic embryos	NaCl (up to 12.5)	Two cycles	R_0, R_1
Brassica napus	Microspore-derived embryos	NaCl (6 – 7))	One step	R_0
Vigna radiata	Cotyledons from *in vitro* seedlings	NaCl (up to 150 mM)	Two steps	R_0, R_1
Alfalfa	Immature ovary derived callus	NaCl (10)	One step	R_0, R_1, R_2
Tobacco	Protoplasts (haploid)	NaCl (11.7)	Short term	$R_0 \times$ wild type $\times F_1$
Flax	Callus	Sulphate salt (26.8)	One step	R_0, R_1, R_2
Coleus	Leaf discs from *in vitro* multiplied shoots	NaCl (5.3)	One step	R_0, R_1

(R111) resistant to drought stress. A novel hybrid was developed by crossing R111 with several sterile lines, indicating that selection of somaclonal variants is an effective method for creating new sources of genetic variation. Drought-tolerant rice lines were obtained by *in vitro* selection of seed induced callus on a media containing polyethylene glycol as a selective agent, which simulated the effect of drought in tissue culture conditions (Adkins *et al.*, 1995)

4. *Herbicide resistance*: Through *in vitro* selection several cell lines resistant to herbicides have been isolated and a few have been regenerated into complete plants. Among the important achievements are tobacco, soybean, wheat, maize, etc. resistant to various herbicides such as glyphosate, sulfonylurea, imidazolinones, etc. Some of the examples have been reported in Table 10.6.

5. *Insect resistance:* Zemetra *et al.* (1993) used *in vitro* selection technique for generation of somaclonal variants for Russian wheat aphid (*Diuraphis noxia*) in wheat. Calli from susceptible wheat cultivar "Stephens" were exposed to an extract from aphid. Resistance to aphid was observed in both R_2 and R_3 generations.

6. *Seed quality:* Recently, a variety Bio L 212 of lathyrus (*Lathyrus sativa*) has been identified for cultivation in central India which has been developed through somaclonal variation and has low ODAP, a neurotoxin (Mehta and Santha, 1996), indicating the potential of somaclonal variations for the development of varieties with improved seed quality.

7. *Alien gene introgression:* The increase in genome rearrangement during tissue culture provides a new opportunity for alien gene introgression, which can help widen the crop germplasm base, particularly by culturing immature embryos of wide crosses where crop and alien chromosomes cannot replicate through meiosis. Introgression of genes may be better achieved by imposing one or more cell culture cycles on interspecific hybrid material. The enhanced somatic genome exchange is likely to produce regenerants where part of the alien genome has been somatically recombined into the chromosomes of crop species.

BASIS OF SOMACLONAL VARIATION

Karyotype changes: Variant plants with altered chromosome number have been reported by several workers. Polyploidy is the most frequently observed chromosome abnormality, e.g. aneuploidy in oats (McCoy, 1982), potato (Sree Ramulu *et al.*, 1985), and barley (Kole and Chawla, 1993). The change in chromosome number in a variant plant is commonly associated with reduced fertility and with altered genetic ratios in the progeny of self-fertilized plants.

Changes in chromosome structure: In contrast to gross changes in chromosome number, more cryptic chromosome rearrangements may be responsible for genetic variation. Translocations have been reported in potato, ryegrass, oats, etc. Published work also suggests that chromosome deletions, duplications, inversions, and other minor reciprocal and nonreciprocal rearrangements occur among

Table 10.6 Some examples of somaclonal variants for herbicide resistance

Crop	Herbicide	Reference
Glycine max	Imazethapyr	Tareghyan *et al.* (1995)
Nicotiana tabacum	Glyphosate	Stefanov *et al.* (1996)
Gossypium hirsutum	Sulfonylurea, Imidazolinone	Rajasekaran *et al.* (1996)
Zea mays	Sethoxydim, cycloxydim	Landes *et al.* (1996)
Zea mays	Glyphosate	Forlani *et al.* (1992)
Triticum aestivum	Difenzoquat	Bozorgipour and Snape (1997)
Beta vulgaris	Imidazolinone	Wright *et al.* (1998)
Datura innoxia	Chlorosulfuron	Rathinasabapathi and King (1992)

regenerated plants, e.g. ryegrass, barley. Chromosome irregularities such as breaks, acentric and centric fragments, ring chromosomes, and micronuclei were observed in garlic somaclones. Cytological abnormalities such as multilobed nuclei, multinucleate cells, abnormal anaphase, and mixoploidy were observed in SC_3 generation of barley.

Single-gene mutations: Single-gene mutations have been detected by several workers (Edallo *et al.*, 1981; Oono, 1978). Recessive single-gene mutations are suspected if the variant does not appear in R_0 or SC_1 generation but the self-fertilized R_1 or SC_2 progeny segregate in 3:1 Mendelian ratio for the trait of interest. This type of analysis has been reported for maize, *Nicotiana sylvestris*, rice, wheat, etc.

Cytoplasmic genetic changes: The most detailed work on this has been done on maize. Many studies have implicated mitochondria as the determinants of male sterility and *Drechslera maydis* T toxin sensitivity in maize. Gengenbach *et al.* (1977) and Brettell *et al.* (1979) regenerated plants that were resistant to fungal T-toxin. The original lines used had all the Texas male sterile cytoplasm, which results in plants being normally highly susceptible to this disease. However, when selection pressure was applied by subculturing callus cultures in the presence of toxin, some regenerants were found to be not only resistant to the toxin but also fertile. When restriction analysis of the mitochondrial DNA (mtDNA) was evaluated, it was evident that significant changes had occurred in the mtDNA of plants derived from cell cultures. Molecular analyses based on restriction endonuclease digest patterns of mitochondrial and chloroplast genomes of regenerated potato plants derived from protoplasts have revealed significant change in mitochondrial genomes but not in chloroplast genomes (Kemble and Shepard, 1994). It was suggested that the variation results from substantial DNA sequence rearrangements (e.g. deletion, addition and intramolecular recombinations) and cannot be explained by simple point mutations. However, point mutations can also occur in the genomes of organelles of regenerated plants. Variations in chloroplast DNA have been detected in tomato and wheat somaclones.

Mitotic crossing over: Mitotic crossing over (MCO) can also account for some of the variation detected in regenerated plants. This can include both symmetric and asymmetric recombination. MCO may account for the recovery of homozygous recessive single gene mutations in some regenerated plants. Somatic cell sister chromatid exchange, if asymmetric, can also lead to deletion and duplication of genetic material and hence variation.

Gene amplification and nuclear changes: Studies have shown that nuclear genes are affected by tissue culture stress. De Paepe *et al.* (1982) have shown that heritable quantitative and qualitative changes can be observed in the nuclear DNA content of the doubled haploid *Nicotiana sylvestris* obtained from pollen cultured plants. These plants contain generally increased amounts of total DNA and an increasing proportion of highly repeated sequences. Both AT- and GC-rich fractions are amplified in tissue-culture-derived plants. Durante *et al.* (1983) showed similar amplification of AT- and GC-rich fractions in DNA from *Nicotiana glauca* pith explants within hours of culture. These studies suggest the presence of a differential replication process during the early stages of dedifferentiation. It has been seen that plant cells like those of other eukaryotes can increase or decrease the quantity of a specific gene product by differential gene amplification and diminution. For instance, ribosomal RNA gene amplification and diminution are now known to be widespread in wheat, rye, and tobacco, and in flax, ribosomal DNA is known to alter directly in response to environmental and cultural pressures. Reduction in ribosomal RNA genes (rDNA) has been found in potato plants regenerated from protoplasts. Additionally, both structural rearrangement within rDNA and methylation of nucleotide sequences of rDNA have been observed (Brown, 1989; Harding, 1991).

Transposable elements: Several authors

have speculated that transposable elements may also be responsible for somaclonal variation. Heterozygous light green (Su/su) somaclones with a high frequency of colored spots on the leaf surface have been detected for a clone of *N. tabacum* (Lorz and Scowcroft, 1983) and for a *N. tabacum* + *N. sylvestris* somatic hybrid. The somatic hybrid has an unstable pattern of inheritance that would be consistent with an unstable gene. A causal relationship between genetic instability possibly related to tissue-culture-induced transposition, and somaclonal variation was speculated.

Recently, genomic changes in a maize line and mutations in a tobacco line because of activation of transposable elements under *in vitro* tissue culture conditions have been demonstrated (Peschke and Phillips, 1992). It has also been shown that several types and copies of nonactive transposable elements, including retrotransposable elements, are present in the genome of potato (Flavell *et al.*, 1992). Thus, it is likely that a few active transposable elements may also be present in potato and can be responsible for generating some degree of somaclonal variation in regenerated potato plants.

Disadvantages: The following are the disadvantages associated with somaclonal variation.

1. Uncontrollable and unpredictable nature of variation and most of the variations are of no apparent use.
2. The variation is cultivar-dependent.
3. The variation obtained is not always stable and heritable. The changes occur at variable frequencies.
4. Not all the changes obtained are novel. In majority of cases, improved variants have not been selected for breeding purposes.

GAMETOCLONAL VARIATION

The variation observed among plants regenerated from cultured gametic cells is termed gametoclonal variation as compared to somaclonal variation that is somatic-tissue-derived variation. Gametes are products of meiosis, governed by Mendel's laws of segregation and independent assortment. There are three genetic differences which can be pointed as evidence for gametoclonal and somaclonal variations being distinct.

i. Both dominant and recessive mutants induced by gametoclonal variation will be expressed directly in haploid generated plants, since only a single copy of each gene is present. Hence, regenerated gametoclones (R_0) can be analyzed directly to identify new variants.

ii. Recombinational events that are recovered in gametoclones would be the result of meiotic crossing over and not mitotic crossing over.

iii. To use gametoclones, chromosome number must be doubled. The most frequently used method to double chromosomes is treatment with colchicine.

Gametoclones can be produced through the culture of either male or female gametic cells, or their derivatives. The most successful and commonly used method for a wide range of species is culture of anthers, or isolated microspores. Anther and microspore culture technology has been well established in many plant species for the production of doubled haploid plants and also for *in vitro* screening of microspores against stresses. Since growth of cells in culture results in single gene mutations, culture of microspores could result in recessive mutations and these can be *in vitro* selected against various stresses. Hence, it is likely that gametoclonal variants can be detected in the R_0 generation, while evaluation of R_1 progeny must be completed to detect the full spectrum of variants. It has been observed that mutant cells probably do not regenerate as well as wild type cells. Hence, the mutation spectrum and frequency obtained from regenerated haploids may be strikingly different from that of regenerated diploid tissue.

When referring to gametoclonal variation,

Table 10.7 Variation among plants regenerated from gametophytic tissue culture

Crop	Characteristics
Brassica napus	Time to flower, glucosinolate content, leaf shape and color, flower type, pod size and shape
Hordeum vulgare	Plant height, yield, fertility, days to maturity
Nicotiana sylvestris	Leaf color, leaf shape, growth rate, DNA content, repeated sequences of DNA
N. tabacum	TMV resistance, root knot resistance
	Time of flowering, plant size, leaf shape, alkaloid content, number of leaves
	Altered temperature sensitivity
Oryza sativa	Seed size, seed protein level, plant height, level of tillering
	Waxy mutant
Saintpaulia ionantha	RuBSase activity

one must recognize four distinct sources of variation.

i) New genetic variation induced by cell culture procedures.

ii) Variation resulting from segregation and independent assortment.

iii) New variation at the haploid level induced by the chromosome doubling procedure.

iv) New variation induced at the diploid level resulting in heterozygosity.

In crops such as barley, wheat, rice, *Brassica* spp., tobacco, potato, etc., gametoclonal variants have been generated and used to accelerate and enrich the breeding program. In fact, the first report on selection of disease resistant mutants by Carlson (1973) utilized haploid tissue of tobacco. Now, there have been several reports of variation in plants regenerated from gametophytic tissue and some have been documented in Table 10.7.

Further reading

Chawla, H.S. 1998. *Biotechnology in Crop Improvement.* International Book Distributing Co., Lucknow.

Evans, D.A., Sharp, W.R., Ammirato, P.V. and Yamada, Y. 1983. *Handbook of Plant Cell Culture.* Macmillan Publishing Company, New York.

Larkin, P.J. and Scowcroft, W.R. 1981. Somaclonal variation – a novel source of variability from cell cultures for plant improvement. *Theor. Appl. Genet.,* **60:** 197-214.

Pierik, R.L.M. 1987. *In vitro Culture of Higher Plants.* Martinus Nijhoff Publishers, Dordrecht.

Reinert, J. and Bajaj, P.P.S. (eds.). 1977. *Applied and Fundamental Aspects of Plant Cell, Tissue and Organ Culture.* Springer-Verlag, Berlin.

Vasil, I.K. 1984. *Cell Culture and Somatic Cell Genetics of Plants: Laboratory Procedures and Their Applications.* Academic Press, Inc., New York.

11

Germplasm Storage and Cryopreservation

The aim of germplasm conservation is to ensure the availability of useful germplasm at any time. Breeding programs rely heavily on locally adapted ancient plant varieties and their wild relatives as sources of germplasm. Further, the continuing search for high yielding varieties of crop plants with resistance to pathogens and pests warrants the availability and maintenance of large collections of germplasm. In species which are propagated through seeds, it is economical to preserve the seeds. These seeds are dried to a water content of 5–8% and then stored at a temperature of –18°C or lower and low humidity. Sometimes it is not feasible to store seeds because

i) some plants do not produce fertile seeds;
ii) some seeds remain viable only for a limited duration;
iii) some seeds are very heterozygous and therefore, not suitable for maintaining true to type genotypes; and
iv) seeds of certain species deteriorate rapidly due to seed borne pathogens.

In case of vegetatively propagated species, preservation of germplasm heavily taxes manpower and land resources. As *in vitro* techniques are becoming more important in crop improvement through the use of somatic cell genetics, the material produced *in vitro* will have to be conserved *in vitro*. The *in vitro* system is extremely suitable for storage of plant material, since in principle it can be stored on a small scale, disease free and under conditions that limit growth. Germplasm storage *in vitro* is crucial for the future development and safety of agriculture.

The conservation of plants *in vitro* has a number of advantages over *in vivo* conservation. These are as follows:

i) *In vitro* culture enables plant species that are in danger of being extinct to be conserved.
ii) *In vitro* storage of vegetatively propagated plants can result in great savings in storage space and time.
iii) Sterile plants that cannot be reproduced generatively can be maintained *in vitro*.
iv) It is possible *in vitro* to efficiently reduce growth, which decreases the number of subcultures necessary.
v) If a sterile culture is obtained, often with great difficulty, then subculture *in vitro* is the only safe way of ensuring that it remains sterile.

There are two main approaches for *in vitro* storage of germplasm.

CRYOPRESERVATION

Cryopreservation (Gr., Kryos = frost) means "preservation in the frozen state". In practice, this is generally meant to be storage at very low temperatures, e.g. over solid carbon dioxide (–79°C), in low temperature deep freezers (–80°C or above), in vapor phase of nitrogen (–150°C) or in liquid nitrogen (–196°C). Generally the plant material is frozen and maintained at the temperature of liquid nitrogen (–196°C). At this temperature, the cells stay in a completely inactive state. The theoretical basis of freeze preservation is the transfer of water present in the cells to the solid state. While pure water becomes ice at 0°C, the cell water needs a much lower temperature because of freezing point depression by salts or organic molecules. The lowest temperature at which water has been demonstrated to be liquid is –68°C. Cryopreservation of plant cells, organs and tissues received widespread importance mainly due to the emphasis placed in recent years on *in vitro* manipulations with cultured cells. Storage at very low temperature considerably slows

down or halts metabolic processes and biological deterioration. Two major considerations have to be taken into account for freezing of specimens. (i) Susceptibility or degree of freeze tolerance displayed by a given genotype to reduced temperatures. (ii) The formation of ice crystals within the cells. The cryopreservation of plant cell culture and eventual regeneration of plants from them has been shown in Fig. 11.1, and it involves the following steps:

1. Raising sterile tissue cultures
2. Addition of cryoprotectants and pretreatment
3. Freezing
4. Storage
5. Thawing
6. Determination of survival/viability
7. Plant growth and regeneration

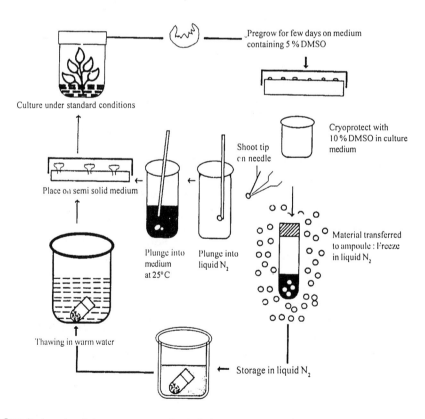

Fig. 11.1 Standard protocol for cryopreservation of shoot tips.

Raising sterile tissue cultures

The morphological and physiological conditions of the plant material influence the ability of an explant to survive freezing at −196°C. Different types of tissues can be used for freezing such as the apical and lateral meristems, plant organs (embryos, endosperm, ovules, anther/pollen), seeds, cultured plant cells, somatic embryos, protoplasts, calluses, etc. In general, small, richly cytoplasmic, meristematic cells survive better than the larger, highly vacuolated cells. Cell suspensions have been successfully frozen in a number of species. These should be in the late lag phase or exponential phase. Callus tissues, particularly among tropical species, are more resistant to freezing damage and have also been used. A rapidly growing stage of callus shortly after one or more weeks of transplantation to the medium is best for cryopreservation. Old cells at the top of callus and blackened area should be avoided. Cultured cells are not the ideal system for cryopreservation. Instead, organized structures such as shoot apices, embryos or young plantlets are preferred. Plant meristems have been quite often used for preservation. Plant meristems when aseptically isolated are incubated for few days in growth media before proceeding to the next step. Water content of cells or tissues for cryopreservation should be low because with low freezable water, tissues can withstand extremely low temperatures.

Addition of cryoprotectants and pretreatment

The tissue is prepared for freezing by treating with various cryoprotectants to withstand the freezing conditions of −196°C. The effect of temperature on plants depends equally on genotype and environment, as well as physiological conditions. While genotype for practical purposes is an unchangeable factor, the other two factors can be altered according to circumstances. The freeze tolerance of a culture can be improved by timing the harvest of cells (in exponential phase) or by hardening process (growing the plants for shorter duration of one week at a lower temperature). Culturing of shoot apices for a brief period under optimum conditions before freezing has proved beneficial. Growing plants at 4°C for 3–7 days before taking explant also increases the survival of specimens. There are two potential sources of cell damage during freeze preservation.

I. Formation of large ice crystals inside the cells, leading to the rupture of organelles and the cell itself.

II. Intracellular concentration of solutes increases to toxic levels before or during freezing as a result of dehydration.

Addition of cryoprotectants controls the appearance of ice crystals in cells and protects the cells from the toxic solution effect. A large number of heterogeneous groups of compounds have been shown to possess cryoprotective properties with different efficiencies. These are glycerol, dimethyl sulfoxide (DMSO), glycols (ethylene, diethylene, propylene), acetamides, sucrose, mannose, ribose, glucose, polyvinylpyrrolidone, proline, etc. Most frequently used are DMSO, sucrose, glycerol and proline. Cryoprotectants depress both the freezing point and the supercooling point of water i.e. the temperature at which the homogeneous nucleation of ice occurs thus retarding the growth of ice crystal formation. DMSO has proved to be an excellent cryoprotectant because of its good properties. It has a low molecular weight, is easily miscible with the solvent, non-toxic at low concentrations, easily permeable into the cells and easily washable from the cells. Generally a concentration of 5 – 10% for DMSO and 10–20% for glycerol is adequate for most materials. In instances where application of a single cryoprotectant does not result in higher survival, a mixture of cryoprotectants has proven beneficial.

In practice, the material suspended in the culture medium and treated with a suitable cryoprotectant is transferred to sterile polypropylene cryovials or ampoules with a screw cap. A dilute solution of the cryoprotectant (e.g. 5–10%

DMSO) is added gradually or at intervals to the ampoules containing the tissue material to prevent plasmolysis and protects cells against osmotic shocks. After the last addition of cryoprotectant, but before cooling, there should be an interval of 20–30 min. The ampoules containing the cryoprotectants are then frozen.

Freezing

The type of crystal water within stored cells is very important for survival of the tissue. For this reason, three different types of freezing procedures have been developed. Various types of cryostats and freezing units are available by which different cooling rates can be easily regulated.

(a) *Rapid freezing*: The plant material is placed in vials and plunged into liquid nitrogen, and a decrease of –300 to –1000°C/min or more occurs. This method is technically simple and easy to handle. The quicker the freezing is done, the smaller the intracellular ice crystals are. A somewhat slower temperature decrease is achieved when the vial containing plant material is put in the atmosphere over liquid nitrogen (–10 to –70°C/min). Dry ice (CO_2) instead of nitrogen can also be used in a similar manner. It is also possible that ultra-rapid cooling may prevent the growth of intra-cellular ice crystals by rapidly passing the cells through the temperature zone in which lethal ice crystal growth occurs. Rapid freezing has been employed to cryopreserve shoot tips of carnation, potato, strawberry and *Brassica napus* and somatic embryos.

(b) *Slow freezing*: In this method, the tissue is slowly frozen with a temperature decrease of 0.1–10°C/min from 0°C to –100°C and then transferring to liquid nitrogen. Survival of cells frozen at slow freezing rates of –0.1 to –10°C/min may involve some beneficial effects of dehydration, which minimizes the amount of water that freezes intracellularly. Slow cooling permits the flow of water from the cells to the outside, thereby promoting extracellular ice formation instead of lethal intracellular freezing. It is generally agreed that upon extracellular freezing the cytoplasm will be effectively concentrated and plant cells will survive better when adequately dehydrated. Slow freezing is generally employed to freeze the tissues for which computer-programed freezers have been developed. Successful examples of cryopreservation of meristems include diverse species as peas, strawberry, potato, cassava, etc.

3. *Stepwise freezing:* This method combines the advantages of both rapid and slow methods. A slow freezing procedure down to –20 to –40°C, a stop for a period of time (approximately 30 min) and then additional rapid freezing to –196°C is done by plunging in liquid nitrogen. A slow freezing procedure initially to –20 to –40°C permits protective dehydration of the cells. An additional rapid freezing in liquid nitrogen prevents the growing of big ice crystals in the biochemically important structures. The stepwise freezing method gives excellent results in strawberry and with suspension cultures.

Storage

Storage of frozen material at the correct temperature is as important as freezing. The frozen cells/tissues are immediately kept for storage at temperature ranging from –70°C to –196°C. A liquid nitrogen refrigerator running at –150°C in the vapor phase or –196°C in the liquid phase is ideal for this purpose. The temperature should be sufficiently low for long-term storage of cells to stop all metabolic activities and prevent biochemical injury. Long-term storage is best done at –196°C. Thus as long as regular supply of liquid nitrogen is ensured in liquid nitrogen refrigerator, it is possible to maintain the frozen material with little further care. The injury caused to cells could be due to freezing or storage. The latter occurs primarily when the cells are not stored at sufficiently low temperatures.

Thawing

Thawing is usually carried out by putting the ampoule containing the sample in a warm water bath (35–45°C) for thawing. Frozen tips of the sample in tubes or ampoules are plunged into the warm water with a vigorous swirling wrist action just to the point of ice disappearance. It is important for survival of the tissue that the tubes should not be left in the warm bath after the ice melts. Just at the point of thawing, quickly transfer the tubes to a water bath maintained at room temperature (20–25°C) and continue the swirling action for 15 sec to cool the warm walls of the tubes. Then continue with washing and culture transfer procedures. It is necessary to avoid excessive damage to the fragile thawed tissues or cells by minimizing the amount of handling at pre-growth stage.

Determination of survival/viability

Regrowth of plants from stored tissues or cells is the only realistic test of survival of plant materials. However, at any stage, the viability of frozen cells can be determined. Cell viability tests *viz.* FDA staining, growth measurements by employing parameters like cell number, packed cell volume, dry and fresh weight, mitotic index have already been discussed in Chapters 7 and 9 on cell suspension and protoplast isolation. Other staining methods of triphenyl tetrazolium chloride (TTC) and Evan's blue have been described.

TTC method of staining

In this method, cell survival is estimated by the amount of formazan produced as a result of reduction of TTC. This reaction results in a pink color. The procedure involves the following steps.

1. Buffer solution: 78% $Na_2HPO_4 \cdot 2H_2O$ solution 0.05 M (8.9 g/l): 22% KH_2PO_4 solution 0.05 M (6.8 g/l).
2. TTC solution: 0.18 M TTC dissolved in buffer solution.
3. About 150 mg of cell sample is put into 3 ml of TTC solution and incubated for 15 h at 30°C.
4. The TTC solution is drained off and the cells washed with distilled water.
5. Cells are centrifuged and extracted with 7 ml of ethanol (95%) in a water bath at 80°C for 5 min.

6. The extract is cooled and made to 10 ml volume with 95% ethanol.
7. The absorbance (pink color) is then recorded with a spectrophotometer at 530 nm.

The amount of formazan produced by the frozen cells is expressed as a percentage (survival) of formazan produced by the control cell suspension.

Evan's blue staining

One drop of 0.1% solution of solid Evan's blue in culture medium is added to one drop of cell suspension on a microscopic slide and observed under light microscope. Living cells remain unstained while dead cells are stained blue.

It should be noted that viability of cells based on staining reaction alone would be insufficient and misleading and cannot replace the plant regrowth experiments.

Plant growth and regeneration

Major biophysical changes occur in the cell during freezing and thawing. The freshly thawed cell is strongly prone to further damage and requires appropriate nursing. Regrowth of cryopreserved materials would depend upon the manner in which specimens are handled, such as plating the cells without washing away the cryoprotectants, incorporation of special additives in regrowth medium and the physical environmental conditions the cells are exposed to during the early stages of regrowth.

SLOW GROWTH METHOD

Cells or tissues can be stored at non-freezing conditions in a slow growth state rather than at the optimum level of growth. This technique is comparable to the traditional technique of Japanese gardeners known as Bonsai, in which trees are kept alive in small pots over hundreds of years. Growing processes are reduced to a minimum by limitation of a combination of factors like temperature, nutrition medium, hormones, etc. By this method, water in the tissue is maintained in the liquid condition, but all biochemical processes are delayed. All cultures in slow growth are static and therefore require only shelf space in suitable environments. This approach can be applied to tissue cultures mainly by following methods:

(a) *Temperature:* The tissues (shoot tips) can be cultured at low temperatures of 1–4°C. The explants such as callus, shoot tip cultures are established and maintained for 3–4 weeks under standard growth conditions. The explants are then transferred to lower temperatures of 1–4°C for cultures normally grown at 20–25°C and to 4–10°C for cultures normally grown at 30°C. A 16 h photoperiod with reduced lighting (~50 lux) may be beneficial.

(b) *Low oxygen pressure:* The oxygen pressure can be regulated by atmospheric or partial pressure.

(c) *Hormones*: It is possible to use plant growth regulators like succinic acid 2,2-dimethyl hydrazide (B-9), cycocel (CCC) and abscisic acid (ABA) to delay culture growth. These chemicals cause reduced growth, mainly shorter internodes of the plants.

(d) *Osmotic inhibitors*: Cultures can be grown on a medium supplemented with 3–6% (w/v) mannitol or high levels of sucrose as an osmotic inhibitor which slows down the growth.

By this method, virus free strawberry plants were maintained for 6 years at 4°C. Grape plants have been stored for 15 years at 9°C by yearly transfer to fresh medium. This method holds great promise in the industry employing micropropagation techniques. During the period of low demand for a particular genotype the cultures may be simply shelved in refrigerator to avoid frequent subculturing.

APPLICATIONS

1. *Conservation of genetic material:* Large number of plant species have been successfully maintained by means of cryopreservation of cultured embryos, tis-

Table 11.1 A list of tropical plants cryopreserved in various forms

Cell suspensions	Somatic embryos
Berberis dictyophilla	*Citrus sinensis*
Capsicum annum	*Coffea arabica*
Glycine max	*Daucus carota*
Nicotiana plumbaginifolia	
Nicotiana tabacum	
Oryza sativa	
Zea mays	
Protoplasts	**Pollen embryos**
Glycine max	*Arachis hypogea*
Nicotiana tabacum	*Citrus* spp.
Zea mays	*Nicotiana tabacum*
	Atropa belladona
Callus	**Zygotic embryos**
Gossypium arboreum	*Camellia sinensis*
Oryza sativa	*Manihot esculenta*
Saccharum spp.	*Zea mays*
Capsicum annum	*Hordeum vulgare*
	Triticum aestivum
Meristems	
Arachis hypogea	
Cicer arietinum	
Solanum tuberosum	

sues, cells or protoplasts. A list of cryopreserved tropical plants has been shown in Table 11.1.

2. *Freeze storage of cell cultures:* A cell line to be maintained has to be subcultured and transferred periodically and repeatedly over an extended period of time. It also requires much space and manpower. Freeze preservation is an ideal approach to suppress cell division and to avoid the need for periodical subculturing.

3. *Maintenance of disease free stocks:* Pathogen-free stocks of rare plant materials could be frozen, revived and propagated when needed. This method would be ideal for international exchange of such materials.

4. *Cold acclimation and frost resistance:* Tissue cultures would provide a suitable material for selection of cold resistant mutant cell lines, which could later differentiates into frost resistant plants.

Further reading

Bhojwani, S.S. and Razdan, M.K. 1990. *Plant Tissue Culture: Applications and Limitations.* Elsevier Science Publishers, Amsterdam.

Pierik, R.L.M. 1987. *In vitro Culture of Higher Plants.* Martinus Nijhoff Publishers, Dordrecht.

Reinert, J. and Bajaj, P.P.S. (eds.). 1977. *Applied and Fundamental Aspects of Plant Cell, Tissue and Organ Culture.* Springer-Verlag, Berlin.

Thorpe, T.A. 1981. *Plant Tissue Culture: Methods and Applications in Agriculture.* Academic Press Inc., New York.

Vasil, I.K. 1984. *Cell Culture and Somatic Cell Genetics of Plants: Laboratory Procedures and Their Applications.* Academic Press, Inc., New York.

PART II

Genetic Material and Its Organization

12

Genetic Material

Cell is the basic unit of life. All living organisms including bacteria, plants, and animals are built from cells. Generally cells are very small, with diameters much less than 1 mm, so they are invisible to the naked eye. In the simplest cells, bacteria, a cell wall surrounds a very thin fatty acid containing outer plasma membrane, which in turn surrounds a superficially unstructured inner region. Within this inner region is located the bacterial DNA that carries the genetic information. These nuclei-free bacteria and their close relatives are known as prokaryotes. Thus, in prokaryotes, nuclear material is not separated from the cytoplasm by a discrete membrane. These also lack an extensive cellular architecture: membranous organelles are absent and the genetic material is not enclosed in a distinct structure (Table 12.1). In virtually all cells other than bacteria, the inner cellular mass is partitioned into a membrane-bound spherical body called nucleus and an outer surrounding cytoplasm. In the nucleus is located the cellular DNA in the form of coiled rods known as chromosomes. Cells that contain a nucleus and with other distinctive membranous organelles such as mitochondria, vesicles, and Golgi bodies are referred to as eukaryotes. In the cells of eukaryotes, a nuclear membrane separates the genetic material from the cytoplasm, which is then further subdivided by other distinct membranous structures.

Eukaryotes can be unicellular or multicellular and comprise all macroscopic forms of life such as plants, animals, and fungi. Plant cells differ from animal cells in the presence of outer cell wall surrounding the cell membrane and the chloroplasts, whereas animal cells contain centrosomes, which are involved in cell division and lysosomes.

The essence of a cell is its ability to grow and divide to produce progeny cells, which are likewise capable of generating new cellular molecules and replicating themselves. Thus, cells are in effect tiny factories that grow by taking up simple molecular building blocks. Considering the complexity of a living organism, one is struck by the relatively small number of organic components used in the formation of substances. Cells contain molecules of great variety, but a living cell is composed of restricted set of elements, six of which (C, H, N, O, P, S) make up more than 99% of its weight. Certain simple combinations of atoms such as methyl ($-CH_3$), hydroxyl ($-OH$), carboxyl ($-COOH$), and amino ($-NH_2$) groups occur repeatedly in biological molecules. Each such group has distinct chemical and physical properties. Molecules can be broadly grouped into two categories.

1. *Small organic molecules*: These are compounds with molecular weight in the 100–1000 range containing up to 30 or so carbon atoms. Molecules of this kind are usually found free in solution in the cytoplasm where

Table 12.1 Characteristics of pro- and eukaryotic organisms

Feature	Prokaryote	Eukaryote
Structural unit	Bacteria, blue green algae	Unicellular organisms, Multicellular plants and animals
Size	<1–2 × 1–4 μm	>5 μm in width or diameter
Location of genetic material	Nucleoid, chromatin body, or nuclear material	Nucleus, mitochondria and chloroplasts
Genetic material	Naked DNA	DNA complexed with proteins
Chromosomes	1	>1
Nuclear membrane	Absent	Present
Histone proteins	Absent	Present
Nucleolus	Absent	Present
Cell wall	Mainly consist of carbohydrates and amino acid complex	Mainly consist of cellulose, hemicellulose, pectin, etc.
Membrane-bound organelles: mitochondria, endoplasmic reticulum, etc.	Absent	Present
Ribosome	70S structure	80S structure
Site of electron transport	Cell membrane	Organelles
Genetic exchange mechanisms	Conjugation, transformation, transduction	Gamete fusion
Cytoplasmic movement	Absent	Present
Cell division	Binary fission	Mitosis
Life-cycle	Short	Long

they form a pool of intermediates from which large molecules called macromolecules are made. Some of the small organic molecules are sugars, amino acids and fatty acids. At least 750 different types of small molecules are found.

2. *Macromolecules:* These are invariably polymeric molecules that are formed by joining together specific types of small molecules. A large polymeric molecule, or more than one polymer associated into a large molecule can be referred to as macromolecule. Some of the macromolecules such as carbohydrates, proteins, and nucleic acids are most important.

Three simple molecules—sugars, amino acids and nucleotides—which form macromolecules have been briefly described.

SUGARS

The simple monosaccharide sugars are molecules composed of carbon, hydrogen, and oxygen, which participate in the proportion $(CH_2O)_n$ and for this reason they are called carbohydrates (hydrates of carbon). All sugars contain hydroxyl groups (–OH) and either an aldehyde or ketone group. The hydroxyl group of one sugar can combine with aldehyde or ketone group of a second sugar with the elimination of water to form a disaccharide. The addition of more monosaccharides in the same way results in oligosaccharides and polysaccharides.

Glucose, a hexose monosaccharide, is the principal food component of many cells. Units of glucose are capable of polymerizing to form large polysaccharide molecules such as cellulose and starch in plants. Starch is the major source of food for humans. Cellulose is the most abundant molecule on earth, as it is the main constituent of plant fiber and wood. Besides, glucose, ribose and other simple sugars, derivatives of sugars such as uronic acids, amino sugars, or acetylated sugars may also form polymers. These polysaccharides play different role in organisms.

AMINO ACIDS

Amino acids are chemically varied but they all contain a carboxyl acid group and amino group both linked to a single carbon atom. Proteins are long linear polymers of amino acids joined head to tail by a peptide bond between the carboxyl acid group of one amino acid and amino group of the next amino acid. There are twenty amino acids in proteins, each with a different side chain attached to α-carbon atom. All amino acids contain nitrogen atoms; therefore, the proteins represent a class of nitrogenous compounds. The properties of amino acid side chains in aggregates determine the properties of proteins they constitute. This results in diverse structure and sophisticated functions of proteins. Specific terms commonly used to refer to different levels of protein structure are given below:

Primary structure: The sequence of amino acids linked by peptide bonds along the polypeptide chain is the primary structure.

Secondary structure: An ordered structure of a protein in which the polypeptide chain is folded in a regular manner with a characteristic helical or pleated structure. Much of the folding is the result of linking carbonyl and amide groups of the peptide chain backbone by hydrogen bonds. In many proteins the hydrogen bonding produces a regular coiled arrangement called the α helix.

Tertiary structure: A final structure of a protein in three-dimensional form from a single polypeptide chain. The final conformation of a protein from a single polypeptide, which is bent or tightly folded and compact is determined by a variety of interactions including hydrogen bonds, salt bonds, hydrophobic or nonpolar bonds and van der Walls forces. It is by virtue of their tertiary structure that proteins adopt a globular shape.

Quaternary structure: Used for proteins that contain more than a single polypeptide chain to indicate the spatial relationships among the separate chains or subunits. It means how the two or more chains are arranged in relation to each other. Most larger proteins contain two or more polypeptide chains, between which there are usually no covalent linkages. Proteins with two or more polypeptide chains are known as oligomeric proteins. A well-known example of an oligomeric protein is hemoglobin.

NUCLEOTIDES

Nucleotides are subunits of DNA (deoxyribonucleic acid) and RNA (ribonucleic acid). The basic control and functioning of the organism is exercised by DNA and RNA. There are no molecules as important as nucleic acids, as they carry within their structure the hereditary information that determines the structure of proteins. The instructions that direct cells to grow and divide are encoded by them. In prokaryotes, DNA is present as a naked molecule, but in eukaryotes, DNA is complexed with proteins to form chromosomes. In some viruses, RNA is the genetic material. These DNA and RNA consist of basic monomer units called nucleotides, which are linked by phosphodiester bridges to form the linear polymers. The nucleotides in turn consist of three molecular compounds: a phosphate radical, a ribose sugar and a nitrogenous base (Fig. 12.1).

Five-carbon sugars are called pentoses. RNA contains the pentose D-ribose, whereas 2-deoxyribose is found in DNA. In both cases it is a five-membered ring known as *furanose*: D-ribofuranose for RNA and 2-deoxy-D-ribofuranose for DNA. When these ribofuranoses are found in nucleotides, their atoms are numbered as 1′, 2′ 3′ and so on to distinguish them from the ring atoms of nitrogenous bases.

Nitrogenous bases are ring compounds, which can combine with H^+. They are of two types: (1) Pyrimidines: These are single derivatives of a six-membered pyrimidine ring (C: cytosine, T: thymine, and U: uracil), and various pyrimidine derivatives such as dihydrouracil are present as minor constituents in certain RNA molecules, and (2) Purines: They have a second five-membered ring fused to a

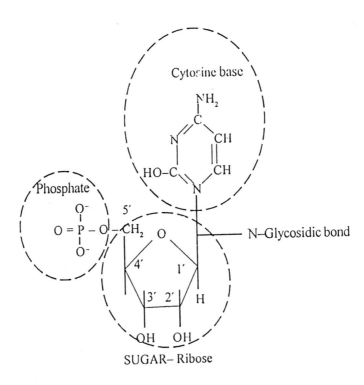

Fig. 12.1 A nucleotide.

six-membered ring (A: adenine, G: guanine). Adenine (6-amino purine) and guanine (2-amino-6-oxy purine) are the common purines found in both DNA and RNA. Other naturally occurring purine derivatives include hypoxanthine, xanthine, and uric acid. Hypoxanthine and xanthine are found rarely in nucleic acids whereas uric acid, the most oxidized state for a purine derivative, is never found in nucleic acids. The nitrogenous bases are numbered in a clockwise fashion in the usual system for organic compounds (Fig. 12.2).

Structural formulas of nucleotides

One may find that the structural formulas of bases are written in different forms. This is a result of their so-called tautomeric behavior. For instance, guanine can be written in two forms, as shown in Fig. 12.2.

Nomenclature of nucleoside and nucleotide compounds

When the ribose and nitrogenous base complex is separated from the phosphate group, the remaining structure is called a nucleoside. So a phosphonucleoside is a nucleotide

Base + Sugar = Nucleoside
Base + Sugar + Phosphate = Nucleotide

So nucleosides are compounds formed when a base is linked to a sugar by a glycosidic bond (Fig. 12.1). Glycosidic bonds involve the carbonyl carbon atom of the sugar, which in cyclic structures is joined to the ring O atom. Such carbon atoms are called anomeric. In nucleosides, the bond is an N-glycoside because it connects the anomeric C-1′ to the N-1 of a pyrimidine or to the N-9 of a purine.

Nucleotides may contain several phosphate

Fig. 12.2 The nitrogenous bases involved in DNA and RNA formation.

residues, but are present in DNA and RNA only as monophosphates. The letter M in the compound name signifies monophosphate, D means di-, and T is triphosphate. The conventional nomenclature is shown in Table 12.2. Nucleosides are named by adding the ending -*idine* to the root name of a primidine or -*osine* to the root name of a purine. The common nucleosides are thus cytidine, uridine, thymidine, adenosine, and guanosine. Because thymine was originally thought to occur only in DNA, the use of the term *deoxythymine* is considered to be redundant. Both TMP and dTMP are currently used. The nucleoside formed by hypoxanthine and ribose is inosine.

Polynucleotides

The next stage in building up the structure of a DNA molecule is to link the individual nucleotides together to form a polymer. This polymer is called a polynucleotide and is formed by attaching one nucleotide to another by way of a phosphate group. The nucleotide monomers are linked together by joining the phosphate (α) group attached to the 5′ carbon of one nucleotide to the 3′ carbon of the next nucleotide in the chain. Normally a polynucleotide is built up from nucleoside triphosphate subunits, so during polymerization the iPP (β and γ phosphates) are cleaved off. The hydroxyl group attached to the 3′ carbon of the second nucleotide is also lost. The linkage between the nucleotides in a polynucleotide is called a phosphodiester bond: phospho indicating the presence of a phosphorus

atom and diester referring to the two ester (C–O–P) bonds in each linkage. To be precise, this should be called a 3′–5′ phosphodiester bond so that there is no confusion about which carbon atoms in the sugar participate in the bond (Fig. 12.3). Polymers of ribonucleotides are named ribonucleic acid or RNA whereas deoxyribonucleotide polymers are called deoxyribonucleic acid (DNA). C-1′ and C-4′ in deoxyribonucleotides are involved in furanose ring formation and because there is no 2′-OH, only 3′ and 5′ hydroxyl groups are available for internucleotide phosphodiester bonds.

In polynucleotides, the two ends of the molecule are not the same. The top of the polynucleotide ends with a nucleotide in which the triphosphate group attached to the 5′-carbon has not participated in a phosphodiester bond and the β and γ phosphates are still in place. This end is called the 5′ or 5′-P terminus. At the other end of the molecule the unreacted group is not the phosphate but the 3′-hydroxyl. This end is called the 3′ or 3′-OH terminus. The chemical distinction between the two ends means that polynucleotides have a direction, which can be looked on as 5′→ 3′ or 3′→ 5′. The direction of the polynucleotide is very important in molecular biology.

There are no chemical restrictions on the order in which the nucleotides can join together to form an individual DNA polynucleotide. At any point in the chain the nucleotide could be A, G, C, or T. Consider a polynucleotide just 10 nucleotides in length: it could have any one of

Table 12.2 Nomenclature of compounds

Base	Nucleoside	Nucleotide
RNA		
Adenine (A)	Adenosine	Adenylate (AMP)
Guanine (G)	Guanosine	Guanylate (GMP)
Uracil (U)	Uridine	Uridylate (UMP)
Cytosine (C)	Cytidine	Cytidylate (CMP)
DNA		
Deoxyadenine (A)	Deoxyadenosine	Deoxyadenylate (dAMP)
Deoxyguanine (G)	Deoxyguanosine	Deoxyguanylate (dGMP)
Deoxythymidine or Thymine (T)	Deoxythymidine or Thymidine	Deoxythymidylate or Thymidylate (dTMP)
Deoxycytosine (C)	Deoxycytidine	Deoxycytidylate (dCMP)

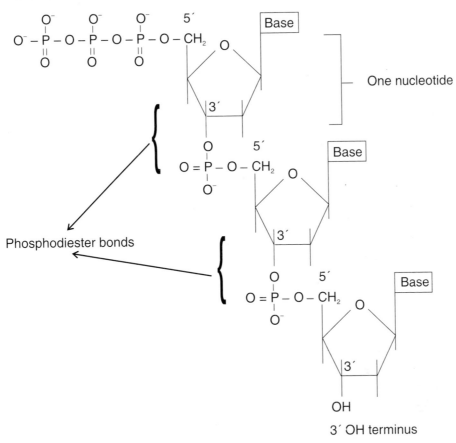

5′ P terminus

One nucleotide

Phosphodiester bonds

3′ OH terminus

Fig. 12.3 Structure of a short polynucleotide showing phosphodiester bonds.

$4^{10} = 1048576$ different sequences. Now imagine the number of different sequences possible for polynucleotide 1000 nucleotides in length, or one million. It is the variability of DNA that enables the genetic material to exist in an almost infinite number of forms.

RNA is also a polynucleotide but with two important differences from DNA

1. The sugar in RNA is ribose.
2. RNA contains uracil instead of thymine. Three other nitrogenous bases adenine, guanine, and cytosine are found in both DNA and RNA. In DNA, the fourth base is thymine instead of uracil.

RNA in the cell usually exists as a single polynucleotide. The structure of RNA polynucleotide is exactly similar to DNA. DNA is invariably in the form of two polynucleotides wrapped around one another to form the famous double helix.

Significance of differences between DNA and RNA

Due to the chemical differences between DNA and RNA, the greater stability of DNA has been postulated.

DNA contains thymine instead of uracil. The key observation is that cytosine deaminates to form uracil at a finite rate *in vivo*. Because C in one DNA strand pairs with G in the other strand, whereas U would pair with A in the DNA. Conversion of C to U would result in a heritable change of a CG pair to a UA pair. Such a change in nucleotide sequence would constitute a mutation in the DNA. To prevent this, a cellular proof reading mechanism exists, and U arising from deamination of C is treated as inappropriate and is replaced by C. Now, if DNA normally contained U rather than T, the repair system of the host could not really distinguish U formed by C deamination from U correctly paired with A. However, the U in DNA is 5-methyl U, or as it is commonly known, thymine. That is, the 5-methyl group on T labels it as if to say this U belongs; do not replace it.

The ribose 2′-OH group of RNA is absent in DNA. Consequently, the ubiquitous 3′-O of polynucleotide backbones lacks a vicinal hydroxyl neighbor in the DNA. This difference leads to a greater resistance of DNA to alkaline hydrolysis. In other words, RNA is less stable than DNA because its vicinal 2′-OH group makes the 3′-phosphodiester bond susceptible to nucleophilic cleavage.

Shorthand notation for polynucleotide structures

In a polynucleotide structure, repetitious uniformity exists in which the chain can be visualized as running from 5′ to 3′ along the atoms of one furanose and thence across the phosphodiester bridge to the furanose of the next nucleotide in line. Thus, this backbone can be portrayed by the symbol of a vertical line representing the furanose and a slash representing the phosphodiester link (Fig. 12.4). The diagonal slash runs from the middle of a furanose line to the bottom of an adjacent one to indicate 3′- (middle) to 5′- (bottom) carbon atoms of neighboring furanoses joined by the phosphodiester bridge. The base attached to each furanose is indicated above it by one-letter designations: A, C, G, or T (U). The convention in all notations of nucleic acids structure is to

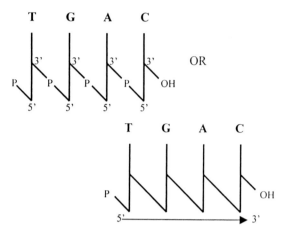

Fig. 12.4 Notation for polynucleotide structres. Furanoses are represented by lines, and phosphodiesters by diagonal slashes.

read the polynucleotide chain from the 5′ end to the 3′ end.

It is known that bases are not part of the sugar–phosphate backbone, but instead serve as distinctive side chains, as the R groups of amino acids along a polypeptide backbone. A simple notation of these structures is to list the order of bases in the polynucleotide using single capital letters—A, G, C, and T (or U). Occasionally, a lowercase 'p' is written between each successive base to indicate the phosphodiester-bridge as in GpApGpCpTpA. A 'p' preceding the sequence indicates that the nucleic acid carries a PO_4 on its 5′ end as in pGpApGpCpTpA; a 'p' terminating the sequence denotes the presence of a phosphate in the 3′-OH end as in GpApGpCpTpApAp.

A more common method representation of nucleotide sequence is to omit the 'p' and write only the order of bases such as GAGCTA. This notation assumes the presence of phosphodiesters joining adjacent nucleotides. The presence of 3′- or 5′ phosphate termini, however, must still be specified as in GAGCTAp for a 3′ PO_4 terminus. To distinguish DNA from RNA sequences, DNA sequences are typically preceded by a lowercase 'd' to denote deoxy as in d-GAGCTA.

GENETIC MATERIAL

Mendel's experiments in 1865-66 on peas led him to conclude that traits of the pea were under the control of two distinct factors (later named genes), one coming from the male parent and the other from the female parent. About the same time, it became known that heredity is transmitted through the egg and sperm, and since a sperm contains little cytoplasm in comparison to the amount of material in its nucleus, the obvious conclusion was that nucleus carries the hereditary determinants of a cell. Soon afterwards, the chromosomes within the nucleus were made visible through the means of special dyes. The cells of members of a given species were seen to contain a constant number of chromosomes. A few years later, the number of chromosomes in the sex cells of male and female gametes were shown to be exactly *n* (the haploid number), half the *2n* (diploid number). The chromosome number is reduced to *n* during gamete formation and fertilization restores the *2n* chromosome number. Chromosomes thus behave exactly as they should if they were the bearers of Mendel's genes.

Several lines of indirect evidence have long suggested that DNA contains the genetic information of the living organisms. Results obtained from several different experiments showed that most of the DNA is located in chromosomes, whereas RNA and proteins are abundant in the cytoplasm. Moreover, a correlation exists between the amount of DNA per cell and the number of sets of chromosomes per cell.

Discovery of DNA

The discovery of the double helical structure of DNA by James Watson and Francis Crick at Cambridge in 1953 was one of the greatest triumphs in biology as a whole. But the structure of polynucleotide was already known when Watson and Crick came to the scene. Besides, certain other observations made at that time helped Watson and Crick to elucidate its structure. These have been briefly mentioned.

Friedrich Miescher (1868) isolated cells that contained an unusual phosphorus containing compound, which he named "nuclein" (today's nucleoprotein).

Miescher (1872) reported that sperm heads, isolated from salmon sperm, contained an acidic compound (today's nucleic acid) and a base which he named "protamine."

Altman (1889) described a method for preparing protein-free nucleic acids from animal tissue and yeast. At this stage it was recognized that nucleic acids in the organisms are enmeshed with proteins.

Chargaff revealed a simple mathematical relationship between the proportion of bases in any one sample of DNA. The data of Chargaff showed that the four bases commonly found in DNA (A, C, G, and T) do not occur in equimolar amounts and that the relative amounts of each vary from species to species. However, certain pairs of bases, namely, adenine residues, equal the number of thymine, and the number of guanine equals the number of cytosine, i.e. A = T and G = C (Chargaff, 1951). It means total purines (A+G) will equal the total pyrimidine (T+C), i.e. there is a 1: 1 ratio. These findings are known as *Chargaff's rules:* [A] = [T]; [C] = [G]; [pyrimidines] = [purines].

By the early 1950s, X-ray studies on DNA by Wilkins, Franklin, and others indicated a well-organized, multiple-stranded fiber about 22 Å in diameter that was also characterized by the presence of groups spaced 3.4 Å apart along the fiber and a repeating unit every 34 Å.

Taking into account the facts known at that time, Watson and Crick in 1953 proposed a double helix structure of DNA as shown in Fig. 12.5 (Watson and Crick, 1953). According to them,

1. DNA double helix comprises two polynucleotide chains.

2. The nitrogenous bases are stacked on the inside of the helix, while the sugar–phosphate forming the backbone of the molecule is on the outside.

3. The two polynucleotide chains interact by

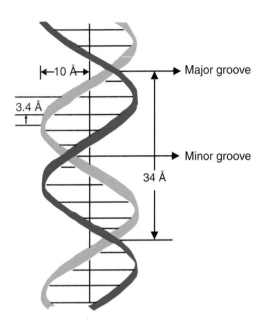

Fig. 12.5 The Watson-Crick model of DNA double helix.

hydrogen bonding. There is a very strict combining rule for hydrogen bonding

4. Ten base pairs occur per turn of the helix. The pitch of the helix is 34 Å, meaning that the spacing between the adjacent base pairs is 3.4 Å. The helix diameter is 20 Å.

5. The two strands of the double helix are antiparallel, with one polynucleotide chain running in 5′→3′ direction and the other chain in 3′→5′ direction. Only antiparallel polynucleotides form a stable double helix.

6. The double helix has two different grooves: major and minor. Thus, the helix is not absolutely regular. This feature is important in the interaction between the double helix and proteins involved in DNA replication and expression.

7. The double helix is right handed.

8. The two polynucleotide chains base pairs are complementary. The sequence of one determines the sequence of the other. This complementary base pairing is of fundamental importance in molecular biology. The rule is that A base pairs with T and forms two

hydrogen bonds, and G base pairs with C and forms three hydrogen bonds. Other base pairs are not normally permitted in the double helix as they will be either too large (purine–purine) or too small (pyrimidine–pyrimidine), or will not align in a manner allowing hydrogen bonding to occur (as with A–C and G–T).

Double helix is a stable structure

Several factors account for the stability of the double helical structure of DNA

i. Both internal and external hydrogen bonds stabilize the double helix. In DNA, H-bonds form between A:T and C:G bases, whereas polar atoms in the sugar-phosphate backbone form external H bonds with surrounding water molecules.

ii. The negatively charged phosphate groups are all situated on the exterior surface of the helix in such a way that they have minimal effect on one another and are free to interact electrostatically with cations in solution, such as Mg^{++}.

iii. The core of the helix, which consists of base pairs, is bonded together with hydrophobic interactions and van der Walls forces besides H bonding.

DNA replication

If genetic information is to be faithfully inherited from generation to generation, the nucleotide sequence of DNA must be faithfully copied from the original parental DNA molecule to the two daughter molecules. The clue as to how this duplication is achieved lies in the double helical structure of the DNA molecules. Since complementary DNA strands are held together by hydrogen bonding, their separation can be readily achieved and when separated, each strand serves as a template for the synthesis of a complementary copy. After replication, the original (parental) DNA duplex will give rise to two identical daughter duplex molecules.

Following DNA replication, one strand of a daughter molecule has been conserved from the

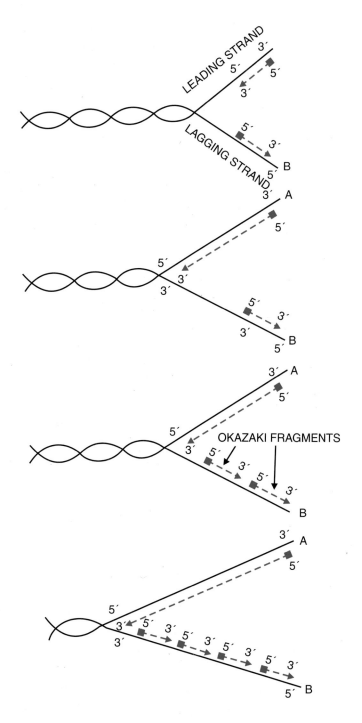

Fig. 12.6 DNA replication showing sequence of unwinding and replication at the fork of a DNA double helix.

parental duplex, while the other strand has been newly synthesized. This mode of replication is called semiconservative. Although semiconservative replication requires strand separation, the disruption of the double helical structure occurs only transiently and only a small part of a chromosome is single stranded at any given time. DNA replication is extremely complex and involves the coordinated participation of many proteins with many different enzymatic activities.

When a DNA molecule is being replicated, only a limited region is in a non base-paired form. The breakage in base pairing starts at a distinct position called the replicating origin and gradually progresses along the molecule in both the directions. The important region at which the base pairs of the parent molecule are broken and the new polynucleotides are synthesized is called the replication fork.

An enzyme able to synthesize a new DNA polynucleotide is called a DNA polymerase. In 1959, Arthur Kornberg and his colleagues isolated from *E. coli* an enzyme capable of DNA synthesis on a polynucleotide template and named it as polymerase I. But later studies revealed that DNA polymerase III and not DNA polymerase I is the primary enzyme involved in DNA replication. There are two main properties of DNA polymerase III.

1. It can synthesize DNA only in the 5'→3' direction.
2. It cannot initiate DNA synthesis unless there is already a short double-stranded region to act as a primer.

During DNA replication there is a breakage of base pairing between the two strands of the parent DNA molecule, which is carried out by helicase enzyme in conjunction with single-strand binding proteins (SSBs). This prevents the two strands from immediate reannealing. The result is the replication fork that provides the template on which DNA polymerase III can work (Fig. 12.6).

Since DNA polymerase III can synthesize only in the 5'→3' direction, it means that the template strands must be read in the 3'→5' direction. Thus, one strand is called the leading strand, where new polynucleotides can be synthesized continuously. However, the second strand where DNA strand cannot be copied continuously is called a lagging strand. This strand has to be replicated in sections: a portion of the parent helix is dissociated and a short stretch of lagging strand replicated, a bit more of the helix dissociated and another segment of the lagging strand replicated and so on. These are called Okazaki fragments, which are 100 to 1000 nucleotides in length.

DNA polymerase III cannot initiate DNA replication. To initiate DNA replication, RNA polymerase first synthesizes a short primer of 6 to 30 ribonucleotides on the template DNA strand. Once replication has been initiated, then DNA polymerase III continues DNA replication. On the lagging strand, DNA polymerase III can synthesize DNA only for a short distance before it reaches the RNA primer at the 5' end of the next Okazaki fragment. At this point, DNA polymerase III stops and DNA polymerase I carries out the functions of removing the ribonucleotides of the primer of the adjacent Okazaki fragment and replacing them with deoxyribonucleotides. When all the ribonucleotides are replaced, DNA polymerase I either stops or possibly carries on a short distance into the short region of the Okazaki fragment, continuing to replace nucleotides before it dissociates from the new double helix. Finally the adjacent Okazaki fragments, which are separated by a single gap between neighboring nucleotides, are sealed by ligase enzyme, which forms a phosphodiester bond.

Further reading

Garret, R.H. and Grisham, C.M. 1999. *Biochemistry.* Saunders College Publishing, Orlando, Florida.

Lewin, B. 1995. *Genes V.* Oxford University Press, Oxford, UK.

Snustad, D.P., Simmons, M.J. and Jenkins, J.B. 1997. *Principles of Genetics.* John Wiley & Sons, New York.

13

Organization of DNA and Gene Expression

When the double helix of DNA was found, it was believed that many distinct molecules are present in a chromosome. But this picture has suddenly changed due to the realization that long DNA molecules are fragile and easily breakable into smaller fragments. By the late 1950s, when great care was taken to prevent shearing of DNA, it was found that DNA molecules contained as many as 200,000 base pairs. But now it is known that a single DNA molecule of *E. coli* contains more than 4 million base pairs. Likewise, a single DNA molecule is thought to exist within each of the chromosome of higher plants. These chromosomes contain an average some 20 times more DNA than present in the *E. coli* chromosome. A chromosome is thus properly defined as a single genetically specific DNA molecule to which are attached large number of proteins that are involved in maintaining chromosome structure and regulating gene expression.

DIFFERENT FORMS OF DNA

DNA molecules can have more than one conformation in solution, depending on the pH and ionic environment. Under physiological conditions, the dominant form of DNA is the B-form. The rare A-DNA exists only in a dehydrated state. The A-form differs from the B-form in relatively minor ways. A-DNA is still a double helix, but more compact than the B-form of DNA. The A form has 11 base pairs per turn of the helix and

a diameter of 23Å. The pitch or distance required to complete one helical turn in B-DNA is 3.4 nm whereas that in A-DNA it is 2.46 nm. The base pairs in A-DNA are no longer nearly perpendicular to the helix axis, but instead are tilted 19° with respect to this axis. Successive base pairs occur every 0.23 nm along the axis, as opposed to 0.332 nm in B-DNA. The B-DNA is thus longer and thinner than the short, squat A-form, which has its base pairs displaced around, rather then centered on the helix axis (Table 13.1). It has been shown that relatively dehydrated DNA fibers can adopt the A-conformation under physiological conditions, but it is still unclear whether DNA ever assumes this form *in vivo*. However, double-helical DNA:RNA hybrids probably have an A-like conformation. The 2'-OH in RNA sterically prevents double-helical regions of RNA chains from adopting the B-form helical arrangement. Importantly, the double-stranded region in RNA chains assumes an A-like conformation, with their bases strongly tilted with respect to the helix axis. Since 1953, other forms of DNA have also been found: C, D, E, F, and Z, each with a slightly different conformation to the double helix. There is no strong evidence that C, D, E, and F conformations exist *in vivo*.

Z-DNA exists in vivo and is the only left-handed form of the helix. It takes its name from the zigzag path that the sugar-phosphate backbone follows. Z-DNA was first recognized by Alexander Rich and his associates in X-ray

Table 13.1 A comparison of the structural properties of A, B and Z-DNA

Character	A-DNA	B-DNA	Z-DNA
Overall proportions	Short and broad	Longer and thinner	Elongated and slim
Rise per base pair	2.3 Å	3.32 Å ± 0.19 Å	3.8 Å
Helix packing diameter	25.5 Å	23.7 Å	18.4 Å
Helix rotation sense	Right handed	Right handed	Left handed
Base pairs per helix repeat	1	1	2
Base pairs per turn of helix	~11	~10	12
Mean rotation per base pair	33.6 Å	35.9 ± 4.2°	−60°/2
Pitch per turn of helix	24.6 Å	33.2 Å	45.6 Å

Table adapted from Dickerson, R.L. *et al.* (1982). Cold Spring Harbor Symposium on Quantitative Biology **47**: 14

analysis of synthetic deoxynucleotide dCpGpCpGpCpG, which crystallized into an antiparallel double helix of unsuspected conformation (Fig. 13.1a). The alternating pyrimidine–purine (Py–Pu) sequence of this oligonucleotide is the key to its unusual properties. Because an alternate nucleotide assumes different conformations, the repeating unit on a given strand in the Z-helix is the dinucleotide; that is for any number of bases, *n*, along one strand, *n*–1 dinucleotides must be considered. For example, a GpCpGpC subset of sequence along one strand is com-

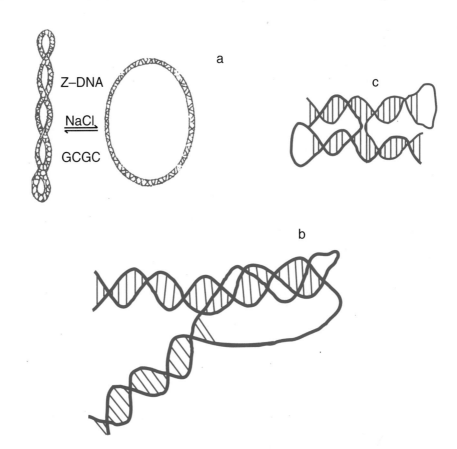

Fig. 13.1 DNA conformations. a: Z-DNA; b: the triple helix conformation of DNA; c: cruciform DNA.

prised of the successive dinucleotide units GpC, CpG, and GpC. In B-DNA, the nucleotide conformations are essentially uniform and the repeating unit is the mononucleotide. It follows that the CpG sequence is distinct conformationally from the GpC sequence along the alternating copolymer chains in the Z-double helix. The conformational alterations going from B to Z realign the sugar-phosphate backbone along a zigzag course that has a left-handed orientation, thus the designation Z-DNA. In any GpCpGp subset, the sugar-phosphate of GpC forms the horizontal "zig" whereas the CpG backbone forms the vertical "zag." The mean rotation angle circumscribed around the helix axis is −15° for a CpG step and −45° for a GpC step (giving − 60° for the dinucleotide repeat). The minus sign denotes a left handed or counterclockwise rotation about the helix axis. Z-DNA is more elongated and slimmer than B-DNA.

Segments of DNA in which purine and pyrimidine residues alternate are able to form DNA under certain conditions, and Z-DNA may be stabilized *in vivo* by binding Z-specific proteins. One possible biological role for left-handed DNA might be based on the fact that small regions of Z-DNA relieve the superhelical strain in a negatively supercoiled DNA molecule, in the same way that small regions of single-stranded DNA can. These local regions of Z-DNA may bind to regulatory proteins that are specific for Z-DNA.

The Z form can arise in sequences that are not strictly alternating Py–Pu. For example, the hexanucleotide ^{m5}CGAT^{m5}CG, a Py–Pu–Pu–Py–Py–Pu sequence containing two −methylcytosine (^{m5}C), crystallizes as Z-DNA. Indeed, the *in vivo* methylation of C at position 5 is believed to favor a B to Z switch because, in B-DNA, these hydrophobic methyl groups would protrude into the aqueous environment of the major groove and destabilize its structure. In Z-DNA, the same methyl group can form a stable hydrophobic patch. It is likely that the Z-conformation naturally occurs in specific regions of cellular DNA, which otherwise is predominantly in B-form. Further-

more because methylation is implicated in gene regulation, the occurrence of Z-DNA may affect the expression of genetic information.

DNA SUPERCOILING–TERTIARY STRUCTURE OF DNA

In vivo, the majority of DNA molecules does not have free ends, but is present as closed circular structures. When a closed circular duplex is twisted around its axis, supercoils appear in the DNA, rather like the coiling seen when a rubber band is twisted. When B-DNA is exposed to forces that twist it in a direction opposite to the twist of the double helix, it forms negative supercoils. This allows the DNA molecules to relieve the twisting force by loosening the winding of two strands around each other. Supercoiling can only occur in a closed circular structure, and the greater the number of supercoils, the greater the torsion in the closed duplex. It has become clear that the degree of supercoiling in cells is strongly controlled by specific enzymes (topoisomerases and gyrases), which are capable of either adding or subtracting super-coiled twist in the DNA. A DNA molecule cannot form a supercoil if one or both of the DNA strands are broken. If this occurs, the DNA supercoil unwinds to a relaxed form. Supercoils are classified in two ways. Negative supercoils twist the DNA around its axis opposite to the direction of the turns of right-handed helix. Positive supercoils twist the DNA in the same direction as the right-handed helix. It is thought that only negatively supercoiled DNA exists *in vivo*, although positively supercoiled DNA may be generated *in vitro*.

Linking number

The basic parameter characterizing supercoiled DNA is the linking number (L). This is the number of times the two strands are intertwined, and, provided both strands remain covalently intact, L cannot change. In a relaxed circular DNA duplex of 400 bp, L is 40 (assuming 10 bp per turn in B-DNA). The linking number for relaxed

DNA is usually taken as the reference parameter and is written as L_0. L can be equated to twist (T) and write (W) of the duplex, where twist is the number of helical turns and write is the number of supercoils:

$$L = T + W$$

In any closed circular DNA duplex that is relaxed, $W = 0$. A relaxed circular DNA of 400 bp has 40 helical turns, and $T = L = 40$. This linking number can only be changed by breaking one or both strands of the DNA, winding them tighter or looser, and rejoining the ends.

Values of T and W for various positively and negatively supercoiled circular DNAs are given below:

Positively supercoiling

$T = $ 0, $W = 0$ then $L = 0$
$T = $ +3, $W = 0$ then $L = +3$
$T = $ 2, $W = +1$ then $L = +3$
$T = $ +2, $W = +1$ then $L = +3$

Negative supercoiling

$T = $ 0, $W = 0$ then $L = 0$
$T = $ −3, $W = 0$ then $L = −3$
$T = $ −2, $W = −1$ then $L = −3$
$T = $ −2, $W = −1$ then $L = −3$

Negative supercoils favor local unwinding of the DNA, allowing processes such as transcription, DNA replication, and recombination. Negative supercoiling is also thought to favor the transition between B-DNA and Z-DNA, and moderate the interactions of DNA binding proteins involved in gene regulation.

CRUCIFORM—TERTIARY STRUCTURE OF DNA

Besides the conformations, DNA may also form unusual structures that appear to be important in regulating gene expression. These include a triple-helix conformation and a cross-shaped structure called cruciform DNA (Fig. 13.1 b and c). Triple helices are characterized experimentally by their susceptibility to digestion *in vitro* with S1 nuclease (an enzyme that specifically degrades regions of single stranded DNA). Cruciform DNA

takes its name from a locally formed cross-shaped segment. A DNA segment is only able to fold into a cruciform if the DNA sequence contains an inverted repeat or palindromic sequence. Palindromes are words, phrases or sentences that are the same when read backward or forward, such as "radar," "Madam, I'am Adam," or "Malayalam." Inverted repeat sequences (palindromes) are symmetrically repeated DNA sequences that are same in both chains of a double helix when read in the 5′ to 3′ direction. The example shows a double-stranded 16 base pair inverted repeat with the center of symmetry (dyad axis) indicated by a dash.

5′ TATACGAT-ATCGTATA 3′
3′ ATATGCTA-TAGCATAT 5′

In naturally occurring DNA sequences, the two halves of the inverted repeat are often separated by a noninverted repeat sequence. Inverted repeat sequences are present in both prokaryotic and eukaryotic genomes, and the cruciform structures have been postulated to be the sites recognized or induced by regulatory DNA-binding proteins.

EUKARYOTIC DNA ORGANIZATION INTO NUCLEOSOMES

A typical human cell is 20 μm in diameter. Its genetic material consists of 23 pairs of dsDNA molecules in the form of chromosomes. Thus, the average length of DNA would be 3×10^9 bp /23 or 1.3×10^8 bp. At 0.34 nm/bp in B-DNA, this represents a DNA molecule 5 cm long. If we consider all the 46 dsDNA molecules together it would be approximately 2 m of DNA that must be packaged into a nucleus of perhaps 5 μm in diameter. Hence, it is very clear that DNA must be condensed by a factor of 10^5.

In eukaryotes, the DNA molecules are tightly complexed with proteins to form chromatin, which is a nucleoprotein fiber with a beaded appearance. The key to the bead-like organization of chromatin is its histone proteins, which are divided into five main classes: H1, H2A, H2B,

H3, and H4. They come together to form nucleosome units containing an octomeric core composed of two molecules each of the H2A, H2B, H3, and H4 components Fig. 13.2. Around each octomeric histone core, 146 bp of B-DNA in a flat, left handed superhelical conformation makes 1.65 turns. Histone 1, a three-domain protein serves to seal the ends of the DNA turns to the nucleosome core and to organize the additional 60 bp of DNA that link consecutive nucleosomes.

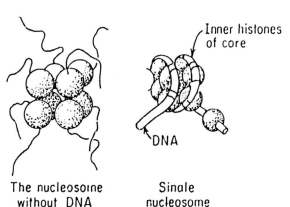

The nucleosome without DNA

Single nucleosome

Inner histones of core

DNA

Fig. 13.2 The core of nucleosome formed by eight histone molecules, made up of two molecules each of H2A, H2B, H3 and H4 joined at their globular portions with the ends of the proteins free (left). A single nucleosome with its inner core of histones, the DNA wrapped around the core and bound by the free ends of the histones, and a stretch of linker DNA to which is bound a single molecule of H1.

A higher order of chromatin structure is created when the nucleosomes, in their characteristic beads on a string motif, are wound in the fashion of a solenoid having six nucleosomes per turn. The resulting 30-nm filament contains about 1200 bp in each of its solenoid turns. Interactions between the respective H1 components of successive nucleosomes stabilize the 30-nm filament. This filament then forms long DNA loops of variable length, each containing, on an average, between 60,000 and 150,000 bp. Electron microscopic analysis of human chromo-

some 4 suggests that 18 such loops are then arranged radially about the circumference of a single turn to form a miniband unit of the chromosome. According to this model, approximately 10^6 of these minibands are arranged along a central axis in each of the chromatid of human chromosome 4 that forms at mitosis. Despite many studies, much of the higher order structure of chromosomes remains a mystery. Also found within chromatin are a large number of different DNA-binding proteins. These proteins are likely to play a variety of specific roles in regulating gene expression.

DNA CONTENT

The nuclear genome is the largest in the plant cell, both in terms of picograms of DNA and in the number of genes encoded. Nuclear DNA is packaged into chromosomes along with histones and non-histone proteins, all of which play an important role in gene expression. These various components are held together to form chromatin. The DNA encodes the genetic information, and the proteins are involved in controlling the packaging of DNA and regulating the availability of DNA for transcription. One process that is basic to all development is the replication of nuclear genetic material, and its equal distribution between two daughter cells during mitosis. The DNA contained in each of the daughter cells must be identical to that contained in the original cell.

The amount of DNA in the haploid genome of an eukaryotic cell is known as its C-value where the 1C amount is an unreplicated haploid DNA content. The nuclear DNA content as C-value varies widely in higher plants from 0.15 to over 200 pg. *Arabidopsis thaliana* has 0.15 pg/haploid genome to some members of the Loranthaceae (mistletoes), which have over 100 pg/haploid genome. Even the smallest plant genome is larger than that found in *Drosophila melanogaster* (165,000 kb) and contains much more DNA than is required to specify all the proteins synthesized during the course of development. The nuclear content and number of base

Table 13.2 Nuclear DNA content of plant species

Species	Common name	pg DNA/ 2C	Mbp/1C
Arabidopsis thaliana	Arabidopsis	0.30	145
Arachis hypogaea	Peanut/ groundnut	5.83	2813
Avena sativa	Oats	23.45	11315
Brassica campestris ssp *oleifera*	Turnip rape	0.97–1.07	468–516
Brassica juncea	Brown mustard	2.29	1105
Carica papaya	Papaya	0.77	372
Cicer arietinum	Chick pea	1.53	738
Citrus sinensis	Orange	0.76	367
Daucus carota	Carrot	0.98	473
Glycine max	Soybean	2.31	1115
Gossypium hirsutum	Cotton	4.39, 4.92	2118, 2374
Helianthus annus	Sunflower	5.95–6.61	2871–3189
Hordeum vulgare	Barley	10.10	4873
Lens culinaris	Lentil	8.42	4063
Lycopersicon esculentum	Tomato	1.88–2.07	907–1000
Oryza sativa	Rice	0.87–0.96	419– 463
Saccharum officinarum	Sugarcane	5.28–7.47	2547–3605
Solanum tuberosum	Potato	3.31–3.86	1597–1862
Sorgum bicolor	Sorghum	1.55	748
Triticum aestivum	Wheat	33.09	15966
Tulipa sp.	Garden tulip	51.2, 63.6	24704, 30687
Vigna mungo	Black gram/urd	1.19	574
Zea mays	Maize	4.75–5.63	2292–2716

1C: unreplicated haploid genome; 2C: Diploid genome;
1 picogram (pg) = 965 million or mega base pairs (Mbp)

pairs of some crop species have been given in Table 13.2 adapted from Arumuganathan and Earle (1991).

All higher eukaryotes contain significantly more DNA than what appears to be necessary to encode their structures and functions. For example, the human haploid genome has a C-value of about 3×10^6 kb, yet estimates of the number of genes in the human genome, calculated by a variety of means, suggest that there are only about 30,000 genes. If one takes 3 kb as a generous estimate for the coding sequence of an average gene, then there exists at least a 10-fold excess of DNA in the genome over that required to produce a human being. A comparison of values from other organisms show that pea, for example, contains 1000 times as the DNA of *E. coli*, which probably contains about 4000 genes. Thus, if the entire pea DNA were expressed in the conventional sense, it would represent four million genes. Although it is difficult to estimate the number of genes

necessary to account for the differences in biological complexity between pea and *E. coli*, a value as high as this seems extraordinary.

Estimates of the haploid DNA content of higher plant nuclei have also been made by studying the kinetics of DNA renaturation. Kinetic analysis shows that a large proportion of the DNA is composed of highly repetitive sequences and is unlikely to represent genes coding for proteins. Most of the protein-coding regions are represented only once per haploid genome and are therefore often referred to as unique or single copy sequences. The proportion of higher plant DNA present as single-copy sequences ranges from 20% to 80%, with a general tendency for plants with large genomes to have higher proportion of repetitive DNA.

Much of the genome comprises sequences that are repeated many times. The number of repeats can vary from a few to millions. It has been convenient to classify repeated sequences broadly according to their repetition frequency.

Sequences that are repeated at about 10^4 to 10^6 copies per haploid genome are highly repetitive and other repeated sequences are called moderately repetitive. Much of the highly repetitive DNA is composed of simple sequences of 2 to 200 bp repeated often in large tandem arrays, and these are known as satellite DNA. Satellites usually consist of long tandem arrays of thousands of related sequences. In many cases the repeats are short tandem repeats of oligonucleotides (2–6 nucleotides tandem repeats as microsatellites or tandem repeats of 6–200 bp as minisatellites), but there are examples of repeats of several thousand base pairs also. As observed by microscopy, many satellites are located around the centromeres and telomeres, where they appear as constrictions or densely staining regions called heterochromatin in the metaphase chromosome. They are usually not transcribed. Most eukaryotic genomes also contain discrete segments of DNA called transposons that are capable of moving from one part of a genome to another.

Plants with a small genome and a low proportion of repetitive DNA have short somatic and meiotic cycles and a short generation time. A good example is *Arabidopsis thaliana,* which is popular in plant molecular biology because of its low DNA content and five-week generation time. Amongst cereals, rice has a small genome of 415 Mbp compared to the wheat genome, which has 16 billion base pairs. In general, horticultural plants have a small genome size and this should facilitate molecular studies of these crops as is being done on rice.

Denaturation

When duplex DNA molecules are subjected to conditions of pH, temperature, or ionic strength that disrupt hydrogen bonds and result in the separation of two strands of DNA, is referred to as denaturation. If temperature is the denaturing agent, the double helix is said to melt. The course of dissociation of two strands of DNA can be followed spectrophotometrically because the relative absorbance of DNA solution at 260 nm increases as much as 40% as the bases unstuck. This absorbance increase or hyperchromic shift is because the aromatic bases in DNA interact via their π electron clouds when stacked together in the double helix. Because the UV absorbance of the bases is a consequence of π electron transitions, and because the potential for these transitions is diminished when the bases stack, the bases in duplex DNA absorb less 260 nm radiation than expected for their numbers. Unstacking alleviates the suppression of UV absorbance. The rise in absorbance coincides with strand separation, and the midpoint of the absorbance increase is termed the melting temperature, T_m. DNAs differ in their T_m values because they differ in relative G+C content of a DNA. The higher the G+C content of a DNA, the higher its melting temperature, because G:C pairs are held by three H bonds whereas A:T pairs have only two H bonds.

T_m is also dependent on the ionic strength of solution: lower the ionic strength, lower the melting temperature. At 0.2 M Na$^+$, T_m = 69.3 + 0.41 (% G + C). Ions suppress the electrostatic repulsion between the negatively charged phosphate groups in the complementary strands of the helix, thereby stabilizing it. DNA in pure water melts even at room temperature. At high concentration of ions, T_m is raised and the transition between helix and coil is sharp.

At pH values greater than 10, extensive deprotonation of the bases occurs, which destroys the hydrogen bonding potential and denatures the DNA double helix. Similarly, at pH values below 2.3, extensive protonation of bases occurs and disrupts the pairing of two strands of DNA. Alkali is generally preferred because it does not hydrolyze the glycosidic linkages in the sugar-phosphate backbone. Small solutes such as formamide and urea also act as denaturants for DNA as they readily form H bonds at temperatures below T_m, if present in a sufficiently high concentration to compete effectively with the H-bonding between the base pairs.

DNA renaturation

Denatured DNA will renature to reform the duplex structure if the denaturing conditions are removed (i.e. solution is cooled, pH is returned to neutrality, or the denaturants are diluted out). Renaturation requires reassociation of the DNA strands into a double helix, a process known as reannealing. This occurs when complementary bases in the two strands come together and the helix can be zippered up. The general nature of the eukaryotic genome can be assessed by the kinetics with which denatured DNA reassociates. Reassociation between complementary sequences of DNA occurs by base pairing, which is a reversal of denaturation by which they were separated.

Renaturation rate and DNA sequence complexity–*Cot* curves

The renaturation rate of DNA is an excellent indicator the sequence complexity of DNA. For example, T4 DNA contains about 2×10^4 base pairs, *E. coli* possesses 4.64×10^6 bp, whereas humans have 3.3×10^9 bp. *E. coli* DNA is considerably more complex as compared to T4 phage DNA in that it encodes more information. Expressed in another way, for any given amount of DNA (in grams), the sequences represented in an *E. coli* sample are more heterogeneous, i.e. more dissimilar from one another, than those in an equal weight of phage T4 DNA. Therefore, it will take the *E. coli* DNA longer to find out its complementary partners and reanneal. This renaturation of DNA can be analyzed quantitatively. The rate of reaction is governed by the equation

$$\frac{dC}{dt} = -kC^2,$$

where C is the concentration of single-stranded DNA at time t, and k is a second-order reassociation rate constant. Starting with an initial concentration of DNA, C_0 at time $t = 0$, the amount of single-stranded DNA remaining after time t is

$$\frac{C}{C_0} = \frac{1}{1 + k \cdot C_0 t}$$

where the unit of C is moles of nucleotides per liter and that for t is seconds.

This equation shows that the parameter controlling the reassociation reaction is the product of DNA concentration (C_0) and time of incubation (t) usually described as the **Cot**.

A useful parameter is derived by considering the conditions when the reaction is half complete at time $t_{1/2}$. Then,

$$\frac{C}{C_0} = \frac{1}{2} = \frac{1}{1 + k \cdot C_0 t_{1/2}}$$

Yielding

$$C_0 t_{1/2} = -1/k$$

The value required for half reassociation ($C_0 t_{1/2}$) is called $Cot_{1/2}$. Since the $Cot_{1/2}$ is the product of the concentration and time required to proceed halfway, a greater Cot implies a slower reaction. $Cot_{1/2}$ is given is moles nucleotides $\times$ seconds/liter as Cot value.

The reassociation of DNA usually is followed in the form of a cot curve, which plots the fraction of DNA that has reassociated ($1 - C/C_0$) against the log of the Cot. Figure 13.3 gives Cot curves for several genomes. The rate of reassociation can be followed spectrophotometrically by the UV absorbance decrease as duplex DNA is formed. Note that relatively more complex DNAs take longer to renature as reflected by their greater $Cot_{1/2}$. The form of each curve is similar, but the Cot required in each case is different. It is described by the $Cot_{1/2}$.

The $Cot_{1/2}$ is directly related to the amount of DNA in the genome. It means as the genome becomes more complex, there are fewer copies of any particular sequence within a given mass of DNA. For example, if the C_0 of DNA is 12 pg, then the bacterial genome whose size is 0.004 pg will contain 3000 copies of each sequence, whereas the eukaryotic genome of size 3 pg will contain only 4 copies of each sequence in an initial concentration of 12 pg of DNA. Thus the same absolute concentration of DNA measured in moles of nucleotides per liter (the C_0) will provide a concentration of each eukaryotic sequence.

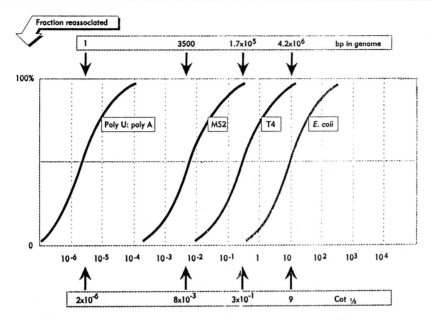

Fig. 13.3 *Cot* curves for different genomes. The form of each curve is similar but the *Cot* required in each case is different.

Since the rate of reassociation depends on the concentration of complementary sequences, for the eukaryotic sequences to be present at the same relative concentration as the bacterial sequences, it is necessary to have more DNA or to incubate the same amount of DNA for longer time. Thus the $Cot_{1/2}$ of the eukaryotic reaction is much greater than the $Cot_{1/2}$ of the bacterial reaction.

The $Cot_{1/2}$ of a reaction therefore indicates the total length of different sequences that are present. This is described as the complexity, usually given in base pairs.

The renaturation of the DNA of any genome should display a $Cot_{1/2}$ that is proportional to its complexity. Thus, the complexity of any DNA can be determined by comparing its $Cot_{1/2}$ with that of a standard DNA of known complexity. Usually, *E. coli* DNA is used as a standard. Its complexity is taken to be identical with the length of the genome (implying that every sequence in the *E. coli* genome of 4.2×10^6 bp is unique. So it can be written as

$$\frac{Cot_{1/2} \text{ (DNA of any genome)}}{Cot_{1/2} \text{ (}E. coli \text{ DNA)}} =$$

$$\frac{\text{Complexity of any genome}}{4.2 \times 10^6 \text{ bp}}$$

FLOW OF GENETIC INFORMATION: CENTRAL DOGMA

How is the genetic information held within DNA translated into protein products? In 1958, Francis Crick postulated that the biological information contained in the DNA of the gene is transferred first to RNA and then to protein. Crick also stated that the information is unidirectional, proteins cannot themselves direct the synthesis of RNA and RNA cannot direct the synthesis of DNA (Crick, 1958). However, the latter part of central dogma was shaken in 1970 when Temin and Baltimore (Baltimore, 1970; Temin and Mizutani, 1970) independently discovered that certain viruses do transfer biological information from RNA to DNA by an enzyme reverse transcriptase, which synthesizes DNA from an RNA template. This central dogma representing the flow of information from DNA to RNA is not unidirectional but it can be from RNA to DNA also (Fig. 13.4).

Fig. 13.4 Central dogma: ◄·········· indicates that reverse transcription was added later to the central dogma proposed by Crick in 1958.

ORGANIZATION OF GENES IN DNA MOLECULES

In the cells of higher organisms, all the genes are carried by a small number of chromosomes, each of which contains a single DNA molecule. Each of these DNA molecules must carry several hundreds of genes. **Gene** is a unit of inheritance. **Gene** can be defined as a segment of genome with a specific sequence of nucleotides, which has a specific biological function. Genes transcribe in to mRNA and are translated in to polypeptides. Ribosomel RNA (rRNA) and transfer RNA (tRNA) genes are transcribed but not translated. These are noncoding RNAs, which have various functions in the cell.

The part of a protein coding gene that is translated into protein is called open reading frame (ORF). The ORF is read in the 5′ to 3′ direction along the mRNA. The ORF starts with an initiation codon and ends with a termination codon (Fig. 13.5). The majority of genes are spaced more or less randomly along the length of a DNA molecule. In some cases, genes are grouped into distinct clusters. The clusters are made up of genes that contain related units of biological information. Examples of such cluster of genes are operons and multigene families.

Operon

Operons are fairly common in the organization of genes in bacteria. An operon is a cluster of genes coding for a series of enzymes that work in concert in an integrated biochemical pathway. For example, the lactose operon of *E. coli* is a cluster of three genes, each coding for one of three enzymes involved in conversion of disaccharide lactose into monosaccharide units of glucose and galactose (Fig. 13.6).

Multigene family

These are found in many organisms. A multigene family is also a cluster of related genes, but here the individual genes have identical or similar nucleotide sequences and therefore contain identical or similar pieces of biological information. There are simple multigene families in which all of the genes are exactly or virtually the same,

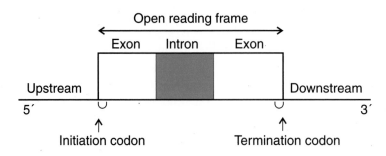

Fig. 13.5 Key features of an eukaryotic gene.

Fig. 13.6 Lactose operon of *E. coli* showing different genes and products.

for example, all the 5S rRNA genes are clustered into a single multigene family. Complex multigene families are made up of similar but nonidentical genes, for example, globin polypeptides of vertebrates. In humans, four globin polypeptides 2a and 2b combine with a heme cofactor to make a single molecule of hemoglobin.

PLANT GENE STRUCTURE AS A DISCONTINUOUS GENE

A number of plant structural genes have been analyzed in a variety of species. Figure 13.7 represents a typical nuclear gene with some of its sequences. Several investigators in 1971 discovered that the biological information carried by the genes is split into distinct units separated by intervening segments of DNA. The sections containing biological information or coding sequences are called exons, and the intervening sequences are referred to as introns. These discontinuous genes, also called split or mosaic genes, are now known to be common in higher organisms. These introns are transcribed but not represented in mature mRNA and are hence not translated. No introns have been found in rRNA genes but they have been demonstrated in a number of plant structural genes. In higher organisms, a gene may contain no introns or may have as many as hundred. Often, introns are much longer than exons.

Introns can be detected by comparing the nucleotide sequences of DNA to the mRNA using genomic clones. Cloned genomic DNA is annealed with RNA and the products examined under an electron microscope. Since the intron sequences are not contained in the RNA, they form single-stranded loops in the heteroduplex molecule.

The number of introns is highly variable. A number of plant genes have no introns at all, for example, zein storage genes of maize, soybean

Fig. 13.7 Diagrammatic representation of a typical nuclear gene.

protein genes, one rare class leaf gene and most *cab* genes (nuclear genes encoding chlorophyll *a/b* binding protein of the photosystem). Plant genes with two or three introns are the soybean actin gene, soybean leghemoglobin gene, and soybean glycinine genes, a rare class leaf gene and genes for the small subunit of RuBP carboxylase. There are nine introns in the maize *Adh* genes and five introns in the phytochrome gene.

When intervening sequences are present, all plant genes are identical in the first and last two bases of the introns GT and AG respectively. Conservation of the two dinucleotides at the intron–exon junctions in all eukaryotic genes suggests that similar RNA splicing mechanisms are involved.

CONTROL SEQUENCES

In prokaryotes, the leader sequence of bacterial transcript contains a short sequence called the Shino-Dalgarno sequence (SDS). It is complementary to a polypyrimidine sequence at the 3' end of 16s rRNA. Base pairing between these sequences is the mean by which the bacterial ribosome selects the correct AUG codon for initiation (Fig. 13.8). Such type of sequences are not found in eukaryotes.

Initiation of transcription occurs when RNA polymerase binds to a specific DNA sequence called the promoter at the beginning of an operon that occurs just upstream of genes. The maximum rate of transcription from a particular segment of DNA depends on the sequence of bases in its promoter. The actual DNA sequence at a promoter site varies from gene to gene, but all promoters have a consensus sequence, which in *E. coli* has two distinct components, a −35 box

Fig. 13.8 Shino–Dalgarno sequences in bacterial transcript.

and −10 box. The −10 box is also called Pribnow box after the scientist who first characterized it.

| −35 box | 5′ TTGACA 3′ |
| −10 box | 5′ TATAAT 3′ |

The negative numbers refer to the position of each box relative to the site at which transcription starts, which is designated as +1. The +1 position therefore corresponds to the site at which the first nucleotide is polymerized into the encoded RNA molecule.

These consensus sequences were determined by sequencing the DNA of a large number of *E. coli* promoters and identifying those bases that predominate at certain positions relative to known transcriptional start sites. The sequence of any one promoter may differ from the consensus sequence and certain variations give rise to promoters with reduced efficiency of RNA polymerase binding. This in turn leads to lower levels of transcription and is an important factor in the control of gene expression. The sequences of a few of the promoters recognized by the *E. coli* RNA polymerase are shown in Table 13.3.

In eukaryotes, the RNA polymerase attach-

Table 13.3 The sequences of a few of the promoters recognized by the *E. coli* RNA polymerase

Gene	−35 sequence	−10 sequence
E. coli consensus	TTGACA	TATAAT
Lac operon	TTTACA	TATGTT
Trp operon	TTGACA	TTAACT
TRNA gene	TTTACA	TATGAT

ment site is most important, which is usually just upstream of the gene to be transcribed. This attachment site is equivalent to the *E. coli* promoter, but is not the only nucleotide sequence that has a role in transcription.

TATA box

Upstream from the cap site are one or two short sequence elements that are common to many eukaryotic genes that are involved in normal transcription. Of the sequences, the TATA –box or Goldberg Hogness box is the best character- ized. This sequence normally occurs –25 to –30 nucleotides upstream of the cap site in almost all plant and animal genes. This sequence is required for correct expression of eukaryotic genes *in vitro* and *in vivo*.

AGGA box

Another sequence that may be involved in regulation of transcription of some eukaryotic genes is the consensus sequence "CAAT" box or GG (CT) CAATCT. This sequence is often found –40 and –180 nucleotides upstream of the cap site. The sequence of these elements is much more variable but often includes the sequence CAAT, termed the CAAT box. In plants, this sequence is referred to as the AGGA box.

Other regulatory elements

Enhancer sequences are another class of elements that show high levels of transcription. These sequences were originally discovered in viral genes. These have also been found in plants. Enhancers are located within a few hundred nucleotides upstream of the start of transcription. Although most enhancer elements are found in flanking sequences on the 5′ ends of genes, they are also found to occur elsewhere. An enhancer core sequence such as GGTGTGGAAAG, or more generally GTGGT/AT/ AT/AG (T/A means that either T or A is found in this position), can be identified by sequence comparisons, although it is clear that sequences other than the core also are important. Enhancer- like sequences in plants stimulate the tissue specific expression of genes.

TYPES OF RNA MOLECULES

The RNA molecules produced by transcription can be grouped into three major classes according to function. These are ribosomal or rRNA, transfer or tRNA, and messenger or mRNA. Besides, heterogeneous nuclear RNA and small nuclear RNA are also present. Ribosomal and transfer RNAs are the end products of gene expression and perform their roles in the cell as RNA molecules. Messenger RNA undergoes the second stage of gene expression, acting as the intermediate between a gene and its final expression product, a polypeptide.

Ribosomal and transfer RNAs are often re- ferred to as stable RNAs, indicating these mol- ecules are long lived in the cell, in contrast to mRNA, which has a relatively rapid turnover rate. Messenger RNA is involved in transcription and both types of stable RNA—rRNA and tRNA—are involved in translation of mRNA information to polypeptide synthesis. Details of these RNAs have been described before the actual process of protein synthesis. The aim is to describe various sequences involved in gene regulation

Messenger RNA and processing

Messenger RNA acts as the intermediate be- tween the gene and the polypeptide translation product. mRNA molecules are not generally long lived in the cell. Most bacterial mRNAs have a half-life of only a few minutes, whereas eukary- otic cells mRNA molecules have longer half lives (e.g. 6 h for mammalian mRNA). In prokaryotes, the mRNA molecules that are translated are direct copies of the gene. In eukaryotes, most mRNAs undergo a fairly complicated series of modifications and processing events before translation occurs:

1. Chemical modifications to the two ends of the mRNA molecules,
2. Removal of introns, and
3. RNA editing.

Chemical modifications to the two ends of the mRNA molecules

That the eukaryotic mRNA undergoes modification became clear when fractions present in the nucleus and cytoplasm were compared. The nucleus can also be divided into two regions: the nucleolus, in which rRNA genes are transcribed, and the nucleoplasm, where other genes including those for mRNA, are transcribed. The nucleoplasmic RNA fraction is called heterogeneous nuclear RNA or hnRNA, the name indicating that it is made up of a complex mixture of RNA molecules, some over 20 kb in length. The mRNA in the cytoplasm is also heterogeneous, but its average length is only 2 kb. If the mRNA in the cytoplasm is derived from hnRNA, then modification and processing events, including a reduction in the length of the primary transcripts, must occur before mRNA leaves the nucleus.

A RNA molecule synthesized by transcription and subjected to no additional modification will have at its 5′ end the chemical structure pppNpN. . ., where "N" is the sugar-base component of the nucleotide and "p" represents a phosphate group. Note that a 5′ terminus will be a triphosphate.

It was found that the 5′ end of eukaryotic mRNAs are blocked by the addition of 7-methyl-Gppp caps (7 methyl guanosine residues joined to the mRNAs by triphosphate linkages) that are added during the synthesis of the primary transcript (Fig. 13.9).

Some eukaryotic mRNAs may undergo further modification of the 5′ end with the addition of methyl groups to one or both of the next two nucleotides of the molecule. The 5′ caps may guide ribosomes to begin translation at the correct AUG codon.

A second modification of most eukaryotic RNAs is the addition of a long stretch of up to 250 A nucleotides to the 3′ end of the molecule, producing a poly(A) tail:

$$...pNpNpA(pA)_npA$$

Polyadenylation does not occur at the 3′ end of the primary transcript. Instead, the final few nucleotides are removed by a cleavage event that occurs between 10 and 30 nucleotides downstream of a specific polyadenylation signal (consensus sequences 5′– AAUAAA– 3′) to produce an intermediate 3′ end to which the poly(A) tail is subsequently added by the enzyme poly(A) polymerase.

The reason for polyadenylation is not known, although several hypotheses have pointed that the length of poly(A) tail determines the time in which the mRNA survives in the cytoplasm before being degraded.

Removal of introns

Eukaryotic genes contains introns, and transcription produces a faithful copy of the template strand. However, the introns must be removed and exon regions of the transcript attached to one another before translation occurs. This process is called splicing. Different types of introns have different splicing mechanisms. All the introns found in nuclear genes that are transcribed into mRNA fall into GT-AG class of introns. This designation GT-AG refers to a particular class in which introns possess the same dinucleotide sequences at their 5′ and 3′ ends. These sequences are GT for the first two nucleotides of the intron (GT is the sequence in the DNA and GU is the transcript) and AG for the last two nucleotides. In fact, the GT-AG rule represents only a component of slightly longer consensus sequences found at the 5′ and 3′ splice junctions of GT-AG introns. The consensus sequences are

$$\downarrow$$
5′ splice site   5′ – AGGTAAGT – 3′
$$\downarrow$$
3′ splice site   5′ – PyPyPyPyPyPyNCAG – 3′

These are slightly more complicated consensus sequences. 'Py' indicates a position where either of the two-pyrimidine nucleotides (C or U) can occur, and 'N' represents any nucleotide. The consensus sequence for the 3′ splice site is therefore six pyrimidines in a row, followed by any nucleotide, followed by a C and then the AG. The actual sequences in an intron may be slightly

Fig. 13.9 Processing of eukaryotic messenger RNA.

different, but the GT and AG components are always present.

In outline, the splicing reaction involves the following steps:

1. Cleavage occurs at the 5′ splice site.
2. The resulting free 5′ end is attached to an internal site within the intron to form a lariat structure.
3. The 3′ splice site is cleaved and the two exons joined together.

RNA editing

In RNA editing, the nucleotide sequence of a mRNA is altered by inserting new nucleotides or by deleting or changing existing ones.

Ribosomal RNA

Ribosomal RNA molecules are components of ribosomes that act as factories for protein synthesis. During translation, ribosomes attach to mRNA molecules and migrate along them, synthesizing polypeptides. Ribosomes are made up of rRNA molecules and proteins, and are numerous in most cells. They comprise about 80% of the total cell RNA and about 10% of the total protein. The eukaryotic ribosomes are large, having a total molecular mass of 4,220,000 daltons and a size of 32 nm × 22 nm. Sizes of ribosomes are determined by velocity sedimentation. The sedimentation coefficient is expressed as S-value (S = Svedberg unit). Prokaryotic ribosomes have a sedimentation coefficient of 70S, whereas eukaryotic ribosomes are variable, but on average have a sedimentation coefficient of about 80S. A typical eukaryotic ribosome has two subunits of 40S and 60S. The large subunit contains 3 rRNAs (28S, 5.8S and 5S) and 49 polypeptides. The small subunit has a single 18S rRNA and 33 polypeptides.

Ribosomal RNAs characteristically contain a number of specially modified nucleotides including pseudouridine residues, ribothymidylic acid, and methylated bases.

Transfer RNA

Each cell contains many different types of tRNA molecules. Transfer RNAs are the adapter molecules that read the nucleotide sequences of mRNA transcript and convert it into a sequence of amino acids. Transfer RNA molecules are relatively small, mostly between 74 and 95 nucleotides, for different molecules in different species. Almost every tRNA molecule in every organism can be folded into a base paired structure referred to as a clover leaf (Fig. 13.10). This structure is made up of the following components:

1. The acceptor arm formed by a series of usually seven base pairs between nucleotides at the 5′ and 3′ ends of the molecule. During protein synthesis, an amino acid is attached to the acceptor arm of the tRNA.
2. The D or DHU arm, so named because the loop at its end almost invariably contains the unusual pyrimidine dihydrouracil.

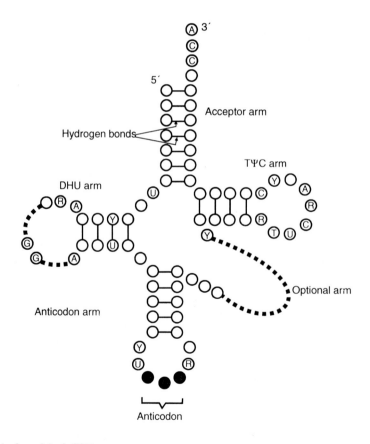

Fig. 13.10 Cloverleaf model of tRNA.

3. The anticodon arm plays the central role in decoding the biological information carried by the mRNA.
4. The extra, optional or variable form that may be a loop of just 3–5 nucleotides, or a much larger stem loop of 13–21 nucleotides with up to 5 base pairs in the stem.
5. The TΨC arm, named after the nucleotide sequence T, Ψ, C (Ψ = nucleotide containing pseudouracil, another unusual pyrimidine base), which its loop virtually always contain.

In both prokaryotes and eukaryotes, tRNAs are transcribed initially as precursor tRNAs, which are subsequently processed to release the mature molecules. A pre-tRNA molecule is processed by a combination of different ribonucleases that make specific cleavages at the 5′ (RNase P) and 3′ (RNase D) ends of the mature tRNA sequence. Eukaryotic tRNA genes are also clustered, and in addition occur in multiple copies. The processing of eukaryotic pre-tRNAs includes the following:

a) The three nucleotides at the 3′ end may be added after transcription. All tRNAs have 5' – CCA – 3′ trinucleotide sequence at the 3' end.

b) Certain nucleotides undergo chemical modification. The most common types are methylation of the base or sugar component of the nucleoside, for example, guanosine to 7-methyl guanosine; base rearrangements (interchanging the position of atoms in the purine or pyrimidine ring, e.g. uridine to pseudouridine; double bond saturation (uridine to dihydrouridine); deamination (removal of an amino group, e.g. guanosine to inosine); sulphur substitution (uridine to 4-thiouridine); addition of complex groups (guanosine to queosine).

Over 50 types of chemical modification have been discovered, each catalyzed by a different tRNA-modifying enzyme. Each tRNA molecule is able to form a covalent linkage with its specific amino acid at the end of the acceptor arm of the tRNA clover leaf by a group of enzymes called aminoacyl tRNA synthetases, and the process is called aminoacylation or charging.

Small nuclear RNA

Small nuclear RNAs or snRNAs are a class of RNA molecules found in the nucleus of eukaryotic cells. They are neither tRNA nor small rRNA molecules, although they are similar in size to these species. They contain from 100 to 200 nucleotides, and some of the bases are methylated and modified as in tRNA and rRNA. The snRNA does not exist as naked RNA, but is found as stable complexes with specific proteins forming small nuclear ribonucleoprotein particles or snRNPs, which are about 10S in size. It is present in 1 to 10% of the number of ribosomes. The snRNPs are important in the processing of eukaryotic gene transcripts (hnRNA) into mature mRNA for export from the nucleus to the cytoplasm.

TRANSCRIPTION

The first stage of gene expression, where the template strand of the gene directs the synthesis of a complementary RNA molecule, is known as transcription. The RNA molecule that is synthesized is called a transcript. The notation used to describe the nucleotide sequence of a gene and the actual process of transcription in prokaryotes and eukaryotes is explained.

Nucleotide sequences

To illustrate aspects of gene structure and expression, a standard procedure is followed for nucleotide sequences. Usually, the sequence presented will be that of the nontemplate strand of the gene in the 5′→3′ direction. This may seem illogical, as we know that biological information is carried by the template strand and not by the complement. However, it is the sequence of the RNA transcript (which is the same as that of nontemplate strand) that the molecular biologist is really interested in, because it is from this sequence that the important conversion to amino acids is made by way of the genetic code (Fig. 13.11). For this reason, when detailing a gene sequence, the convention is to describe the nontemplate strand.

The template is generated by transcription in the form of mRNA that is identical to one strand of DNA duplex. Thus, the two strands of DNA are as follows:

The DNA strand that bears a sequence identical to the mRNA (except for possessing T instead of U) is called the *coding strand, sense strand*, or *nontemplate strand*.

The other strand of DNA, which directs the synthesis of mRNA via complementary base-pairing is called the *template strand* or *antisense strand*. The antisense strand has a 3'→5' direction, which directs the synthesis of mRNA in the 5'→3' direction.

The underlying chemical reaction in transcription is the synthesis of an RNA molecule. This occurs by polymerization of ribonucleotide subunits in the presence of a DNA template, which directs the order in which the individual ribonucleotides are polymerized into RNA. During polymerization, the 3' OH group of one ribonucleotide reacts with the 5' P group of the second to form a phosphodiester bond. The transcript is therefore synthesized in the 5'→3' direction. Remember that in order to base pair, complementary polynucleotides must be antiparallel: this means that the template strand of the gene must

be read in the 3'→5' direction. The enzyme that catalyzes the reaction is called DNA-dependent RNA polymerase (usually referred to as RNA polymerase).

TRANSCRIPTION IN PROKARYOTES

RNA polymerase in *E. coli*

A single enzyme species catalyzes the synthesis of all types of *E. coli* RNA molecules (messenger, transfer, and ribosomal). This is in contrast to the finding in eukaryotic cells where several different RNA polymerases have been identified.

The complete *E. coli* RNA polymerase complex is composed of five protein subunits: two alpha (α) subunits, one beta (β) and one beta prime (β') subunit, and a sigma (σ) subunit. Collectively, the complex is known as the holoenzyme, and is represented as $\alpha_2\beta\beta'$ σ. The $\alpha_2\beta\beta'$ core of the enzyme is required for the polymerase activity. The σ subunit is required solely for DNA sequence recognition at a promoter site.

The actual process of transcription can be divided into three phases: initiation, elongation, and termination.

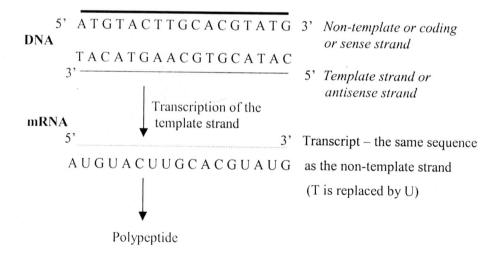

Fig. 13.11 Representation of a nucleotide sequence.

Initiation

Initiation of transcription occurs with the binding of RNA polymerase to a DNA molecule at a specific position, just in front (upstream) of the gene. These attachment points are called promoters. A promoter is a short nucleotide sequence that is recognized by an RNA polymerase as a point at which to bind to DNA in order to begin transcription. The actual DNA sequence at a promoter site varies from gene to gene, but all promoters have a consensus sequence, which for *E. coli* has two distinct components a −35 box and −10 box. This has been explained earlier.

The σ subunit attaches to the promoter sequence (following recognition by σ), while the other subunits cooperate in catalyzing RNA synthesis during transcription. Once transcription has been initiated, the σ subunit dissociates from the rest of the enzyme after approximately eight ribonucleotides have been polymerized into RNA, leaving the core enzyme to continue transcription. At initiation, the RNA polymerase complex has an elongated shape and covers up to 80 bp of DNA; during movement along the DNA the core enzyme covers approximately 30 bp of DNA. The optimal rate of polymerization is estimated to be approximately 40 nucleotides per second at 37°C.

Elongation

During the elongation stage of transcription the RNA polymerase moves along the DNA, reading the template information in the 5′→3′ direction. A RNA–DNA hybrid is thus formed in the unwound regions. It must be noted that requirement of all double-stranded nucleic acid structures is that antiparallel nature be maintained. As the RNA polymerase moves further along the DNA template, the newly synthesized RNA strand is displaced and the DNA duplex reforms behind the transcriptional complex. Three points should be noted:

1. The transcript is not longer than the gene. Position +1 is the point where synthesis of the RNA transcript actually begins, but it is not the start of the gene itself. Almost invariably the RNA polymerase enzyme transcribes a leader segment before reaching the gene. The length of the leader sequence varies from gene to gene. Similarly, when the end of the gene is reached, the enzyme continues to transcribe a trailer segment before termination occurs.

2. Only a small region of the double helix is unwound at any one time. Once elongation has started, the polymerized portion of the RNA molecule can gradually dissociate from the template strand, allowing the double helix to return to its original state. The open region or transcription bubble contains between 12 and 17 RNA–DNA base pairs. This is especially important because unwinding a portion of the double helix necessitates overwinding the adjacent areas and there is a limited molecular freedom for this to occur.

3. The rate of elongation is not constant. It has been observed that occasionally the enzyme will slow down, pause and then reaccelerate, or even go so far as to reverse over a short stretch, removing ribonucleotides from the end of the newly synthesized transcript before resuming its forward course.

Termination

Termination signals for *E. coli* genes vary immensely, but all possess a common feature: they are complementary palindromes. The complementary palindromes exert their influence on transcription by enabling a stem loop to form in the growing RNA molecule. Some stem loop terminators are followed by a run of 5 to 10 As in the DNA template, which after transcription into Us will leave the transcript attached to the template by a series of weak A–U base pairs. It is thought that RNA polymerase pauses just after the stem loop structure and that the A–U base pairs then break, so that the transcript detaches from the template.

Other terminators do not have the run of As. These are called Rho-dependent terminators. Rho is a large protein and is not a true subunit of the RNA polymerase. Rho attaches to the growing RNA molecule and when the polymerase pauses at the stem loop structure, it disrupts the base pairing between the DNA template and RNA transcript.

Termination of transcription results in dissociation of the RNA polymerase enzyme from the DNA and release of the RNA transcript. The core enzyme is now able to reassociate with a σ subunit and begin a new round of transcription on the same or a different gene.

TRANSCRIPTION IN EUKARYOTES

Eukaryotic RNA polymerases

Eukaryotic cells contain three different RNA polymerases, each with a distinct functional role. Ribosomal RNA (rRNA) is transcribed off the rDNA gene by RNA polymerase I. The synthesis of messenger RNA is catalyzed by RNA polymerase II, with the subsequent addition of poly(A) tails carried out by poly(A) polymerase. Transfer RNA and a variety of smaller nuclear and cytoplasmic RNAs are made by RNA polymerase III. The existence of three different types of eukaryotic RNA polymerases suggests the existence of three different types of promoters that can be regulated independently. Eukaryotic RNA polymerases are composed of many proteins that interact both with each other and with multiple sites on the DNA.

Initiation. Initiation of transcription in eukaryotes depends on recognition and binding of RNA polymerase II to a promoter sequence. This attachment site is not the only nucleotide sequence that has a role in transcript initiation. Other possible sequences spread out over several hundred base pairs will also be involved in the binding. Besides this, certain transcription factors (TFs) are also involved. The eukaryotic core promoter element is called a "TATA box" because it contains the attachment site for RNA polymerase II. The attachment site has the following consensus sequence

5′–TATAAAT –3′

The TATA box

The TATA box is usually located about 25 nucleotides upstream of the position at which transcription starts, and is therefore also called −25 box. RNA polymerase attaches to TATA box, the process being mediated by a set of proteins called transcription factors (Fig. 13.12). These TFs have been designated TFIIA, TFIIB, and so on, distinguishing them from TFI and TFIII transcription factors that work with RNA polymerase I and III respectively. The first step in this attachment is the binding of TFIID to the TATA box, which is probably mediated by TFIIA. Once TFIID has attached to the DNA template, the transcription complex starts to form. TFIIB binds next, followed by RNA polymerase II itself, and then the two final transcription factors TFIIE and TFIIF. Transcription complex is now ready to begin synthesizing RNA.

Termination

Identification of termination signals for RNA polymerase has proven difficult because the 3′ ends of the transcripts are removed immediately after synthesis. But it is clear that the trailer regions may be very long, extending 1000 to 2000 bp downstream of the gene. This led to the suggestion that there are in fact no set termination points downstream of genes transcribed by RNA polymerase II. Possibly one of the transcription factors detaches from the complex at the end of the gene, destabilizing the complex and leading it to fall off the template at some later point.

GENETIC CODE AND TRANSLATION

Translation is the process by which the information in the mRNA molecule is decoded into a polypeptide chain. This process is carried out by ribosomes and aminoacyl transfer RNAs. The central fact about translation is that the sequence of amino acids in the polypeptide being synthesized is specified by the sequence of nucleotides in the mRNA molecule being translated. The rule

Upstream sequences
influence RNA polymerase
attachment

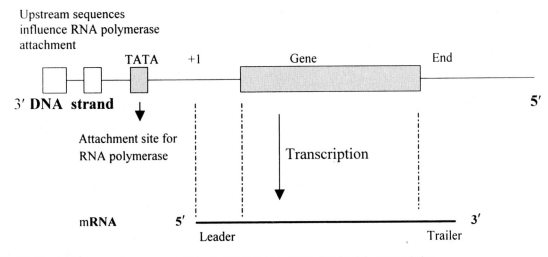

Fig 13.12 A diagrammatic representation of different sequences involved in transcription.

that determines which sequence of nucleotides specifies which amino acids is embodied in the genetic code.

Genetic code

The problem of exactly how the genetic code works was the major preoccupation of biologists when the double helix structure of DNA was postulated. Since proteins are constructed from a population of 20 different amino acids, the code that specifies them must have at least 20 different possible words. In order to decipher this coded information, two basic assumptions were made, which were subsequently confirmed.

1. *Colinearity between gene and protein:* It means the order of nucleotides in the gene directly correlates with the order of amino acids in the corresponding polypeptide.
2. Each code word or codon is a triplet of nucleotides that codes for a single amino acid. Since there are only four different nucleotide bases in DNA (A, G, C, T) and correspondingly in mRNA (A, G, C, U), a one-base code could only specify four different amino acids. A two-base code could specify 16 amino acids (4 × 4), but this is still not sufficient. A three base code is capable of specifying all 20 amino acids since there are

64 possible combinations (4 × 4 × 4). Information in mRNA is indeed translated in units of three nucleotides, and these units are called triplet codons. The assignment of particular codons to particular amino acids is called the genetic code, and is shown in Table 13.4.

Of the 64 possible codons, 61 specify a particular amino acid, with the remaining 3 codons (UAA, UAG, UGA) being used as stop signals which indicate to the translation machinery that protein synthesis is complete. The codon AUG (which specifies methionine) is also known as a start codon since in almost all cases the start site for translation is determined by an AUG in the mRNA. Note that only two amino acids (methionine and tryptophan) have a single codon; all other amino acids have more than one.

The genetic code has three important features:

1. *The code is degenerate:* All amino acids except methionine and tryptophan have more than one codon, so that all possible triplets have a meaning. Most synonymous codons are grouped into families (GGA, GGU, GGG and GGC, for example all code for glycine).
2. *The code contains punctuation codons:* Three codons, UAA, UAG and UGA, do not

Table 13.4 Genetic code

UUU	Phe	UCU	Ser	UAU	Tyr	UGU	Cys
UUC	Phe	UCC	Ser	UAC	Tyr	UGC	Cys
UUA	Leu	UCA	Ser	UAA	Stop	UGA	Stop
UUG	Leu	UCG	Ser	UAG	Stop	UGG	Trp
CUU	Leu	CCU	Pro	CAU	His	CGU	Arg
CUC	Leu	CCC	Pro	CAC	His	CGC	Arg
CUA	Leu	CCA	Pro	CAA	Gln	CGA	Arg
CUG	Leu	CCG	Pro	CAG	Gln	CGG	Arg
AUU	Ile	ACU	Thr	AAU	Asn	AGU	Ser
AUC	Ile	ACC	Thr	AAC	Asn	AGC	Ser
AUA	Ile	ACA	Thr	AAA	Lys	AGA	Arg
AUG	Met	ACG	Thr	AAG	Lys	AGG	Arg
GUU	Val	GCU	Ala	GAU	Asp	GGU	Gly
GUC	Val	GCC	Ala	GAC	Asp	GGC	Gly
GUA	Val	GCA	Ala	GAA	Glu	GGA	Gly
GUG	Val	GCG	Ala	GAG	Glu	GGG	Gly

Key to amino acid abbreviations: Ala: alanine; Arg: arginine; Asn: asparagine; Asp: aspartic acid; Cys: cysteine; Gln: glutamine; Glu: glutamic acid; Gly: glycine; His: histidine; Ile: isoleucine; Leu: leucine; Lys: lysine; Met: methionine; Phe: phenylalanine; Pro: proline; Ser: serine; Thr: threonine; Trp: tryptophan; Tyr: tyrosine; Val: valine.

code for amino acids, and if present in the middle of a heteropolymer cause protein synthesis to stop. These are called termination codons and one of the three always occurs at the end of a gene at the point where translation must stop. Similarly, AUG virtually always occurs at the start of a gene and marks the position at which translation should begin. AUG is called the initiation codon and codes for methionine. Thus, most newly synthesized polypeptides will have this amino acid at the amino terminus, though it may be subsequently removed by posttranslational processing of the protein.

3. *The code is not universal:* When the genetic code was completed in 1966 it was assumed to be universal, that is, it was used by all organisms. It was therefore surprising to learn in 1979 that human mitochondrial genes use a slightly different code in which UGA (normally a stop codon) codes for tryptophan. AGG and AGA are stop codons and AUA codes for methionine. Other rare differences have also come to light, but it is true to say that the code shown in Table 13.4 is followed by virtually all nuclear genes.

TRANSLATION

There are three major steps in the synthesis of polypeptide
1. Chain initiation
2. Chain elongation
3. Chain termination

Chain initiation

The first step in translation is the attachment of the small 40S subunit of ribosome (30S in prokaryotes) to the mRNA molecule. The ribosome-binding site ensures that translation starts at the correct position. The correct attachment point for ribosomes in *E. coli* has the following consensus sequence

5′   A G G A G G U   3′

An eukaryotic mRNA does not have an internal ribosome binding site equivalent to the above consensus sequence. Instead, the small subunit of the ribosome recognizes the cap structure as its binding site and therefore attaches to the extreme 5′ end of the mRNA molecule. The translation process itself starts when an aminoacylated tRNA base pairs with an initiation codon that has been located by the small subunit of ribosome. The initiator tRNA is charged with methionine, because methionine is the amino acid coded by AUG. The initiation process involves the ordered formation of a complex which constitute an initiator factor (eIF2), GTP, the initiating methionine tRNA (met-tRNA), and the small 40S subunit ribosomal unit. This complex along with other initiation factors binds to the 5′ end of mRNA and then moves along the mRNA until it reaches the first AUG codon (Fig. 13.13).

Chain elongation

Once the initiation complex has formed, the large subunit joins the complex, which requires the hydrolysis of GTP and the release of initiation factors. Two separate and distinct sites are now available at the ribosome to which tRNAs can bind. The first of these called the peptidyl or P-site is at that moment occupied by the aminoacylated t-RNA met which is still base paired with the initiation codon. The aminoacyl or A-site is positioned over the second codon of the gene and at first it is empty. Elongation begins when the correct aminoacylated tRNA enters the A-site and base pairs with the second codon. This requires the elongation factor eEF1, and a second molecule of GTP is hydrolyzed to provide the required energy. Now both sites of the ribosome are occupied by aminoacylated tRNAs and the two amino acids are in close contact. A peptide bond is formed between the carboxyl group of methionine and the amino group of the second amino acid. The reaction is catalyzed by peptidyl transferase. Peptidyl transferase acts in conjunction with tRNA deacylase, which breaks the link of met-tRNA. This results in a dipeptide attached to the tRNA present in the A-site.

Translocation occurs when the ribosome slips along the mRNA by three nucleotides so that aa-aa-tRNA enters the P-site, expelling the now uncharged $tRNA^{met}$ and making the A-site vacant again. The third aminoacylated tRNA enters the A-site and the elongation cycle is repeated. Each cycle requires hydrolysis of a molecule of GTP and is controlled by the elongation factor eEF2.

After several cycles of elongation, the start of the mRNA molecule is no longer associated with the ribosome, and a second round of translation can begin by binding of 40S subunit and forming a new initiation complex. The end result is a polysome, a mRNA being translated by several ribosomes at once.

Chain termination

Termination of translation occurs when a termination codon (UAA, UAG, or UGA) enters the A-site. There are no tRNA molecules with anticodons able to base pair with any of the termination codons. A termination factor eRF recognizes all the termination codons in eukaryotes and it requires hydrolysis of a GTP molecule. The ribosome releases the polypeptide and mRNA and subsequently dissociates into subunits, which enter the cellular pool before becoming involved in a new round of translation. The mechanism of protein synthesis is essentially the same for both plants and animals. Ribosomes from one species will faithfully translate mRNA from the other.

Post-translation modifications

Primary translation products undergo a variety of modifications vital to the production of a fully functional protein. These post-translational processing steps are varied and more important in eukaryotes. One reason for this is the existence of more membrane-bound compartments into which proteins must be directed.

Membrane-bound polysomes synthesize proteins that are either secreted from the cell, or form part of a membrane, or are packaged into specialized organelles (e.g. seed storage protein bodies). It has been postulated that translocated proteins carry an N-terminal extension of about 20 amino acids, termed as signal peptide. It binds to a receptor in the membrane as soon as it is synthesized and emerges from the ribosome. Signal peptides have been found in the primary translation products of a number of mRNAs, e.g. zein from maize. These signal peptides are to be removed. Besides, glycosylation occurs during

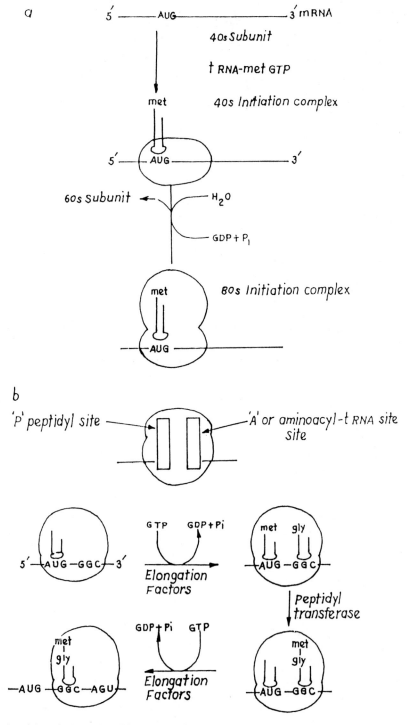

Fig. 13.13 Initiation (a) and elongation (b) process of polypeptide chain synthesis.

passage of proteins through a membrane. Many proteins remain inactive until their polypeptide chains are cleaved by proteases. In some cases, the original single polypeptide chain is not a functional product; fragments produced by proteolytic cleavage are needed. Other types of processing involve the addition of moieties (carbohydrates, lipids, cofactors and phosphate groups) to the amino acid residues.

In conclusion, it can be seen that there are many steps in eukaryotes, each of which must function properly for the successful expression of a gene. Thus, there are many potential points at which gene expression may be controlled.

Further reading

Adams, R.L.P., Knowler, J.T. and Leader, D.P. 1992. *The Biochemistry of the Nucleic Acids*, 11th ed. Chapman & Hall, London

Garret, R.H. and Grisham, C.M. 1999. *Biochemistry*. Saunders College Publishing, Orlando, Florida.

Lewin, B. 1995. *Genes V*. Oxford University Press, Oxford, UK.

Snustad, D.P., Simmons, M.J. and Jenkins, J.B. 1997. *Principles of Genetics*. John Wiley & Sons, New York.

PART III

Recombinant DNA Technology

14

Basic Techniques

The last few years have seen the rapid development of new methodologies in the field of molecular biology. The initial impetus for rDNA technology came about in the early 1970s with the simultaneous development of techniques for the transformation of *E. coli*, cutting and joining of DNA molecules, and monitoring of the products of cutting and joining reaction. New techniques have been regularly introduced and the sensitivity of older techniques greatly improved upon. To explain the processes of gene cloning and gene manipulations, understanding of certain basic techniques is essential. These techniques are discussed in this chapter.

Agarose gel electrophoresis

Electrophoresis through agarose gels is the standard method for the separation, identification, and purification of DNA and RNA fragments ranging in size from a few hundred to 20 kb. Besides agarose, polyacrylamide gel electrophoresis is also used for the same purpose, but is preferred for small DNA fragments. The technique of agarose gel electrophoresis is simple, rapid to perform, and capable of resolving DNA fragments that cannot be separated adequately by other procedures such as density gradient centrifugation. Furthermore, the location of DNA within the gel can be determined directly by staining with low concentrations of intercalating fluorescent ethidium bromide dye under ultraviolet light. If necessary, these bands of DNA

can be recovered from the gel and used for a variety of cloning purposes.

Agarose is a linear polymer extracted from seaweed. Agarose gels are cast by melting the agarose in the presence of desired buffer until a clear transparent solution is obtained. The melted solution is poured into a mould and can be made into a variety of shapes, sizes, and porosities (4–10%) and the gel is allowed to harden. On hardening, the agarose forms a matrix, the density of which is determined by the concentration of agarose. When an electric field is applied across the gel, DNA, which is negatively charged at neutral pH, migrates towards the anode. The rate of migration depends on a number of parameters: (1) molecular size of DNA, (2) agarose concentration, (3) conformation of DNA, and (4) composition of electrophoresis buffer. Larger molecules migrate more slowly than the smaller molecules do because they have to find their way through the pores of the gel. By using gels of different concentrations it is possible to resolve a wide range of DNA molecules. The electrophoretic mobility of DNA is also affected by the composition and ionic strength of electrophoresis buffer. In the absence of ions, electrical conductivity is minimal and DNA migrates very slowly. In buffers of high ionic strength, electrical conductance is very efficient. Several different buffers are available, *viz.* TAE (Tris-acetate), TBE (Tris-borate), TPE (Tris-phosphate) and alkaline buffer, at a pH range of

7.5–7.8. Electrophoresis is normally carried out at room temperature. Gels are stained with intercalating fluorescent ethidium bromide dye and as little as 0.05 µg of DNA in one band can be detected as visible fluorescence under ultraviolet light. Ethidium bromide is a powerful mutagen and toxic, and therefore proper care should always be taken while handling the solution, and the staining solution should always be decontaminated after its use.

Pulsed field gel electrophoresis

In recent years the means to analyze entire genomes have increased enormously, culminating in the construction of physical maps of the genomes of *E. coli, Bacillus subtilis, Mycobacterium tuberculosis, Caenorhabditis elegans,* and the sequencing of the yeast and human genomes. In plants the progress has not been as fast, though the situation is improving rapidly. *Arabidopsis* and rice are being used as model systems and have been completely sequenced.

The pulsed field gel electrophoresis (PFGE) technique was conceived by Schwartz and Cantor (1984) to allow the electrophoretic separation of DNA molecules several megabases in size. This was necessary for the development of megabase mapping and cloning techniques because all linear double-stranded DNA molecules larger than a certain size migrate through agarose gels at the same rate. Agarose gel electrophoresis is the standard technique for separating nucleic acids; however, conventional agarose gels cannot resolve fragments >20 kb. This is because large DNA fragments are unable to migrate through the gel matrix by the same sieving mechanism that small fragments do. Ordinary agarose gel electrophoresis fails to separate extremely long DNA molecules because the steady electric field stretches them out so that they travel end-first through the gel matrix in a snake-like fashion. This mode of migration is known as *reptation*. This DNA can no longer be sieved according to its size. The greater the pore size of the gel, the larger the DNA that can be sieved. Thus, agarose gels cast with low

concentrations of agarose (0.1–0.2%) are capable of resolving extremely large DNA molecules. However, such gels are quite fragile and must be run very slowly. Even then, they are incapable of resolving linear DNA molecules larger than 750 kb in length. Pulsed field gel electrophoresis is a variation of agarose gel electrophoresis, which takes advantage of the electrophoretic property of large DNA fragments. PFGE resolves DNA molecules of 100–1000 kb by intermittently changing the direction of electric field in a way that causes large DNA fragments to realign more slowly with the new field direction than do smaller molecules. In the PFGE method, pulsed alternating, orthogonal electric fields are applied to a gel, which forces the molecules to reorient before continuing to move snake-like through the gel. This reorientation takes much more time for larger DNA molecules, so that progressively longer DNA molecules move more and more slowly. Large DNA molecules become trapped in their reptation tubes every time the direction of electric field is altered and can make no further progress through the gel until they have reoriented themselves along the new axis of the electric field. The larger the DNA molecule, the longer the time required for this alignment (Fig. 14.1a). DNA molecules whose reorientation times are less than the period of electric pulse will therefore be fractionated according to size. In conventional electrophoresis, a single constant electric field is employed, which limits the separation range to 50 kb. Separation in PFGE is achieved by applying two perpendicularly oriented electric fields in an alternating mode, which enabled Schwartz and Cantor initially to separate electrophoretically individual yeast chromosomes ranging in size from 230 kb to 2000 kb. Thus, the gap in resolution between standard techniques in molecular biology and the 100–5000 kb size range covered by genetic analysis was closed.

The understanding of PFGE is still incomplete, and several models to explain the behavior of DNA molecules during electrophoresis have been proposed. The separation of molecules is

Electric field alternates 120° every 90
seconds for 18 to 24 hours at 14° C

Fig. 14.1 Pulsed field gel electrophoresis. (a) very large DNA molecules that are migrating parallel to the electric field in one
direction takes more time (T) to reorient in an alternating electric field than do smaller molecules. (b) CHEF
pulsed field gel systems use a hexagonal gel box that alters the angle of the fields relative to the agarose gel.
EtBr–Ethidium bromide; S1 S2 and S3–Sample lanes; Mkr–Marker lane. Concept of the figure taken from
Miesfeld: *Applied Molecular Genetics.* Copyright © 1999 John Wiley and Sons. This material is used by
permission of John Wiley and Sons, Inc.

basically achieved by forcing them to reorient
periodically to changing field conditions. This
change of direction is quicker by smaller
molecules, which obtain a larger net mobility and
are thus separated from larger molecules. Since
its introduction, PFGE has been rapidly accommodated to the analysis of chromosomal DNA
from a wide variety of organisms. Currently,
techniques have advanced to allow the separation and analysis of DNA molecules up to 12
megabases.

Two basic types of PFGE systems have

been developed. The simplest is called field
inversion PFGE, which works by periodically
reversing the direction of the electric field by
using a standard gel box apparatus. Each time
the direction of the field is inverted, larger
molecules take longer to reverse the direction
than do smaller molecules. Net movement in one
direction is accomplished by keeping the time of
field inversion in the forward direction greater
than in the reverse orientation. A second type of
PFGE is based on alternating the angle of the
electric field. The original PFGE design for

alternating angle was developed by Schwartz and Cantor and was based on a 90°-field alteration. However, this created an uneven DNA migration path through the gel. Chu *et al.* (1986) improved the basic PFGE design by changing the shape of the gel box so that multiple electric fields could be produced by placing electrodes around the perimeter of a hexagonal box (Fig. 14.1b). This type of PFGE gel box uses alternating angles of 120° and produces a contoured clamped homogeneous electric field or CHEF. Commercially available CHEF gel box systems have been designed that control the angle of the alternating electric fields, and the pulsed time in each direction, to optimize the resolution capacity for any range of DNA sizes.

An essential element in all PFGE studies is the availability of DNA preparations containing molecules that are several million base pairs in size. These large DNA molecules are extremely sensitive to shearing and nuclease activity. Therefore, for isolating high molecular weight DNA, cells are lysed *in situ* in an agarose plug or a bead. Intact cells or nuclei are resuspended in molten low-melting-temperature agarose and solidified in blocks whose size matches the thickness of the loading slot of the gel. Following cell lysis, a protease inhibitor is soaked into the agar block to inactivate proteinase K, and DNA is then digested with restriction enzymes prior to placing the agar block into the well of the PFGE gel. Gel electrophoresis is initiated, and the DNA fragments migrate out of the agar block and directly into the gel.

The limit of resolution of pulsed field gel electrophoresis depends on several factors:

Pulse time: This is the most critical factor in all PFGE systems, and is the time the electric field is applied in each direction. The longer the pulse time, the more the separation is extended to higher molecular weight DNAs.

Temperature: Mobility of DNA molecules increases with temperature. In PFGE, the temperature is usually maintained at 10 15°C. When higher temperatures are used, separation over a somewhat larger range can be obtained

with the separation window moving to larger molecules at the expense of more diffuse bands.

Field strength: Electric field strength also has a strong influence on the separation range. Depending on the desired separation range, field strengths vary from 0.5 to 10 V/cm. Above a critical value, which depends on DNA size, molecules becomes trapped within the agarose matrix and no longer migrate. Trapping can occur after various times of electrophoresis and leads to a strong base smearing, especially for molecules in the megabase size range. Very high field strength, that above 10 V/cm, should be avoided, as bands become too diffuse. A good compromise is usually a field strength of 5–7 V/cm.

Field angle: The angles of the two electric fields to the gel is an important factor. Smaller angles, between 95 and 120 increases the speed at which separation is attained by a factor of 2 to 3. Especially for separating very large molecules in the magabase size range this may reduce the run times by several days.

Agarose: The agarose concentration is kept at 1% (w/v). When higher agarose concentration is used separation occurs over a smaller range but with sharper bands and a decrease of the overall mobility.

DNA concentration: The mobility of DNA molecules in PFGE is sensitive to the concentration of DNA, with high concentrations leading to lower mobilities. This effect may be substantial and results in size estimates that differ by as much as 50 kb.

DNA topology: The behavior of nonlinear DNA molecules strongly deviates from linear molecules. Typically, circular DNA molecules (supercoiled and relaxed) show a much lower mobility in PFGE than do linear molecules.

One has to keep in mind, however, that changing one parameter will influence the effect of other parameters. In the original Schwartz and Cantor PFGE apparatus, the speed and migration direction of DNA molecules was dependent on the position at which the DNA is loaded on the gel. Consequently, DNA migrates in a complex path and skews towards the edges of the gel, making size estimates difficult. These problems were solved by several modifications of the electrode configuration, resulting in the development of variant forms.

Field inversion gel electrophoresis

The field inversion gel electrophoresis (FIGE) apparatus was designed by Carle et al. (1986). In this apparatus, there is periodic inversion of a single electric field. The field is uniform in both directions and the forward pulse is slightly longer than the reverse pulse, resulting in the movement of the DNA along a perfectly straight track. FIGE resolves DNA fragments up to 2000 kb in length.

Vertical pulse field gradient gel electrophoresis

The apparatus designed by Gardiner et al. (1986) uses a vertical gel apparatus with platinum wire electrodes positioned on opposite sides of the gel. The DNA moves first toward one set of electrodes and then toward the other as the electric fields are switched. The net result of these zigzag movements is a straight line from the loading well toward the bottom of the gel. Because all the lanes in the gel are exposed to equivalent electric fields, there is no horizontal distortion of the DNA bands. The size of the molecules resolved at a given voltage is a function of the angle between the electrodes and the pulse time. Typically the electrodes are arranged at a 90° angle and 10-sec pulses are used to separate molecules between 50 and 450 kb in length and 50–60 sec pulses are used to fractionate molecules larger than 100 kb in length.

Contour-clamped homogeneous electric field gel electrophoresis

In contour-clamped homogeneous electric field gel electrophoresis (CHEF), electric field is generated from multiple electrodes that are arranged in a square or hexagonal contour around the horizontal gel and are clamped to predeter-

mined potentials. A square contour generates electric fields that are oriented at right angles of 120° or 60° depending on the placement of the gel and polarity of electrodes. By using a combination of low field strengths (1.3 V/cm), low concentration of agarose (0.6%), long switching intervals (1 h), and extended periods of electrophoresis (130 h), it is possible to resolve DNA molecules up to 5000 kb in length (Chu et al., 1986; Vollrath and Davis, 1987).

Detection: The standard technique of ethidium bromide is used for staining the gels. For detection of minor species of DNA, the gels are usually destained in water for extended periods of time (24–48 h) before photography.

In plant molecular biology, PFGE has already found widespread applications in the following aspects:

1. Determination of size of genomic DNA;
2. Characterization of the large-scale organization of telomeric and sub-telomeric regions of rice, tomato, and barley;
3. Construction of long-range restriction maps of centromeric and non-centromeric regions;
4. Organization of gene clusters such as the pea leghemoglobin genes, the wheat α-amylase genes, the patatin genes of potato and tomato, and the tandemly repeated tomato 5S RNA genes;
5. Study of the higher order chromatin structure in maize;
6. Assessment of *in vivo* damage and repair of DNA following exposure to radiation; and
7. Map-based cloning to isolate genes for which only a variant phenotype and genetic linkage map position are known.

Polyacrylamide gel electrophoresis

In native polyacrylamide gel electrophoresis (PAGE), proteins are applied to a porous polyacrylamide gel and separated in an electric field. When proteins are placed in an electric field, molecules with a net charge, such as proteins, will move toward one electrode or the other, a phenomenon known as electrophoresis. There are different types of PAGE. In native polyacrylamide gel electrophoresis, the molecules separate in an electric field on the basis of their net charge and size of the protein, since the electrophoresis separation is carried out in a gel, which serves as a molecular sieve. Small molecules move faster through the pores as compared to large molecules.

The gels are made of polyacrylamide, which is chemically inert. It is readily formed by the polymerization of acrylamide. Choosing an appropriate concentration of acrylamide and the crosslinking agent, methylene bisacrylamide, can control the pore sizes in the gel. The higher the concentration of acrylamide, the smaller is the pore size of the gel. The gel is usually cast between the two glass plates of 7–20 cm² separated by a distance of 0.5–1.0 mm. The protein sample is added to the wells in the top of the gel, which are formed by placing a plastic or Teflon comb in the gel solution before it sets (Fig. 14.2). A bromophenol blue dye is mixed with the protein sample to aid its loading on to the gel. Because the bromophenol blue dye is a small molecule, it also migrates quickly through the gel during electrophoresis, thus indicating the progress of electrophoresis. In the simple native PAGE, buffer is same in upper and lower reservoirs and in the gel with a pH of approximately 9, such that most proteins have net negative charges and migrate toward the anode in the lower reservoir. An electric current (~300 V) is applied across the gel from top to bottom for a period to move the protein through the gel. When blue indicator of bromophenol dye reaches at the bottom of the gel, electric current is switched off and the gel is removed from the electrophoresis apparatus. The gel is stained and the protein bands within it are visualized.

Fig. 14.2 Native polyacrylamide gel electrophoresis.

A modification of simple PAGE is SDS-PAGE (sodium dodecyl sulfate-PAGE). In SDS-PAGE, the protein sample is treated with a reducing agent such as 2-mercaptoethanol or dithiothreitol to break all disulfide bonds. It is then denatured with SDS, a strong anionic detergent, which disrupts nearly all the noncovalent interactions and covers them with an overall negative charge. Approximately one molecule of SDS binds via its hydrophobic alkyl chain to the polypeptide backbone for every two amino acid residues, which gives the denatured protein a large net negative charge that is proportional to its mass. The SDS-protein mixture is then mixed with the bromophenol blue dye and applied in to the sample wells, and electrophoresis is performed as for native PAGE. As all the proteins now have an identical charge to mass ratio, they are separated on the basis of their mass. The smallest proteins move farthest. Thus, if proteins of known molecular mass are electrophoresed alongside the samples, the mass of unknown proteins can be determined. SDS-PAGE is a rapid, sensitive, and widely used technique from which one can determine the degree of purity of a protein sample, the molecular mass of unknown sample, and the number of polypeptide subunits with a protein.

Isoelectric focusing

In isoelectric focusing (IEF), proteins are separated by electrophoresis in a pH gradient in a gel. They separate on the basis of their relative content of positively and negatively charged groups. Each protein migrates through the gel until it reaches the point where it has no net charge—this is its isoelectric point (pI); here, the protein's net charge is zero and hence it does not move in an electric field. In IEF, a polyacrylamide gel is used that has large pores so as not to impede protein migration and contains a mixture of polyampholytes (small multicharged polymers that have many pI values). If an electric field is applied to the gel, the polyampholytes migrate and produce a pH gradient. To separate proteins by IEF, they are electrophoresed through such a gel. Each protein migrates through the gel until it reaches a position at which the pH is equal to its pI (Fig. 14.3). If a protein diffuses away from this position, its net charge will change as it moves into a region of different pH and the resulting electrophoretic forces will move it back

(a)

(b)

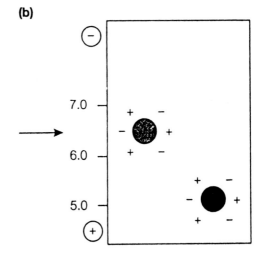

Fig. 14.3 Isoelectric focusing. (a) Before applying an electric current, (b) after applying an electric current the proteins migrate to a position at which their net charge is zero (isoelectric point, pI).

to its isoelectric position. In this way each protein is focused into a narrow band (as thin as 0.01 pH unit) about its pI.

2-Dimensional gel electrophoresis

Isoelectric focusing can be combined with SDS-PAGE to obtain very high-resolution separations by a procedure known as two-dimensional gel electrophoresis. In this technique, the protein sample is first subjected to isoelectric focusing in a narrow strip of gel containing polyampholytes. This gel strip is then placed on top of an SDS polyacrylamide gel and electrophoresed to produce a two-dimensional (2D) pattern of spots in which the proteins have been separated in the horizontal direction on the basis of their pI, and in the vertical direction on the basis of their mass. The overall result is that proteins are separated on the basis of their size and charge. Thus two proteins that have very similar or identical pIs and produce a single band by isoelectric focusing will produce two spots by 2D gel electrophoresis (Fig. 14.4). Similarly, proteins with similar or identical molecular masses, which would produce a single band by SDS-PAGE, also produces two spots because of the initial separation by isoelectric focusing. This 2D gel electrophoresis has enormous use in proteomics study.

Staining of proteins

The most commonly used protein stain is the dye Coomassie brilliant blue. After electrophoresis,

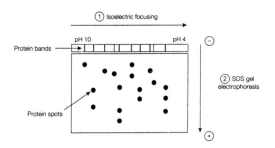

Fig. 14.4 Two-dimensional gel electrophoresis. The protein sample is first subjected to isoelectric focusing in one direction and then to SDS-PAGE in the second direction.

the gel containing separated proteins is immersed in an acidic alcoholic solution of the dye. This denatures the proteins, fixes them in the gel so that they do not wash out, and allows the dye to bind to them. After washing away the excess dye, the protein bands are visible as discreet blue bands. A more sensitive stain is soaking the gel in a silver salt solution. However, this technique is rather difficult to apply.

Nucleic acid blotting

Blotting describes the immobilization of sample (nucleic acids, proteins) on to a solid support, generally nylon or nitrocellulose membranes. In nucleic acid blotting, the blotted samples are nucleic acids (DNA or RNA) on nylon or nitrocellulose membranes, which are then used as targets for hybridization with labeled probes. Labeling of probe nucleic acids and hybridization have formed the basis for a range of experimental techniques that help understand the organization and expression of genetic material.

The blotting method was originally developed by Southern (1975), who demonstrated that DNA restriction fragments that had been electrophoretically fractionated on agarose gels could be transferred to a solid support (nitrocellulose) and detected as discreet bands following hybridization to a complementary DNA probe. When the blotting method is applied to RNA, it is termed *northern blotting*. *Western blot* analysis refers to the transfer of proteins to membranes and their detection with antibody probes.

Southern blot analysis

Southern (1975) gave the method for detecting DNA fragments in an agarose gel by blotting on a nylon or nitrocellulose membrane followed by detection with a probe of complementary DNA or RNA sequence. The blotting technique has not changed over the years since its first description, except for the availability of increasingly sophisticated blotting membranes, kits for labeling, and apparatus for electrophoresis transfer. Although membranes originally used for DNA transfer were made of nitrocellulose, nylon-based membranes used nowadays perform much better due to their higher nucleic acid binding capacity and physical strength. In Southern blotting, agarose gel electrophoresis is performed. (The procedure of agarose gel electrophoresis has already been explained.)

Procedure: An outline of a procedure for Southern hybridization is shown in Fig. 14.5. Genomic DNA is digested with one or more restriction enzymes, and the resulting fragments are separated according to size by electrophoresis on an agarose gel. The agarose gel is mounted on a filter paper wick. A piece of nylon membrane slightly larger than the size of gel is placed over it, followed by 2–4 pieces of prewet filter papers. Thus the membrane is sandwiched between the gel and a stack of filter papers and paper towels (Fig. 14.6). Air bubbles between each layer are removed. A 10-cm stack of dry filter paper towels cut to size is placed on top

DNA digested with restriction enzyme(s)

↓ Agarose gel electrophoresis

Agarose gel with DNA fragments

↓ 1. Depuration treatment
2. Neutralizing solution

Agarose gel equilibrated

Nylon/nitrocellulose membrane
(capillary action)

↓

DNA transferred to nylon/nitrocellulose membrane

↓ Fixation of DNA to membrane
(Baking at 80°C/UV crosslinking)

Membrane with DNA fragments

↓ Labeled probe

Hybridization of membrane with labeled probe

↓

Detection of DNA fragments on the membrane

Fig. 14.5: A general scheme for Southern hybridization.

of the pyramid and a glass plate and weight of around 0.5 kg placed on top. Strips of parafilm or plastic wrap are placed around the gel to prevent direct contact between the stack of paper towels and the wick, which can cause a short circuit in the flow of the transfer solution.

DNA in the agarose gel is denatured *in situ*. For efficient blotting, after electrophoresis the gel is generally pretreated. The electrophoresed DNA is exposed to a short depurination treatment (0.25 M HCl), followed by alkali treatment. It shortens the DNA fragments at depurinated sites and denatures the fragments so that the DNA is single stranded and accessible for probing. The gel is then equilibrated in neutralizing solution prior to blotting. The use of positively charged nylon membranes removes the need for extended gel pretreatment.

DNA fragments from the gel get transferred to a membrane. The fragments are transferred from the gel to membrane by the capillary transfer method. DNA fragments are carried from the gel in a flow of liquid and deposited on the surface of the solid support of membrane. The liquid is drawn through the gel by capillary action, which is established and maintained by stack of dry and absorbent paper towels. This transfer is allowed to proceed for 12 h or longer. The paper towels and the 3 MM filter papers above the gel and membrane are removed and the membrane laid on a dry sheet of 3 MM filter papers. The membrane is dried on filter paper, followed by baking in an oven at 80°C for 2 h. DNA can also be fixed to the membranes by the UV crosslinking method. It is based on the formation of crosslinks between a small fraction of thymine residues in the DNA and positively charged amino groups on the nylon membrane.

Following the fixation step, the membrane can then be used for the hybridization labeling reaction. Probes can be labeled single-stranded DNA, oligodeoxynucleotides, or an RNA that is a complementary sequence to the blotted DNA. Labeling of probe can be done by radioactive or nonradioactive methods, which has been discussed in Chapter 17. For maximum

Fig. 14.6 Agarose gel electrophoresis and Southern blotting.

hybridization of probe nucleic acid sequence to the blotted DNA the perfect conditions are chosen because unlike solution hybridizations, membrane hybridizations tend not to proceed to completion. In membrane hybridization some of the bound nucleic acid is embedded and is inaccessible to the probe. Prolonged incubations may not generate any significant increase in detection sensitivity. The composition of hybridization buffer can greatly affect the speed of reaction and the sensitivity. Dextran sulphate and other polymers act as volume extruders to increase the rate and extent of hybridization. Dried milk, heparin and detergents such as SDS have been used to depress non-specific binding of the probe. Formamide can be used to depress the melting temperature of the hybrid so that reduced temperatures of hybridization can be used.

After hybridization, the membrane is washed to remove unbound probe and regions of hybridization are detected as per the protocol for a particular probe used. If a radioactive probe sequence has been employed, the regions of hybridization are detected by the autoradiographic method.

Northern blot analysis

Northern hybridization is the blotting of RNA separated by gel electrophoresis on nylon or nitrocellulose membranes followed by hybridization with nucleic acid (RNA or DNA) sequences. The procedure of northern blotting is similar to Southern blotting except that RNA rather than DNA is blotted on the membrane. Further, in Southern hybridization, DNA has to be denatured before blotting, which is not needed in northern hybridization. Since RNA was not able to bind nitrocellulose, initially northern blotting was carried out on chemically reactive paper, where RNA sequences are bound covalently. The RNA was blotted or immobilized on diazotized cellulose [diazobenzyloxymethyl (DBM) cellulose)] paper (Alwine *et al.*, 1977). Subsequently *o*-aminophenylthioether (APT) cellulose was developed, which was easier to prepare and more stable than DBM cellulose. Subsequently, various other methods were developed where it was shown that RNA denatured by various chemical treatments binds tightly to nitrocellulose and hybridizes to radioactive probes with high efficiency. Thus, presently nitrocellulose membranes are used for blotting of RNA under appropriate conditions because of the convenience of using these membranes. Attachment of denatured RNA to nitrocellulose is presumed to be noncovalent but is essentially irreversible. It is therefore possible to hybridize sequentially RNA immobilized on nitrocellulose to a series of radioactive probes without significant loss of the bound nucleic acid.

Proteins blotting

Western blotting: This refers to blotting of electrophoresed protein bands from a SDS-polyacrylamide gel on to a membrane (nylon or nitrocellulose) and their detection with antibody probes (Fig. 14.7). The sample is electrophoresed on SDS-PAGE, which separates the proteins on the basis of size, resulting in a series of bands down the gel. Because the gel matrix does not allow large antibodies to enter rapidly, the sample proteins must be transferred to a solid support. The gel is overlayed with a nitrocellulose or nylon sheet, and an electric field is applied so that proteins migrate from the gel to the sheet where they become bound. The nitrocellulose membrane thus has an exact image of the pattern of proteins that was in the gel. This particular form of blotting is called western blotting. The western blot is incubated with a protein such as casein to bind to nonspecific protein binding sites and hence prevent spurious binding of antibody molecules in subsequent steps. This step blocks nonspecific binding sites. The membrane is then incubated in a solution containing the labeled antibody that recognizes the protein of interest (primary antibody level). Unreacted antibody is washed away and those protein bands that have bound the antibody become visible and are identified. The method of visualization depends on how the antibody was labeled. If it has been labeled with ^{125}I, then autoradiography is carried out to detect the radioactive protein bands. The second approach is that the primary antibody is not labeled but the membrane with first antibody is detected by incubating with second antibody that recognizes the first antibody. For example, if the first specific antibody was raised in rabbits, the second antibody could be a goat anti-rabbit antibody. The second antibody is labeled and its binding is detected by autoradiography (if radio-labeled) or it could be conjugated to an enzyme that generates a colored product or fluorescent tag. The advantage of the second method is that a single preparation of labeled second molecule can be employed as general detector for different probes.

Dot blot technique

Southern and northern blotting are widely used in molecular biology, but these techniques require extensive purification of nucleic acid samples and the entire process is time consuming and

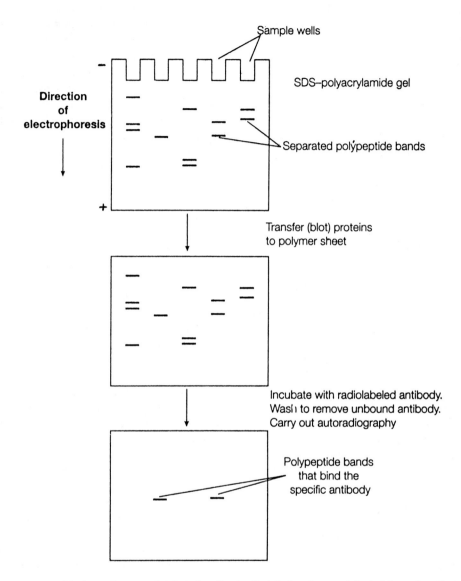

Fig. 14.7 Western blotting using a radiolabeled antibody. Protein bands are detected by autoradiography.

expensive, although indispensable for structural analysis of nucleic acids. However, if only the detection and quantitation of any given sequence is required, it is possible to shorten the procedure to avoid purification steps, electrophoresis, and blotting of the gel.

The purified nucleic acid sample (DNA or RNA) may be applied directly to nitrocellulose filter paper either by "dotting" it on with a pipette or an apparatus is used for placing the spots on the membrane through slots made in the equipment. Subsequent steps involve hybridization and detection. As the sample is normally applied in a circular form (either as a drop or oblong slot) the exposure of the membrane filter to detection procedure will be visualized as a "dot" or "slot" hence the "dot and slot blot." Double-stranded DNA does not bind efficiently to the filters and it is therefore necessary to denature the DNA by 0.4 M NaOH.

The major applications of this technique are: (1) rapid detection of specific sequences, and (2) determination of the relative amounts of any given species or sequence of RNA or DNA in a complex sample. The intensity of hybridization of a nucleic acid sample is quantified by densitometer scanning of autoradiographic signals. This helps in comparing quantities of any given nucleic acid sequence in many samples.

Autoradiography

Autoradiography is a process of photographic image formation, caused by exposure of film to emissions from radioisotopes. As the image density will be at a maximum close to the source of radiation, autoradiography allows both the detection and localization of radioactive molecules in a gel or on the surface of a filter or thin layer chromatography (TLC) plate placed in contact with a sheet of film. Techniques that require autoradiography include protein synthesis *in vitro*, colony hybridization, Southern blotting, DNA sequencing, dot blotting, etc.

The power of autoradiography lies not only in its very high sensitivity but also in its ability to discriminate between labeled and unlabeled molecules. For example, high sensitivity is needed for the detection of radioactively labeled "probes," hybridized to complementary sequences of nucleic acids which are immobilized on sheets of nitrocellulose by blotting processes or colony hybridization. One of the most dramatic examples of autoradiography is the sequencing of DNA, in which the sequence is actually read from an autoradiograph of the sequencing gel.

The procedure used for autoradiography depends on the radioisotope being employed, the levels of radioactivity being incorporated, the medium in which the radioactive compounds are found, and on the degree of resolution and speed of analysis required. In every case, photographic film is used to detect the radiation emitted by isotopes, and it is because of the nature of photographic process that there is no single, universally acceptable method of autoradiography. A basic understanding of this process can be useful in deciding which method of autoradiography will produce the optimum results.

Photographic film consists of a support, such as plastic sheet, coated with a thin layer of an emulsion that contains microscopic crystals of a silver halide suspended in gelatin. If these crystals are treated with a reducing agent or developer, they will be converted to metallic silver. However, the activation energy for this reduction is greatly decreased if there are already some silver atoms in the crystals, and in such cases the crystal is converted to silver far more rapidly. Once conversion of Ag$^+$ to Ag atom begins in a crystal, the process is autocatalytic, accelerating as more Ag atoms are formed. Clearly with limited development, only those crystals that already contain some silver atoms will be further reduced to metallic silver. This provides the basis of the photographic process since exposure of the emulsion to radiation will result in the formation of a few silver atoms in some of the crystals, thus producing a latent image, which can be revealed by limited development. After development, and before further exposure to light, any undeveloped silver halide is removed by being dissolved in a fixer solution such as sodium thiosulphate, thus preventing any further formation of silver.

The exact mechanism is not known, but it appears that radiation absorbed by a crystal of silver halide can promote an electron into a "conduction band," allowing it to move through the crystal until trapped in a sensitivity speck. Once trapped, the electron can combine with a mobile interstitial Ag$^+$ ion to form Ag, and this then act as an efficient trap for the next electron produced in the crystal, which will also combine with an Ag$^+$ ion. Formation of the first Ag is a reversible process, but two or more silver atoms forms a stable speck in the crystal. Hence the latent image is stable only if it is composed of at least two Ag atoms per crystal. Absorption of either β or γ emissions by a silver halide will result in the formation of several Ag atoms, and so a latent image is formed.

In direct autoradiography, the sample present in a gel or filter is placed in contact with a sheet of X-ray film, and left to form a latent image. However, this process will work if emitted radiations can reach and be absorbed by the film. This method is ideal for use with ^{14}C or ^{35}S, particularly when the compounds are on a surface (e.g. nitrocellulose filter or TLC), since β emissions can penetrate into the X-ray film, where they be absorbed by the photographic emulsion. However, this cannot be used for detecting tritiated compounds since they produce very low-energy β-emissions and can rarely penetrate the protective coating of X-ray film. When direct autoradiography is used with ^{32}P or ^{125}I, unfortunately they are so highly penetrating that majority of the emissions pass completely through the film without being absorbed and hence without forming a latent image.

The procedure for direct autoradiography is very simple. The sample (a dried gel, membrane) is placed in firm contact with a sheet of "direct" type X-ray film in the dark. This is done using a commercial cassette, where the sample and film can be sandwiched between two sheets of perplex, glass, or aluminium taped or clipped together. The assembly is then left at room temperature for several hours or days (depending on the level of radioactivity in the sample) to allow adequate exposure of the film. Development and fixing of the film are then carried out according to the manufacturer's instructions. If the sample is a nondried gel, e.g. sequencing gel, it is necessary to carry out exposure at −70°C to stop any diffusion of molecules within the gel.

Indirect autoradiography (fluorography) is employed when either low-energy radioactive compounds (e.g. ^3H) or high-energy compounds (^{32}P or ^{125}I) are used. The principle of fluorography is that the energy of emissions is converted by a scintillant into light, which then causes the

formation of a latent image in the film. For low-energy emission compounds ($-^3H-$), if the sample is gel or membrane, it is impregnated with a solution of scintillant. The most commonly used scintillant is 2,5-diphenyloxazole (PPO). After drying under vacuum, the impregnated gel is covered with a sheet of thin plastic film (Saran wrap or cling film) and placed in contact with a sheet of preexposed screen or medical X-ray film, in the dark. The gel and film are sealed in a cassette and left at −70°C for exposure of the film. Since the latent image is formed by light rather than by β-emissions, the process is known as fluorography.

When radioactive compounds that emit high-energy emissions are used, the gels or membranes are not impregnated with scintillant (PPO). In such cases, "intensifying screens," which contain a dense solid scintillant (usually calcium tungstate) are used. The intensifying screens are placed against the film, on the far side from radioactive sample. Any emissions passing through the photographic emulsion are absorbed by the screen and converted to light, effectively superimposing a photographic image upon the direct autoradiographic image.

It has been reported that small amounts of radioactivity are under-represented with the use of fluorography and intensifying screens. This problem can be overcome by a combination of preexposing a film to an instantaneous flash of light (preflashing) and exposing the autoradiograph at −70°C. Preflashing provides many of the silver halide crystals of the film with a stable pair of silver atoms and −70°C temperature increases the stability of a single silver atom increasing the time available to capture a second photon.

E. coli transformation

The process of transferring exogenous DNA into cells called "transformation." Thus it refers to any application in molecular biology in which purified DNA is actively or passively imported into host cells using nonviral methods. There are basically two general methods for transforming bacteria.

Log phase *E. coli* culture

↓ Centrifuge

Resuspend bacterial pellet in ice cold 50 mM CaCl₂ solution

↓

Chill on ice for 30 min

↓

Aliquot competent cells in vials (200 µl)
(Chill on ice)

↓ Plasmid DNA (*Amp*ᴿ)

Chill on ice

↓ Heat shock at 42°C for 2 min

Plating on LB media containing ampicillin

↓

Ampicillin resistant colonies
(Transformed *E. coli* cells)

Fig. 14.8 Calcium chloride method of *E. coli* transformation.

1. *Chemical transformation method:* The chemical method utilizes CaCl₂ and heat shock to stimulate the entry of DNA into cells (Fig. 14.8). Cohen *et al.* (1972) showed that CaCl₂-treated *E. coli* cells are effective recipients for plasmid DNA. Almost any strain of *E. coli* can be transformed with plasmid DNA with varying efficiency. But it is known that many bacteria contain restriction systems, which can influence the efficiency of transformation. Thus it is essential to use a restriction-deficient strain of *E. coli* as a transformable host. Bacterial cells in the log phase which are competent (cells that are in a state in which they can be transformed by DNA in the environment) are used for transformation. Cells are incubated with DNA on ice in a solution containing CaCl₂, followed by a brief heat shock at 42°C. These cells are allowed to recover for a short time in the growth media and then plated on agar plates containing appropriate antibiotics. In most cases it is possible to obtain ~10^6–10^8 antibiotic-resistant colonies per microgram of

circular plasmid DNA using this method. The mechanism of bacterial transformation is not known, but some studies have suggested that the DNA enters the cell through protein pores that are activated by both calcium and heat. Numerous modifications of the original CaCl₂ procedure have been developed in an attempt to increase the transformation efficiency of various *E. coli* strains. For example, addition of hexamine cobalt chloride or dimethyl sulfoxide to the CaCl₂ solution was found to increase transformation efficiencies by 10 to 100-fold.

2. *Electroporation method:* It uses a short pulse of electric charge to facilitate DNA uptake (Fig. 14.9). Bacterial cells are suspended in

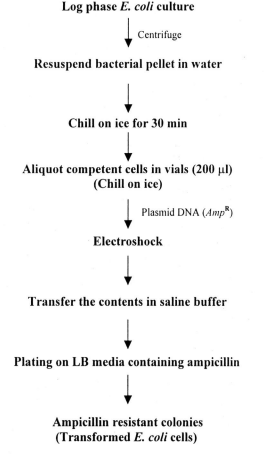

Log phase *E. coli* culture

↓ Centrifuge

Resuspend bacterial pellet in water

↓

Chill on ice for 30 min

↓

Aliquot competent cells in vials (200 µl)
(Chill on ice)

↓ Plasmid DNA (*Amp*ᴿ)

Electroshock

↓

Transfer the contents in saline buffer

↓

Plating on LB media containing ampicillin

↓

Ampicillin resistant colonies
(Transformed *E. coli* cells)

Fig. 14.9 Electroporation method of *E. coli* transformation.

a low ionic strength solution in a special cuvette and electric shock is given. Electroporation was first applied to animal cells, but lately by increasing the field strength using high voltage power sources, electroporation parameters were found that worked well for bacteria. The mechanism of DNA transformation is poorly understood but it is thought that short pulse (~3 msec) of electric charge (2.5 kV, ~200 ohms, ~30 microfaradays) may cause a transitory opening in the cell wall or perhaps activate membrane channels. The major advantages of electroporation are that high transformation frequencies of 10^8–10^{10} transformants/ mg plasmid DNA are routinely obtained, and once optimized, electroporation works well with most bacterial strains.

Protocol

Agarose gel electrophoresis

Agarose gel electrophoresis is a simple and highly effective method for separating, identifying, and purifying DNA fragments. A general procedure is given here since it is routinely used in molecular biology experiments.

1. Seal the ends of a suitable plastic gel-casting platform with an adhesive tape.
2. Prepare (0.6–0.8%) agarose mix depending on the experimental requirements in 1× TAE electrophoresis buffer and boil with gentle stirring till a homogeneous clear solution is formed. Cool to 55°C.
    TAE buffer (1×)
    40 mM Tris-acetate (pH 8.0)
    1 mM EDTA (pH 8.0)
3. Pour agarose into the gel mould and place the comb; allow to set for at least 30 min.
4. Remove the comb carefully and pull the tape off the gel-casting platform.
5. Keep the gel in an electrophoresis unit

containing 1× TAE buffer. The gel should be completely submerged in the buffer.
6. Add 1/10th volume of the tracking dye (10× stock) to the DNA sample. Mix by tapping and spin for 2–3 sec in a microfuge.
    Tracking dye (10× stock)
    0.1 M EDTA pH 8.0
    0.01 M Tris-Cl pH 8.0
    Bromophenol blue (0.25%)
    Glycerol (50%)
7. Load the DNA sample carefully. Load a suitable molecular weight marker in the first as well as the last lane along with the DNA samples. Attach the electrodes. (Make sure that the negative black terminal is at the same end of the unit as the wells.)
8. Turn on the power supply and run at 20 mA power current. Run for an appropriate time by looking at the tracking dye.
9. Disconnect the current after an appropriate time. Remove the gel from electrophoresis unit and immerse it in an ethidium bromide stain (10 mg/ml) for 20–30 min. (Gloves should be worn while handling ethidium bromide.)
    Ethidium bromide (10 mg/ml)
    Dissolve 100 mg ethidium bromide in 10 ml distilled water and store at 4°C. Working staining solution is 20 µl ethidium bromide stock (10 mg/ml) to 200 ml distilled water.
10. View the gel on a UV transilluminator and photograph the gel.

Further reading

Hames, B.D., Hooper, N.M. and Houghton, J.D. 1998. *Instant Notes in Biochemistry*, Bios Scientific Publishers Ltd., Oxford.

Miesfeld, R.L. 1999. *Applied Molecular Genetics*. Wiley-Liss, New York.

Old, R.W. and Primrose, S.B. 1995. *Principles of Gene Manipulation*. Blackwell Science, UK.

Sambrook, J., Fritsch, E.F. and Maniatis, T. 1989. *Molecular Cloning: A Laboratory Manual*. Cold Spring Harbor Laboratory Press, USA.

Walker, J.M. and Gaastra, W. (Eds.). 1987. *Techniques in Molecular Biology*. Croom Helm, Australia.

15

Gene Cloning:
Cutting and Joining DNA Molecules

The term *gene cloning* can be defined as the isolation and amplification of an individual gene sequence by insertion of that sequence into a bacterium where it can be replicated. Cloning a fragment of DNA allows indefinite amounts to be produced from even a single original molecule. Cloning technology involves the construction of novel DNA molecules by joining DNA sequences from different sources. The product is often described as recombinant DNA. The construction of such composite or artificial DNA molecules has also been termed *genetic engineering* or *gene manipulation* because of the potential for creating novel genetic combinations by biochemical means. A precise legal definition of gene manipulation has been given as follows: The formation of new combinations of heritable material by the insertion of nucleic acid molecules, produced by whatever means outside the cell, into any virus, bacterial plasmid or any other vector system so as to allow their incorporation into a host genome in which they do not naturally occur but in which they are capable of continued propagation.

The basic events in gene cloning have been described in the following steps and shown in Fig. 15.1.

1. Isolation of the gene of interest.
2. A fragment of DNA (gene of interest) to be cloned is incorporated into a small replicating (usually circular) DNA molecule called a *vector*. This can be an *E. coli* plasmid, a virus, cosmid, etc. The vector with an incorporated gene is called a *recombinant vector* or *molecule*.
3. The recombinant vector is introduced into a host cell by transformation.
4. Cells that have acquired recombinant DNA molecule are selected.
5. Recombinant DNA molecule is multiplied within the host cell to produce a number of identical copies of the cloned gene.

In gene cloning, a fragment of DNA that codes for a required gene product has to be removed from the host organism and moved to a vector molecule (e.g. plasmid, phage, cosmid) to construct recombinant DNA molecule. This involves cutting DNA molecules at specific sites and joining them together again in a controlled manner. In order to understand the process of gene cloning, i.e. to insert a piece of foreign DNA into a vector for construction of recombinant DNA molecule, one must understand the various vectors, mechanism of cutting and joining DNA molecules at specific sites, the use of agarose gel electrophoresis as a method for monitoring the cutting and joining reaction, and a means of monitoring transformation of *E. coli*. Thus, a gene cloning experiment has four essential components, which have been dealt in details to explain

Fig. 15.1 Fundamental steps of a cloning experiment. The cloning experiment involves (1) the production and preparation of vector and foreign (passenger) DNA, (2) the ligation of vector and passenger DNA so as to construct a recombinant DNA molecule, (3) the transformation of suitable host cells, and (4) the selection of those cells which have acquired recombinant DNA molecules.

the process of gene cloning.

1. Enzymes for cutting and joining the DNA fragments into vector molecules.
2. Cloning vehicles or vectors.
3. DNA fragments, i.e. gene libraries.
4. Selection of a clone of transformed cells that has acquired the recombinant chimaric DNA molecule.

ENZYMES FOR CUTTING – RESTRICTION ENDONUCLEASES

Present-day DNA technology is totally dependent on the ability to cut DNA molecules at specific sites with restriction endonucleases. Before 1970, there was no method available for cutting a duplex DNA molecule. It became apparent that related phenomenon of host-controlled restriction and modification might lead towards a solution to this problem when it was discovered that restriction involves specific endonucleases. What is host-controlled restriction and modification? Restriction means the identification of incoming DNA to the cell (e.g. bacteria) and its destruction by cleaving into pieces if it is recognized as foreign DNA. Modification is the protection of the cell's own DNA by methylation of certain bases so that host DNA is not cleaved. The phenomenon of restriction and modification has been explained by the behavior of λ phage on two *E. coli* host strains C and K (Fig. 15.2). In the bacteria/bacteriophage system, when the incoming DNA is a bacteriophage genome to the bacteria, the effect is to reduce the efficiency of plating (EOP), i.e. to reduce the number of plaques formed in plating test.

If a stock preparation of phage lambda is grown on *E. coli* strain C and is then titred on C and K strains of *E. coli* . The titres observed on these two strains differed with lower titer on K strain of *E. coli*. The phage is said to be restricted by the second host strain *E. coli* K. If those phages that result from infection on *E. coli* K are collected and replated on *E. coli* K, then they are no longer restricted. But if they are again first grown on *E. coli* C and then on *E. coli* K, they are again restricted when plated on *E. coli* K. Thus the efficiency with which the phage plates on a particular host strain depends on the strain on which it was last propagated. This nonheritable change conferred upon the phage by the second host strain (*E. coli* K) that allows it to be replated on that strain without any restriction is called modification. It was observed by radioactive labeling experiments that when restricted phages adsorb on the surface of restrictive hosts and inject their DNA, it is degraded soon. The endonuclease responsible for this degradation is called restriction endonuclease or restriction enzyme. The restrictive host must protect its own DNA from degradation, and its DNA must be

modified. The modification involves methylation of certain bases at a very limited number of sequences within the DNA, which constitutes the recognition sequences for the restriction endonucleases. Thus, the restriction enzyme consists of two enzymatic elements: nuclease and methylase. The genes that specify host-controlled restriction and modification systems may reside on the host chromosome itself or may be located on a plasmid or prophage such as P1. Different enzymes recognize different but specific sequences, which range from 4 to 8 base pairs.

During construction of a recombinant DNA molecule, the circular vector must be cut at a single point into which the fragment of DNA to be cloned can be inserted. Not only must each vector be cut just once, but also all the vector molecules must be cut at precisely the same position. Restriction endonucleases are enzymes that make site-specific cuts in the DNA. The first enzyme was isolated from *Haemophilus influenzae* in 1970. Now, many types of restriction endonucleases have been isolated from a wide variety of bacteria, which calls for a uniform nomenclature and classification. According to the recommendation of Smith and Nathans (1973), each enzyme is represented by a three-letter code in italics. For example, *Hae* for *Haemophilus aegypticus* and *Hin* for *Haemophilus influenzae*. Sometimes a four-letter code is used when the enzymes are isolated from different serotypes of a species. The letter code denotes the origin of enzyme. For example, *Hinf* for *Haemophilus influenzae* serotype f. If more than one enzyme is isolated from a single origin, then these are

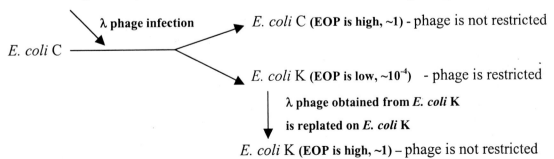

Fig. 15.2 Host controlled restriction and modification of λ phage on *E. coli* strain C and K analyzed by efficiency of plating.

denoted by Roman numerals (I, II, III, etc.). For example, *Eco*RI for the first enzyme of *Escherichia coli* serotype R.

There are three types of restriction endonucleases (Yuan, 1981).

Type I endonucleases

1. They are complex nucleases, e.g. *Eco*K and *Eco*B.
2. They function simultaneously as an endonuclease and a methylase and require ATP, Mg++, and S-adenosyl methionine as cofactors.
3. They are single multifunctional enzymes with three different subunits: a restrictive subunit, a modification subunit, and a specificity subunit, which determines the recognition specificity.
4. The recognition site of class I enzymes is 15 bp in length, which can be methylated at the adenine position in the recognition site, and the cleavage site is approximately 1000 bp away from the 5' end of the sequence TCA located in the 15-bp recognition site. DNA molecules with recognition sequences methylated at both adenine residues on either strand are resistant to type I endonucleases. Cleavage occurs when the recognition site is unmethylated.
e.g.

*Eco*K 5' – A A*C N N N N N N G T G C –3'
3' – T T G N'N'N'N'N'N'N'C A.C G –5'

* Show methylation at adenine position; N— unspecified or any base base; N'—corresponding complementary base to N. Heteroduplex molecules consisting of one modified and one unmodified strand will always be methylated by the enzyme. Unmodified DNAs are restricted, i.e. cleaved.

5. These enzymes exhibit high ATPase activity. For each phosphodiester cleavage, more than 10,000 molecules of ATP are hydrolyzed.
6. These enzymes show specificity for the recognition site but not for their cleavage

site. This produces a heterogeneous population of DNA fragments. In view of this, it is very undesirable to use in the gene cloning techniques.

Type II endonucleases

1. They are simple enzymes that consist of a single polypeptide. They have separate endonuclease and methylase activities.
2. They recognize a specific nucleotide sequence and cut a DNA molecule at this sequence and nowhere else. Most of these enzymes recognize a hexanucleotide sequence, but others recognize tetra-, penta-, or octanucleotide sequences on the DNA. For example, the restriction endonuclease *Pvu*I isolated from *Proteus vulgaris* cuts DNA at the hexanucleotide CGATCG, whereas a second enzyme *Pvu*II isolated from the same bacterium cuts at a different hexanucleotide CAGCTG.
3. They are very stable and only require Mg++ as cofactors.
4. The majority of the known recognition sequences have an axis of rotational symmetry (palindromic sequences), i.e. the sequences read the same in either direction in opposite strands. DNA sequences are written from left to right in the 5'→3' direction. For instance, the recognition sequence of *Eco*RI is as follows.

5'—G A A T T C —3'
3'—C T T A A G —5'

5. Many restriction endonucleases make a simple double-stranded cut in the middle of the recognition sequence, resulting a blunt or flush end. *Pvu*II, *Alu*I, *Hae*III, etc. are examples of blunt-end cutters (Fig. 15.3).
6. Other restriction endonucleases do not cut the two DNA strands at exactly the same position. Instead, the cleavage is staggered usually by 2 or 4 nucleotides, so that the resulting DNA fragments have short single-stranded overhangs at each end. These are called sticky or cohesive ends, as base

Recognition sequence   Cleavage pattern

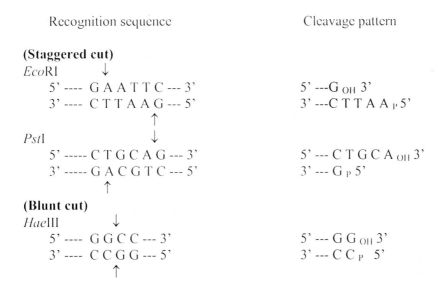

(Staggered cut)

*Eco*RI

5' ---- G A A T T C --- 3'   5' ---G $_{OH}$ 3'
3' ---- C T T A A G --- 5'   3' ---C T T A A $_P$ 5'

*Pst*I

5' ----- C T G C A G --- 3'   5' --- C T G C A $_{OH}$ 3'
3' ----- G A C G T C --- 5'   3' --- G $_P$ 5'

(Blunt cut)

*Hae*III

5' ---- G G C C --- 3'   5' --- G G $_{OH}$ 3'
3' ---- C C G G --- 5'   3' --- C C $_P$ 5'

Fig. 15.3 Type II restriction endonuclease cleavage patterns.

pairing between them can stick the DNA molecules back together again. The staggered cut may result in overhanging ends either at 5' (*Eco*RI) or 3' (*Pst*I). The 3'-end always carries a hydroxyl group whereas the 5'-end always carries a phosphate group (Fig. 15.3). The cleavage patterns of some restriction enzymes are shown in Table 15.1.

7. The DNA ends of one molecule generated by a given enzyme with a given recognition sequence can form complementary base pairs with any other DNA molecule, provided they possess the same ends. The cleavage products of *Eco*RI is shown in Fig. 15.4. Hence, these enzymes are widely used in gene technology.

Table 15.1 Catalog of some restriction endonucleases with their recognition sequences

Enzyme	Producing organism	Recognition sequence[a]
Staggered cut (sticky ends)		
*Bam*HI	*Bacillus amyloliquefaciens* H	G↓GATCC
*Bgl*II	*Bacillus globigi*	A↓GATCT
*Eco*RI	*Escherichia coli* R	G↓AATTC
*Hind*III	*Haemophilus infuenzae* Rd	A↓AGCTT
*Pst*I	*Providencia stuartii*	CTGCA↓G
*Sal*I	*Streptomyces albus* G	G↓TCGAC
*Taq*I	*Thermus aquaticus*	T↓CGA
*Pst*I	*Providencia stuartii*	CTGCA↓G
Blunt ends		
*Alu*I	*Arthrobacter luteus*	AG↓CT
*Hae*III	*Haemophilus aegypticus*	GG↓CC
*Hpa*I	*Haemophilus parainfuenzae*	GTT↓AAC
*Pvu*II	*Proteus vulgaris*	CAG↓CTG
*Sau*3A	*Staphylococcus aureus* 3A	↓GATC
*Sma*I	*Serratia marcescens*	CCC↓GGG

↓ indicates the site of enzyme cut within the specified nucleotide sequence.
[a]The sequence shown is of one strand in the 5' → 3' direction.

Fig. 15.4 Cleavage pattern of *Eco*RI and joining of restriction fragments due to complementary base pairs.

8. Some restriction endonucleases possess identical recognition and cleavage sites that are called isoschizomers, e.g. *Hin*dIII and *Hsu*I

$$5'—A\ A\ G\ C\ T\ T\ —3'$$
$$3'—T\ T\ C\ G\ A\ A\ —5'$$

9. Some enzymes possess the same recognition site but differ in the cleavage site. *Sma*I and *Xma*I have recognition site as 5'—C C C G G G—3', but their cleavage sites are different (Fig. 15.5).

10. Some restriction endonucleases with different recognition sequences may produce the same sticky ends, that is, generate identical overlapping termini. Examples are *Bam*HI (recognition sequence GGATCC), *Bgl*II (recognition sequence AGATCT), and *Sau*3A (recognition sequence GATC). All these produce GATC sticky ends. Fragments of DNA produced by cleavage with either of these enzymes can be joined to each other, as each fragment will have a complementary sticky end.

The number of type II enzyme cleavage sites in a DNA molecule depends on the size of DNA, its base composition, and the GC content of the recognition sequences. The probability P for a particular tetranucleotide consisting of all four bases to occur in a DNA molecules containing 50% AT and 50% GC bp is $(1/4)^4$. Tetranucleotide cleavage occurs at every 256 bp distance and hexanucleotide at every 4096 bp.

Star activity is another property shown by type II restriction endonucleases, i.e. the enzyme changes its specificity. It arises mainly due to the change in reaction conditions from the optimum, e.g. increasing the pH, lowering the concentration of NaCl, or replacing magnesium by manganese, and also due to the methylation of DNA. For example, *Eco*RI normally recognizes and cleaves the sequence 5'—G A A T C—3'. When the reaction condition changes, it recognizes a tetranucleotide sequence 5'—A A T T—3'.

Type III endonucleases

1. They have two different subunits. One unit is responsible for site recognition and modification while the other is responsible for nuclease action.
2. They require ATP and Mg++ as cofactors.
3. The recognition sites are asymmetric. These nucleases cleave DNA at specific sequences, which are nonpalindromic. Cleavage takes place by nicking one strand in the immediate vicinity of the recognition site,

$$\downarrow$$
*Sma*I 5' — C C C G G G 3' ——> 5' C C C$_{OH}$ 3'
3' — G G G C C C 5' ——> 3' G G G$_P$ 5'
$$\uparrow$$

$$\downarrow$$
*Xma*I 5'— C C C G G G 3' ——> 5' C$_{OH}$ 3¢
3' – G G G C C C 5' ——> 3' G G G C C$_P$ 5'
$$\uparrow$$

Fig. 15.5 Cleavage pattern of type II enzymes with same recognition site but differs in cleavage site.

usually 25–27 bp downstream of the recognition site. Two sites in opposite orientations are necessary to break the DNA duplex. Recognition sequences of some of the type III enzymes are given below:

$$5'— G \ A \ C \ G \ C \ (N)_5 \ — 3'$$
$$3'— C \ T \ G \ C \ G \ (N)_{10} \ — 5'$$
*Hga*I

$$5'— G \ A \ A \ G \ A \ (N)_8 \ — 3'$$
$$3'— C \ T \ T \ C \ T \ (N)_7 \ — 5'$$
*Mbo*II

$$5'— G \ G \ A \ T \ G \ (N)_9 \ — 3'$$
$$3'— C \ C \ T \ A \ C \ (N)_{13} \ — 5'$$
*Fok*I

4. In contrast to the staggered ends generated by type II endonucleases, the single-strand ends produced by type III endonucleases always differ from each other and thus cannot be recombined at random.
5. They lack ATPase activity, and S-adenosyl methionine is not required.
6. They produce a relatively homogenous population of DNA fragments, unlike fragments obtained by cleavage of type I enzymes.
7. They are not used for gene cloning because cleaved products are not uniform.

Other restriction enzymes

Some restriction enzymes have characteristics different from those of the above-mentioned three types. For example, the *Eco*571 consists of a single polypeptide that has both modification and nuclease activities, with the nuclease activity stimulated by S-adenosyl methionine. It has other properties of type II. Thus, this system should be reclassified from type II to type IV.

Another system *Mcr* (modified cytosine reaction) also falls outside the type I–IV classification. It was found that DNA from some bacterial and eukaryotic sources was restricted in some of the commonly used host strains of *E. coli*. The phenomenon is caused by methyl cytosine in the DNA, and was called modified cytosine reaction. There are two *Mcr* systems: (*Mcr*A) encoded by a prophage-like element and

(*Mcr*BC) encoded by two genes, *mcr*B and *mcr*C, which are located very close to the genes encoding the *E. coli* K type I system. The recognition site for *Mcr*BC is

$$R^mC \ (N_{40–80}) \ R^mC, \quad \text{where } R = A \text{ or } G$$

Cleavage occurs at multiple sites in both strands between the methyl cytosines and it requires GTP. The *Mcr*BC system is analogous to the *Rgl*B (restricts glucoseless phage) system discovered in T-even phages. It was observed that wild type T-even phages have glucosylated 5-hydroxymethyl cytosine instead of cytosine in their DNA. If glucosylation of 5-hydroxymethyl cytosine is prevented by mutation of the phage or host genes, the phage is restricted. It was not originally clear why *E. coli* should possess a system for restricting DNA containing 5-hydroxymethyl cytosine, since this is an uncommon form of DNA. But it is now apparent that the *Rgl*B (= *Mcr*BC) system that restricts DNA containing 5-hydroxymethyl cytosine, methyl cytosine or 4-methyl cytosine.

JOINING DNA MOLECULES

The enzymes used to join DNA fragments are called DNA ligases. When the vector DNA and foreign DNA are both cut with the same restriction enzyme, the overlapping ends of the vector and foreign DNA are compatible and complementary. These DNA fragments and vector molecules are mixed together, which forms complementary base pairs between overlapping terminal single-stranded DNA sequences. Ligases act on DNA substrates with 5' terminal phosphate groups and form the phosphodiester bond between the two DNA sequences (vector molecule and the DNA to be cloned) to join them together. This is the final step in construction of a recombinant DNA molecule. This process is called *ligation.*

DNA ligase

E. coli and bacteriophage T4 encode the enzyme DNA ligase, which seals single-stranded nicks between adjacent nucleotides in a duplex DNA chain. The reaction catalyzed by the enzymes of *E. coli* and T4 phage infected *E. coli* is very similar, but they differ in their cofactor requirements. The T4 enzyme requires ATP, whereas the *E. coli* enzyme requires NAD⁺. Both of these catalyze the formation of a phosphodiester bond between adjacent 3' hydroxyl and 5' phosphate termini in DNA (Fig. 15.6). In each case, the cofactor is split and forms an enzyme–AMP complex. The complex binds to the nick and makes a covalent bond in the phosphodiester chain.

Bacteriophage T4 DNA ligase is often used in gene cloning because it can join DNA molecules of annealed cohesive DNA termini or nicks as well as join blunt-ended double-

Fig. 15.6 Action of DNA ligase. An enzyme–AMP complex binds to a nick bearing 3' OH and 5' P groups. The AMP reacts with the phosphate group. Attack by the 3' OH group on this moiety generates a new phosphodiester bond, which seals the nick. Reproduced from *Principles of Gene Manipulation*, R.W. Old and S.B. Primrose, by kind permission of Blackwell Science, Osney Mead, Oxford, U.K.

stranded DNA molecules, whereas *E. coli* DNA ligase is ineffective in ligating blunt-ended DNA fragments.

DNA-modifying enzymes

Kinase

Bacteriophage T4 polynucleotide kinase catalyzes the transfer of γ-³²P phosphate of ATP to a 5' terminus of DNA or RNA.

$$5'—G_{OH}\ 3' \qquad\qquad \gamma.\overset{32}{\ }_{P\text{-}ATP}\quad 5'—G_{OH}\ 3'$$
$$3'—CpTpTpApA_{OH}\text{--}5' \longrightarrow 3—CpTpTpApAp^*\text{---}5'$$
Polynucleotide
kinase

Two types of reactions are commonly used. In the forward reaction, the gamma-phosphate is transferred to the 5' terminus of dephosphorylated DNA (i.e. the DNAs that lack the required phosphate residue). In the exchange reaction, an excess of ADP causes bacteriophage T4 polynucleotide kinase to transfer the terminal 5' phosphate from phosphorylated DNA to ADP; the DNA is then rephosphorylated by transfer of a radiolabeled gamma phosphate from γ³²P-ATP. Thus, this enzyme of T4 kinase has phosphorylation activity as well as phosphatase activity. This enzyme is used in radiolabeling the 5' termini for sequencing by the Maxam-Gilbert technique (see DNA sequencing methods) and for other uses requiring terminally labeled DNA. It is also used in phosphorylating synthetic linkers and other fragments of DNA that lack terminal 5' phosphates in preparation for ligation.

Alkaline phosphatase

It catalyzes the removal of 5' phosphate groups from the DNA and thus modify the termini of DNA.

$$5'—G_{OH}\ 3' \qquad\qquad \text{Alkaline}\quad 5'—G_{OH}\ 3'$$
$$3'—CpTpTpApAp\text{--}5' \longrightarrow 3'—CpTpTpApA_{OH}{}^*\text{--}5'$$
phosphatase

By treatment with alkaline phosphatase, both recircularization and plasmid dimer formation are prevented because ligase cannot join the ends (Fig. 15.7).

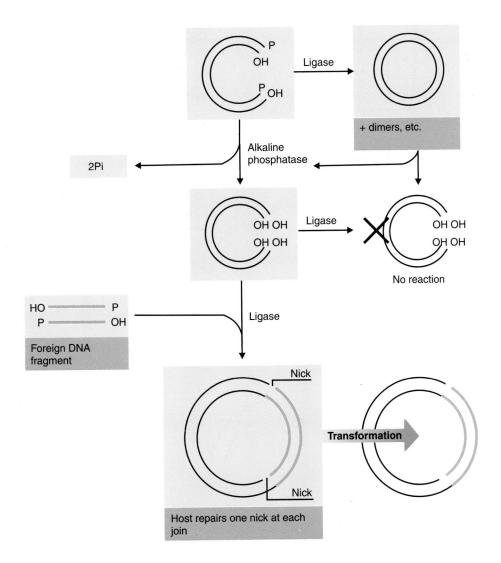

Fig. 15.7 Action of alkaline phosphatase to prevent recircularization of vector plasmid without insertion of foreign DNA. Reproduced from *Principles of Gene Manipulation,* R.W. Old and S.B. Primrose, by kind permission of Blackwell Science, Osney Mead, Oxford, U.K.

DNA polymerase

DNA polymerase I of *E. coli*, also known as DNA-dependent DNA polymerase, directs the synthesis of complementary nucleic acids using single-stranded DNA as a template. DNA synthesis requires a preexisting DNA or RNA primer with a 3′ hydroxyl to initiate *de novo* synthesis. The *in vitro* enzymatic synthesis of DNA is very important in genetic manipulation experiments. It converts single-stranded DNA to the double-stranded form, for example, by converting single-stranded tails on restriction fragments to blunt flush ends. DNA polymerase I is susceptible to protease action and splits into two fragments. The larger of these two fragments has polymerase and 3′→5′ exonuclease activity but lacks

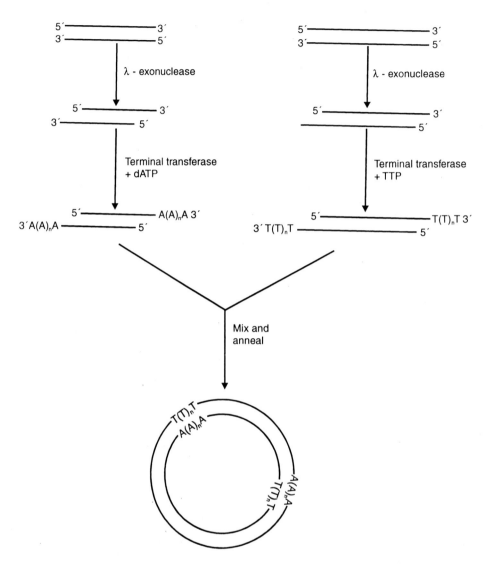

Fig. 15.8 Use of terminal transferase enzyme to add complementary homopolymer tails to two DNA molecules.

$5' \rightarrow 3'$ exonuclease activity, and is employed instead of the intact enzyme, known as **Klenow polymerase.**

$5'—G_{OH}\,3'$ Klenow
$3'—CpTpTpApAp—5'$ polymerase
              $\xrightarrow{}$
        dATP, dTTP

$5'—GpApApTpT_{OH}\,3'$
$3'—CpTpTpApAp\;\;5'$

Terminal transferase

This enzyme has been purified from calf thymus, and is known as terminal deoxynucleotidyl transferase. This enzyme can add oligodeoxynucleotide tails to the 3' ends of DNA duplexes. Thus, it provides the means by which the homopolymeric extensions can be synthesized, known as *homopolymeric tailing*. For example, if single deoxynucleotide triphosphate (dATP) is provided, it will repeatedly add nucleotides to the 3'-OH termini of a population of DNA molecules. Thus by adding oligo d(A) sequences to 3' ends of one population and oligo d(T) to the 3' ends of another population of DNA molecules, the two types of DNA molecules can anneal to form mixed dimeric circles. Jackson *et al.* (1972) applied for the first time the homopolymer tailing method when they constructed a recombinant in which a fragment of phage λ DNA was inserted into SV40 DNA using dA·dT homopolymers (Fig. 15.8). This has been extensively used in the construction of recombinant plasmids. An example of using dC·dG tailing is shown in Fig. 17.5.

S1-Nuclease

It converts cohesive ends of duplex DNA to blunt or flush ends or trimming away single-stranded ends. It is used when annealing of two incompatible ends requires overlapping ends to be removed, as shown below:

$$5'—G_{OH}\ 3' \quad \xrightarrow{\text{S1 Nuclease}} \quad 5'—G_{OH}\ 3'$$
$$3'—CpTpTpApAp–5' \qquad\qquad 3'—Cp\ 5'$$

λ-Exonuclease

This enzyme is used for 5' end modification. Thus, it removes nucleotides from the 5' ends of duplex DNA to create an improved substrate for terminal transferase.

$$5'—GpApApTpT_{OH}\ 3' \quad \xrightarrow{\text{λ-Exonuclease}} \quad 5'—GpApApTpT_{OH}\ 3'$$
$$3'—CpTpTpApAp–5' \qquad\qquad 3'—Cp\ 5' + Np5'$$

Exonuclease III

This enzyme is used for 3' end modification. It removes nucleotides from the 3' end of duplex DNA to create an improved substrate for manipulation.

$$5'—GpApApTpT_{OH}3' \quad \xrightarrow{\text{Exonuclease III}} \quad 5'—G_{OH}\ 3' + Np\ 3'$$
$$3'—CpTpTpApAp–5' \qquad\qquad 3'—CpTpTpApAp—5'$$

Bal 31 Nuclease

It can simultaneously degrade both the 3' and 5' strands of a DNA molecule, which results in DNA fragments that are shortened and possess blunt ends at both the termini. It is a highly specific single-stranded endodeoxyribonuclease

$$5'—GpApApTpT_{OH}3' \quad \xrightarrow{\text{Bal 31 nuclease}} \quad 5'\ pA_pA_{OH}\ 3'$$
$$3'—CpTpTpApAp–5' \qquad\qquad 3'\ TpTp\ 5'$$

Oligo and mono nucleotides

Linkers and adaptors

Recombinant DNA and gene cloning experiments take advantage of the ability of restriction enzymes to cut DNA at particular positions. If suitable enzyme sites are not available for manipulation on the DNA, then these cleavage sites can be added as linker or adaptor molecules. *Linkers* are synthetic single-stranded oligonucleotides, which self-associate to form symmetrical double-stranded molecules containing the recognition sequence for a restriction enzyme. For example, the oligonucleotide dCCGAATTCGC will self-associate to give a duplex structure containing the *Eco*RI recognition sequence (indicated by asterisks).

$$5'\ C\ C\ G^*A^*A^*T^*\ T^*\ C^*\ G\ G\ 3'$$
$$3'\ G\ G\ C\ T\ T\ A\ A\ G\ C\ C\ 5'$$

These nucleotides can be ligated to the blunt end of any DNA fragment and then cut with the specific restriction enzymes to generate DNA fragments with cohesive termini (Fig. 15.9).

Adaptor molecules are chemically synthesized DNA molecules within preformed cohesive ends. They are required when the target site for the restriction enzyme used is present in within the DNA sequence to be cloned. Consider a blunt-ended foreign DNA. A synthetic adaptor molecule is ligated to the foreign DNA. The adaptor is used in the 5' hydroxyl form to prevent self-polymerization (Fig. 15.10). The foreign DNA plus ligated adaptors is phosphorylated at the 5' terminus and ligated to the vector previously cut with a particular restriction enzyme.

PROTOCOL

15.1: Restrictive digestion of plasmid DNA

1. Pipette the following components in a clean micro centrifuge tube kept on ice. Add the restriction enzyme at the end. (Always make

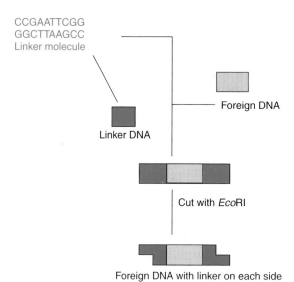

```
CCGAATTCGG
GGCTTAAGCC
Linker molecule
```

Linker DNA

Foreign DNA

Cut with *Eco*RI

Foreign DNA with linker on each side

Fig. 15.9 The use of linker for construction of DNA fragments with cohesive termini.

```
5' HO AATTCGGG OH 3'
3'      HOGCCC P  5'
Adaptor molecule
```

Adaptor

Foreign DNA

Foreign DNA with adaptor on each side

Fig. 15.10 The use of adaptors for construction of DNA fragments with cohesive termini.

a control in which DNA is not added.)
Appropriate 10x buffer
DNA sample (~ 0.5 µg)
2 – 3 Enzyme units (0.5 to 1µl)
Water to final volume of 10µl

2. Mix gently by vortexing or by finger tapping.
3. Incubate the reaction mixture at the recommended temperature (for most of the enzymes it is 37°C) for 2 h.

4. Keep it at −40°C till further analysis
5. Separate the DNA fragments on 0.8% mini agarose gel using 1xTBE buffer

   10xTBE buffer

   Tris base   108 g
   Boric acid   55 g
   0.5M EDTA (pH 8.0)   40 ml
   Make the volume to one liter
6. Add 2 µl of DNA loading dye in the digested sample and mix it well by finger tapping.
7. Load the DNA sample carefully. Load a suitable molecular weight marker (1-kb ladder) in the first and as well as the last lane along with the DNA samples. Attach the electrodes (make sure that the negative black terminal is at the same end of the unit as the wells).
8. Turn on the power supply and run at 20 mA power current. Run for an appropriate time by looking at the tracking dye.
9. Disconnect the current after an appropriate time. Remove the gel from electrophoresis unit and immerse it in an ethidium bromide stain. Working staining solution is 20 µl ethidium bromide stock (10 mg/ml) to 200 ml distilled water. Keep the gel in the staining solution for 20–30 min. (Gloves should be worn while handling ethidium bromide.)
10. View the gel on UV transilluminator and photograph the gel.

Further reading

Chawla, H.S. 1998. *Biotechnology in Crop Improvement*. International Book Distributing Co., Lucknow.

Grierson, D. and Covey, S.N. 1988. *Plant Molecular Biology*. Blackie, Glasgow.

Maniatis, T., Fritsch, E.F. and Sambrock, J. 1982. *Molecular Cloning: A Laboratory Manual*. Cold Spring Harbor Laboratory Press, USA.

Mantell, S.H., Mathew, J.A. and McKee, R.A. 1985. *Principles of Plant Biotechnology: An Introduction to Genetic Engineering in Plants*. Blackwell Scientific Publications, Oxford.

Old, R.W. and Primrose, S.B. 1985. *Principles of Gene Manipulation: An Introduction to Genetic Engineering*. Blackwell Scientific Publications, Oxford.

Sambrook, J., Fritsch, E.F. and Maniatis, T. 1989. *Molecular Cloning: A Laboratory Manual*. Cold Spring Harbor Laboratory Press, USA.

Winnacker, E.L. 1987. *From Genes to Clones: Introduction to Gene Technology*, VCH, Weinheim, Germany.

16

Gene Cloning: Vectors

One of the most important elements in recombinant DNA (rDNA) technology or in gene cloning is the vector. A random DNA or cDNA segment or specific gene is linked into a small replicating (usually circular) DNA molecule called vector to form rDNA molecule, which can be propagated in suitable host cells to a large number, is a cloning vector. There are different types of cloning vectors for use with different types of host cells. The largest number exists for *E. coli* and the best known of these is the plasmid vector. Besides, there are cosmids, phages, phagemids, yeast artificial chromosomes, transposons, bacterial artificial chromosomes, etc. A few of the vectors have been described.

A vector should have the following features:
1. It must contain a replicon that enables it to replicate in host cells.
2. It should have several marker genes, which help to differentiate the transformed cells from the nontransformed cells and also from the transformed cells, which contain recombinant DNA molecules, e.g. genes for ampicillin and tetracycline resistance.
3. It should have a unique cleavage site within one of the marker genes so that insertion of foreign DNA into the marker gene leads to its inactivation and identification of recombinant DNA molecule.
4. For the expression of cloned DNA, the vector DNA should contain suitable control elements, such as promoters, terminators, and ribosome binding sites.

Biology of *E. coli* K-12

Stanley Cohen and Herb Boyer chose *Escherichia coli* strain K-12 for their molecular biology experiments in the early 1970s because it was easy to grow and amenable to metabolic studies. During the last 30 years, *E. coli* K-12 has proved to be an innocuous biological host for the propagation of recombinant DNA molecules. The attenuated *E. coli* K-12 strain does not thrive outside of the laboratory environment and it is unable to compete against the more genetically robust *E. coli* serotypes normally found in the human intestine.

E. coli is a gram-negative rod-shaped bacterium about 2.5 μm long, which contains flagella and a genome of 4,639,221 bp encoding at least 4000 genes. The natural habitat of *E. coli* is the intestine, and it is able to grow in simple media containing glucose, ammonia (or peptide digests), phosphate and salts (Na^+, K^+, Ca^{2+}, and Mg^{2+}). *E. coli* contains both an outer and inner membrane separated by a protein-rich periplasmic space. The outer membrane contains proteins that function in the import and export of metabolites and also provide binding sites for infectious bacteriophage. Like all gram-negative bacteria, *E. coli* have no nuclear membrane and the chromosome is a large circular duplex molecule with a membrane attachment site and a single origin of replication. *E. coli* cells undergo cell division every 20 minutes when grown under optimal aerobic conditions, and they are facultative anaerobes, which allows them to be grown to high density using fermentation devices. Single *E. coli* colonies are isolated on standard bacterial agar plates and individual strains can be stored anaerobically in agar stabs or frozen as glycerol stocks.

    E. coli K-12 is a safe, nonpathogenic bacterium, but·specific laboratory practices are always warranted and specific biosafety guidelines must be followed for recombinant DNA research. *E. coli* K-12 as a standard host used in gene cloning experiments has several added properties, which has greatly increased the efficiency.

i) Bacterial restriction modification system poses the biggest problem for gene cloning. The *E. coli* K-12 restriction system (*Eco*K) requires three proteins encoded by *hsd*R, *hsd*M, and *hsd*S genes that function together as a protein complex for both cleaving and methylation of DNA. If the recognition system $5' - AAC(N)_5GTGC - 3'$ is unmethylated, it cleaves the DNA at a variable distance from this recognition site. *E. coli* K-12 strains with mutations in the endogenous restriction modification system (*hsdR⁻*) are used.

ii) *E. coli* K-12 strains used for gene cloning contain mutations in the *rec*A gene, which encodes a recombination protein. *rec*A⁻ enhances the biosafety of *E. coli* K-12 strains because *rec*A⁻ strains are sensitive to ultraviolet light.

iii) Most *E. coli* K-12 strains contain mutations in *end*A gene, which encodes DNA-specific endonuclease I. Loss of this endonuclease greatly increases the plasmid DNA yields and improves the quality of DNA that is isolated using standard biochemical preparations.

PLASMIDS

Plasmids are double-stranded, closed circular DNA molecules, which exist in the cell as extrachromosomal units. They are self-replicating, and found in a variety of bacterial species, where they behave as accessory genetic units. Plasmids replicate and are inherited independently. There are three general classes of plasmids: (i) virulence plasmids, which encode toxin genes, (ii) drug resistance plasmids, which confer resistance to antibiotics, and (iii) plasmids, which encode genes required for bacterial conjugation. Basically all molecular cloning methods involve the manipulation of specific DNA fragments that utilize some aspects of different plasmids. Plasmids range in size from 1 to 200 kb and depend on host proteins for maintenance and replication functions. Most of the plasmid-encoded genes that have been characterized impart some growth advantage to the bacterial host. These are present in a characteristic number of copies in a bacterial cell, yeast cell or even in organelles (mitochondria) of eukaryotic cells. These plasmids can be single copy plasmids that are maintained as one plasmid DNA per cell or multicopy plasmids that are maintained as 10–20 genomes per cell (Table 16.1). There are also plasmids that are under relaxed replication control, thus permitting their accumulation in very large numbers (up to 1000 copies per cell). Most plasmids used in cloning have a relaxed mode of replication. The best-studied plasmid replication system is that of *ColE1*, which is a small *E. coli* plasmid encoding antibacterial proteins called colicins. *ColE1* replication has been shown to involve transcription of a DNA sequence near the origin, which creates the RNA primer required for initiation of DNA synthesis. These plasmids are used as

Table 16.1 Replicons carried by the currently used plasmid vectors

Plasmid	Replicon	Copy number
pBR322 and its derivatives	pMB1	15–20
pUC	pMB1	500–7
pACYC and its derivatives	P15A	10–12
pSC101 and its derivatives	pSC101	~15
ColE1	ColE1	15–20

cloning vectors due to their increased yield potential.

Most plasmid vectors in current use carry a replicon derived from the plasmid pMB1, which was originally isolated from a clinical specimen (its close relative is *ColE1* replicon). Under normal conditions of growth, 15–20 copies of plasmids carrying this replicon are maintained in each bacterial cell. The pMB1 (or *ColE1*) replicon does not require plasmid-encoded functions for replication; instead, it relies entirely on long-lived enzymes supplied by the host. Replication occurs unidirectionally from a specific origin and is primed by an RNA primer.

Replication of plasmid DNA is carried out by a subset of enzymes used to replicate the bacterial chromosome. Usually, a plasmid contains only one origin of replication together with its associated *cis*-acting control elements. The control of plasmid copy number resides in a region of the plasmid DNA that includes the origin of DNA replication.

The F plasmids encode the gene required for bacterial conjugation. This extrachromosomal genetic element is ~95 kb, and was named because of its role in bacterial fertility. The F plasmid is present as one copy per cell and encodes ~100 genes, some of which are required for F plasmid transfer during conjugation (*tra* genes). As a result of integration into *E. coli* chromosome and then an imprecise excision resulting in the reformation of a circular plasmid, the F plasmid is able to transfer portions of the *E. coli* genome into another bacterium via conjugation. F plasmid containing *E. coli* strains are required in molecular cloning involving M13 filamentous bacteriophage. The F plasmid carries the sex pili gene, which encodes the cell-surface-binding site for M13 phages.

Frequently, plasmids contain some genes advantageous to the bacterial host, for example, resistance to antibiotics, production of antibiotics, degradation of complex organic compounds, and production of colicins, and enterotoxins. Antibiotics are antimicrobial agents that can be classified based on their mode of action. Bacte-

rial antibiotic-resistance genes encode proteins that inactivate these agents either by preventing their accumulation in the cell or by inactivating them once they have been imported. By utilizing these antibiotic-resistance genes as dominant genetic markers in plasmid cloning vectors, it is possible to select for *E. coli* cells that have maintained high copy replication of plasmid DNA molecules. Table 16.2 lists examples of commonly used antibiotics and the corresponding bacterial antibiotic resistance genes that have been exploited for specific cloning experiments.

Discussions about the danger involved in cloning foreign DNA into bacteria have led to requirements for cloning vectors regarding containment. As genetically engineered plasmids should not be able to spread in nature by means of conjugation, cloning vectors should not be self-transmissible plasmids. Plasmids are self-transmissible when they contain transfer (*tra*) genes, which allow the production of pili involved in conjugal transfer of plasmid DNA. If a second plasmid, which does not contain these *tra* genes, is also present in the same cell, it will also be transferred as long as it contains the so-called mobilization (*mob*) genes. This plasmid is then said to have been mobilized. Laboratory-used plasmids, therefore, have their *tra* and *mob* genes deleted to prevent the possibility of self-transmission or mobilization of recombinant plasmids should they escape from the laboratory environment.

A plasmid vector used for cloning is specifically developed by adding certain features:

i. Reduction in size of vector to a minimum to expand the capacity of vector to clone large fragments. Since the efficiency of transformation of bacterial cells drops drastically when plasmids larger than 15 kb are used, the size of cloning vector should be small, preferably 3–4 kb. In this way foreign DNA fragments of 10–12 kb can be accommodated.

ii. It should contain an origin of replication that operates in the organism into which the cloned DNA is to be introduced.

Table 16.2 Antibiotics and antibiotic resistance genes used in gene cloning

Antibiotic	Mode of action	Resistance gene	Application
Ampicillin	Inhibits cell wall synthesis by disrupting peptido-glycan crosslinking	β-lactamase (*amp*^{R}) gene product is secreted and hydrolyzes ampicillin	*Amp*^{R} gene is included on plasmid vector as positive selection markers
Tetracycline	Inhibits binding of aminoacyl tRNA to the 30S ribosomal subunit	*tet*^{R} gene product is membrane bound and prevents tetracycline accumulation by an efflux mechanism	*tet*^{R} gene product is a positive selection marker on some plasmids
Kanamycin	Inactivates translation by interfering with ribosome function	Neomycin or aminoglycoside phospho transferase (*neo*^{R}) gene product inactivates kanamycin by phosphorylation	*Neo*^{R} gene is a positive selection marker on plasmids
Bleomycin	Inhibits DNA and RNA synthesis by binding to DNA	The *bla*^{R} gene product binds to bleomycin and prevents it from binding to DNA	*Bla*^{R} gene is a positive selection marker used in plasmids
Hygromycin B	Inhibits translation in prokaryotes and eukaryotes by interfering with ribosome translocation	Hygromycin-B-phospho transferase (*hph* or *hyg*^{R}) gene product inactivates hygromycin B by phosphorylation	*Hyg*^{R} gene is used as a positive selection marker in eukaryotic cells that are sensitive to hygromycin B
Chloramphenicol	Binds to the 50S ribosomal subunit and inhibits translation	Chloramphenicol acetyl transferase (*CAT* or *CM*^{R}) gene product metabolizes chloramphenicol in the presence of acetyl CoA	*CAT*/*CM*^{R} gene is used as a selectable marker, and as transcriptional reporter gene of promoter activity in eukaryotic cells.

iii. Introduction of selectable markers.

iv. Introduction of synthetic cloning sites termed *polylinker, restriction site bank*, or *polycloning sites* that are recognized by restriction enzymes. This polycloning site is usually present inside a marker gene so that with the insertion of foreign DNA it will inactivate that marker gene and the recombinants can be selected.

v. Incorporation of axillary sequences such as visual identification of recombinant clones by histochemical tests, generation of single-stranded DNA templates for DNA sequencing, transcription of foreign DNA sequences *in vitro*, direct selection of recombinant clones and expression of large amounts of foreign proteins.

Some of the plasmid vectors have been described below.

pBR322

This was one of the first artificial cloning vectors

to be constructed, and undoubtedly the most widely used cloning vector up to now (Bolivar and Rodriguez, 1977a; 1977b). It is a 4.36-kb double-stranded cloning vector. This plasmid vector has been put together from fragments originating from three different naturally occurring plasmids. Figure 16.1a indicates all the salient features of this vector. It contains *ColE1 ori* of replication with relaxed replication control. Generally there are 15–20 molecules present in a transformed *E. coli* cell, but this number can be amplified by incubating a log phase culture of *ColE1* carrying cells in the presence of chloramphenicol. Like *ColE1*, pBR322 and all its derivatives are amplifiable, which makes it very simple to isolate large amounts of these plasmids. pBR322 contains two antibiotic-resistance genes, one is the ampicillin-resistance (*amp*^{R}) gene coding for a β-lactamase (which modifies ampicillin into a form that is nontoxic to *E. coli*) and the other is the tetracycline-resistance (*tet*^{R}) gene (a set of genes coding for enzymes that detoxify tetracycline). The β-lactamase gene originates from the transposon *Tn*1, originally located on the R plasmid R1 drd 19. Due to the way in which pBR322 was constructed, this gene is no longer transposable. The gene determining tetracycline resistance was derived from the naturally occurring plasmid pSC101, a low-copy-number plasmid which in itself has also been used as a cloning vector. The presence of these two resistance genes indicates that a bacterium transformed with pBR322 will be able to grow on a medium containing ampicillin and tetracycline, whereas the cells lacking the plasmid will not be able to grow.

The complete nucleotide sequence of pBR322 has been determined. The plasmid contains 20 unique recognition sites for restriction enzymes. Six of these sites, i.e. *Eco*RV, *Bam*HI, *Sph*I, *Sal*I, *Xma*III and *Nru*I, are located within the gene coding for tetracycline resistance; two sites for *Hind*III and *Cla*I lie within the promoter of the tetracycline resistance gene; and three sites for *Pst*I, *Pvu*I and *Sca*I lie within the β-lactamase gene. Cloning of a DNA fragment into any of these 11 sites results in the insertional inactivation of either one of the antibiotic resistance markers. The gene of interest is spliced into *tet*^{R} gene cluster, and then the *E. coli* cells are transformed. Thus, three types of cells are obtained:

1. Cells that have not been transformed and so contain no plasmid molecules and will be *amp*^{S} *tet*^{S}.
2. Cells that have been transformed with pBR322 but without the inserted DNA fragment or gene will be *amp*^{R} *tet*^{R}. These are transformed cells.
3. Cells that contain a recombinant DNA molecule, that is, the DNA fragment has been inserted into the pBR322 at *tet*^{R} gene cluster. These cells will lose tetracycline resistance because the fragment has inserted in the middle of tetracycline resistance gene cluster. These are recombinants and will be *amp*^{R} *tet*^{S}.

The bacteria are spread on an agar medium and colonies grown. By checking for growth on agar media modified with ampicillin, one can exclude untransformed cells, but both transformants and recombinants will produce colonies. Replica plates are made and then grown on agar with ampicillin and tetracycline. Colonies that do not grow on tetracycline agar are recombinants whereas both transformants and recombinants will grow on ampicillin agar. By comparing the replica plates, recombinants can be picked up from agar ampicillin plates. Plasmids can take up an insert of up to 10 kb.

Like *ColE1*, pBR322 is not self-transmissible, but it can be mobilized by other self-transmissible plasmids such as the F-factor.

pACYC184

This is a small 4.0 kb cloning vector developed from the naturally occurring plasmid p15A (Fig. 16.1b). It contains the genes for resistance to chloramphenicol and tetracycline. The chloramphenicol gene has a unique *Eco*RI restriction site and the tetracycline gene has a number of restriction sites as mentioned in pBR322. Insertional inactivation of one of the antibiotic-resistance genes facilitates the recognition of strains carrying plasmids with cloned DNA fragments. Like pBR322, it is not self-transmissible, and can be amplified. It has been a very popular cloning vector since it is compatible with the *ColE1*-derived plasmids of the pBR322 family and can therefore be used in complementation studies.

pUC vectors

The name pUC is derived because it was developed in the University of California (UC) by Messings and his colleagues (Norrander *et al.*, 1983). These plasmids are of 2.7 kb and possess the *ColE1 ori* of replication. These vectors contain *amp*^{R} gene and a new gene called *lac*Z, which was derived from the *lac* operon of *E. coli* that codes for β-galactosidase enzyme (Fig.16.1c).

The Lac operon A number of bacterial operons in *E. coli* serve to illustrate various

Fig. 16.1 Plasmid vectors. a – pBR322; b – pUC18/19; c – pACYC184.

principles of prokaryotic gene regulation. But the *lac* operon has application in molecular biology experiments also. The *lac* operon encodes three enzymes: *lacZ* gene encodes β-galactosidase, which cleaves lactose into glucose and galactose; *lacY* encodes permease; and *lacA* encodes transacetylase. When glucose levels are low and lactose levels are high, transcription of the *lac* operon is maximally induced through a mechanism of transcriptional derepression and activation, which requires *lac* repressor and catabolite activator protein (CAP). The lac repressor protein is encoded by *lacI* gene. Binding of the lac repressor to the operator sequence is extremely tight and specific, which cause transcriptional repression through a mechanism involving repressor interference with RNA polymerase binding to the *lac* promoter. When an effector molecule such as isopropyl thiogalactoside (IPTG), a nonmetabolizable allolactose analog, binds to the lac repressor protein, its DNA binding affinity for the *lac* operator decreases and transcription of the *lac* operon is derepressed. The β-galactosidase activity of the *lac* operon has been used as a tool for identification of clones. The β-galactosidase activity can be measured in live cells by using the chromogenic substrate 5-bromo-4-chloro-3-indolyl-β-galactoside (X-gal). In wild type *E. coli* cells, cleavage of X-gal produces a blue color that can be visualized as a blue colony on agar plates. The X-gal assay can be used to identify *lac* operon mutants, which appear as white colonies against a blue background. Another way the X-gal assay has been exploited is by taking advantage of β-galactosidase mutants, which function through a mechanism called α complementation. Characterization of these mutants showed that β-galactosidase activity could be reconstituted in *E. coli* cells that expressed both nonfunctional N-terminal (α fragment) and C-terminal (ω fragment) β-galactosidase polypeptides. α complementation has been used to monitor insertional cloning into plasmids containing restriction sites within the ~150 amino acid coding sequence of the α fragment. When these plasmids are grown in *E. coli* strains expressing the ω fragment on an F′ plasmid, it is possible to score antibiotic-resistant cells that contain the plasmid cloning vector and are either blue (no gene insertion) or white (coding sequences are disrupted). The advantage of using α complementation for DNA cloning is that plasmid vectors are limited by the size of insert DNA that can be accommodated, and the α fragment (~450 bp) is small compared to the entire *lacZ* gene (~3500 bp).

This enzyme splits lactose into glucose and galactose. A recombinant pUC molecule is constructed by inserting new DNA into one of the restriction sites that are clustered near the start of the *lacZ*. This means that

1. Cells harboring normal pUC plasmids are ampicillin resistant and able to synthesize b-galactosidase (transformant but non-recombinant). When an inducer of *lac* operon IPTG (isopropyl thio-galactoside) and X-gal (5-bromo-4-chloro-3-indolyl-β-D-galactopy-ranoside), a substrate for β-galactosidase, is put in the media, it will be broken down by the enzyme into a product. It will give deep-blue colonies.

2. Cells harboring recombinant pUC plasmids are ampicillin resistant but unable to synthesize β-galactosidase. When IPTG and X-gal are put in the media, it will give white colonies because the substrate cannot be broken down. Recombinants can therefore be selected by the color of colonies and there will be no need of replica plating as done in the case of pBR322.

Cells plated + agar + ampicillin + IPTG + X-gal = Nonrecombinant **– Blue colonies**
Cells plated + agar + ampicillin + IPTG + X-gal = Recombinant **– White colonies**

pUN121

This is a 4.4-kb vector derived from pBR322 (Nilsson *et al.*, 1983). This vector has been developed for rapid selection of bacteria containing a recombinant plasmid. This is required when

genomic libraries are constructed, which involves the construction of several thousands of clones and handpicking of individual clones with inserted DNA would be extremely tedious. Direct-selection vectors have therefore been constructed. In these vectors (Fig. 16.2), the original promoter region of the tetracycline-resistance gene is exchanged for the P$_R$ promoter of bacteriophage (λ). Transcription from this promoter is suppressed by the protein encoded by the lambda *cI* gene, which was also cloned on pUN121. Cells harboring pUN121 will therefore have the phenotype *amp^{R}tet^S* as the *cI* gene

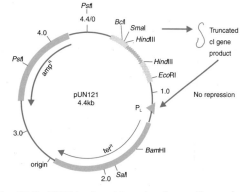

Fig. 16.2 pUN121 plasmid vector (a positive selection vector). a – When no foreign DNA is inserted in one of the restriction sites in the *cI* gene, the *cI* gene product is synthesized and represses the P$_L$ promoter. The cells harboring such plasmids are sensitive to tetracycline. b – Plasmids with foreign DNA inserted in the *cI* gene produce a truncated repressor. The P$_L$ promoter is derepressed and the cells become tetracycline resistant.

product suppresses expression of the tetracy-cline-resistance gene. The *cI* gene contains unique sites for the restriction enzymes *Eco*RI, *Hind*III, *Bcl*I, and *Sma*I. Cloning of foreign DNA into these restriction sites will cause insertional inactivation of the *cI* gene, abolishing the suppression of P$_R$ promoter. Cells harboring such a plasmid will become *tet*^{R}. Plating of trans-formed cells on plates containing tetracycline will therefore only result in growth of cells harboring the pUN121 vector with inserted DNA fragments.

Yeast plasmid vectors

Yeast (*Saccharomyces cerevisiae*) is a single-celled eukaryotic microbe, which can be cultured and manipulated using the standard techniques applied to bacteria. The genetic material of yeast is packaged into chromosomes within a mem-brane-enclosed nucleus and partitioned at cell division by mitosis and meiosis. In 1976, Struhl *et al.* reported that a fragment of yeast DNA when cloned into *E. coli* restored histidine-independent growth to strains carrying the *hisB* mutation. Similarly, Ratzkin and Carbon (1977) ligated fragments of yeast genome into plasmid *ColE1* and obtained a small number of leucine-independent transformants from a *leuB E. coli* strain. In each case, the yeast chromosomal segment carried the gene for the yeast enzyme equivalent to that absent in the bacterial strain. Thus, the yeast *HIS3* gene was able to be expressed in a bacterial cell and produce the yeast version of the enzyme imidazole glycerol phosphate dehydratase and thereby complement the mutation in the bacterial *hisB* strain. (With yeast genes it is customary to indicate a wild-type allele in capital letters and the mutant allele in lower case letters.) Other yeast genes isolated in this way include *LEU2*, *URA3*, *TRP1*, and *ARG4*. In general, genes from eukaryotic organ-isms rarely function in bacterial cells, but the success with yeast genes arises from the lack of introns in most such genes coupled with the ability of some yeast promoters to function in bacteria.

The wild-type yeast genes thus isolated

provided essential markers in subsequent at-tempts to introduce exogenous DNA into yeast cells. Hinnen *et al.* (1978) demonstrated that the yeast *LEU2* gene carried by the plasmid *ColE1* could be taken up by *leu2* yeast sphaeroplasts and transform them to a leucine-independent phenotype. In this case the transformation resulted from the integration of the added DNA into a yeast chromosome. Thus, a variety of approaches have resulted in a cloning system that makes use of many of the powerful techniques developed for *E. coli*.

Vector systems

Most yeast vectors developed share the follow-ing common features:

1. Most of the vectors are derived from bacterial plasmids and retain both the ability to replicate in bacteria and selection markers suitable for use in bacterial systems. As plasmid amplification is not high in yeast as compared to that in *E. coli*, this is a most useful feature. Thus, initial plasmid construc-tion is done in *E. coli* before transfer to yeast. Such vectors permitting cloning in two different species are referred to as *shuttle vectors.*

2. In most cases, the selection markers employed are nutritional markers such as *LEU2* or *HIS3*. In some cases, such markers are not suitable. For example, Brewing yeasts are polyploid and it is hence not possible to obtain auxotrophic mutants. In such cases, resistance to copper or to the drug chloramphenicol have been used.

There are different classes of yeast vectors (Fig. 16.3) and the main distinguishing factors are the replication and transmission of the vector once inside the yeast cells.

Yeast integrating plasmids (YIps). These are not actually yeast plasmid vectors, because they are unable to replicate within a yeast cell without integration into a chromosome. Such vectors contain a yeast marker as the only addition to the bacterial plasmid. As integration is essential for a successful transformation, the

VECTOR TYPE STRUCTURE FEATURES

Integrating
YIp

Must integrate
Low frequency transformation
Stable

X⁺

Episomal
YEp

Autonomous
High frequency transformation
High copy number
Moderately stable

X⁺

Replicating
YRp

Autonomous
High frequency transformation
~10 per cell
Very unstable

X⁺ ARS

Centromeric
YCp

Autonomous
High frequency transformation
1 per cell
Stable

X⁺ ARS CEN

Linear
YLp

TEL ARS X⁺ TEL   Similar to YRp

Mini-chromosome

CEN ARS X⁺ TEL   Similar to YCp
TEL   Very stable

LEGEND

———————— Bacterial plasmid DNA

Yeast chromosomal DNA with selective marker

Yeast 2 μ DNA

λ DNA

Fig. 16.3 Features of yeast vectors.

frequencies are very low, with a typical yield of one transformant per μg of DNA. Here the plasmid vector integrates into the host chromosomal DNA. Integration is due to homologous recombination between the yeast gene carried by the YIp as the selectable marker and the inactive version of the gene present in one of the yeast chromosomes. Thus, the *LEU2*-carrying plasmid would integrate at the chromosomal *leu2* region as a result of crossover either within the gene itself or the flanking region carried by the plasmid. The product is a yeast chromosome with an integrated bacterial plasmid flanked by a duplication of the gene. Low frequency excision events involving an intrachromosomal crossover between the homologous regions will eliminate either the incoming *LEU2* allele or the original *leu2* allele as well as the bacterial segment. This latter case represents a replacement of a chromosomal allele with an added allele from outside the cell.

An important development in the use of these vectors has been the observation that opening the plasmid within the yeast segment before adding it to the yeast cells greatly increased the transformation frequency. For that purpose, the plasmid is treated with a restriction enzyme that has a unique recognition site within the yeast segment. If a plasmid carried two yeast segments (e.g., *URA3* and *LEU2*), cutting the plasmid within one of these regions will target the integration to that chromosomal locus.

A YIp will not replicate as a plasmid because it lacks autonomously replication sequences (*ARS*); therefore, it cannot be multiplied in large numbers inside the yeast host.

Episomal plasmids – Yeast episomal plasmid (YEp). It is a yeast vector with the ability to replicate autonomously without integration into a yeast chromosome. Yeast vectors with this property have been built around a naturally occurring yeast plasmid, the so-called 2 μ circle. This molecule has no function and hence its name is based on contour length. It is present in most yeast strains with a copy number of 40–50 per haploid genome. The 2-μm plasmid, as it

is called, is 6 kb in size. These are autonomous and show high frequency transformation.

The first 2-μm-based vectors carried the whole of this plasmid and gave high frequency transformation (up to 2×10^4 transformants per μg DNA) combined with reasonable stability levels. Such plasmids are, however, large and carry only a limited number of unique restriction sites. Smaller vectors carrying only a limited region of the 2-μm circle have been developed. Such plasmids are more easily handled in cloning experiments, but the loss of 2-μm sequences has cost in terms of stability. Hence, stability levels acceptable for most purposes can be obtained in *cir*⁺ cells, and the high copy numbers obtained have made these plasmids the main workhorse for yeast cloning.

Yeast replicating plasmids (YRp). This is a vector based on chromosomal elements. Chromosomal sequences with the ability to support autonomous replication of plasmids in yeast were first discovered when it was observed that a region closely linked to *TRP*1 gene would allow high frequency transformation and propagation without the need for integration. Such so-called ARS have been isolated from many regions of the yeast genome and from a range of other eukaryotic organisms. Vectors based on *ARS* are highly unstable once in the yeast cell, but they show high frequency transformation. The copy number in cells retaining the plasmid is high, in the order of 10 per cell.

Yeast centromeric plasmid (YCp). This vector is also based on chromosomal elements. It is an ARS-based plasmid, which carries in addition a yeast chromosomal centromeric DNA sequence. This greatly increased the stability of an ARS-based plasmid. Such plasmids have a copy number of one per cell and are thus useful in situations in which a low copy number is essential, such as cloning genes expressing a product lethal to the cell in excess. ARS centromeric plasmids have also been referred to as mini chromosomes, which are circular

Yeast linear plasmid (YLp). These are ARS centromeric plasmids, but are linear rather than circular. Linear plasmids have been developed by adding telomeric (chromosome end) sequences to the ends of a cut ARS-based plasmid. The original construction made use of telomeres from ciliate protozoan *Tetrahymena*. These linear plasmids are present like a genuine chromosome, in a single copy, and are the most stable yeast plasmid vectors. Linear plasmids cannot replicate in *E. coli* and are hence technically difficult to work with. They have nonetheless attracted much interest as they raise the exciting possibility of putting together complete sets of genes in artificial eukaryotic chromosomes.

Ti plasmids

Most cloning vectors for plants are based on the Ti plasmid, which is not a natural plant plasmid, but belongs to a soil bacterium *Agrobacterium tumefaciens*. This bacterium invades plant tissues, causing a cancerous growth called a crown gall. During infection, a part of the Ti plasmid called the T-DNA is integrated into the plant chromosomal DNA. Cloning vectors based on Ti plasmid have been explained in greater details in Chapter 23. Likewise, there are Ri plasmids present in *Agrobacterium rhizogenes*, which cause hairy root disease in plants.

COSMIDS

Cosmids are plasmid vectors that contains a bacteriophage lamda *cos* site, which directs insertion of DNA into phage particles (Fig. 16.4). The development of cosmid vectors was based on the observation that a ~200-bp DNA sequence in the λ genome called *cos* is required for DNA packaging into the phage particle during lytic infection. Cleavage at *cos* by the λ terminase protein during phage packaging produces a 12-nucleotide cohesive end at the termini of the linear λ genome. Recircularization of the λ genome after bacterial infection is facilitated by base pairing between the complementary cohesive ends. Cosmids were developed to overcome the technical problem of

introducing large pieces of DNA into *E. coli*. Thus, bacteriophage *cos* DNA sequence was introduced that is required for packaging DNA into preformed phage heads. For successful packaging of DNA, there must be two *cos* sequences separated by a distance of 38–52 kb. Cosmid cloning vectors with DNA inserts of 30–45 kb can be packaged in vitro into λ phage particles, provided that the ligated double-stranded DNA contains λ *cos* sequences on either side of the insert DNA. Because the λ phage head can hold up to 45 kb of DNA, the optimal ligation reaction in cosmid cloning produces recombinant molecules with *cos* sequences flanking DNA segments of ~40 kb. Following adsorption of these phage particles on to suitable *E. coli* host cells, the cosmid vector circularizes via the cohesive ends and replicates as a plasmid. Cosmid vectors possess an origin of replication, a selectable genetic marker (antibiotic resistance), and suitable cloning sites. For this reason, cosmids are ideal vectors for genome mapping. Like plasmids, cosmids can multiply in large copy number using bacterial *ori* of replication (*ColE1*) inside the bacterial cell and do not carry the genes for lytic development. The advantages of cosmids are that (a) relatively large size of insert DNA (up to 45 kb) can be cloned; and (b) DNA can be introduced into the host using bacteriophages derived by *in vitro* packaging. The disadvantages are that (a) it is difficult to store bacterial host as glycerol stock; and (b) *in vitro* packaging is needed to maintain cosmids inside the viral heads.

BACTERIOPHAGE VECTORS

Bacteriophages are viruses that infect bacteria. These are usually called phages. The phages are constructed from two basic components: proteins form a coat or capsid within which the nucleic acid genome is enclosed. Phages can be double stranded (T2, T4, T6, λ) or single stranded (φX174, M13). There are both DNA (T2, T4, T6, λ, φX174) and RNA (MS2) phages. Phages contain an origin of replication, but

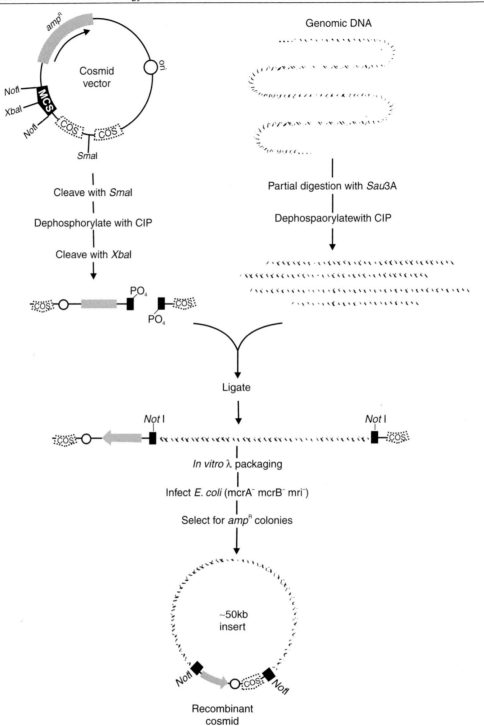

Fig. 16.4 Schematic diagram of cosmid vector and cloning of genomic DNA. Concept of the figure taken from Miesfeld: *Applied Molecular Genetics.* Copyright © 1999 John Wiley and Sons. This material is used by permission of John Wiley and Sons, Inc.

unlike plasmids, phages lack the machinery necessary to actually make proteins. Phages generally gain entry into the host cell by injecting their DNA directly into the cell or by being internalized by host cell processes. In most cases, once the phage genome is inside the host cell, phage proteins and phage DNA are synthesized and infectious particles are released through lysis or membrane budding. The ability to transfer DNA from the phage genome to bacterial hosts during the bacterial infection is known as transfection. In gene cloning, two specially designed phage vectors are used, which are constructed from components of the temperate bacteriophage lambda and the production of single-strand DNA using cells infected with vectors based on filamentous bacteriophage M13.

Biology of bacteriophage λ

λ is a bacteriophage capable of infecting certain strains of *E. coli*, such as *E. coli* K-12. The λ phage particle contains two major structural components called a head and a tail. The λ genome is a double-stranded DNA molecule of 48,503 bp that is packaged within the phage head. The tail proteins are required for phage attachment to the *E. coli* outer membrane and injection of phage DNA into the host cell.

λ infection of *E. coli* can result in phage replication and cell lysis leading to release of ~100 infective phage particles (lytic phase) or λ DNA can integrate into the *E. coli* genome as a prophage and remain dormant, creating a lysogenic host cell.

Having established the basic sequence of events, we can now look at the molecular level. λ DNA exists as a double-stranded linear DNA molecule about 48.5 kb long. At each 5′ end there is a single-stranded region 12 bases in length. These singlestranded regions are complementary to each other and are able to pair with one another and are therefore referred to as cohesive end sites (*cos*). After injection of the DNA into the cell, the cohesive ends pair together by hydrogen bonds to form a circular DNA molecule.

Initially, λ DNA replication begins at the *ori* and proceeds bidirectionally to form a theta (θ) structure. This yields a number of monomeric circular DNA molecules. However, as bidirectional replication continues, a rolling circle mechanism begins to produce multiple-length linear forms of double-stranded λ genomic DNA. Staggered cleavage of this large concatamer at the regularly spaced *cos* sites is mediated by the λ terminase protein. Packing of the newly replicated λ genomic DNA into phage head particles is coincident with *cos* site cleavage.

Factors that control the decision of whether the phage will take a lytic or lysogenic path have been well characterized and shown to reflect transcriptional control of λ encoded genes. The primary event in this genetic switch between lysis and lysogeny is governed by the level of λ encoded proteins required for the lytic pathway. If one set of the λ immediate early genes is preferentially transcribed, then the lytic pathway is favored; however, if a transcriptional repressor protein encoded by λ *cI* gene inhibits this transcription, then lysogeny is favored.

Physical constraints on the amount of DNA that will fit into phage head also control DNA packaging. Although the λ genome is 48 kb, DNA that is at least 38 kb but no more than 52 kb can be packaged provided that *cos* site cleavage occurs.

Often a bacteriophage vector is introduced into the host *E. coli* cells by transfection, which is exactly the same as transformation, but involves phage rather than plasmid DNA. The transfectants are mixed with normal bacteria and poured on to an agar plate, giving a set of plaques on a lawn of bacteria, each plaque being a clone consisting of bacteriophages produced by a single infected cell.

The portions of phage genome that determine lysogenic growth are removed from the DNA, and the phage is made into a DNA which always enters lytic lifecycle and ensures that transfected cells immediately give rise to plaques. With most phage vectors, nonessential region of the genome can be disrupted and deleted without harming the genes for the important replicative and structural proteins of the phage, and the DNA fragments to be cloned are ligated into it.

The utility of λ vectors for gene cloning was significantly enhanced following the development of two major improvements in the general strategy.

First, to overcome low efficiency of CaCl$_2$-mediated DNA transformation (0.1% of the cells), as compared to phage infection (10% of the cells), protein extracts were developed that permitted ligated DNA to be packaged *in vitro* into infectious phage particles. Following the ligation of foreign DNA to the purified λ vector arms, the ligation reaction is then carefully mixed with an enriched protein fraction containing phage-packaging proteins. Through an energy-independent association process involving highly ordered protein–protein interactions, the λ DNA is cleaved at appropriately spaced *cos* sites by λ terminase and packaged into infectious particles. Commercially available packaging extracts are commonly used to achieve phage titers of ~10^7 plaque forming units/microgram vector DNA.

A second improvement in λ cloning systems led to an increase in the ratio of recombinant to nonrecombinant phage in the pool of infectious particles. In the first generation of *Eco*RI-based λ vectors because of uncleaved λ vector or religation of vector without an insert, resulted in high numbers of nonrecombinant phages. Therefore, several genetic strategies were devised to select genetically against nonrecombinant λ genomes. Two of these selection schemes are illustrated in two different λ-phage-based cloning vectors. One involves small DNA insertions (usually cDNA) into the λ *cI* gene, which controls the lysogenic pathway (λgt10, λgt11), and the other relies on replacement of vector "stuffer" DNA encoding two λ genes that promote lysogeny (*red, gam*), with large fragments of foreign DNA (EMBL3, EMBL4).

Bacteriophage cloning vectors based on λ are λgt10, λgt11, EMBL3 (European Molecular Biology Lab), EMBL4, Charon, etc.

λ *phage cloning vectors*

The bacteriophage λ contains double-stranded DNA of length 48.5 kb, which is a linear molecule with short complementary single-stranded projections of 12 nucleotides at its 5′ ends. These cohesive termini are referred to as *cos* sites, which allow the DNA to be circularized after infection of the host cell. Derivatives of bacteriophage have been developed as cloning vectors (Fig. 16.5). The main advantage of a λ-based vector is that large fragments of up to 25 kb can be inserted as compared to plasmid vectors, which can take up 10 kb fragments. When a bacteriophage DNA is digested with restriction enzyme, it generates three fragments: a 19.6-kb left arm carrying the genes for the lambda head and tail proteins, a 12- to 14-kb central fragment carrying the *red* and *gamma* genes under pL control; and a 9- to 11-kb right arm carrying the lambda replication and lysis genes. Genomic DNA is partially digested with restriction endonucleases to produce a population of DNA fragments, having sizes from 15 to 20 kb. The size-selected fragments are ligated to vector arms. Viable phages containing recombinant DNA are obtained by in vitro packaging. In the replacement vector, "stuffer" DNA of λ vector encoding two λ genes *red* and *gam* that promotes the lysogeny, is replaced with large fragments of foreign DNA. Lack of *red/gam* genes permits lytic growth. A permanent collection of recombinant phage is then established after amplifying the phages through several generations of growth on a strain. A number of strategies are used to identify recombinant plaques. Some vectors carry the *lacZ* gene and are therefore plated on agar, IPTG, X-gal plates (as explained earlier in pUC plasmid), whereas others rely on differences in plaque morphology.

Advantages of this vector are as follows: (i) The size of cloned DNA is relatively large (9–25 kb); (ii) The multiplication of cloned DNA is very easy by infecting a suitable host with bacteriophages containing recombinant DNA; (iii) There is direct selection of clones containing recombinant DNA during *in vitro* packaging; and

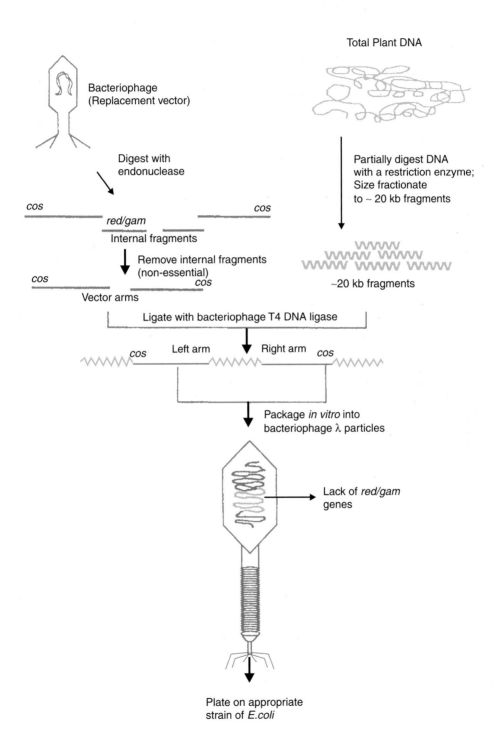

Fig. 16.5 Genomic DNA cloning in bacteriophage λ.

(iv) Storage of clones is easy.

λ*gt10* and λ*gt11*. These are insertion vectors. These are modified lambda phages designed to clone cDNA fragments. λgt10 is a 43-kb double-stranded DNA for cloning fragments that are only 7 kb in length. The insertion of DNA inactivates the *cI* (repressor) gene, generating a *cI*⁻ bacteriophage (Fig. 16.6). Nonrecombinant λgt10 is *cI*⁺ and forms cloudy plaques on appropriate *E. coli* host, whereas recombinant cI ⁻ λgt10 forms clear plaques. Further, in an *E. coli* strain carrying an *hfl*A150 mutation, only cI ⁻ phage will form plaques because *cI*⁺ will form lysogens and will not undergo lysis to form any plaques. Thus, recombinant λgt10 plaques can be easily selected. λgt11 is a 43.7-kb double-stranded phage for cloning DNA segments that are less than 6 kb in length. Foreign DNA can be expressed as a β-galactosidase fusion protein. Recombinant λgt11 can be screened using either nucleic acid or antibody probes. The recombinant λgt11 becomes *gal*⁻ whereas nonrecombinant λgt11 remains *gal*⁺, so that an appropriate *E. coli* host with recombinant phage will form white or clear colonies. Nonrecombinant phage (*gal*⁺) forms blue colonies, thus permitting screening in the presence of IPTG and X-GAL.

EMBL 3 and EMBL 4. The EMBL (European Molecular Biology Lab) vectors are replacement vectors used for cloning large (up to 20 kb) fragments of genomic DNA to develop genomic libraries. λ DNA containing the *red* and *gam* genes in λ EMBL3 is replaced with DNA fragments generated by *Bam*HI. Fragments are substituted at the site of *trp*E gene. These

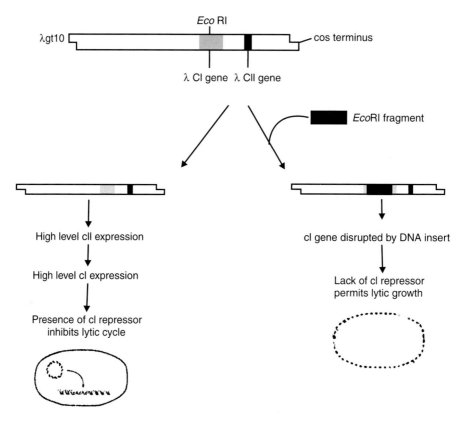

Fig. 16.6 Insertional cloning in λgt10 phage vector.

vectors are particularly useful because three enzyme sites are present in the cloning site (*Eco*RI, *Bam*HI, and *Sal*I). The genomic DNA can be digested partially with *Sau*3A1 (tetracutter) to generate large number of fragments and can be cloned at *Bam*HI site since the cohesive ends produced by *Sau*3A1 and *Bam*HI are same. These fragments produced by *Sau*3A1 can be cloned at *Bam*HI site. The cloned fragments can be excised back using *Eco*RI or *Sal*I enzymes. Recombinants can be selected on *E. coli* containing the P2 bacteriophage lysogen. Nonrecombinant λEMBL3 vector DNA is not large enough to be efficiently packaged. It is never used as an insertion vector. EMBL 3 and EMBL 4 have polylinkers with reverse orders of restriction sites with respect to each other.

Advantages of this vector are as follows: (i) The cloned DNA is relatively large in size (9–25 kb); (ii) The multiplication of cloned DNA is very easy by infecting a suitable host with bacteriophages containing recombinant DNA; (iii) There is direct selection of clones containing recombinant DNA during in vitro packaging; and (iv) Storage of clones is easy.

Charon: Charon bacteriophages are replacement (substitution) vectors designed for cloning large fragments (9–20 kb) of DNA. Charon 40 is unique in that its stuffer consists of small DNA fragments polymerized head-to-tail, which can be digested by a single restriction enzyme *Nae*. The charon bacteriophage can be used to screen recombinant genomic DNA libraries by *in vivo* recombination. Charon 4A vector is used chiefly as a substitution vector to construct genomic libraries of eukaryotic DNA. Typically, fragments of genomic DNA of size 7–20 kb obtained by partial digestion with *Eco*RI can be cloned. The *Eco*RI sites used for cloning are located in the *lacZ* gene and the *bio* gene. This can be used as insertion vector if digested with *Xba*I, which will create single cut and up to 6 kb fragments can be cloned. There are other charon series vectors, such as charon 34 and 35.

M13 PHAGES

M13 is a filamentous *E. coli* phage. Joachim Messing and his colleagues developed cloning vectors from the M13 isolate of an *E. coli* filamentous phage in the early 1980s, primarily as a means to obtain single-stranded DNA for use as templates in dideoxy DNA sequencing reactions. Filamentous bacteriophages infect *E. coli* through the bacterial sex pili normally used for bacterial conjugation. Proteins required for sex pili formation are encoded on the *E. coli* F plasmid, and therefore this genetic element must be present in the host cell (also referred to as a male cell). It is a single-stranded circular DNA of 6407 nucleotides. A single nonessential region has been located in the M13 genome within the intergenic region between genes II and IV, which comprises 507 nucleotides. This is the basis for development of cloning vectors in the M13 system. Through modifications of M13 DNA, the M13mp series of cloning vectors was developed. Cloning vectors of the M13 mp series have a *lacZ* gene that complements the *gal* host giving blue colonies. On transformation, only white or clear plaques are obtained, thus permitting selection of recombinant M13mp plaques. M13mp8 and M13mp9 have the same arrangement of genes, but differ in the direction of polycloning sites. This permits sequencing from both ends of the double-stranded DNA molecule using Sanger's method of dideoxy chain termination reaction. The advantages associated with this vector are that about 100–200 copies of RF (replicative form) accumulate in the cell, the DNA is obtained from the virus as single strands, and this avoids the technical difficulties involved in the separation of single-stranded DNA from double-stranded DNA using alkaline CsCl gradient. Since the intergenic region is very small, the cloning range is very less.

PHAGEMIDS

These are plasmid vectors having both bacterial

as well as phage *ori* of replication, for example, pBluescript II KS. M13-based cloning vectors have disadvantages: (i) It is difficult to obtain sufficient amounts of the double-stranded DNA (RF type) required for recombinant DNA reactions, and (ii) the rolling circle mechanism of M13 replication can be affected by insert sequence and size, causing under representation of certain recombinant molecules following infection. To overcome these problems, hybrid M13 phage/plasmid cloning vectors called phagemids were constructed in the early 1990s. The original phagemid vectors were derived from the high copy pUC18/pUC19 plasmids first developed by Messing.

pBluescript II KS is a pUC-based plasmid and is 2958 bp long with *lacZ* gene precedes from polycloning site (Fig. 16.7). The main features of this vector are as follows:

1. Multiple cloning site bordered by T3 and T7 promoters to read in opposite directions on the two strands to direct the *in vitro* synthesis of RNA.

2. Selection of the recombinant DNA molecules through blue-white colonies on X-gal plates through a *lacl* gene in upstream portion of *lacZ* gene.

3. The fl (+) and fl (−) origins of replication from filamentous phage yields sense (+) and antisense (−) strands. The f1 origin sequence allows the production of single-stranded circular DNA following infection of the host bacterial cell with M13 helper phage.

4. Modified *ColEl ori* of replication, which results in high copy plasmid replication.

5. The ampicillin resistant gene (*amp*R) helps in selection of cells containing phagemids. These vectors can thus be used to generate single-stranded DNA *in vivo* or RNA *in vitro* that is complementary to either of the two strands of foreign DNA inserted in to the polycloning site.

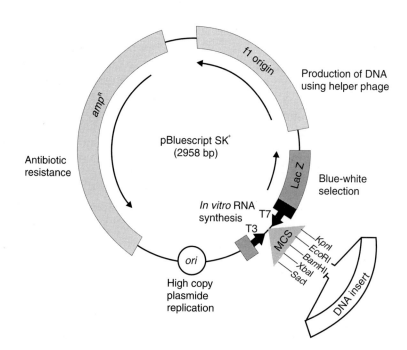

Fig. 16.7 pBluescript SK+ phagemid cloning vector. Concept of the figure taken from Miesfeld: *Applied Molecular Genetics.* Copyright © 1999 John Wiley and Sons. This material is used by permission of John Wiley and Sons, Inc.

YEAST ARTIFICIAL CHROMOSOME (YAC)

Very simple gene cloning vectors, such as plasmids and λ vectors, can accommodate an insert DNA of size up to 10 kb and 25 kb respectively, whereas cosmids can accommodate 4 DNA segments of up to 40 kB. The development of yeast artificial chromosomes (YAC) by David Burke *et al.* (1987) extended the cloning range to several thousand base pairs (i.e. about 1000–2000 kb) by using a totally new approach. YAC vectors use DNA sequences that are required for chromosome maintenance in *S. cerevisiae*. A typical YAC consists of centromere elements (*CEN*4) for chromosome segregation during cell division, two telomere sequences (*TEL*) for chromosome stability, an origin of replication and one or more growth selectable markers (*URA*3) to select positively for chromosome maintenance. In essence, the YAC is a minichromosome. Normal eukaryotic chromosomes are always linear. In yeast, the sequence elements required for circular plasmid vectors to be replicated and stably maintained in an extrachromosomal state have been identified as ARS. CEN (centromere) sequences ensure segregation and telomeres seal the chromosome ends. Linear plasmids must carry telomeres, which are a tandem array of simple satellite-like repetitive sequences. For cloning purposes, circular YAC is digested with restriction enzymes. From the resulting fragments, the fragments that contain *his* gene are deleted and the other fragments are utilized for cloning insert DNA in the YAC. The most commonly used YAC for genome mapping of higher eukaryotes is pYAC4 (Fig. 16.8). The left and right arms have all the essential yeast chromosomal elements to enable replication in a yeast host, as well as markers *TRP* and *URA* on the left and right arm respectively, which permits selection for transformants. The two arms can be ligated to *Eco*RI ends of a large fragment of genomic DNA insert to create a recombinant chromosome. This molecule can then be maintained as a linear chromosome in yeast, and is referred to as a yeast artificial chromosome. Yeast transformants containing recombinant YAC molecules can be identified by red/white color selection using a yeast strain that has the *ADE2*-1 ochre mutation, which is suppressed by the *SUP*4 gene product.

Transformed yeast containing recombinant YAC molecules – **Red colony**
(red color is due to inactivation of *SUP*4 gene by DNA insertion into *Eco*R1 site)

Non-transformed yeast – **white colony**

The plasmid vector can be grown in bacteria because it contains a *ColE1* plasmid origin and an antibiotic resistance gene (*amp*^{R}). The advantage of this kind of vector is that only few clones are required to cover a particular genome because YAC can accommodate large inserts of DNA.

BACTERIAL ARTIFICIAL CHROMOSOME (BAC)

BAC is another cloning vector system in *E. coli* developed by Mel Simon and his colleagues as an alternative to YAC vectors. BAC vectors are maintained in *E. coli* as large single copy plasmids that contain inserts of 50–300 kb. BAC vectors contain F-plasmid origin of replication (*ori*S), F plasmid genes that control plasmid replication (*rep*E) and plasmid copy number (*par*A, B, C), and the bacterial chloramphenicol acetyltransferase (*CM*^{R}) gene for plasmid selection. The methods used to construct a BAC library are essentially the same as those for a standard plasmid library, except that the insert DNA must be prepared by preparative pulse field gel electrophoresis. pBeloBAC11 is an example of a BAC vector developed in the Simon lab (Fig. 16.9). High molecular weight DNA is partially digested with the enzyme *Hin*dIII, size fractionated by PFGE, and ligated to the *Hin*dIII digested and phophatase-treated pBeloBAC 11 vector. *E. coli* cells are electroporated with the ligated material and white colonies (DNA insertion into *lac*Z α subunit gene) are isolated on the basis of chloramphenicol resistance. The insert DNA can be excised with the rare cutting enzyme *Not*I

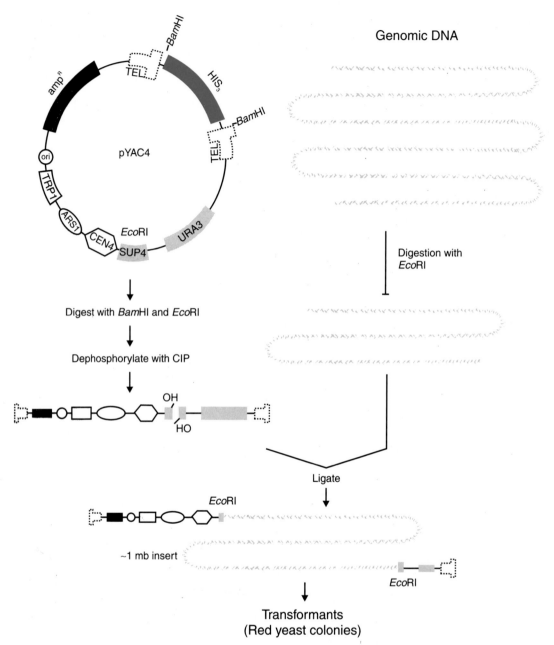

Genomic DNA

Digestion with
*Eco*RI

Digest with *Bam*HI and *Eco*RI

Dephosphorylate with CIP

Ligate

~1 mb insert

*Eco*RI

*Eco*RI

Transformants
(Red yeast colonies)

Fig. 16.8 A schematic diagram of pYAC4 vector and cloning of genomic DNA. Concept of the figure taken from Miesfeld: *Applied Molecular Genetics*. Copyright © 1999 John Wiley and Sons. This material is used by permission of John Wiley and Sons, Inc.

(8 bp recognition site). The plasmid can also be linearized at the *cos*N site by lambda terminase to facilitate end labeling for restriction enzyme mapping. The bacteriophage T7 and SP6 tran-

scriptional promoter sequences flanking the *Hind*III cloning site provide a means to generate end specific RNA probes for library screening.

DNA segments from the bithorax gene of

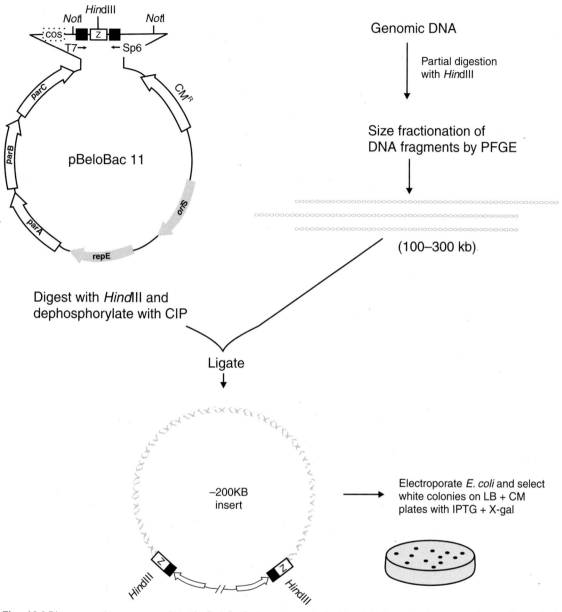

Fig. 16.9 Diagrammatic representation of pBeloBAC11 vector and cloning of high molecular weight DNA. Concept of the figure taken from Miesfeld: *Applied Molecular Genetics*. Copyright © 1999 John Wiley and Sons. This material is used by permission of John Wiley and Sons, Inc.

Drosophila have already been cloned using this vector. Advantages of this vector are as follows: (1) The clones are stable over many generations and (2) BAC can be transformed into *E. coli* very efficiently through electroporation, thus avoiding the packaging extract. The disadvantages are that it lacks positive selection for clones containing inserts and very low yield of DNA.

P1 PHAGE VECTORS

The P1 cloning system was developed as an

alternative to YAC and cosmid vectors for cloning of high molecular weight DNA fragments. The vector, which is 30.3 kb, can accommodate an insert DNA of size up to 95 kb. With 1–2 µg of vector arms and 2–4 µg of size-fractionated DNA fragments, which have *Sau*3A1 sticky ends, greater than 10^5 clones can be generated using the P1 system. In spite of *cos* site in vector, the packaging of P1 phage DNA into the viral head is determined by a 162-bp DNA sequence known as *pac* site. P1 phage head can accommodate DNA of length up to 110 kb.

pAD10SacBII, and pAD10 are the most common P1 phage vectors. The former pAD10 Sac BII vector and its elements are shown in Fig. 16.10. It contains a *ColE1* replicon, P1 *pac* site, an 11-kb *Bam*HI-*Sca*I stuffer fragment from adenovirus DNA, kanamycin resistant (*kan*) gene from a transposon *Tn903*, and an ampicillin resistant gene (*amp*). The P1 vector also contains P1 *lox*P recombination site. The cells containing this vector DNA can be induced by IPTG, but the cells containing this vector will die if there is sucrose in the media, which helps in the selection of transformed cells. There are T7 and Sp6 promoters and rare cutting sites for *Sal*I, *Sfi*I, and *Not*I endonucleases, which permits recovery of the insert. The host bacterial cell should have P1 recombinase gene for multiplying the cloned DNA as a plasmid. It offers certain advantages: (1) Multiple copies of genome are produced from the representative library; (2) Particular DNA cloned can be produced in large quantities very easily; (3) The cloning process is very easy; and (4) Physical dissection of cloned DNA is easy.

P1-DERIVED ARTIFICIAL CHROMOSOME (PAC)

This cloning system was developed by Ioannou *et al.* (1994). The constructed vector incorporates features of both P1 and F factor systems and can be transformed into *E. coli* by electroporation. In a PAC vector, inserts of size 100–300 kb can be cloned. It is devoid of

Fig. 16.10 pAD10sacBII vector and its elements.

problems such as chimarism and instability of cloned DNA.

Shuttle vectors

These are plasmids capable of propagating and transferring (shuttling) genes between two different organisms, one of which is typically a prokaryote (*E. coli*) and the other a eukaryote (e.g. yeast). Shuttle vectors must have unique origins of replication for each cell type as well as different markers for selection of transformed host cells harboring the vector. Shuttle vectors have the advantage that eukaryotic genes can be cloned in bacterial hosts, yet the expression of these genes can be analyzed in appropriate backgrounds.

Expression vectors

These are engineered so that any cloned insert can be transcribed into RNA, and, in many instances, even translated into proteins. cDNA expression libraries can be constructed in specially designed vectors derived from either plasmids or bacteriophage lambda. Proteins encoded by the various cDNA clones within such expression libraries can be synthesized in the host cells, and if suitable assays are available to identify a particular protein, its corresponding cDNA clone can be identified and isolated. Expression vectors designed for RNA expression or protein expression, or both, are available.

PROTOCOLS

16.1. Isolation of plasmid DNA: mini preparation

1. Preparation of bacterial culture: Plasmids can be isolated from their bacterial cultures. The Luria-Bertinin (LB) agar medium with antibiotics can be used.
2. The medium has the following constituents:

Tryptone	10 g/l
NaCl	10 g/l
Yeast extract	5 g/l

   Adjust pH to 7.0 with 1N NaOH; add agar 20 g/l for solid media.
3. Dissolve all the constituents in water and adjust pH. For solid medium add 20 g/l agar. The medium is aliquoted 50–100 ml in each conical flask. The medium is autoclaved at 15 p.s.i. pressure, 121°C for 20 min. The flasks are cooled to 50°C and appropriate antibiotics are added (50 mg/100 ml) depending upon the plasmid selectable marker genes.
4. Streak glycerol stock of bacterial culture on solid media plates containing appropriate antibiotics and incubate at 37°C.
5. After 16 h of incubation, pick a single colony from the petri dish and inoculate in 3 ml volume of LB broth containing appropriate antibiotics. Incubate the cultures for 16 h at 37°C on a shaker at a speed of 120–150 rpm.

Mini preparation of plasmid DNA

1. Distribute the bacterial culture in two sterile eppendorf tubes (1.5 ml). Spin for 30 sec and pour off the supernatant, completely leaving the bacterial pellet as dry as possible.
2. Suspend the pellet in 100 ml of Solution 1 by vortexing.

Solution 1

Glucose	(50 mM)
EDTA	(10 mM)
Tris-Cl	(25 mM)

pH 8.0, stored at 4°C

3. Leave on ice for 5–10 min.
4. Add 200 µl freshly prepared Solution 2 and gently mix by inverting

Solution 2: Alkali lysis buffer

10N NaOH	20 µl
10% SDS	100 µl
Distilled water	880 µl

5. Leave on ice for 10 min.
6. Add 150 µl of Solution 3 (ice cold) to the tube and gently mix by inverting. Incubate on ice for 15 min.

Solution 3: Potassium acetate buffer (pH 5.5) for 100 ml

Potassium acetate (5M)	60 ml
Glacial acetic acid	11.5 ml
Water	28.5 ml

7. After incubation spin at 15,000 rpm for 10 min at 4°C. Collect the supernatant carefully in a fresh tube.
8. Add equal volume of phenol:chloroform in a 1:1 ratio and spin to remove the aqueous phase.
9. Precipitate the DNA with two volumes of ethanol at room temperature. Invert the tubes 10–15 times and incubate at –20°C for 20 min.
10. Centrifuge at 15,000 rpm for 5 min at 4°C. Collect the pellet and dry.
11. Rinse the pellet with 70% ethanol, decant ethanol and dry the pellet.
12. Dissolve the pellet in appropriate volume (50 µl) of Tris-EDTA (TE) buffer.

Tris EDTA buffer pH 8.0
Tris (10 mM) pH 8.0
EDTA (1 mM) pH 8.0

13. Estimate the concentration of DNA: DNA has a maximum absorbance at about 260 nm. Based on extinction coefficient an optical density (OD) of 1.0 at 260 nm corresponds approximately to 50 mg/ml of double stranded

DNA. The ratio of OD_{260}/OD_{280} provides the estimate of purity. A typically pure preparation of DNA has a ratio of approximately 1.8. If there is a contamination with protein or phenol, the ratio will be less than this value. The steps for UV quantification of DNA are as follows:

i. To measure the concentration of DNA, dilute it appropriately with TE buffer (either 1:50 or 1:100).

ii. Standardize the spectrophotometer using TE buffer as blank.

iii. Measure the absorbance of the sample at 260–280 nm.

iv. Calculate the DNA concentration of the sample as follows:
Concentration of DNA (μg/ml) = OD_{260} × 50 × dilution factor
Concentration of DNA (μg/ml) = OD_{260} × 50 × dilution factor/1000

16.2. Isolation of genomic DNA by SDS–Proteinase K method

1. Any plant tissue can be used for DNA extraction. While collecting material it is advisable to avoid brown plant tissue material.

2. Grind tissue with liquid nitrogen in a pre-cooled mortar and pestle.

3. Take 3.8 ml extraction buffer (600 ml of proteinase K (1 mg/ml) and 600 μl of 5% SDS) in a centrifuge tube. Add the ground tissue in the above buffer solution and homogenize it.
Extraction buffer (pH 8.0)
Tris base   0.05 M   6 g/l
EDTA (pH 8.0)   0.03 M   87.7 g/l
Total volume 1000 ml

4. Transfer the homogenized mixture to a tube and incubate at 37°C for 1 h.

5. Add 1 g sodium perchlorate to the mixture and slightly roll the tubes for homogenization. Centrifuge at 8,000 rpm for 5 min.

6. Transfer the supernatant into another tube and add 9 ml ethanol perchlorate. A precipitate of DNA material will be observed.
Ethanol perchlorate solution: Dissolve 40 g $NaClO_4 \cdot H_2O$ in 320 ml absolute ethanol, then add 120 g $NaClO_4 \cdot H_2O$ and 80 ml distilled water.

7. Scoop out DNA into a tube.

8. Add 4 ml of phenol saturated with TE buffer (pH 8.0) and 4 ml of chloroform.

9. Spin at 5000 rpm for 3 min and transfer the aqueous phase to a fresh tube.

10. Repeat the phenol chloroform procedure as in steps 8 and 9.

11. Precipitate the DNA with two volumes of ethanol at room temperature. Invert the tubes 10–15 times and incubate at –20°C for 20 min.

12. Centrifuge at 15,000 rpm for 5 min at 4°C. Collect the pellet and dry.

13. Rinse the pellet with 70% ethanol, decant ethanol and dry the pellet.

14. Dissolve the pellet in appropriate volume (250 ml) of Tris-EDTA (TE) buffer.
Tris EDTA buffer   pH 8.0
Tris (10 mM)   pH 8.0
EDTA (1 mM)   pH 8.0

15. Estimate the concentration of DNA: Follow the Step 13 of protocol 16.1.

Further reading

Chawla, H.S. 1998. *Biotechnology in Crop Improvement*. International Book Distributing Co., Lucknow.

Gupta, P.K. 1996. *Elements of Biotechnology*. Rastogi Publishers, Meerut.

Kingsman, .S.M. and Kingsman, A.J. 1988. *Genetic Engineering: An Introduction to Gene Analysis and Exploitation in Eukaryotes*. Blackwell Scientific Publication, Oxford.

Maniatis, T., Fritsch, E.F. and Sambrock, J. 1982. *Molecular Cloning: A Laboratory Manual*. Cold Spring Harbor Laboratory Press, USA.

Mantell, S.H., Mathew, J.A. and McKee, R.A. 1985. *Principles of Plant Biotechnology: An Introduction to Genetic Engineering in Plants*. Blackwell Scientific Publications, Oxford.

Miesfeld, R.L. 1999. *Applied Molecular Genetics*. Wiley-Liss, New York.

Old, R.W. and Primrose, S.B. 1985. *Principles of Gene Manipulation: An Introduction to Genetic Engineering*. Blackwell Scientific Publications, Oxford.

Sambrook, J., Fritsch, E.F. and Maniatis, T. 1989. *Molecular Cloning: A Laboratory Manual*. Cold Spring Harbor Laboratory Press, USA.

Walker, J.M. and Gaastra, W. (Eds.). 1987. *Techniques in Molecular Biology*. Croom Helm, Australia.

17

Gene Cloning: cDNA and Genomic Cloning and Analysis of Cloned DNA Sequences

In order to clone a DNA sequence that codes for a required gene product, the gene has to be removed from the organism and cloned in a vector molecule. Libraries of such cloned fragments are made. A *gene library* is a random collection of cloned fragments in a suitable vector that ideally includes all the genetic information of that species. This is sometimes also called the shotgun collection because for obvious reasons an appropriate selection has to be made. There are two ways by which gene libraries are made, referred to as complementary (cDNA) and genomic DNA libraries. Cloning of DNA fragments obtained from cDNA and genomic DNA have been discussed in terms of cDNA and genomic DNA cloning.

COMPLEMENTARY DNA (cDNA) LIBRARIES AND CLONING

For many plants and animals, a complete genome library contains such a vast number of clones that the identification of the desired gene is a mammoth task. cDNA libraries are used when it is known that the gene of interest is expressed in a particular tissue or cell type. Since all the genes are not expressed at the same time, relatively few clones will be needed to make a cDNA library that contains copies of all the genes that are active in the tissue or cells from which the mRNA is obtained. It will be a relatively easy task to screen the cDNA library for a clone carrying a particular expressed gene. mRNA cannot itself be ligated into a cloning vector, but it can be converted into DNA by cDNA synthesis and then ligated into a cloning vector.

The strategy of cDNA cloning involves the copying of mRNA transcripts into DNA, which are then inserted into bacterial plasmids and then placed into bacteria by transformation. At this stage it should be known that mRNA used for cDNA preparation is a processed transcript and not the original one transcribed from DNA. The processing of mRNA has been explained in Chapter 13. Almost all RNA transcripts are processed to form the mature RNA.

The primary transcript formed from the DNA in eukaryotes is known as heterogeneous nuclear (hn) RNA. This hnRNA is often modified in the nucleus, which involves a number of steps before it reaches cytoplasm for translation. A cap structure is added to the 5' end, which consists of a 7-methyl guanosine residue, and a tail of adenine residues is added to the 3' end poly(A) tail. The final processing step is the removal (or splicing out) of intervening sequences (introns), which are transcribed along the rest of the gene. Endonucleases specifically cleave the mRNA molecule, and the coding regions (exons) are joined to form a continuous coding sequence. After completion of these processing steps, the mature mRNA molecule is transported into the cytoplasm where it can participate in protein synthesis.

The principal steps involved in cDNA cloning are summarized in Fig. 17.1.

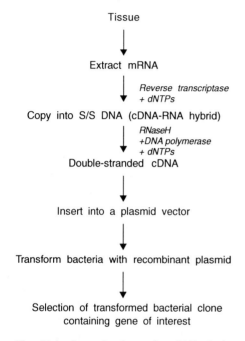

Tissue

↓

Extract mRNA

↓ *Reverse transcriptase + dNTPs*

Copy into S/S DNA (cDNA-RNA hybrid)

↓ *RNaseH +DNA polymerase + dNTPs*

Double-stranded cDNA

↓

Insert into a plasmid vector

↓

Transform bacteria with recombinant plasmid

↓

Selection of transformed bacterial clone containing gene of interest

Fig. 17.1 General scheme for cDNA cloning.

Isolation or extraction of mRNA

When it is known that the gene of interest is expressed in a particular tissue or cell type, a crude extract of the tissue is prepared and then freed from proteins, polysaccharides, and other contaminants. Many eukaryotic mRNAs can be further purified from the total or polysomal fraction using the technique of oligo-dT cellulose chromatography. It is known that mRNAs have poly(A) (adenosine residues) tail at their 3' end. Under appropriate conditions this tail will bind to a string of thymidine residues immobilized on cellulose and then poly(A)$^+$ fraction can be eluted. Two or three passages of the poly(A)$^+$ fraction through such a column produces a fraction highly enriched for mRNA.

The mRNA fraction will contain many different mRNA sequences, but certain techniques can be employed to enrich for a particular mRNA species. At the beginning itself, careful selection of a tissue may be made so that a particular mRNA is abundantly produced for its isolation, for example, the use of bean cotyledons harvested during their phase of storage of protein synthesis. Having prepared a mRNA fraction, it is necessary to check that the sequence of interest is present. This is done by translation of mRNA *in vitro* and identification of appropriate polypeptides in the products obtained.

Synthesis of first strand of cDNA

The mRNA fraction is copied into the first strand of DNA as single-stranded, using RNA-dependent DNA polymerase, also known as reverse transcriptase (Fig. 17.2). This enzyme, like other DNA polymerases, can only add residues at the 3'-OH group of an existing primer, which is base paired with the template. For cloning of cDNAs, the most frequently used primer is oligo-dT, 12-18 nucleotides in length, which binds to the poly(A) tract at the 3' end of mRNA molecules. The RNA strand of the resulting RNA–DNA hybrid is destroyed by alkaline hydrolysis prior to second-strand synthesis.

Synthesis of second strand of cDNA

The second strand of cDNA can be synthesized by two methods.

1. *Self priming: c*DNA: The mRNA hybrid so obtained is denatured so that a second strand can be synthesized on the single strand of cDNA by the Klenow fragment of DNA polymerase I, which is self-priming (Fig. 17.3). Due to unknown reasons, the single-stranded DNA has a transitory hairpin structure at the 3' end, which can be used to prime the synthesis of the second strand of cDNA by the Klenow fragment of *E. coli* DNA polymerase 1. The hairpin loop and any single-stranded overhang at the other end are then digested away with the single-

strand-specific S1 nuclease. The final product is a population of double-stranded, blunt-ended DNA molecules complementary to the original mRNA fraction.

2. *Replacement synthesis:* The cDNA:mRNA hybrid is used as a template for a nick translation reaction (Fig. 17.4). RNase H produces nicks and gaps in the mRNA strand of the hybrid, creating a series of RNA primers that are used by *E. coli* DNA polymerase I during the synthesis of the second strand of cDNA. The reaction has three advantages: (1) It is very efficient, (2) it can be carried out directly using the products of the first strand reaction, and (3) it eliminates the need to use nuclease S1 to cleave the single-stranded hairpin loop in the double stranded cDNA. Most cDNA libraries are now being constructed using a replacement reaction to synthesize the second strand of cDNA.

Cloning of cDNA

The most commonly used procedure for cloning cDNAs involves the addition of complementary homopolymeric tracts to double-stranded cDNA and to the plasmid vector (Fig. 17.5). A string of cytosine residues are added to the cDNA using the enzyme terminal transferase, to form oligo-dC tails on the 3' ends. Similarly, a plasmid is cut open at a unique restriction endonuclease site and tailed with oligo-dG. The vector and the double-stranded cDNA are then joined by hydrogen bonding between the complementary homopolymers to form open circular hybrid molecules capable of transforming *E. coli.*

Introduction to host cells

The recombinant plasmids are used to transform bacteria, usually the *E. coli* K-12 strain. *E. coli* cells treated with calcium chloride will take up plasmid molecules from the surrounding medium and the host cell will repair any gaps in the recombinant plasmid. If the plasmid has

been chosen carefully, it is possible to select transformed from nontransformed bacteria on the basis of antibiotic resistance. Many cloning plasmids contain two antibiotic resistance genes, one of which is destroyed during cloning. In the case of pBR322, cloning into unique *Pst*I site destroys ampicillin resistance but leaves tetracycline resistance intact (Fig. 17.5). Bacteria transformed with a recombinant plasmid will be sensitive to ampicillin but resistant to tetracycline. This simple selection tells the investigator which colonies carry a cDNA copy of some sort.

Clone selection

The antibiotic resistance selection already carried out has identified which clones carry a recombinant plasmid, but there will be thousands of different inserts. The cloning procedure usually begins with a whole population of mRNA sequences. The most difficult part of cDNA is selecting which of these clones carries the sequence of interest. If the gene is expressed, then the simplest selection is to screen for the presence of the protein, either by bacterial phenotype it produces or by the protein detection methods usually based on immunological or enzymological techniques. More often than not the protein is not expressed, and other methods such as nucleic acid hybridization are used. Identification of the gene is discussed after the genomic DNA cloning. The procedure is same for both the cDNA and genomic cloning.

GENOMIC CLONING

Genomic libraries and clones are necessary in addition to cDNA clones because cDNA clones are generated from processed mRNA, which lack introns and other sequences surrounding the gene. From the biotechnological point of view, if genes are to be modified and returned to plants, it is likely that genomic sequences will be more useful.

The genome of plants is remarkably com-

Fig. 17.2   Synthesis of the first strand of cDNA using an oligo (dT) primer and reverse transcriptase.

Fig. 17.3   Synthesis of double-stranded cDNA by the self-priming method.

Fig. 17.4 Synthesis of double-stranded cDNA by replacement synthesis method.

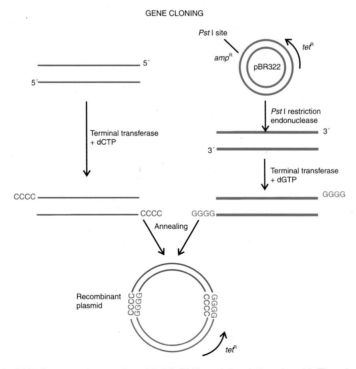

Fig. 17.5 Cloning of cDNA fragment into a plasmid. (–)cDNA and (___) the plasmid. The plasmid, pBR 322 carries ampicillin resistance (amp^R) tetracycline resistance (tet^R). After insertion of foreign DNA the ampicillin resistance gene is destroyed and bacterial clones carrying recombinant plasmids can be selected on the basis of drug resistance

plex, and a particular fragment of interest comprises only a small fraction of the whole genome. Therefore, construction of a useful and representative recombinant genomic library is dependent on the generation of a large population of clones. This is necessary to ensure that the library contains at least one copy of every sequence of DNA in the genome. A general strategy to construct genomic library from random fragments of DNA in phage is shown in Fig. 17.6.

Isolation of DNA

For genomic cloning, nuclear DNA must be isolated from the tissue in a high molecular weight form, largely free from organelle DNA and RNA. The first step for construction of genomic library is isolation of genomic DNA. Different methods are available for the isolation of DNA, but proteinase K, SDS method, and the CTAB method are most commonly employed. Preparation of high molecular weight DNA with minimum amount of shearing and nonspecific damage and whose size range is 100–150 kb is adequate for production of genomic DNA libraries. The size of large molecular weight DNA can be checked by pulsed field gel electrophoresis. DNA preparation can be made RNA free by digestion with ribonucleases.

Partial digestion

The high molecular weight genomic DNA is partially digested with a restriction enzyme, such as *Sau*3A, to obtain fragments with desired range of 9–25 kb, with about 30% of the genomic DNA remaining undigested.

Vectors for cloning

Plasmids are usually used for cDNA cloning, but they accommodate a relatively small piece of DNA due to problems with transformation using very large plasmids. It is generally desirable to clone larger pieces of genomic DNA as a large clone is more likely to contain an intact copy of a gene plus some of the flanking sequences. Cloning large pieces also brings the

number of different clones needed to cover the entire genome down to a manageable number. Thus, genomic cloning is often carried out by using bacteriophage λ and cosmids as vectors, which can accommodate relatively large DNA molecules. Bacteriophage λ has high cloning and packaging efficiency and is easy to handle and store as compared to plasmid vectors. The phage genome is double stranded and 50 kb in length. It is amenable to genetic manipulation and the structure of the genome is well understood. Since its first use as a cloning vector in 1974, many different vectors have been constructed that enable replacement of part of the λ genome with foreign DNA. Part of the central genome is nonessential, so it can be replaced without any detriment to phage reproduction. This type of vector is termed as a replacement vector. Here, the vector is cleaved at two different sites and a portion of the vector is removed and foreign DNA can be inserted into this. The replacement vectors such as EMBL3 (European molecular biology lab) and the lambda DASH and lambda FIX vectors contain a stuffer fragment that is replaced by the genomic DNA and will hold 9–23 kb of inserted DNA. There are insertion vectors also, which are derivatives of wild type phage with a single target site at which DNA can be inserted. These are also used in which the foreign DNA is directly inserted into the digested arms of the vector which generally accommodate up to 12-kb fragments depending on the vector, for example, λgt10, λgt11 and lambda ZAP (see Fig. 16.6). However, these insertion vectors are not commonly used for genomic cloning because the insert size is small.

Cosmids can accept approximately 45 kb of foreign DNA, almost twice as much as bacteriophage vectors (see Fig. 16.4). However, genomic DNA libraries are considerably more difficult to construct and maintain in cosmids than in bacteriophages. In the past few years, it has become obvious that certain genes are too large to be accommodated as a single fragment

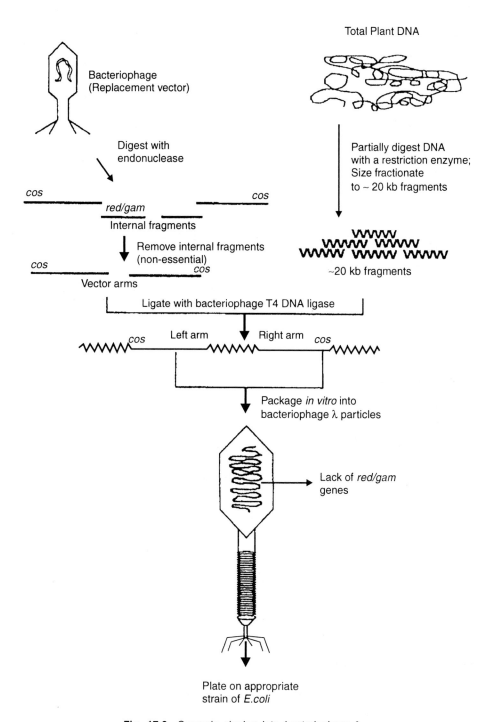

Fig. 17.6 Genomic cloning into bacteriophage λ.

in cosmids also. For this, yeast artificial chromosomes (YACs) and bacterial artificial chromosome (BAC) as vectors are used for cloning (see Figs. 16.8 and 16.9).

Ligation of fragment to a vector

The nonessential region of the phage genome is removed with the same restriction enzyme used for digestion of DNA and the vector arms are purified. These vector arms and the genomic fragments are annealed together. The genome has cohesive ends so as well as the foreign DNA, the genome will anneal to form a concatamer. Proper ligation conditions must be chosen for producing the recombinant clones. The quality of insert DNA and vector DNA is crucial to the success. The usual cause for failure is that the DNA has defective ends and therefore cannot ligate properly.

Packaging

The resulting bacteriophage lambda of the ligations are packaged by using packaging extracts, which contain head and tail proteins of the phage and endonucleolytic cleavage enzymes. The recombinant λ molecules can be packaged into head and tail structures in the test tube. These particles can then be added directly to a culture of bacteria. Proteins required for packaging are purified from infected cells and added to vectors that have been linearized and joined to each other through their terminal *cos* sites. The phage particles are then used to infect *E. coli* [(e.g. XL1 Blue MRF (P2)] for initial selection of phage particles containing recombinant DNA only. Then these phage particles are further infected on *E. coli* XL1 Blue MRF for amplification of the phage particle containing recombinant DNA, and clone selection can subsequently be carried out.

To calculate how representative the constructed library is, the following equation can be used for 99% probability of isolating a fragment of interest:

$$I \times N = 4.6 \times G$$

where I is the size of the average cloned fragment (in base pairs), N is the number of clones screened, and G is the target genome size (in base pairs). This assumes that the cloned fragments were prepared by completely random cleavage of the target DNA.

Once cloned DNA has been produced, hybridization techniques can be applied to identify individual clones. However, as cDNA cloning of a gene is frequently carried out first, a specific probe is often available. These probes can simplify the selection of genomic clones from the very large number produced. Different methods to analyze and identify cloned genes are discussed.

IDENTIFICATION AND ANALYSIS OF CLONED GENES

Once the gene is cloned, the DNA must be analyzed to yield information about gene structure and function. A number of procedures can be employed to attempt identification of a clone carrying the desired gene. Most of these are based on a technique of *hybridization probing.* It is known that any two single-stranded nucleic acids have the potential to base pair with one another. If two strands of nucleic acids are not exactly complementary to each other, they will form an unstable hybrid structure because only a small number of individual interstrand bonds will be formed (Fig. 17.7). However, if the polynucleotides are complementary, then extensive base pairing will occur to form a stable double-stranded molecule. This pairing can occur between single-stranded DNA molecules to form a double helix between single-stranded RNA molecules and also between combinations of one DNA and one RNA strand.

Probes

Nucleic acid hybridization is a fundamental tool in molecular cloning. Virtually every aspect of cloning, characterization, and analysis of genes involves hybridization of one strand of nucleic acid to another. This brings us to the aspect of probes for nucleic acid detection. A probe is a molecule (DNA or RNA segment) that recognizes the complementary sequence in DNA or RNA molecules by molecular hybridization and thus allows identification and isolation of these specific DNA sequences from an organism. The success of colony or plaque hybridization as a means of identifying a particular recombinant clone depends on the availability of a DNA molecule that can be used as a probe. In practice, the nature of the probe is determined by the information available about the desired gene.

The following probes that can be used for hybridization:

i. *Homologous probes*: These contain at least part of the exact nucleic acid sequence of the desired gene.

ii. *Partially homologous probes*: A similarity in the nucleotide sequence is seen when two genes for the same protein, but from different organisms are compared. These similarities mean that a single-stranded probe prepared from the gene of one organism will form a hybrid with the gene from the second organism. Although the two molecules will not be entirely complementary, enough base pairs will be formed to produce a stable structure. This strategy is called *heterologous probing*, and has proved very useful in screening genomic libraries for genes that have already been identified in other organisms.

iii. *Synthetic oligonucleotide probes*: The gene that is being sought will be one that codes for a protein. The amino acid sequence of this protein is determined. From this amino acid sequence it is possible to use the genetic code to predict the nucleotide sequence of the relevant gene. This prediction will always be an approximation. Once the sequence of the gene has been predicted, a mixture of oligonucleotides that will hybridize to the gene can be synthesized. Usually a region of five or six codons where the prediction is fairly unambiguous is chosen, and a set of synthetic oligonucleotides including all the possible sequences is used as a probe. One of these oligonucleotides will have the correct sequence and so will hybridize to the gene being sought.

The advantages of the hybridization method are as follows:

i. It does not require expression of the inserted sequence.

ii. It can be applied to any sequence, provided a suitable radioactive probe is available.

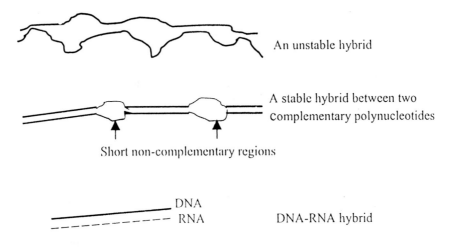

An unstable hybrid

A stable hybrid between two Complementary polynucleotides

Short non-complementary regions

DNA
RNA

DNA-RNA hybrid

Fig. 17.7 Nucleic acid hybridization.

iii. Several identical prints can be easily made from a single phage plate, allowing the screening to be performed in duplicate and hence with increased reliability.

iv. This method allows a single set of recombinants to be screened with two or more probes.

Approaches for identification of genes

Colony and plaque hybridization

Grunstein and Hogness (1975) developed a screening procedure to detect DNA sequences in transformed bacterial colonies or bacteriophage plaques by hybridization *in situ* with a radioactive probe (Fig. 17.8). Depending on the specificity of the probe, this can lead to the rapid identification of one colony amongst many thousands. The colonies to be screened are first replica plated onto a nitrocellulose or nylon membrane by placing on the surface of an agar plate prior to inoculation. A replica agar plate is retained as a reference set. The colonies are grown and the nitrocellulose membrane disc carrying the colonies removed. The membrane bearing the colonies is removed and treated with an alkali so that the bacterial colonies are lysed and their DNAs are denatured. The membrane disc is then treated with proteinase K to remove proteins and leave the denatured DNA bound to the nitrocellulose, for which it has a high affinity, in the form of a DNA print of the colonies. The single-stranded DNA molecules can then be bound tightly to the membrane by a short heat treatment at 80°C or by UV irradiation. The molecules will be attached to the membrane through their sugar-phosphate backbones, so the bases will be free to pair with complementary nucleic acid molecules The probe must now be labeled, denatured by heating, and applied to the membrane in a solution of chemicals that promote nucleic acid hybridization. After a period to allow hybridization to take place, the membrane is washed to remove unbound probe molecules, dried, and the positions of bound probe detected. The nature of probes that can be used has been explained earlier. During hybridization, any colony carrying a sequence complementary to the probe will become radiolabeled and can be identified by autoradiography. Any clone showing a positive result can be picked from the master plate and grown to provide DNA for further analysis.

Immunological detection

Immunological detection of clones synthesizing a foreign protein has also been successful where the inserted gene sequence is expressed. A particular advantage of the method is that genes that do not confer any selectable property on the host can be detected, but it does require that specific antibody is available. In this method transformed cells are plated on agar. A replica plate must also be prepared because subsequent procedures kill these colonies. The bacterial colonies are then lysed to release the antigen from positive colonies. Nitrocellulose filters imprinted with detritus of bacterial lysis are soaked in a solution containing the antibody (unlabeled). The antigen complexes with the bound IgG antibody. The filter is removed and exposed to ^{125}I labeled IgG. The ^{125}I-IgG can react with the bound antigen via antigenic determinants at sites other than those involved in the initial binding of antigen to the IgG coated filter (Fig. 17.9). Positively reacting colonies are detected by washing the filter and making an autoradiographic image. The required clones can then be recovered from the replica plate. Nowadays, the secondary ligand ^{125}I-IgG is available, which is covalently linked to an enzyme whose activity can be detected histochemically (e.g. alkaline phosphatase). The key to the success lies in the quality of the antibody and it is essential that antibody efficiently recognize the denatured protein.

Southern blot analysis of cloned genes

The method was originally developed by Southern (1975), who demonstrated that DNA restriction fragments that had been electrophoretically

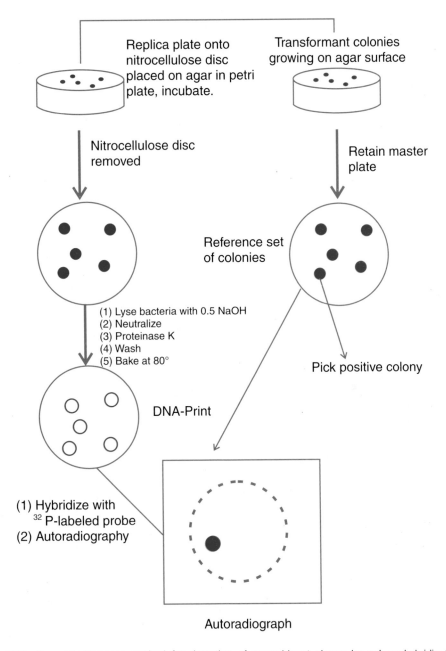

Fig. 17.8 Grunstein-Hogness method for detection of recombinant clones by colony hybridization.

fractionated on agarose gels could be transferred to a solid support (nitrocellulose) and detected as discreet bands following hybridization to a complementary DNA probe. The procedure of agarose gel electrophoresis has been explained in Chapter 14.

Blotting procedure. Genomic DNA is digested with one or more restriction enzymes

Fig.17.9 Antigen–antibody complex formation in the immunochemical detection method.

and the resulting fragments are separated according to size by electrophoresis on an agarose gel. The DNA is denatured in situ and transferred from the gel to a membrane. The procedure of Southern blotting has already been explained in Chapter 14. The membrane to which DNA has been transferred from the gel can then be used for hybridization labeling reaction.

Detection of nucleic acid sequences

There are both radioactive and nonradioactive methods for labeling of nucleic acids to be used as probes for detection of nucleic acid sequences.

Radioactive labeling

During the 1970s, methods became available to introduce radioactivity into nucleic acids in vitro. Before this period, radioactivity could be incorporated into nucleic acids by metabolic labeling, in which radioactive precursors usually ^{32}P were introduced into the cells that were synthesizing the DNA of interest. This procedure required large quantities of radioactivity and involved laborious methods for purification of nucleic acids. Further, this could be applied to small number of DNAs, e.g. viral genomes. But later methods use phosphorylation by bacteriophage T4 polynucleotide kinase in which the γ phosphate of ATP is transferred to 5' hydroxyl terminus in DNA or RNA or the use of *E. coli* DNA polymerase 1 to replace unlabeled nucleotides in double-stranded DNA with alpha ^{32}P labeled nucleotides. The following methods are used for *in vitro* labeling of nucleic acids.

i. *Nick translation:* This method is employed for labeling of double-stranded DNA probes. Nicks in the DNA can be introduced at widely separated sites by very limited treatment with DNase 1. At such a nick, there is a free 3' OH group and DNA polymerase of *E. coli* will incorporate residues at 3'OH terminus that has been created (Fig. 17.10). DNA polymerase 1 has a 5'→3' exonuclease activity and removes or hydrolyzes nucleotides from the 5' side of the nick. The simultaneous elimination of nucleotides from the 5' side and the addition of nucleotides to the 3' side results in movement of the nick along the DNA, i.e. nick translation. If the four deoxynucleoside triphosphates are radiolabeled (e.g. ^{32}P-dNTPs), the reaction progressively incorporates the label into a duplex that is unchanged except for translation of the nick along the molecule. Because the original nicking occurs virtually at random, a DNA preparation is effectively labeled uniformly to a degree depending on the extent of nucleotide replacement and specific radioactivity of the labeled precursors. Generally, the reaction is performed with only one of the four deoxynucleoside triphosphates (e.g. alpha ^{32}P-dCTP) in labeled form and three unlabeled dNTPs.

A typical reaction mixture contains plasmid DNA (probe DNA to be labeled), nick translation buffer, one labeled and other unlabeled dNTPs, DNase 1, *E. coli* DNA polymerase 1, and water to make up the volume. The reaction mixture is incubated for 2–4 h at 14°C. Reaction is stopped by adding EDTA. Normally, the radiolabeled DNA is separated from the unincorporated dNTPs on a Sephadex column and the probe is ready for detection and hybridization studies.

ii. *Random primed radiolabeling of probes:* This method is used to generate probes from denatured, closed circular DNA or denatured, linear, double stranded DNA especially the restriction fragments purified by gel electrophoresis (Feinberg and Vogelstein, 1983; 1984). In this method, DNA is not nicked but primers are used for synthesis and extension of DNA strand. The purified DNA is mixed with primers and is first denatured by boiling and then synthesis is carried out using the Klenow fragment of *E. coli* DNA polymerase.

This enzyme lacks 5'→3' exonuclease activity; therefore, the radioactive product is synthesized exclusively by primer extension rather than by hydrolysis of nucleotides from the nick at 5' terminus in nick translation method.

In this method, the reaction mixture consisting of probe DNA to be labeled is mixed with random primers and buffer, and DNA is denatured by heating at 100°C for 5 min and then put on ice. Then Klenow fragment of *E. coli* DNA polymerase 1 and one labeled and three unlabelled dNTPs are added and volume is made up with water. The mixture is incubated from 30 min to 16 h at 37°C. The labeled DNA is separated from unincorporated nucleotides as described in the nick translation method.

iii. *Probe preparation by PCR:* Advantages of this method are that very small amounts of template material can be used, e.g. specific parts of whole plasmids can be labeled, all of the probe DNA can be labeled, and that the probe has a defined length (Memelink *et al.*, 1994). The procedure is same as described in PCR reactions. A reaction mixture contains template DNA (linearized), amplification buffer, *Taq* DNA polymerase, primers, one labeled and other unlabelled dNTPs, and the volume is made up with water. Forty cycles are run in the PCR thermal cycling apparatus. (For more details refer to Chapter 18 on PCR.) The probe is separated from the unincorporated nucleotides as described for nick translation.

Nonradioactive labeling

Nonradioactive methods to label nucleic acids to be used as probes are used more often. Nonradioactive detection methods typically require shorter exposure times to detect the hybridization signal. At least three different types of nonradioactive labeling and detection systems are widely used. These include the horseradish peroxidase system, the digoxigenin–anti-digoxigenin system, and the biotin–streptavidin system. All three systems may then be detected with either chromogenic (colorimetric) or chemiluminescent substrates for the respective enzymes involved.

Horseradish peroxidase system. In the horseradish peroxidase (HRP) nonradioactive system, DNA is covalently linked with HRP and then chromogenic or chemiluminescent substrates for horseradish peroxidase are used for detection. Chloronapthol is often used as a chromogenic

5′ ——————————————————— 3′　Nicked DNA Duplex

3′ ——————————————————— 5′

E. coli DNA polymerase I
+ 4 dNTPs
(α ^{32}P-labeled)

PPi

Exonucleolysis

Continued reaction

5′ ——————————— 3′　Labeled, nicked DNA
duplex

3′ ——————————— 5′

Fig. 17.10　^{32}P-labeling of duplex DNA by nick-translation. Asterisks indicate radiolabeled phosphate groups.

substrate. In the presence of peroxide and HRP, chloronaphthol forms an insoluble purple product. The chemiluminescence substrate used for HRP is luminol. Chemiluminescent refers to the emission of light resulting from the transition of a molecule from an excited state to a lower energy state triggered by a chemical reaction with the molecule. HRP catalyzes the oxidation of luminol and emits luminescence, which can be detected by an X-ray film.

Digoxigenin (DIG) labeling system. The digoxigenin–anti-digoxigenin labeling system is

based on the use of digoxigenin (DIG), a cardenolide steroid isolated from *Digitalis* plants, as a hapten. A nucleotide triphosphate analog (digoxigenin [11]-dUTP) containing the digoxigenin moiety can be incorporated into DNA to be used as probes by nick translation or random prime labeling. The DIG-labeled probe is subsequently detected by enzyme-linked immunoassay using an antibody to digoxigenin (anti-DIG) to which alkaline phosphatase has been conjugated. A chromogenic or chemiluminescent substrate for alkaline phosphatase can then be used to detect the DIG-labeled probe. Chromogenic substrates such as 5-bromo-4-chloro-3-indolyl phosphate (BCIP) and nitroblue tetrazolium chloride (NBT) are added to give purple/blue color (Fig. 17.11).

The use of dioxetane chemiluminescent probes has been developed in the last few years. Dioxetane substrates can be used with the digoxigenin or the biotin–streptavidin systems. There are a number of stable 1,2-dioxetane derivatives that emit light when activated by enzymes. The 1,2-dioxetane that is commonly used has a phosphate group attached and can be enzymatically activated by alkaline phosphatase. The alkaline phosphatase can be directly linked to the DNA used as a probe or it can be covalently linked to streptavidin and then bound to a biotin-labeled probe. Alternatively, alkaline phosphatase can be covalently linked to

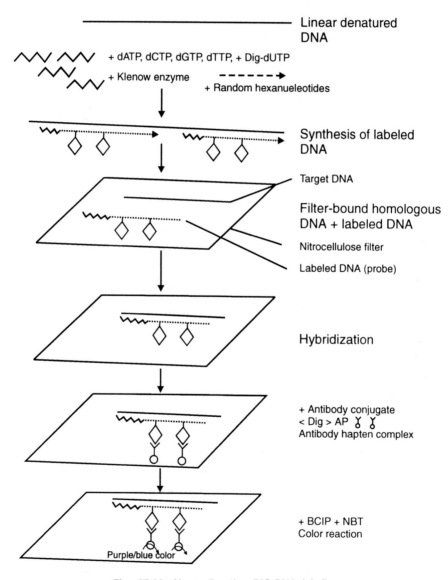

Fig. 17.11 Non-radioactive DIG-DNA labeling.

the antibody directed against a hapten such as digoxigenin. Chemiluminescent substrates for alkaline phosphatase include PPD and CSPD. PPD, also called AMPPD, is 4-methoxy-4-(3-phosphatephenyl) spiro (1,2-dioxetane-3,2' adamantane) (Lumigen, Detriot, MI). The emission of light in this reaction is a two-step process. First, enzymatic dephosphorylation occurs. The alkaline phosphatase cleaves the phosphate group from the protected PPD or AMPPD and an unstable 1,2 dioxetane anion is produced. The anion rapidly breaks down into adamantanone and methyl *m*-oxybenzoate which is in an excited state. When the excited methyl *m*-oxybenzoate decays to its electronic ground state, light is emitted. The wavelength of light emitted is 480 nm. The reaction is shown below. In the presence of excess substrate, the light intensity produced depends upon the concentration of alkaline phosphatase.

Biotin–streptavidin labeling system. It is based on the interaction between biotin and avidin, and is widely used in immunology. A biotin-containing nucleotide analog (biotin-[14]dATP) can be incorporated into DNA by nick translation and random priming methods. After hybridization, the biotin-labeled DNA can be detected by the tight and specific binding of streptavidin that has been conjugated to alkaline phosphatase. A chromogenic or a chemiluminescent substrate for alkaline phosphatase can then be added for detection.

DNA Sequencing

DNA sequencing is an important technique by which the precise order of nucleotides in a segment of DNA can be determined. Rapid and efficient DNA sequencing methods became available in late 1970s and that revolutionized the work on structure of gene. Two different tech-

niques were developed simultaneously by Maxam and Gilbert at Harvard (Maxam and Gilbert, 1977) and by Sanger-Coulson at Cambridge (Sanger *et al.*, 1977).

Sanger-Coulson method. This technique is also known as dideoxy-mediated chain termination method as it involves the use of chain-terminating dideoxy nucleoside triphosphates (ddNTPs) in an enzymatic reaction. 2',3' ddNTPs differ from conventional dNTPs in that they lack a hydroxyl residue at the 3' position of deoxyribose. They can be incorporated by DNA polymerase into a growing chain through their 5' triphosphate groups. However, the absence of a 3'-OH group prevents the formation of a phosphodiester bond with the succeeding dNTP. Further extension of the growing chain is then stopped or terminated. Thus when a small amount of one ddNTP is included along with the four normal dNTPs in a reaction mixture, there is a competition between extension of chain and infrequent but specific termination.

Generally the DNA to be sequenced is a single-stranded template that is cloned into a phage known as M13. This M13 phage infects *E. coli* and has been successfully engineered in the laboratory to contain suitable restriction sites for cloning and to allow easy identification of a phage that is carrying a cloned insert. The M13 viral genome comprises a circular, single-stranded DNA molecule, which is converted into a double-stranded circle inside the bacterial cell. The double-stranded replicative form acts as a template on which the progeny single-stranded molecules are replicated inside the cell, and this single-stranded DNA is packaged into viral protein coats and extruded from the cell to complete the life cycle of the phage. However, it is also possible to use denatured double-stranded DNA template, although the best results are obtained from single-stranded DNA templates.

A short, synthetic oligonucleotide primer is annealed onto the single-stranded template of M13 DNA just upstream of the restriction sites used for cloning (Fig. 17.12). A universal primer that anneals to the vector sequences flanking the target DNA can also be used. When this complex is incubated in vitro with purified DNA polymerase and the four dNTPs, the primed site will act as a starting point for DNA synthesis in the 5′ to 3′ direction through the cloned sequence. If a ddNTP is present, synthesis stops when this molecule is incorporated into the growing chain. Four separate reactions are therefore performed. In each reaction a different ddNTP is present in addition to all four dNTPs (dTTP, dCTP, dGTP, dATP). One of the four dNTPs (usually dATP) is radiolabeled. The ratio of dNTP to ddNTP in the reaction mixture is carefully adjusted so that the dideoxy moiety is incorporated only occasionally

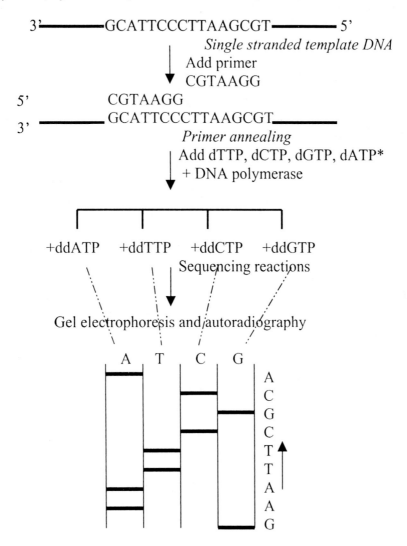

Sequence reads: 5′ GAATTCGCA 3′

Fig. 17.12 DNA sequencing by the dideoxy chain termination method. * indicates the presence of radioactively labeled P or S.

in place of its deoxy homolog. Hence, each of the four reactions produces a population of partially synthesized DNA fragments of different lengths, all sharing a common 5′ end (defined by the primer), all of which are radioactively labeled and all of which are terminated by a dideoxynucleotide. A reaction can be given the name of the base with which the 3′ ends of the synthesized molecules are terminated. For example, all the newly synthesized, radioactive molecules in the T-reaction containing ddTTP will be terminated at a T.

The DNA from these reactions is denatured and the products fractionated according to the size in four different lanes (T, C, G, and A) on a denaturing polyacrylamide gel, which can resolve DNA chain length differences of one nucleotide. Autoradiography of the gel allows the sequence to be read directly. The order of bands on the autoradiograph is determined from the bottom of the gel to the top. This corresponds to the 5′ to 3′ sequence of the strand synthesized in vitro and is therefore complementary to the template-cloned insert. All the bands in the T-track correspond to the position of an A in the template DNA sequence.

Maxam–Gilbert method. This method involves chemical degradation of the original DNA rather than synthesis. The method starts with a single-stranded DNA that is labeled at one end (either its 5′ or 3′ end) with radioactive ^{32}P using either polynucleotide kinase or terminal transferase. A restriction fragment of DNA can be used. From a restriction map, an enzyme is chosen that removes a small piece from one end of the molecule, leaving just one end labeled. Double-stranded DNA can be used if only one strand is labeled at only one of its ends.

The method involves modification of a base chemically and then the DNA strand is cleaved by reactions that specifically fragment its sugar-phosphate backbone only where certain bases have been chemically modified. There is no unique reaction for each of the four bases. However, there is a reaction specific to G only and a purine specific reaction that removes A or

G. Thus the difference in these two reactions is a specific indication of where A occurs (Fig. 17.13). Similarly, there is a cleavage reaction specific for pyrimidine (C + T), which if run in the presence of 1 or 2 M NaCl, works only with C. Differences in these two are thus attributable to the presence of T in the nucleotide sequence (Fig. 17.14). The conditions of chemical cleavage are generally adjusted so that on an average only a single scission occurs per DNA molecule.

In four different reactions, the DNA is then chemically cleaved at specific residues, but the reactions are only partially completed. This generates populations of radiolabeled molecules that extend from a common point (the radiolabeled terminus) to the site of chemical cleavage (Fig. 17.15). The products are a random collection of different sized fragments wherein the occurrence of any base is represented by its unique pair of 5′- and 3′- cleavage products. These products form a complete set, the members of which differ in length by only one nucleotide, and they can be resolved by gel electrophoresis into a ladder, which can be visualized by autoradiography of the gel. Only fragments containing the labeled terminus will be visualized. The sequence can then be deduced and interpreted from the order of the fragments obtained from the different digestions as shown in the figure. The success of this method depends entirely on the specificity of the cleavage reactions. For a number of reasons, the range of this method is less than that of the Sanger method.

In principle, this method can provide the total sequence of a dsDNA molecule just by determining the purine positions on one strand and then the purines on the complementary strand. Complementary base pairing rules then reveal the pyrimidines along each strand, T complementary to where A is, C complementary to where G occurs. With current knowledge, it is possible to read the order of 400 bases from the autoradiogram of a sequencing gel.

The Sanger method required S/S templates, specific primers, and Klenow fragment of *E. coli* DNA polymerase I whereas the Maxam-Gilbert

Cleavage of G

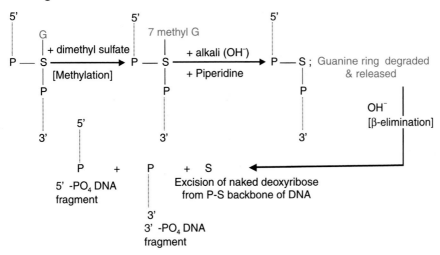

Cleavage of A or G

DNA is treated with acid followed by treatment with dimethyl sulfate which results in methylation of adenine at 3-position as well as guanine at 7 position. Subsequent reaction with OH⁻ and piperidine results in scision of DNA strand and displacement of purines.

Fig. 17.13 Cleavage of purines using dimethyl sulfate in the Maxam-Gilbert method.

Cleavage of pyrimidines (C or T) *reaction for T is shown*

Cleavage of C

If cleavage reaction for pyrimidine is performed in the presence of 1 or 2M NaCl, it works only with C

Fig. 17.14 Cleavage of pyrimidines using hydrazine hydrolysis in the Maxam-Gilbert method.

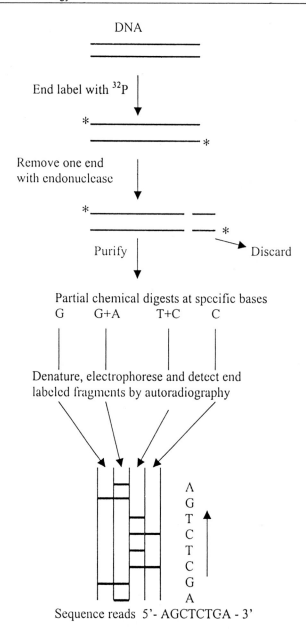

Fig. 17.15 DNA sequencing by the Maxam-Gilbert method. The DNA can be labeled at either the 5′ end (using polynucleotide kinase) or 3′ end (by terminal transferase). Autoradiogram of a hypothetical electrophoretic pattern obtained for four reaction mixtures performed as described in Figs. 17.13 and 17.14 and run in four lanes G, G+A, T+C, and C. Reading this pattern from the bottom up yields the sequence AGCTCTGA. If the 5′ end was ^{32}P labeled, only the 5′ fragments will be evident on the autoradiogram. Assuming 5′ end was labeled the sequence would be AGCTCTGA.

method uses simple chemical reagents. However, with the development of bacteriophage M13 and phagemid vectors, the availability of synthetic primers and improvements to the sequencing reactions, the dideoxy-mediated chain termination method is now used much more extensively than the Maxam-Gilbert method. Nevertheless, the chemical degradation method has one advantage in that the sequence is obtained from the original DNA molecule and not from an enzymatic copy. Therefore, one can sequence synthetic oligonucleotides with the chemical degradation method, analyze DNA modifications such as methylation, and study both DNA secondary structure and the interaction of proteins with DNA by either chemical protection or modification interference experiments. However, because of its ease and rapidity, the Sanger technique is the best choice for DNA sequencing.

High throughput DNA sequencing: The Maxam-Gilbert and Sanger techniques, which are highly conceptual, elegant and efficacious, are in practice time consuming and partly labor intensive, partly because a single radioisotopic reporter is used for detection. Using one reporter to analyze each of the four bases requires four separate reactions and four gel lanes. The resulting autoradiographic patterns, obtained after a delay for exposure and development, are complex and require skilled interpretation and data transcription.

Prober *et al.* (1987) developed a rapid DNA sequencing system based on fluorescent chain terminating dideoxynucleotides. DNA sequence determination using the dideoxy chain termination method results in the synthesis of reaction products that have the same 5′ terminus, but random 3′ ends due to the incorporation of dideoxynucleotides (ddNTPs). This sequencing system is based on the use of a novel set of four chain-terminating dideoxynucleotides, each carrying a different chemically tuned succinyl fluorescein dye tag distinguished by its fluorescent emission. Dideoxy sequencing involves template-directed enzymatic extension of a short oligonucleotide primer in the presence of chain-terminating dideoxyribonucleotide triphosphates (ddNTPs). The fluorescent dyes used are 9-(carboxyethyl)-3-hydroxy-6-oxo-6H-xanthanes or succinyl fluoresceins (SF-xxx, where xxx refers

to the emission maximum in nanometers) (Fig. 17.16). These fluorescent dyes give different colors (e.g. A − red; T − blue; G − green; C − yellow). The wavelengths of the absorption and emission maxima in the succinyl fluorescein system are tuned by changing the substituents $-R_1$ and $-R_2$ as shown in Fig. 17.16. The set of four fluorescence tagged chain-terminating reagents have been designed. These are ddNTPs to which succinyl fluorescein has been attached via a novel acetylenic linker to the heterocyclic base. (These reagents are designated N-SF-*xxx*, where N refers to the ddNTP, SF − succinyl fluoresceins and *xxx* refers to the emission maximum in nanometers.) The linker is attached to the 5 position in the pyrimidines and to the 7 position in the 7-deazapurines. Avian myeblastosis virus reverse transcriptase is used in a modified dideoxy sequencing protocol to produce a complete set of fluorescence-tagged fragments in one reaction mixture. Labeled DNA fragments produced by enzymatic chain extension reactions are separated by polyacrylamide gel electrophoresis, detected and identified as they migrate past the fluorescence detection system specially matched to the emission characteristics of this dye set. A scanning system allows multiple samples to be run simultaneously and computer-based automatic base sequence identifications are made. The DNA sequencing system has been commercialized by Applied Biosystems Incorporated (ABI), and it soon became the workhorse of the Human Genome Project. The ABI system uses four spectrally distinct fluorescent ddNTPs. This strategy makes it possible to perform the entire sequencing reaction in a single tube and to resolve the chain-terminated products in one lane of a sequencing gel. The ABI DNA detection system permits direct real time data acquisition by laser activated dye excitation at a point in the electrophoresis run that maximizes fragment resolution. Using this method, reliable sequence information can be obtained for DNA segments >500 nucleotides long in less than 4 hours. Based on the number of lanes per gel, one ABI instrument

a

SF-505: R₁=R₂=H

SF-512: R₁=H R₂=CH₃

SF-519: R₁=CH₃ R₂ H

SF-526: R₁=R₂=CH₃

Fig. 17.16 Succinyl fluorescein dyes. (A) Chemical structure of the four dyes used to label dideoxynucleotide triphosphates for use as chain-terminators in modified dideoxy DNA sequencing protocols. (B) Normalized absorption spectra of the dyes shown in (A). Absorption coefficient at the maximum for SF-505 is 72,600M⁻¹ cm⁻¹. Spectra were measured in pH 8.2, 50 mM aqueous Tris-HCl buffer. The other dyes have coefficients within 10 percent of this value. Vertical bar (laser) indicates the position of the argon ion laster line at 488 nm used for fluorescence excitation.

can generate ~50,000 nucleotides of sequence information in an 8-hr day. An initial drawback to automated DNA sequencing using a fluorescence-based detection system was the reduced signal-to-noise ratios that occur when small amounts of template are used in the reaction. The solution to this problem was to develop a one-sided PCR amplification strategy, called cycle sequencing, that utilizes a modified *Taq* DNA polymerase and temperature cycling to generate high levels of chain terminated products from a small amount of template. The use of cycle sequencing has greatly extended the automation capabilities of DNA sequencing because it requires much less starting material for each round of sequencing.

Further reading

Chawla, H.S. 1998. *Biotechnology in Crop Improvement.* International Book Distributing Co., Lucknow.

Kingsman, S.M. and Kingsman, A.J. 1988. *Genetic Engineering: An Introduction to Gene Analysis and Exploitation in Eukaryotes.* Blackwell Scientific Publication, Oxford.

Maniatis, T., Fritsch, E.F. and Sambrock, J. 1982. *Molecular Cloning: A Laboratory Manual.* Cold Spring Harbor Laboratory Press, USA.

Mantell, S.H., Mathew, J.A. and McKee, R.A. 1985. *Principles of Plant Biotechnology: An Introduction to Genetic Engineering in Plants.* Blackwell Scientific Publications, Oxford.

Miesfeld, R.L. 1999. *Applied Molecular Genetics.* Wiley-Liss, New York.

Old, R.W. and Primrose, S.B. 1985. *Principles of Gene Manipulation: An Introduction to Genetic Engineering.* Blackwell Scientific Publications, Oxford.

Sambrook, J., Fritsch, E.F. and Maniatis, T. 1989. *Molecular Cloning: A Laboratory Manual.* Cold Spring Harbor Laboratory Press, USA.

Walker, J.M. and Gaastra, W. (Eds.). 1987. *Techniques in Molecular Biology.* Croom Helm, Australia.

18

Polymerase Chain Reaction

The investigation of an organism's genome was greatly enhanced during the early 1970s with the development of recombinant DNA technology. This technology allows for *in vivo* replication (amplification) of genomic DNA regions that are covalently linked with bacterial plasmid or virus clones. In 1985–86, a second major development occurred at Cetus Corporation, USA, where researchers developed an *in vitro* method for the amplification of DNA fragments, referred to as the polymerase chain reaction (PCR). The idea for PCR is credited to Kary Mullis, who along with five other researchers demonstrated that oligonucleotide primers could be used specifically to amplify specific segments of genomic DNA or cDNA. Mullis was awarded the Nobel Prize in chemistry in 1993 for his contribution to the development of PCR. This PCR is based on the features of semiconservative DNA replication carried out by DNA polymerases in prokaryotic and eukaryotic cells. PCR results in the selective amplification of a chosen region of a DNA molecule. In this technique, DNA molecule is not cloned in a bacterial plasmid or virus. The only requirement is that the sequence at the borders of the selected DNA region must be known so that two short oligonucleotides can anneal to the target DNA molecule for amplification. These oligonucleotides delimit the region that will be amplified. In normal DNA replication within a cell, DNA duplex opens up and a small single-stranded RNA primer is synthesized to which DNA polymerase adds further nucleotides. In PCR, a specific region of DNA, which is flanked by two oligonucleotide (deoxy) primers that share identity to the opposite DNA strands, is enzymatically amplified. Amplification of selected region from a complex DNA mixture is carried out *in vitro* by the DNA polymerase I from *Thermus aquaticus*, a bacterium that lives in hot springs.

This amplification is achieved by a repetitive series of cycles involving three steps:
1. Denaturation of a template DNA duplex by heating at 94°C.
2. Annealing of oligonucleotide primers to the target sequences of separated DNA strands at 55–65°C.
3. DNA synthesis from the 3′-OH end of each primer by DNA polymerase at 72°C.

By using repetitive cycles, where the primer extension products of the previous cycle serve as new templates for the following cycle, the number of target DNA copies has the potential to double each cycle (Fig. 18.1).

This technique became important with the discovery of *Taq* DNA polymerase by Kary Mullis. *Taq* DNA polymerase is a thermostable enzyme, which retains activity even after denaturation by heat treatment at 94°C, whereas the DNA polymerase initially used for amplification was the Klenow fragment of *E. coli* DNA polymerase I that could not withstand the denaturation step. Earlier, amplification was

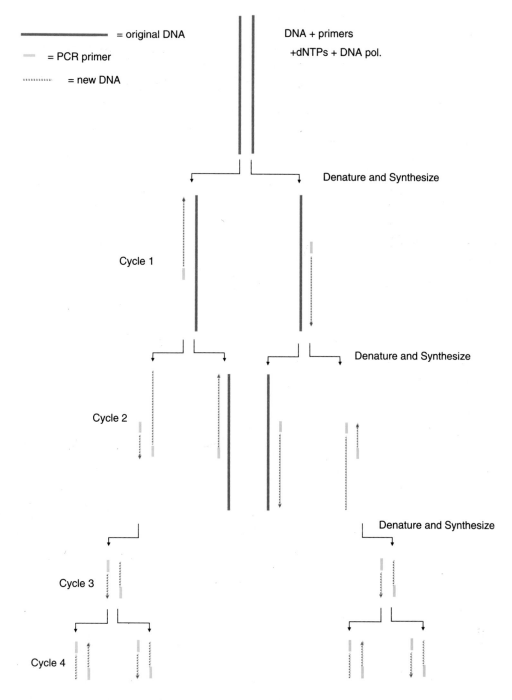

Fig. 18.1 Schematic diagram of the polymerase chain reaction. Only the first two cycles are shown completely. Beginning with the third cycle, the diagram does not show the fate of original DNA and the extension products made from it.

achieved by transferring samples in water baths maintained at different temperatures and adding fresh enzyme during each cycle of amplification. Presently automated thermal cycling apparatus are available that can amplify a million-fold target DNA sequence with just 20 PCR cycles in few hours.

In theory, each amplification cycle should double the number of target molecules, resulting in an exponential increase in the PCR product. However, even before substrate or enzyme becomes limiting, the efficiency of exponential amplification is less than 100% due to suboptimal DNA polymerase activity, poor primer annealing, and incomplete denaturation of the templates. PCR efficiency can be expressed as

PCR product yield = (input target amount) × $(1 + \% \text{ efficiency})^{\text{cycle number}}$

This equation can be used to calculate that ~26 cycles are required to produce 1 µg of PCR product from 1 pg of a target sequence (10^6 amplification) using an amplification efficiency value of 70%. [1 µg PCR product = (1 pg target) × $(1 + 0.7)^{26}$.]

Polymerase chain reaction is useful because of automated instrumentation. A standard PCR machine is basically a thermal cycling instrument. The minimum requirement for a thermal cycler is that it be capable of rapidly changing reaction tube temperature to provide the optimal conditions for each step of the cycle. There are three basic designs, but a design represented by the MJ Research DNA engines™, is the most common. It utilizes the heating/cooling pump principle of the *Pelletier effect*, which is based on heat exchange occurring between two dissimilar surfaces that are connected in series by electric current. Figure 18.2 shows the temperature profile for a typical PCR reaction cycle run on a thermal cycler. The ramp time and temperature accuracy are the most important parameters that must be controlled by the thermal cycler. A hot start is a method that can be used to increase primer specificity during the first round of DNA synthe-

sis. In this procedure, the *Taq* polymerase is held inactive at temperatures below the annealing optimum (<55°C) to avoid extending mismatched primers. The figure also shows a graph depicting the amount of PCR product produced as a function of cycle number. The PCR efficiency equation applies only to the exponential phase, which represents the linear range of the reaction when plotted on a log scale. The plateau effect results from enzyme limitations due to decreased enzyme activity from repeated exposure to 95°C and from stoichiometric limitations of active enzyme molecules, relative to DNA templates, within the time frame of extension period.

Procedure

The polymerase chain reaction is used to amplify a segment of DNA that lies between the two regions of known sequence where two oligonucleotides (deoxy) as primers can bind the opposite strands of DNA due to the complementary nature of base sequences. *Taq* DNA polymerase catalyzes the amplification reaction (Fig. 18.1). The template DNA is first denatured by heating at 94°C. The reaction mixture is then cooled to a temperature that allows the primers to anneal to their target sequences. These annealed primers are then extended (i.e. synthesis of DNA) with *Taq* DNA polymerase. The cycle of denaturation, annealing, and DNA synthesis is repeated many times, because the product of one round of amplification serves as template for the next, each successive cycle essentially doubles the amount of the desired DNA product. The major product of this exponential reaction is a segment of double-stranded DNA whose ends are defined by the 5′ termini of the primers and whose length is defined by the distance between the primers. The products of first round of amplification are heterogeneously sized DNA molecules, whose lengths may exceed the distance between the binding sites of the two primers. In cycle 2, the original strands and the new strands from cycle 1 are separated, yielding a total of four primer sites

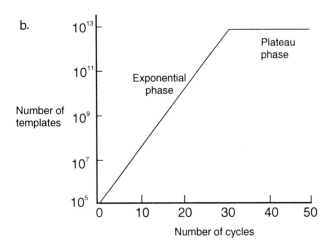

Fig. 18.2 a – Temperature profile of a typical PCR cycle. The dotted lines represent the ramping period which is usually Δ1°C per second with most thermal cyclers. b – Accumulation of PCR target molecules as a function of cycle number. The exponential phase lasts about 30 cycles under standard reaction conditions. The plateau phase results from limiting amounts of enzyme and reduced enzyme activity. Concept of the figure taken from Miesfeld: *Applied Molecular Genetics.* Copyright © 1999 John Wiley and Sons. This material is used by permission of John Wiley and Sons, Inc.

with which primers anneal. The primers that are hybridized to the new strands from cycle 1 are extended by polymerases as far as the ends of the template, leading to a precise copy of the target region. In cycle 3, double-stranded DNA molecules are produced that are precisely identical to the target region. Further cycles lead to exponential doubling of the target region. The original DNA strands and the variable extended DNA strands will continuously be produced from the original template but at a linear rate and become negligible after the exponential increase of target fragments (Table 18.1). The reaction can be continued for 25–45 cycles. Under normal conditions, the amount of *Taq* DNA polymerase becomes limiting after 30 cycles of amplification. If further amplification is required, then a sample of the amplified DNA can be diluted and used as a template for further rounds of synthesis in a fresh PCR reaction.

Table 18.1 PCR amplification of a target fragment with increasing number of cycles

Cycle number	Number of double-stranded target molecules
1	0
2	0
3	2
4	4
5	8
6	16
7	32
8	64
9	124
10	256
11	512
12	1024
13	2048
14	4096
15	8192
16	16,384
17	32,768
18	65,536
19	131,072
20	262,144
21	524,288
22	1,048,576
23	2,097,152
24	4,194,304
25	8,388,608

Components in a polymerase chain reaction

1. *Oligomer primers*: The amplification product of a PCR reaction is determined by the sequence of PCR primers. The primers anneal to complementary sequences on the DNA template and thereby determine the boundaries of the amplified product. Usually primers are used at a concentration of 1µM, which is sufficient for at least 30 cycles of amplifications. It is assumed that the *Taq* DNA polymerase begins to work as soon as priming oligomers become bound to their templates at low temperatures (35–55°C). Once a DNA target has been chosen, there are several rules of thumb for primer design that are important to consider. (a) *Primer length*: Primers for the PCR should be 16–24 nucleotides long, and are generally referred to as primers 1 and 2 for the two ends of target sequence. Primers of this much length maintain specificity (chance of finding the same sequence by chance is only 1 in 4^{20}) and provide sufficient base pairing for stable duplex formation. (b) *Duplex stability*: Both the primers in a PCR reaction should have similar melting temperatures (T_m) to ensure that they will have the same hybridization kinetics during the temperature annealing phase. Primers with an overall G + C content of 45–55% are most common. The T_m of each oligonucleotide can be calculated using an empirical formula that takes into account base composition and the effect of nearest neighbor interactions (the T_m of AAG is different from AGA). For primers with a low GC content, it is desirable to choose a long primer so as to avoid a low melting temperature. (c) *Noncomplementary primer pairs*: The two primers should not share complementarity at the 3′ ends because they will give rise to

primer dimer products. For example, if one primer has the sequence 5′ — ...GGCG — 3′ and the other primer has a terminal sequence 5′ — ...CCGC — 3′, then they can form a short hybrid that will become a substrate for DNA synthesis. Once the primer dimer product is formed, it is a competing target for amplification. (d) *No hairpin loops*: Each primer must be devoid of palindromic sequences that can give rise to stable intrastrand structures that limit primer annealing to the template DNA. (e) *Optimal distance between primers*: This rule is specific for application, but for most diagnostic PCR assays, it is best when the opposing primers are spaced 150–500 bp apart. (f) Sequences with long runs (i.e. more than three or four) of a single nucleotide should be avoided. Presently, a variety of computer algorithms has been developed, which takes all these parameters into account, and the researcher only needs to provide a DNA target sequence.

2. *Amplification buffer:* The standard buffer for PCR reaction contains KCl, TrisCl, and 1.5 mM magnesium chloride. When incubated at 72°C, the pH of the reaction drops by more than a unit. To maintain the pH, presence of divalent magnesium cations is critical.

3. *Deoxyribonucleoside triphosphates*: dNTPs are used at a saturating concentration of 200 µM for each dNTP.

4. *Target sequence*: Template DNA containing the target sequences can be added in a single- or double- stranded form. Linear target sequences are amplified better as compared to closed circular DNAs. Purity of template is flexible and even crude DNA extracts (e.g. pulp of a fossil tooth, a hair follicle) can be used as long as the addition of template does not inhibit the activity of polymerase. The concentration of the target sequence in the PCR reaction is generally in nanograms (5–100 ng).

5. *Taq DNA polymerase*: Two forms of *Taq* DNA polymerase are available: the native enzyme

purified from *Thermus aquaticus* and a genetically engineered form synthesized in *E. coli* (Ampli Taq-™). *Taq* DNA polymerase is the most frequently used and preferred enzyme. The original PCR protocol used the Klenow fragment of *E. coli* DNA polymerase I to perform the primer extension reaction. However, this meant that fresh enzyme had to be added after each round of denaturation because this enzyme is easily heat inactivated. Another problem with *E. coli* DNA polymerase is that the optimal activity level of the enzyme is 37°C, which greatly limits the specificity of the reaction due to degenerate primer annealing at this low temperature. Both these problems were solved by the use of *Taq* DNA polymerase. This enzyme retains activity even after repeated exposure to a temperature of 95°C, and is also fully active at ~75°C, which essentially eliminates degenerate heteroduplex formation. More recently, several other thermostable DNA polymerases have been isolated and characterized that offer advantages for specialized PCR assays (Table 18.2). The biggest difference between these enzymes is their inherent 3′→5′ exonuclease proof reading activity, which is important if the sequence of PCR product must be error free. DNA polymerases with 3′→5′ exonuclease activity are *Tli* from *Thermococcus litoralis* and *Pfu* from *Pyrococcus furiosus*, which have a 10–100 times lower error rate than DNA polymerases lacking this activity. *Taq* DNA polymerase and *E. coli* DNA polymerase I encode a 5′→3′ exonuclease activity. Two other parameters are processivity (number of nucleotides polymerized before template dissociation) and extension rate (nucleotides polymerized/second). Some thermostable DNA polymerases can use RNA templates as a substrate, which can be useful for PCR applications that require a separate cDNA synthesis reaction using viral reverse transcriptase. An enzyme of this is the recombinant form of *Tth* polymerase

Table 18.2 Characteristic features of different thermostable DNA polymerases

Enzyme	Relative efficiency	Error rate	Processivity	Extension rate	3′→5′ exonuclease	5′→ 3′ exonuclease
Taq Pol	88	2×10^{-4}	55	75	No	Yes
Tli Pol (Vent)	70	4×10^{-5}	7	67	Yes	No
Pfu Pol	60	7×10^{-7}	n.d.	n.d	Yes	No
rTth	n.d	n.d	30	60	No	Yes

Note. Relative efficiency – Percent conversion of template to product per cycle; Error rate – Frequency of errors per base pairs incorporated; Processivity – Average number of nucleotides added before dissociation; Extension rate – Average number of nucleotides added per second; n.d. – not determined.

(*rTth*) from *Thermus thermophilus*, which can catalyze high temperature reverse transcription of RNA in the presence of $MnCl_2$. Approximately 2 units of either of the enzymes are required to catalyze a typical PCR. Addition of excess enzyme may lead to amplification of nontarget sequences. Both forms of the polymerase (*Taq* and Ampli *Taq*™) carry a 5′→3′ polymerase-dependent exonuclease activity, but they lack a 3′→5′ exonuclease activity. It means that *Taq* polymerase does not correct the incorporation of mismatched bases.

The reaction is carried out in a sterile 0.5-ml microfuge tube by mixing the components in the order of sterile water, amplification buffer, dNTPs, primers 1 and 2, template DNA and finally the *Taq* DNA polymerase. Total volume can be 50 or 100 μl. The reaction mixture is overlayered with 2–3 drops of mineral oil. This prevents evaporation of the sample during repeated cycles of heating and cooling. But now PCR thermal cyclers are available with heated lid; thus, mineral oil is not required in those reactions. Also, reactions can be carried out in a much smaller volume of 10–25 μl.

The amplified DNA product is detected by agarose gel electrophoresis, followed by staining with ethidium bromide.

Inverse PCR

Conventional PCR is used frequently to amplify DNA segments lying between the two priming oligonucleotides. But PCR can also be used to amplify unknown DNA sequences that lie outside the boundary of known sequences for which primer pairs can be designed. This is referred to as inverse PCR (iPCR). The technique for amplifying unknown flanking DNAs was developed by different research groups (Ochmann *et al.*, 1988; Triglia *et al.*, 1988). In this technique, genomic DNA is digested with restriction enzyme(s) to produce DNA fragments of appropriate size (1 to 4 kb). It must be noted that there should be no sites of cleavage within the core target sequence. The DNA fragments are then ligated to form monomeric circular molecules, followed by PCR using oligomer primers that are designed to extend outwardly from the known core target sequence, i.e. in the direction opposite to those used for normal PCR (Fig.18.3). The major product of the amplification reaction is a linear double-stranded DNA molecule that consists of a head-to-tail arrangement of sequences flanking the original target DNA. Inverse PCR is therefore useful as a method of chromosome crawling to explore the chromosomal sequences that are contiguous to a known segment of DNA. But the length of DNA that can be amplified by PCR is limited to relatively short steps along the chromosome.

REVERSE TRANSCRIPTASE-MEDIATED PCR (RT-PCR)

Combining the process of cDNA synthesis by reverse transcriptase and PCR amplification into a single application leads to reverse transcriptase PCR (RT-PCR). Thus, amplification of double-stranded DNA by PCR is not limited to

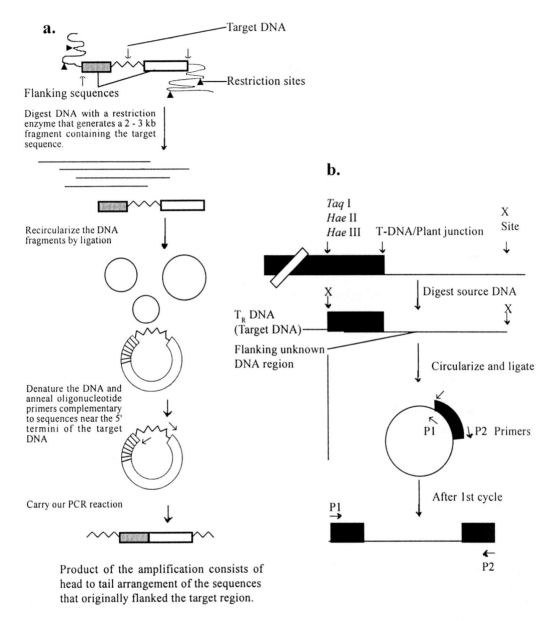

Fig. 18.3 a – Major steps involved in inverse polymerase chain reaction. b – Generalized amplification scheme for unknown DNA flanking the T_R DNA border of pTiT37 by iPCR.

genomic DNA targets, but can also be applied to double-stranded cDNA that has been synthesized from RNA using a reverse transcription reaction. It is possible to detect, and even isolate, very rare mRNA transcripts from cell samples. RT-PCR acts as a means to obtain material for sequence determination and a step in cloning a cDNA copy of the RNA.

As described earlier, there are thermostable DNA polymerases such as *rTth*, which can utilize

RNA templates for cDNA synthesis and thus permit single enzyme RT-PCR. Alternatively, one of the viral reverse transcriptase enzymes (AMV RTase or MMLV RTase) can be used for the cDNA synthesis reaction, and then a second reaction is performed using one of the thermostable DNA polymerases. There are advantages and disadvantages to each of these methods, and the deciding factor often depends on the whether the RT-PCR reaction is being performed for a specialized research application or an automated diagnostic assay. At present, both classes of enzymes (viral RTases and *rTth* DNA polymerase) are used for RT-PCR.

In this section, three basic molecular genetic strategies that utilize RT-PCR as a tool are described.

RACE: Rapid amplification of cDNA ends

RT-PCR can be possibly used to amplify a complete cDNA coding region. But in many applications there may be insufficient information for a straightforward strategy. In general, cDNA synthesis and library screening can lead to the isolation of gene sequences that represent all or part of a processed mRNA transcript. There are always problems when trying to isolate the 5′ end of low abundance transcripts because cDNA libraries often do not contain a full representation of all mRNA sequences. This is especially true of oligo-dT-primed libraries. RACE protocols generate cDNA fragments by using PCR to amplify sequences of a known gene between a single region in the mRNA and either the 3′ or the 5′ end of the transcript. Thus it is necessary to know or to deduce a single stretch of sequence within the mRNA to choose a one gene specific primer complementary to the sequence of cDNA segment and a second "anchored" primer that anneals to a sequence that has been covalently attached to the newly created cDNA terminus. This strategy was first described by Frohman *et al.* (1988).

In the two RACE protocols, extension of the cDNAs from the ends of the transcript to the specific primers is accomplished by using

primers that hybridize either at the natural 3′ poly (A) tail of the mRNA, or at a synthetic poly (dA) tract added to the 5′ end of the first-strand cDNA (Fig. 18.4). In the RACE 3′ end protocol, mRNA is reverse transcribed using an oligo (dT$_{17}$) primer, which has a 17-nucleotide extension at its 5′ end because of the natural 3′ poly(A) tail of the mRNA. This extension, the anchor sequence, is designed to contain the restriction sites for subsequent cloning. Amplification is performed using the anchor 17-mer and a primer specific for the sought-after cDNA.

In the RACE 5′ end protocol, the mRNA is reverse transcribed from a gene-specific primer. The resultant cDNA is then extended by terminal deoxytidyl transferase (TdT) to create a poly (dA) tail at the 3′ end of the cDNA. Amplification is performed with the oligo (dT$_{17}$)/anchor system as used for the 3′ protocol and the specific primer. 5′ RACE is used to amplify uncloned upstream transcript sequences using a gene-specific primer.

Several modifications of the original RACE protocol have been described. One of these, called RNA ligase-mediated RACE (RLM-RACE), involves the use of bacteriophage T4 RNA ligase to attach covalently a single-stranded RNA anchor molecule to the decapped 5′ end of the mRNA. First-strand cDNA synthesis can be performed using a gene-specific primer to produce a pool of cDNAs encoding the anchored primer sequence. Variation of this method is referred to as ligation-anchored PCR (LA-PCR).

Quantitative RT-PCR

It is often important to know relatively steady-state levels of specific gene transcripts as a means to investigate the role of a gene function in a particular cell phenotype. Northern blots and RNase protection assays are the two common ways to measure steady-state mRNA levels. However, both of these assays are labor intensive and may require up to 25 µg of total RNA for each assay. Therefore, most gene expression studies cannot be performed using

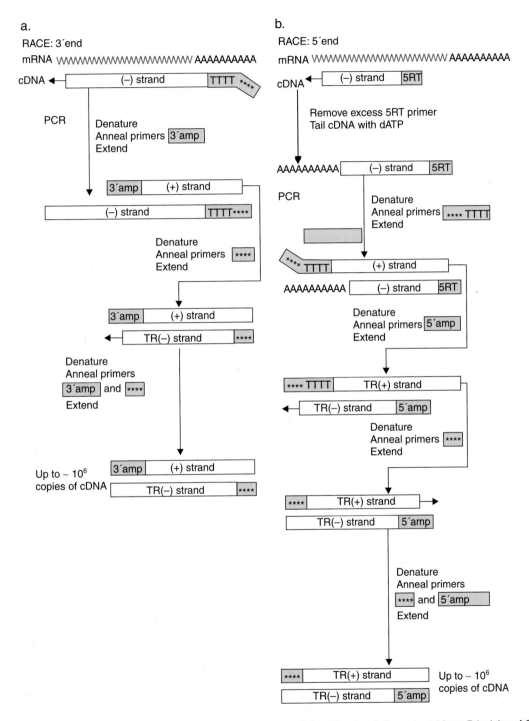

Fig. 18.4 Rapid amplification of cDNA ends (RACE). a – 3′ protocol; b – 5′ protocol. Reproduced from *Principles of Gene Manipulation,* R.W. Old and S.B. Primrose, by kind permission of Blackwell Science, Osney Mead, Oxford, U.K.

conventional RNA-based assays. It is possible to develop a quantitative RT-PCR assay that takes into account variation in the number of target molecules as well as amplification efficiency.

Quantitative RT-PCR measurements require a standard RNA template against which the experimental RNA can be measured. Two types of RNA standards have been used for this purpose. One type is an endogenous gene product that is ubiquitously expressed in all cell types, for example, actin or glyceraldehyde-3-phosphate dehydrogenase (GAPDH), which are the same standards used in northern blots and RNase assays. The other type of standard is an exogenously added cRNA (complementary RNA) that is synthesized by *in vitro* transcription. These exogenous cRNAs are sometimes called RT-PCR "mimics" because they contain the same RT-PCR priming sites and overall sequence as the target RNA, but produce a PCR product that differs from the target RNA by a unique restriction site or a shift in molecular weight. This makes it possible to compare directly the PCR product of the target RNA to the cRNA standard on the same agarose gel. Mimic cRNAs are transcribed from plasmids containing SP6 or T7 bacteriophage promoters and are constructed by *in vitro* mutagenesis of the cloned target cDNA.

There are two ways to use mimic cRNAs to measure the amount of target mRNA in a sample. (1) *Noncompetitive RT-PCR:* This is performed by adding mimic cRNA to the RNA sample at approximately the concentration of the target mRNA. By limiting PCR amplification to the linear range (midexponential phase), usually 25 cycles or less, it is possible to establish a direct correlation between the known amount of cRNA and target mRNA. However, the absolute number of target mRNA molecules in the sample greatly affects the number of cycles needed to detect the product within the linear range. This is a significant problem if this method is used for diagnostic purposes. (2) *Competitive RT-PCR:* It circumvents the problem of variable target copy number by using product ratios that are independent of cycle number. This is done by performing a series of RT-PCR reactions that contain the same amount of sample RNA to which various known amounts of mimic cRNA have been added. The concentration of mimic cRNA that produces the same amount of mimic PCR product as the target RNA in the sample (a mimic: target product ratio of 1.0) represents the point at which both templates are competing equally for the primers. Figure 18.5 illustrates how competitive RT-PCR can be used to quantify the amount of AMG mRNA in a cellular RNA sample that has been doped with increasing amounts of an AMG mimic cRNA. Insertional mutagenesis can be used to construct an AMG mimic cRNA for an internal marker in RT-PCR reactions using AMG gene-specific primers (P1 and P2). The AMG mimic PCR product is 30 bp longer than the normal AMG PCR product. The amount of products generated in reactions can be seen in agarose gel. Since competitive RT-PCR quantification is based on ratios, it is not necessary to restrict PCR amplification to the exponential phase, and therefore maximum sensitivity can be achieved by performing up to 35 cycles.

Amplification of differentially expressed genes

Two techniques–differential screening and subtraction hybridization – can be used to isolate gene transcripts that are present at different levels in two RNA populations. However, these two approaches require large amounts of RNA to synthesize sufficient quantities of an enriched cDNA probe for library screening. In contrast, RT-PCR can be used to generate cDNAs from very small amounts of mRNA. Here, two basic RT-PCR cloning approaches have been described to identify differentially expressed genes.

a) *Differential display reverse transcriptase PCR (DDRT-PCR):* This technique, developed by Liang and Pardee (1992), is based on the principle of two-dimensional gel electrophoresis as a rapid screening procedure for differential display of proteins that were metabolically labeled *in vivo* or *in vitro*.

Fig. 18.5 Competitive RT-PCR uses a mimic cRNA to quantify the amount of target mRNA in an experimental sample based on a ratio of PCR products. a – Flow scheme showing how insertional mutagenesis can be used to construct an AMG mimic cRNA for use as an internal marker in RT-PCR reactions using AMG gene-specific primers (P1 and P2). b – Schematic representation of an agarose gel showing the relative amounts of RT-PCR products generated in reactions using a constant amount of cellular RNA and decreasing amounts of known *in vitro* synthesized AMG mimic cRNA. Concept of the figure taken from Miesfeld: *Applied Molecular Genetics.* Copyright © 1999 John Wiley and Sons. This material is used by permission of John Wiley and Sons, Inc.

The appearance or disappearance of protein spots in the two-dimensional array could provide information about cellular changes in protein synthesis. Liang and Pardee applied the same differential display approach to mRNA by developing a method to array RT-PCR products on standard DNA sequencing gels. DDRT-PCR provides the potential to clone differential RT-PCR cDNA products by physically excising radiolabeled bands from an acrylamide gel. Modified oligo dT primers are used for reverse transcription that anneal to a subset of poly A+ mRNAs owing to differences in dinucleotides at the 3′ end of the primer. Figure 18.6 shows how a 5′ – dT-GC – 3′ primer is used to synthesize cDNA from a subset of RNAs containing the appropriate 3′ dinucleotide adjacent to the poly A tail. The product is then PCR amplified in the presence of ^{33}P-dATP and a known primer. Of these first-strand cDNAs, only a fraction serve as appropriate templates for separate PCR reactions containing a known arbitrary primer. Candidate RT-PCR products are excised from the gel, reamplified with the same primer pair, and used as probes on conventional Northern blots to verify that they correspond to differentially expressed gene transcripts. The false positive rate with DDRT-PCR can be highly variable and therefore it can be best used as a screening procedure rather than a cloning strategy. A similar RT-PCR screening technique called RAP-PCR (RNA arbitrarily primed PCR), described by Welsh and McClellend, is based on a genomic DNA fingerprinting strategy. The primary difference between these two methods is that RAP-PCR uses the same random primer for both reverse transcription and PCR amplification steps, which eliminates bias toward the noncoding 3′ poly A+ sequences.

b) *Suppression subtraction hybridization (SSH)*: Siebert and Larrick (1992) developed a PCR-based strategy to clone differentially expressed gene transcripts. Although it is technically not an RT-PCR approach, it does require the synthesis of double-stranded DNA as a starting point for exponential PCR amplification of gene transcripts present at higher concentration in one of the two mRNA pools. The basis for differential amplification in SSH is two-fold. First, conditions are used that promote rapid reassociation kinetics with excess "driver" cDNA sequences to normalize two "tester" cDNA pools. Double-stranded cDNA is made from mRNA representing tester (containing gene transcripts of interest) and driver (lacks appreciable amounts of desired gene transcript) transcripts (Fig. 18.7). The tester cDNA is first digested with the base cutting restriction enzyme (*Rsa*I) to create blunt-ended termini and then divided into two equal portions, and distinct primers (Adp 1 and Adp 2) are ligated to the two cDNA pools. In the next step, separate rapid reassociation hybridization reactions are performed (rxn 1a and rxn 1b) with each tester cDNA pool and a mass excess of driver cDNA. Following this driver tester hybridization reaction, the two tester pools are mixed without prior denaturation and the unhybridized single-strand low abundance molecules are allowed to anneal under conditions to initiate a second hybridization reaction (rxn 2). In the subsequent PCR amplification step, a pair of adapter primers exponentially amplifies these unique "tester" duplexes. Second, the PCR amplification of nondifferential cDNAs is suppressed because undesirable sequences contain inappropriate priming sites or are capable of forming intrastrand "panhandle" structures that are poor substrates for PCR. The combined PCR effect of exponential amplification of selective tester duplex cDNAs, with the resulting suppression of common sequence cDNAs, can potentially result in a 10^2- to 10^3-fold enrichment of differentially expressed tester cDNA sequences. The amplified cDNAs are used to make an enriched

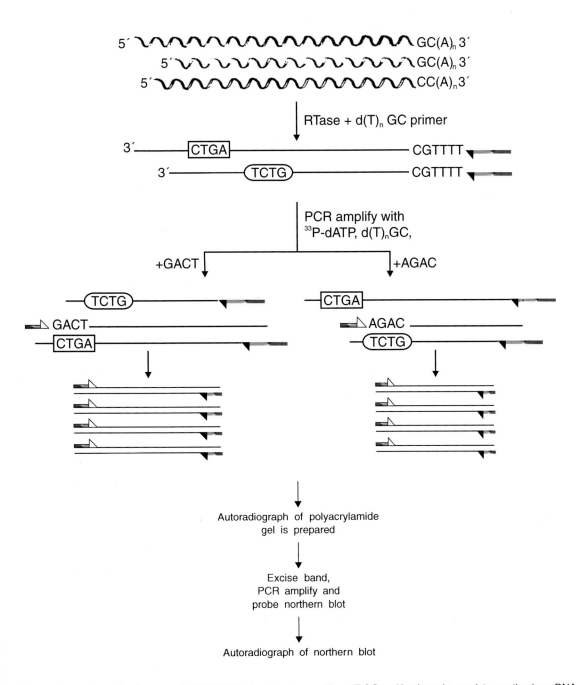

Fig. 18.6 A schematic diagram of DDRT-PCR showing how a 5′ − dT-GC − 3′ primer is used to synthesize cDNA from a subset of RNAs containing the appropriate 3′ dinucleotide adjacent to the poly A tail. Of these first strand cDNAs, only a fraction of will serve as appropriate templates for separate PCR reactions containing a known arbitrary primer. Concept of the figure taken from Miesfeld: *Applied Molecular Genetics.* Copyright © 1999 John Wiley and Sons. This material is used by permission of John Wiley and Sons, Inc.

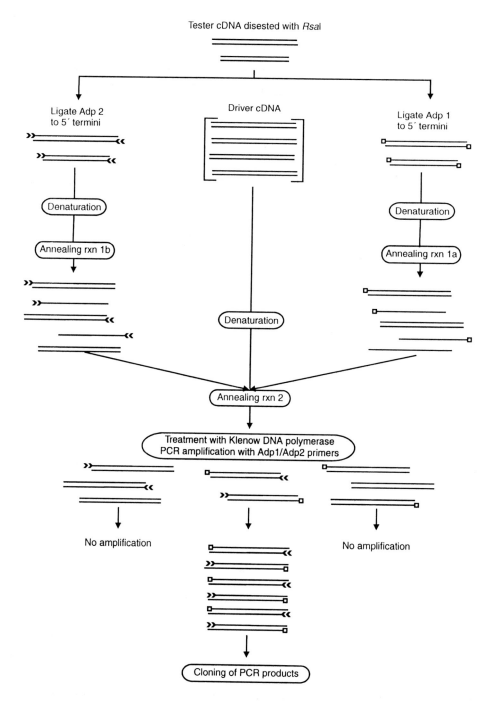

Fig. 18.7 A flow diagram of suppression subtraction hybridization technique. It combines cDNA normalization with selective PCR amplification to achieve enrichment of differentially expressed gene transcripts. Concept of the figure taken from Miesfeld: *Applied Molecular Genetics.* Copyright © 1999 John Wiley and Sons. This material is used by permission of John Wiley and Sons, Inc.

plasmid library, which can be randomly sequenced or screened with a differential cDNA probe.

Cloning of PCR products

There are numerous situations in which a specific DNA segment must be cloned into a plasmid vector. Although PCR generates sufficient material for analysis, when extensive manipulations of the sequence is needed or when large amounts of DNA are required, it is necessary to clone the PCR product. But amplified DNA fragments do not clone efficiently because there are no convenient restriction enzyme sites to facilitate the design of subcloning strategy. Three basic strategies are commonly used to subclone PCR products.

1. *Restriction site addition:* This is done by including the nucleotide sequence of an appropriate restriction enzyme into the 5′ ends of the primer. A PCR primer may be designed which, in addition to the sequence required for hybridization with the input DNA, includes an extra sequence at its 5′ end. Figure 18.8 illustrates the addition of a *Hin*dIII and an *Eco*RI sites to the ends of an amplified DNA fragment. Four nucleotides are placed between the hexanucleotide restriction sites and the extreme ends of the DNA for ensuring that the restriction sites are good substrates for the restriction endonucleases. The extra sequence does not participate in the first hybridization step; only the 3′ portion of the primer hybridizes, but it

Fig. 18.8 Incorporation of extra sequences at the 5' end of a primer. Primer 1 has an extra sequence near its 5′-end which forms a *Hin*dIII site (AAGCTT) and Primer 2 has an extra sequence near its 5′-end which forms an *Eco*RI (GAATTC) site. Each primer has an additional 5′-terminal sequence of four nucleotides so that hexanucleotide restriction sites are placed within the extreme ends of amplified DNA. Reproduced from *Principles of Gene Manipulation,* R.W. Old and S.B. Primrose, by kind permission of Blackwell Science, Osney Mead, Oxford, U.K.

subsequently becomes incorporated into the amplified DNA fragment. The restriction enzyme recognition site does not interfere with template annealing during the first PCR cycle. However, all subsequent cycles contain PCR products that have the restriction enzyme site incorporated into the DNA termini. By digesting the PCR products with the corresponding enzymes, the amplified DNA fragment can be cloned by a standard ligation reaction.

2. *T/A cloning:* This relies on the fact that a proportion of amplified DNA fragments are not truly blunt ended because thermostable DNA polymerases (*Taq* and *Tth*) lack 3′ to 5′ exonuclease activity and have a tendency to add an extra A residue at the 3′ end of each strand. This one base overhang at each end facilitates ligation to the plasmid vectors containing a thymidine nucleotide (TMP or TTP) overhang. Although not as efficient as restriction site addition PCR cloning, T/A cloning is easier because PCR products can be ligated directly to the vector without additional purification steps or enzyme cleavage reactions.

3. *Blunt-end ligation:* This is required if a proofreading thermostable DNA polymerase is used and restriction sites have not been added to the primer. The basic PCR generates double-stranded amplified product. However, the product of the reaction does not contain a large proportion of perfectly blunt-ended molecules. This can be carried out in a combined exonuclease/repair reaction with Klenow polymerase in the presence of all four dNTPs. *Taq* DNA polymerase in general remains tightly bound to the 3′ ends of the DNA, so proteinase K treatment is recommended. Excess primer is removed by preparative gel electrophoresis as a means of purifying the amplified DNA. The DNA fragment can then be blunt ligated into a suitable vector (e.g. pBluescript) linearized with *Sma*I. The chemically synthesized primers, which form the 5′ end of each strand

of the DNA do .not bear 5′-terminal phosphates. Therefore, if the linearized vector has been treated with phosphatase to prevent simple reclosure, it is essential that 5′ phosphate groups be added to the PCR product by using a polynucleotide kinase. Blunt-ligated recombinants can be identified by loss of β-galactosidase activity using a blue-white screening protocol.

Genetic engineering by PCR

In many cases. the use of genetic regions (regulatory and coding DNAs) requires nucleotide modifications. DNA modification techniques are numerous and include the use of linkers, DNA-modifying enzymes to change restriction enzyme sites (Klenow, T4 polymerase, etc.) and single-stranded site-specific modification methods. All these methods are useful, but they have limitations as well as require a considerable amount of time and effort to locate modified product. The use of PCR amplifications to modify a DNA fragment was first discussed by Mullis *et al.* (1986) and first demonstrated by Scharf *et al.* (1986) for the addition of a restriction enzyme site synthesized into the 5′-ends of PCR oligomer primers. Because this technique requires the use of custom-synthesized oligomer primers for each DNA modification, it is referred as custom-PCR (cPCR) engineering (Slightom, 1991). The major strength of cPCR engineering is that most of the complex DNA modifications can be incorporated into the sequence of the synthesized oligomer primers, allowing the engineering process to proceed as a series of straightforward ligation events.

Applications

The usefulness of PCR was quickly recognized by scientists in many different disciplines and it, either directly or after some modifications, has been applied to resolve many problems that were unapproachable by other techniques.

1. It has extensive applications in the diagnosis of genetic disorders such as phenylketonuria, hemophilia, sickle cell anemia, thalassemia,

etc. and the detection of nucleic acid sequences of pathogenic organisms in clinical samples. Sometimes, the RFLP pattern of PCR products in healthy and defective foeti can be studied or the PCR product may also be sequenced to reveal the differences. For detection of pathogens, immunochemistry and histocytochemistry have long been used by pathologists to analyze cell phenotypes in histological specimens. However, a major limitation of these procedures is the inability to detect infectious agents such as viral pathogens and other microorganisms that persist at a very low level in infected cells. When it became clear that PCR could offer increased specificity and sensitivity, researchers began to develop PCR strategies to detect as few as one viral genome in tissue specimens. Two basic PCR approaches have been followed for pathogen detection. The first approach is competitive quantitative PCR, with an added emphasis on low-level detection and strain identification. The second approach is to develop *in situ* PCR (ISPCR) procedures that allow the same level of sensitivity and specificity as solution PCR assays, with the advantage of cell-specific resolution. Similar to solution PCR, the ISPCR amplification parameters must be optimized with regard to temperature, time, and reaction constituents. ISPCR products are detected by using labeled dNTPs (radioactive or biotinylated) in the reaction or by subsequently processing the ISPCR specimen using standard *in situ* hybridization conditions to detect the amplified hybridization target.

2. The genetic identification of forensic samples, even DNA extracted from a hair or a single sperm are sufficient for PCR amplification. Some well-characterized sequences of microsatellites are being used for designing primers, so that DNA finger printing may be achieved through PCR.

3. For the analysis of homologous genes in evolutionary biology.

4. PCR amplification has several plant molecular biology related uses. These include

i. Generation of specific sequences of cloned double-stranded DNA for use as probes.

ii. Generation of probes for genes or cDNAs that have not yet been cloned.

iii. Generation of libraries of cDNA from small amounts of mRNA. The main advantage of this method is that it allows large cDNA libraries to be established from as few as one or two cells.

iv. Generation of large amounts of template DNA for sequencing to be carried out either by the Maxam-Gilbert chemical degradation method or Sanger dideoxy-mediated chain termination method.

v. Identification of genetic mutations: Deletions and insertions at defined loci can be detected by a change in the size of the amplified product. Alternatively, deletions can be recognized by their failure to amplify when one of the priming oligonucleotides maps within the deleted sequence. It is also useful for the identification of point mutations. The PCR technique of *single-strand conformational polymorphism (SSCP)* is the widely used method for detecting single base pair changes in genomic DNA. It is based on the principle that two single-strand DNA molecules of identical length but unlike sequences migrate differentially in a nondenaturing acrylamide gel. This is due to sequence-dependent intramolecular folding reactions that contribute to the overall structure of the molecule. Single-strand conformational polymorphism is done by amplifying a specific DNA segment using a radiolabeled (or fluorescently labeled) primer, or by including trace amounts of ^{32}P-a-dNTP in the PCR reaction. The PCR products are then heat denatured and cooled on ice to promote rapid formation of intrastrand duplexes. Figure 18.9 shows how SSCP can be used to identify mutant alleles in known target DNA sequences. Two other PCR-based approaches for detecting single nucleotide differences are *base* *excision sequence scanning (BESS)*, which uses an enzymatic sequencing method to detect thymidine residues in PCR products, and *artificial mismatch* *hybridization (AMH)*, a method that exploits

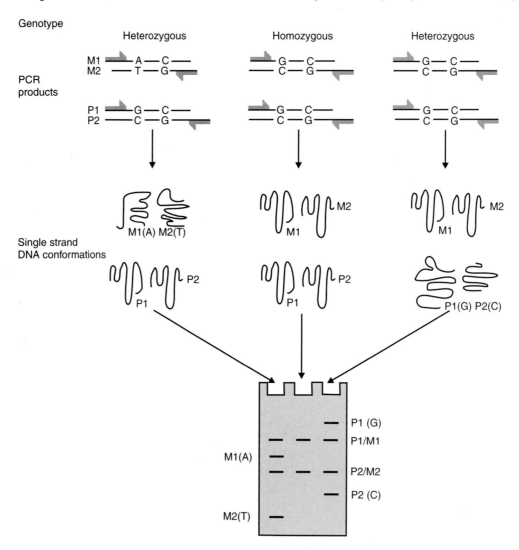

Fig. 18.9 Point mutation in genomic DNA can be rapidly amplified using the PCR technique of a single strand conformational polymorphism (SSCP). In this example, AMG primer pairs were used to detect nucleotide alterations in a specific region of the AMG gene known to cosegregate as a dominant mutation. With DNA from three individuals (two disease heterozygotes and a homozygote normal), PCR amplification was performed with AMG specific primers that flank a region where multiple dominant alleles have previously been identified. Because the primary nucleotide sequences of each of the DNA strands in a heterozygote individual are different, four separate bands are resolved in the corresponding lanes. Maternal (M1, M2) and paternal (P1, P2) strands are shown. Concept of the figure taken from Miesfeld: *Applied Molecular Genetics.* Copyright © 1999 John Wiley and Sons. This material is used by permission of John Wiley and Sons, Inc.

heteroduplex stability assays to detect single base pair mismatches.

vi. PCR detection of a gene transferred into a plant genome: The ability to transfer genes into a plant genome is well established. The determination of those plant tissues that are transformed from those that are not can be accomplished using selectable marker or reporter genes in most cases. However, there are some situations in which the use of selectable or reporter gene is not effective or where a reporter gene is not informative, either because of its lack of expression or due to its incomplete transfer. In the PCR method, genomic DNA is isolated and the PCR target gene; that is, the gene that was used for transformation experiments is selected for amplification. With designing of appropriate primers the gene is amplified and seen on a gel.

vii. PCR technique is employed in various molecular markers like RAPD; microsatellites or simple sequence repeats, amplified fragment length polymorphism (AFLP), etc. Variations observed in the amplification products are used as markers (examples and details have been dealt in Chapter 22).

viii. Identification of unknown plant DNA regions that flank an *Agrobacterium* T-DNA boundary. Although PCR is frequently used to amplify DNA segments lying between the two priming oligonucleotides, it can also be used to amplify unknown DNA sequences that lie outside the boundary of known sequences. This is referred to as inverse PCR (iPCR). Inverse PCR is also useful if the border sequences of a DNA segment are not known and those of a vector are known, then the sequence to be amplified may be cloned in the vector and border sequences of a vector may be used as primers in such a way that the polymerization proceeds in inverse direction, that is, away from the vector sequence flanked by the primers and towards the DNA sequence of inserted segment. Similarly, if the gene sequence is known, one can use its border sequences as primers, for an iPCR to amplify the sequences flanking this gene.

Advantages

1. It can produce large amounts of identical DNA molecules from a minute amount of starting material. In fact, the DNA isolated from a single cell (even plant protoplast) is sufficient for gene detection.
2. The technique is quick and simple.
3. The technique is extremely sensitive.
4. The DNA need not be pure for amplification provided the sample does not contain contaminants that inhibit *Taq* polymerase.

Problems

1. Nucleotide sequence of at least the boundary regions must be known so that oligonucleotide primers can bind and synthesize the DNA. This limits PCR to the study of genes that have already been characterized in part by cloning methods.
2. PCR is a very sensitive technique and is prone to generating false signals, which in many cases are the result of contaminants from previous PCR amplifications being present in the amplification reaction components. False positives can also result from minute contamination by any plasmid or lambda clone that may contain the target gene sequence.

PROTOCOL

16.1. PCR for detection of transgenes

The protocol given below can be taken as a standard "template" but depending on the length (in base pairs) and percentage of G+C, etc. of the primers used to detect the particular transgene, it may be necessary to change the amplifying temperature.

1. Isolate the genomic DNA
2. Add the reagents in 0.2 ml microcentrifuge tube placed on ice

Stock	Final	25 µl
Taq DNA polymerase buffer (10×)	1×	2.5 µl
dNTPs (10 mM each)	0.2 mM each	0.5 µl each
Primer 1 (10 µM)	0.2 µM	0.5 µl
Primer 2 (10 µM)	0.2 µM	0.5 µl
DNA (100 ng/µl)	100 ng	1.0 µl
Taq DNA polymerase (5 units/µl)	1U	0.2 µl

   Sterile water to make up the final volume to 25 µl.
3. Prepare cocktail of components. Always keep one tube with only cocktail and water (without DNA) as control.
4. Mix well and spin briefly.
5. Overlay the reaction with a 25 µl of mineral oil (in those PCR models where hot lid is not provided) to prevent evaporation of reagents at high temperatures.
6. Place the tubes in thermal cycler block and set the machine for following cycling conditions.

Step 1	94°C 2 min
Step 2	(25–30 cycles)
	94°C 1 min
	55°C 1 min
	72°C 1 min
Step 3	72°C 5 min
Step 4	4°C end

After amplification, add 2.5 µl of tracking dye in the sample and store at 4°C.

7. Prepare agarose gel of 1.2%.
8. Load the samples carefully in the wells.
9. Add suitable molecular weight (1 kb ladder) in one of the lanes.
10. Carry out electrophoresis, stain the gel with ethidium bromide and photograph it.

Further reading

Maniatis, T., Fritsch, E.F. and Sambrock, J. 1982. *Molecular Cloning: A Laboratory Manual.* Cold Spring Harbor Laboratory Press, USA.

Miesfeld, R.L. 1999. *Applied Molecular Genetics.* Wiley-Liss, New York.

Old, R.W. and Primrose, S.B. 1995. *Principles of Gene Manipulation.* Blackwell Science, UK.

Saiki, R.K., Gyllenstein, U.B. and Erlich, H.A. 1995. The polymerase chain reaction. In: *Genome Analysis–A Practical Approach* (ed. Davis, K.E.), IRL Press, Oxford.

Slightom, J.L., Drong, R.F. and Chee, P.P. 1995. Polymerase chain reaction: gene detection, inverse PCR, and genetic engineering. *Plant Mol. Bio. Manual.* II Kluwer Academic Publishers, Dordrecht.

19

In vitro Mutagenesis

Present-day knowledge of biological macromolecules originates from the study of spontaneous or induced mutations. Mutants are essential prerequisites for any genetic study also. In the past, isolation and characterization of DNA, RNA, or protein variants had to await the identification of phenotypically deviant host organisms. Because of this and moreover because most mutations are phenotypically silent, there was only a very limited subset of genes or gene products amenable for the study of structure–function relationships. Classically, mutants are generated by treating the test organism with chemical or physical agents that modify DNA. This method though extremely successful suffers from a number of disadvantages. (1) Any gene in the organism can be mutated, and the frequency with which mutants occur in the gene of interest can be very low; (2) Mutants with the desired phenotype when isolated do not guarantee that the mutation has occurred in the gene of interest; (3) Prior to the development of gene cloning and sequencing techniques, there was no way of knowing where in the gene the mutation had occurred and whether it arose by a single base change, an insertion of DNA, or a deletion.

The advent of recombinant DNA methodology offered a change from a passive to a more active mode of study. The basis was laid by the discovery of restriction enzymes, ligases, polymerases, and the enzymes which modify DNA molecules (nucleases, methylases), and by the development of techniques of gene cloning. This enabled one to isolate any DNA in pure form and in large quantities, introduce changes in the DNA, selecting them in the absence of phenotypic expression and studying the effects of these deliberate alterations in the DNA either *in vitro* or after reintroduction *in vivo*.

Sometimes the objective of isolating gene sequences from DNA libraries is simply to obtain inserts that can be analyzed by DNA sequencing, for example, when performing comparative studies of the same gene between different species. However, in many cases, the isolation of novel gene sequences requires the use of functional assays to characterize the cloned DNA. In conventional genetics, a biological comparison between wild type and mutant phenotypes often reveals the normal gene function. This same approach can be used in molecular biology, except that the researcher is able to choose precisely what type of mutations will likely to be most informative.

Site-directed mutagenesis

With the development of techniques in molecular biology, isolation and study of a single gene is now possible and mutagenesis has been refined. Instead of mutagenizing many cells of an organism and then analyzing many thousands of offsprings to isolate a desired mutant, it is now possible to specifically change any given base in

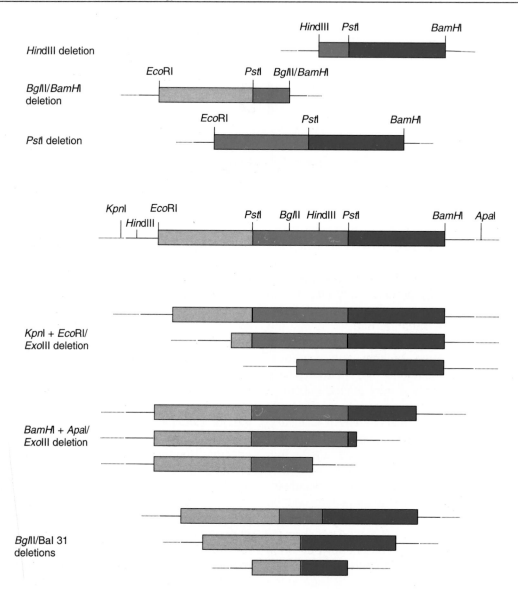

Fig. 19.1 Deletion mutagenesis using restriction enzymes and exonucleases. Concept of the figure taken from Miesfeld: *Applied Molecular Genetics.* Copyright © 1999 John Wiley and Sons. This material is used by permission of John Wiley and Sons, Inc.

a cloned DNA sequence. This technique is known as site-directed mutagenesis. The strategies recently developed for *in vitro* site-directed mutagenesis allow one to create virtually any change at will and this – in principle – enables us to correlate any piece of gene sequence information with gene regulation and product formation. This methodology invariably yields gene mutations, but it is only applicable to fully characterized cloned DNA. The importance of site-directed mutagenesis goes beyond gene structure–function relationships because the technique enables mutant proteins to be generated with very specific changes in particular amino acids. Targeted mutagenesis is therefore virtually impossible without prior sequence determination of the DNA.

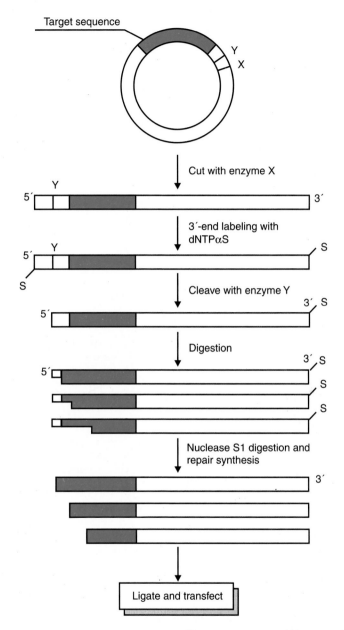

Fig. 19.2 A procedure for making unidirectional deletions in a DNA molecule. X and Y represent recognition sequences for two different restriction endonucleases. Reproduced from *Principles of Gene Manipulation,* R.W. Old and S.B. Primrose, by kind permission of Blackwell Science, Osney Mead, Oxford, U.K.

The type of mutagenesis protocol chosen largely depends on the desired alterations in target DNA. It may involve the creation of single base transitions and/or transversions, deletions, and insertions or precise replacement of whole sequence blocks of variable size. This chapter describes some basics *in vitro* mutagenesis strategies, each of which is used for creating specific types of mutations in cloned DNA sequences.

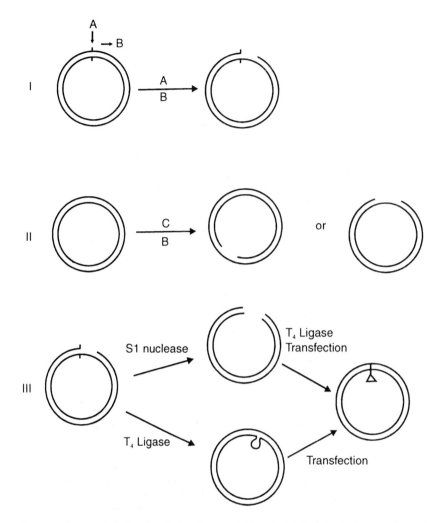

Fig. 19.3 Construction of gapped single-stranded regions and internal deletions in duplex circular DNA. A – Restriction endonuclease in presence of ethidium bromide; B – *Micrococcus luteus* DNA pol I or exonuclease III; C – DNase I in presence of ethidium bromide. Scheme I – Use of restriction enzyme in presence of ethidium bromide to produce a nick at a defined position. Scheme II – Use of DNase I and ethidium bromide to create nick. In both the schemes I and II, nicks are converted to limited gaps by exonucleolytic action of *Micrococcus luteus* DNA pol I or extended gaps by exonuclease III. Scheme III – Two pathways for construction of internal deletions either by sequential action of S1 nuclease and T_4 ligase or by direct gap misligation and *in vivo* repair.

Deletion mutagenesis

In deletion mutagenesis, nucleotides are removed from the termini or internally. Information from restriction maps can be used to design specific deletion that allows convenient recloning of the modified DNA segment. Alternatively, exonucleases can be used to degrade DNA nonspecifically from the termini of a linearized plasmid. The extent of exonuclease treatment, usually as a factor of reaction time, determines the amount of deleted DNA. Fig. 19.1 illustrates how a plasmid insert can be mutated by restriction enzyme mediated deletion mutagenesis using compatible restriction enzyme mediated termini, or by digesting linearized plasmid DNA with exonucleases followed by treatment with Klenow DNA polymerase. Based on the restriction map of a plasmid DNA insert *BgI*II and *Bam*HI digestion results in compatible termini (GATC). Exonuclease III (*Exo*III) degrades one strand of double-stranded DNA from the 3′ end, but only if the DNA termini contain either a 5′ overhang or a blunt end, but not an extended single strand 3′ overhang. This property can be exploited to generate unidirectional deletions by digesting target DNA with two restriction enzymes, one that leaves a 5′ overhang (*Eco*RI or *Bam*HI) and the other a 3′ overhang (*Kpn*I or *Apa*I). *Exo*III deleted DNA must be treated with the single strand nuclease S1 to remove the undigested 5′ strand. *Bal*31 exonuclease degrades double-stranded DNA to yield bidirectional deletions that can be blunt ligated without prior treatment with S1 nuclease.

Unidirectional deletions: Yanisch-Perron *et al.* (1985) and Barcak and Wolf (1986) described a method for making a series of deletions, as shown in Fig. 19.2. This method is based on the principle that α-thiophosphate-containing phosphodiester bonds are resistant to hydrolysis by the 3′→5′ exonucleolytic activity of T4 phage DNA polymerase. Linear duplex DNA molecules blocked at one 3′ terminus with a thiophosphate are then degraded from the other end with the exonuclease. Digestion for different lengths of time followed by treatment with nuclease S1 and ligation allows the preparation and recovery of a set of deletion mutants.

Gap-sealing mutagenesis: This method involves the use of pancreatic deoxyribonuclease or restriction endonucleases in the presence of ethidium bromide to introduce one single-strand scission per closed circular molecule. The nick is then extended into a small gap by the exonucleolytic activity of *Micrococcus luteus* DNA polymerase I or a more extended gap in the presence of exonuclease III. Gapped DNA can be linearized by nuclease S1, and the

shortened molecules can be cyclized by T4 ligase directly or after addition of a synthetic linker molecule (Fig. 19.3).

Linker scanning mutagenesis: McKnight and Kingsbury (1982) reported an elegant method combining deletion and substitution mutagenesis in their study of promoter elements of the thymidine kinase gene of HSV. Two opposing sets of 5′ and 3′ deletion mutants were obtained by sequential exonuclease III and nuclease S1 action and were flanked by synthetic restriction enzyme *Bam*HI-linker sequences. Opposing 5′ and 3′ deletion mutants were sorted out and recombined via the synthetic restriction site. This was done in such a way that the residues in the linker exactly replaced a stretch of nucleotides in the wild-type sequences (Fig. 19.4). For the thymidine kinase gene, an almost ideal set of mutants was obtained, which exhibited no net increase or decrease in length, but now the linker scanned across nearly every individual residue within the region −120 to +20 nucleotides of the transcriptional start. A similar series of linker scanning mutants of the 5′ flankage region of mouse β-globin gene was reported by Charnay *et al.* (1985). The success of this method goes with the initial choice for matching partner molecules from the opposing libraries of unidirectional mutants (5′ and 3′). If the choice is limited, one gets constructs that either have additional residues or small deletions flanking the linker sequence.

Deletion mutagenesis is sometimes used to create novel proteins lacking one or more functional domains, for example, deleting the hormone-binding domain of a steroid receptor to produce a constitutively active transcription factor. Internal deletions of protein coding sequences are more complicated because two-thirds of all deletions result in frame-shift mutations.

Cassette mutagenesis

In cassette mutagenesis, a synthetic DNA fragment containing the desired mutant sequence is used to replace the corresponding sequence in the wild type gene. Wells *et al.* (1985) originally

5′ deletion series ⟶   ⟵ 3′ deletion series

WT sequence

—— Gene sequences

▧ Synthetic linker residues

Fig. 19.4 A typical build up of a series of linker-scanning mutants. The opposing sets of 5′ and 3′ deletion mutants are combined via a linker sequence containing a restriction enzyme recognition site to give DNA segments of lengths similar to WT DNA. WT – Wild type.

used this method to generate improved variants of the enzyme subtilisin. It is a simple method, with an efficiency of almost 100%. But there is an absolute requirement for unique restriction sites flanking the region of interest and the limitation on the number of different oligonucleotide replacements that can be synthesized.

Oligonucleotide-directed mutagenesis

This method is based on the use of a relatively short oligonucleotide (7–20 nucleotides long) containing a base mismatch for defined mutation. The short oligonucleotide is used as a primer, which is strand specific to synthesize one complete mutant DNA strand in an *in vitro* DNA synthesis reaction.

To use this strategy, the nucleotide sequence of the region to be mutated must be known to permit the synthesis of a complementary mutagenic oligonucleotide. An oligonucleotide is designed such that it is complementary to a single-stranded template DNA and cloning the gene in an M13-based vector. However, DNA cloned in a plasmid and obtained in duplex form can also be converted to a partially single-stranded molecule that is suitable for mutagenesis.

The method involves full or partial enzymatic extension of the primer. The basic principle of the method is shown in Fig. 19.5. The reaction proper is started by hybridization of the primer to template DNA followed by enzymatic extension in the presence of all four nucleotides and Klenow DNA polymerase I. This enzyme lacks 5′→3′ exonuclease activity; correction of the mismatch between primer and template, however, may occur due to 3′→5′ exonuclease proofreading activity before the start of the actual priming reaction. This is followed by transfection of the extended duplex into competent *E. coli*. To protect the 5′ end of the primer from being exposed to 5′→3′ exonuclease editing activity *in vivo*, methods are developed to create fully closed circular duplexes in which all nicks are closed by ligation prior to transfection. This can be achieved by extending the primer/template combination to an all the way round circular duplex molecule by using a double primer/template combination to facilitate the full length extension reaction or by cutting and ligation of partially extended regions into a closed duplex.

After transformation of the host *E. coli*, the heteroduplex gives rise to homoduplexes whose sequences are either that of original wild-type

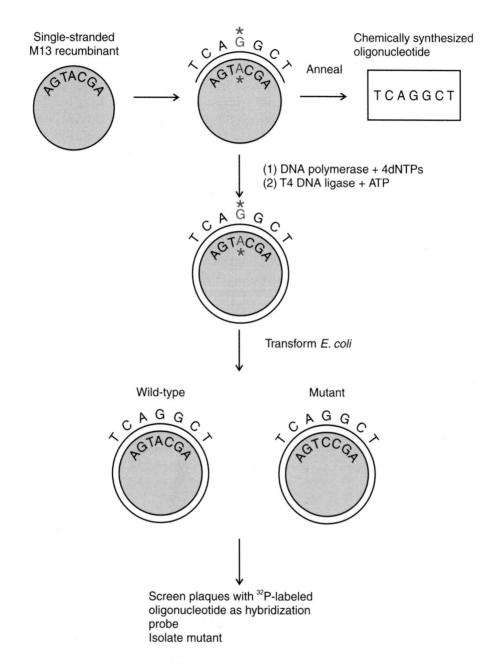

Fig. 19.5 Oligonucleotide-directed mutagenesis. Asterisks indicate mismatched bases. Reproduced from *Principles of Gene Manipulation,* R.W. Old and S.B. Primrose, by kind permission of Blackwell Science, Osney Mead, Oxford, U.K.

DNA or that containing the mutated base. Generally the frequency of mutated clones is rather low. To pick out mutants, the clones can be screened by nucleic acid hybridization with a ^{32}P labeled oligonucleotide as a probe. The mutant can then be sequenced to check that the procedure has not introduced any adventitious changes. This was necessary with early versions of the technique where *E. coli* DNA polymerase was used. The more recent use of high-fidelity DNA polymerases from T4 and T7 phages has minimized the problem of extraneous mutations as well as shortened the time for copying the second strand. Also, these polymerases do not strand displace the oligomer, a process which would eliminate the original mutant oligonucleotide.

The efficiency with which the single primer method yields mutants depends on many factors. The double-stranded heteroduplex molecules that are generated will be contaminated both by any single-stranded nonmutant template DNA that has remained uncopied and by partially double-stranded molecules. The presence of these species considerably reduces the proportion of mutant progeny. They can be removed by sucrose gradient centrifugation or by agarose gel electrophoresis, but this is time consuming and inconvenient.

Following transformation and *in vivo* DNA synthesis, segregation of the two strands of the heteroduplex molecule can occur, yielding a mixed population of mutant and nonmutant progeny. Mutant progeny have to be purified away from parental molecules, and this process is complicated by the cell's mismatch repair system. The major reason for the low yield of mutant progeny is that the methyl directed mismatch repair system of *E. coli* favors the repair of nonmethylated DNA. Thus, it is important during the bacterial *in vivo* replication phase that the directed base pair mismatch not be repaired. In the cells newly synthesized DNA strands that have not been methylated are preferentially repaired at the position of mismatch, thus preventing a mutation. To avoid this, *E. coli* strains carrying the mutations *mut*L, *mut*S, or *mut*H that have specific defects in DNA repair pathways are used. Following transformation, single colonies are isolated and the

Table 19.1 Strategies used to increase the efficiency of oligonucleotide-directed *in vitro* mutagenesis

Descriptive name	Method to remove template strand	Advantages	Disadvantages
dUTP incorporation	M13 template grown in a DNA repair mutant (*dut*-, *ung*-) contains dUTP in place of dATP	No special selections schemes or restriction enzymes required	Requires the use of a M13 phagemid and appropriate *E. coli* strains
Phosphorothioate incorporation	*In vitro* DNA synthesis incorporates phosphorothioates into mutant strand which protects it from enzyme cleavage with *Nci*I	Critical reactions are done *in vitro* and does not require a special *E. coli* strain	Depends on efficient phosphorothioate incorporation and exonuclease III digestion
Gapped duplex DNA	Mutant strand is marked with both gain of chloramphenicol resistance and loss of *amp*^{R} phenotypes	Reversal of two antibiotic resistant phenotypes allows for calculation of mutation frequency	Requires a special vector and depends on efficient *in vitro* heteroduplex formation
Site elimination	Mutant strand is marked with a second oligo-directed mutation in a nonessential enzyme site	Nonessential enzyme site can be anywhere on vector	Appropriate enzyme site is required and differences in oligo-annealing temperature may affect efficiency
amp^s site reversion	Mutant strand is marked with a second oligo-directed reversion of an *amp*^{s} to *amp*^{R} phenotype	Both the *amp*^{R} and *tet*^{R} genes can be used sequentially to obtain mutations	Requires a special *amp*^{R} cloning vector that can be used for reversion
PCR amplification	Methylated template strand is eliminated after PCR using the restriction enzyme *Dpn*I	Requires only one transformation and can be done quickly	PCR reaction can introduce mutations at non-target sites

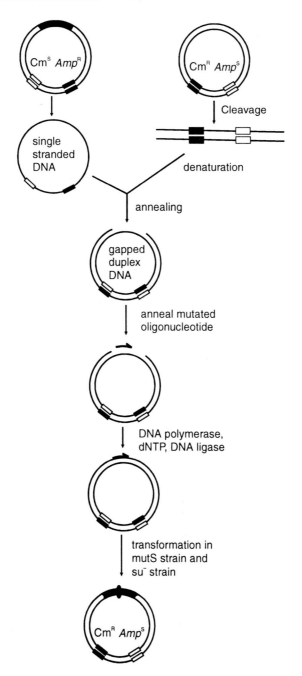

Fig. 19.6 Gapped duplex method of *in vitro* site-specific mutagenesis; Cm–Chloramphenicol; Amp–Ampivillin

recovered plasmid DNA is sequenced to determine if it contains the desired mutation. The efficiency of site-directed mutagenesis could be as high as 80%, depending on the method used to enrich for mutated plasmids (Table 19.1).

A key strategy in the oligonucleotide-directed

Fig. 19.7 dUTP incorporation in to M13 DNA using a phagemid vector to produce single-stranded DNA in an *E. coli* strain defective in the two enzymes uracil-*N*-glycosylase (*ung*) and dUTPase (*dut*). The mutagenic oligo is annealed to uracil containing template DNA (U), and *in vitro* DNA synthesis is performed to generate double-stranded DNA. Following transformation of *ung*⁺, *dut*⁺ *E. coli*, the template strand is degraded and the surviving double stranded phagemid plasmid is isolated and sequenced. In this example, the mutant oligo was used to engineer a novel *Eco*RI site into the DNA insert. Concept of the figure taken from Miesfeld: *Applied Molecular Genetics*. Copyright © 1999 John Wiley and Sons. This material is used by permission of John Wiley and Sons, Inc.

mutagenesis method is to increase the relative number of mutated plasmids in the pooled population. The following strategies are used to preferentially retain the mutated strand.

Gapped duplex method: This is one of the finest methods for the direct selection of the mutant strain. In this strategy, two mutants of a vector each carry an amber mutation in one of the two antibiotic resistance markers (Stanssens *et al.*, 1989). The target fragment to be mutagenized is cloned into one of these mutant vectors (1), which contains the mutation in one resistant marker, e.g. chloramphenicol (*Cm*). Single-stranded DNA is prepared from this construct. The DNA of the second vector (2) containing the mutation in the other marker ampicillin (*Amp*) is cut at the position where the target fragment was cloned into vector 1, denatured, and annealed to the single-stranded DNA of the vector insert hybrid. This results in a duplex molecule that contains a region of single-stranded DNA corresponding to the cloned target fragment, also called a gapped-duplex. The synthetic oligonucleotide containing the desired base changes is annealed to the single-stranded region. Thereafter, the remaining single-stranded regions are filled in and the DNA is transformed into hosts that are deficient in mismatch repair, e.g. a *mut*S strain unable to suppress amber mutations. Selection for clones expressing the resistant marker ensures that only molecules containing the mutated strands are present in the obtained clones (Fig. 19.6).

The dUTP incorporation strategy: This strategy (shown in Fig. 19.7) relies on the degradation and repair of template DNA that has been marked with deoxyuridine prior to the *in vitro* DNA synthesis step. Thomas Kunkel developed the dUTP incorporation strategy by taking advantage of an *E. coli* strain that contains defects in genes responsible for preventing dUTP incorporation into DNA (Kunkel *et al.*, 1987). In this method, M13 phages are grown on a specialized host before mutagenesis, which carries two mutations. One of the mutations is in the *ung* gene, which encodes uracil-N-

glycosylase, an enzyme that normally removes uracil bases from DNA that result from cytosine deamination. The second defect is in the *dut* gene (dUTPase), which prevents the buildup of dUTP pools in the cell. Growing single-stranded M13 phage in an *ung⁻/dut⁻ E. coli* strain results in the phages containing 20–30 residues of uracil in the template DNA. These phages are unable to grow in an *ung⁺* host. The mutagenic oligo is then annealed to uracil containing template DNA (U) and *in vitro* DNA synthesis is performed to generate double-stranded DNA. Following primer annealing and *in vitro* DNA synthesis, a hetero-duplex composed of a uracil-containing strand and a mutant strand is synthesized *in vitro* in the presence of dTTP. This DNA is transformed into a *ung⁺ dut⁺ E. coli* strain. Because the template DNA contains a large number of dU residues in place of dA, it is degraded by the activity of uracil-N-glycosylase and results only in the mutant progeny.

Phosphorothioate method: This strategy is based on the use of nucleotide thioanalog phosphorothioate (e.g. dCTPαS) in the *in vivo* reaction (Taylor and Eckstein, 1985). Since some endonucleases (e.g. *Ava*I, *Ava*II, *Ban*II, *Hin*dII, *Nci*I, *Pst*I and *Pvu*I) cannot cut phosphorothioate containing DNA strand, only the opposite strand is cut. The mutant oligonucleotide is annealed to the single-stranded DNA template and *in vivo* DNA synthesis takes place in the presence of a thionucleotide by DNA polymerase (Fig. 19.8). A heteroduplex is generated in which the mutant strand is phosphorothioated. After sealing the gap in the mutant strand with DNA ligase, the heteroduplex is treated with endonuclease (*Nci*I), which cleaves only the parental strand. Exonuclease III is used to partially degrade the template strand starting from the generated nicks. This strand is subsequently resynthesized with DNA polymerase I, resulting in a homoduplex molecule containing only the mutant sequences on both strands. After transformation, a high yield of mutant clones is obtained. Originally nitrocellulose filters were used with phosphorothioate method to remove single-stranded template

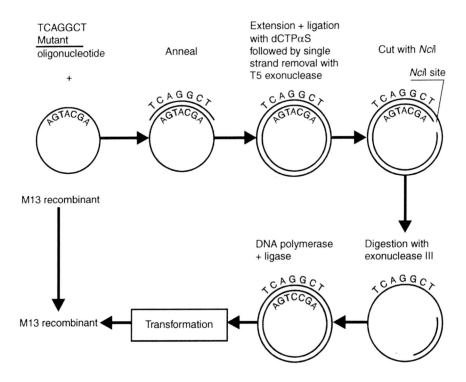

Fig. 19.8 a. Structure of a thionucleotide (dcTP×S). **b.** Phosphorothioate method for *in vitro* strand selection of mutants. Reproduced from *Principles of Gene Manipulation,* R.W. Old and S.B. Primrose, by kind permission of Blackwell Science, Osney Mead, Oxford, U.K.

DNA, but these have been replaced with phage T5 exonuclease, which has single- and double-stranded exonuclease activities with a copurifying single-stranded endonuclease activity. The advantages of this method are as follows:

1. This method permits strand selection *in vitro* and gives very high efficiencies of point mutations, deletions, and insertions.
2. It does not require specialized hosts or phage vectors.

A disadvantage of all primer extension methods is that they require a single stranded-template DNA. The phosphorothioate method has been used with a double-stranded template, but the procedure is very laborious.

Chemical mutagenesis

Chemical modification of DNA *in vitro* is a technique of increasing importance for the simple generation of nucleotide substitutions. Intrinsically mutagenic attack on DNA by chemical treatment is a random process and may therefore at first sight appear nonapplicable. Yet the present state of the art is such that essentially any mutation can be generated by a combination of optimized procedures for mutagenic treatment and highly effective screening. For chemical mutagenesis a variety of chemical reagents have been advocated, including nitrous acid or hydroxylamine (convert C and A to T or G respectively), formic acid (protonates purine ring nitrogen of A and G, leading to depurination and ultimately to transversion and/or transition mutations), hydrazine (splits thymine and cytosine rings, causing all possible base changes), dimethylsulphate at neutral or acid pH (methylates G or A respectively and creates all possible base changes), sodium bisulphite (deaminates C residues, leading to the formation of U residues which after DNA replication causes a C to T change), and methoxylamine (targets C, giving N^4-methoxycytosine and causes change to T). Unlike most chemicals the last two reagents are strictly specific for single-stranded DNA.

Sodium bisulphite in particular has found wide use in protocols developed a few years ago, where duplexes of DNA are formed that leave a short but defined single-stranded region for interaction with the mutagen (Fig. 19.9). In the strategy the region of interest is cloned into a vector designed for generating single-stranded DNA. Such target DNAs can be formed from two nonequal complementary strands by hybridization yielding circular heteroduplexes with a gapped single-stranded region. Other groups have employed molecules that have a defined nick extended to a small gap by 5'→3' exonucleolytic activity of *Micrococcus luteus* DNA polymerase or 3'→5' exonuclease III activity. The circular duplex DNA containing the defined single-stranded region is dissolved in sodium acetate or citrate buffer and a high concentration (0.5–4 M) of reagent is added. After

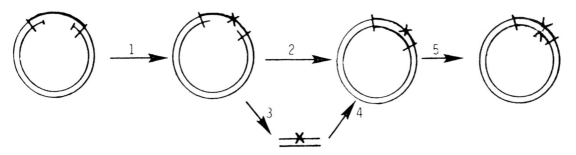

Fig. 19.9 A commonly used strategy for chemical mutagenesis using single-stranded specific reagents. The different steps are: 1 – Treatment with single-stranded specific mutagen; 2 – Reverse transcriptase + dNTPs + T$_4$ ligase; 3 – Reverse transcriptase + dNTPs followed by fragment isolation; 4 – Recloning in nonmutagenized acceptor plasmid; 5 – Selection for mutants.

a few hours of incubation at slightly elevated temperature (25–50°C) the bisulphite is removed by exhaustive dialysis. Bisulphite treated single-stranded DNA is then reconstituted *in vitro* with double-stranded DNA by using a DNA polymerase. For *in vitro* synthesis, DNA is incubated in the presence of all four deoxynucleotide triphosphates to repair the gapped region by DNA polymerase I or Klenow DNA polymerase action, prior to ligation by T4 DNA ligase (some prefer the use of reverse transcriptase instead of DNA polymerase I for gap repair). A fragment comprising the region of interest is then recloned into the original construct, replacing the nonmutagenized homologue. Mutant progeny plasmids or mutant RF or recombinant phage DNA is finally obtained on transfection of the DNA into bacterial host cells.

Meyers *et al.* (1985) developed a protocol that utilizes formic acid or hydrazine, which cause damage to specific bases. The target DNA of rather short length is cloned into a single-stranded vector M13 adjacent to a so-called G–C clamp. This clamp is a short stretch of DNA rich in GC residues introduced in the mutagenesis vector by molecular cloning. It functions to keep the strands of the target DNA segment from complete denaturation during certain steps in the procedure. Single-stranded DNA is generated and treated with chemical mutagens. The mutagenic reagents are removed by ethanol precipitation, and subsequently an oligonucleotide primer is annealed to the insert DNA in the CG clamp region. Next, strand extension with reverse transcriptase in the presence of dNTPs ensures that the region of insert DNA, including the GC clamp of the vector, is double-stranded. This enzyme incorporates random nucleotides at sites of damage with a high frequency. The region of interest is then excised by restriction endonucleases and is recloned, replacing respective wild-type sequences.

Polyacrylamide gel with a linear gradient of denaturant concentration (formamide and urea) is used which depress the melting temperature. First, the insert DNAs plus adherent GC clamps are excised from the pool of DNA obtained from

bacterial transformants run on a preparative gel to separate mutant DNAs from the parental (nonmutant) wild type. The gradient gel causes DNAs that contain nucleotide substitutions to run faster or slower than originally, due to an altered melting behavior. Partially denatured DNAs are eluted from the acrylamide matrix after visualization by ethidium bromide fluorescence and religated again in the vector above containing the M13 origin now to yield only mutant genotypes. The chimeric DNAs are transfected into *E. coli* and individual colonies are prescreened again by restriction analysis of DNA minipreps and analytical gradient gel electrophoresis. Genuine mutants having altered mobility from wild-type DNA are identified and single-stranded DNA from the minipreps of mutant candidates are prepared for dideoxy sequencing to characterize the nucleotide changes.

This approach is attractive for two reasons. First, although it requires a fair number of steps, it yields no nonmutagenized background colonies and distinct genotypes can already be identified prior to characterization by sequencing. Second, depending on the chemical mutagen and the regime of treatment applied, a given DNA segment can really be saturated with transversions and/or transitions.

PCR-mediated *in vitro* mutagenesis

The development of optimal reaction conditions for long-distance PCR made it possible to design site-specific mutagenesis strategies that exploit unique primer designs to direct the amplification of entire plasmids. One example of this is to use inverse PCR to amplify a plasmid with a pair of tail-to-tail oligonucleotides, one of which contains a mismatched nucleotide. Following amplification, the mutated linear product is purified using agarose gel electrophoresis, recircularized, and used to transform *E. coli*. Higuchi *et al.* (1988) have described a method which enables a mutation in a PCR-produced DNA fragment to be introduced anywhere along its length. This method requires four primers and three PCRs. Two primary PCR reactions produce two overlapping DNA fragments, both bearing the same mutation in the overlap region (Fig. 19.10). One PCR uses primer A and primer B whereas the second PCR utilizes primer A′ and primer C. Primers A and A′ are complementary. These two primary reactions generate two overlapping DNA fragments bearing the mutation. The overlap in the sequence allows the fragments to hybridize. These are mixed, denatured, and annealed. One of the two possible hybrids is extended by DNA polymerase to produce a duplex fragment. The other hybrid has recessed 5′ ends and as it is not a substrate for polymerase is lost from the reaction mixture. The third PCR reaction is conducted with primers B and C.

Strategies have also been developed that

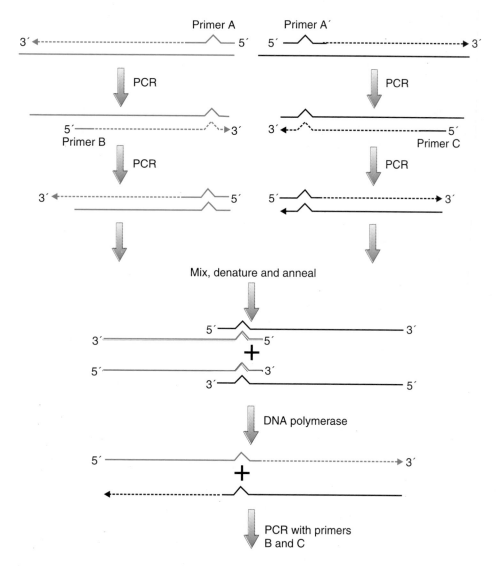

Fig. 19.10 PCR method of site-directed mutagenesis. Primers A and A′ are complementary. Reproduced from *Principles of Gene Manipulation,* R.W. Old and S.B. Primrose, by kind permission of Blackwell Science, Osney Mead, Oxford, U.K.

permit the construction of unique gene fusions using consecutive PCR reactions that use the PCR products of one reaction to prime DNA synthesis during the second amplification (Fig. 19.11). This technique is sometimes called PCR-mediated gene SOEing (splicing by overlap extension). In addition to creating gene fusions, this method also provides a PCR-based approach for inserting nucleotide mutations into the middle of a gene coding sequence. Sarkar and Sommer (1990) described a method that utilizes three oligonucleotide primers to perform two round of PCR. The product of the first PCR is used as a megaprimer for the second PCR.

The advantage of the PCR method of mutagenesis is that it is 100% efficient in generating desired mutations.

The disadvantages of the PCR method are as follows:

1. The PCR product usually needs to be ligated into a vector.
2. *Taq* polymerase copies DNA with low fidelity. Thus the sequence of the entire amplified segment generated by PCR mutagenesis must be determined to ensure that mutation has taken place.

Advantages of site directed mutagenesis

1. It serves as a reliable tool for gene manipulations and has simplified earlier techniques, which were extremely laborious, tedious, and time consuming.
2. It helps to study gene structure and function relationship as it enables mutant proteins to be generated with specific changes in particular amino acids. Such mutations facilitate the study of mechanism catalysis, substrate specificity, stability, etc.
3. The ability to manipulate enzymes has resulted in dramatic advances in understanding biological catalysis and has led to the novel field of protein engineering.
4. Through *in vitro* mutagenesis techniques it is now possible to assess in detail the identity and function of an amino acid by replacing

it with an alternative residue and determining the effect this has on the enzymatic activity.

Random mutagenesis

Random mutagenesis of a gene or gene segment is best carried out using doped oligonucleotides. These are synthesized using oligonucleotide precursors that are contaminated by a small amount of other three nucleotides. Reidhaar-Olson and Sauer (1988) have used this method to probe functionally and structurally important amino acids in a bacteriophage. The principle of this method is that the first strand of the oligonucleotide cassette is synthesized with equal mixtures of all four bases in the first two codon positions (indicated by the letter N) and an equal mixture of G and C in the third position (Fig. 19.12). The resulting population of the base combinations will include codons for each of the 20 naturally occurring amino acids at each of the mutagenized residue positions. When the complementary strand is synthesized, inosine is inserted at each randomized position because it is able to pair with each of the four conventional bases. The two strands are annealed and the mutagenic cassette ligated into the gene of interest.

The advantage of this approach is that it reduces the degeneracy. The limitations of this approach are:

1. It is limited by the length of oligonucleotide that can be synthesized.
2. In a random mutagenesis for even a short oligonucleotide sequence, the resultant mutant library becomes too large to be screened. For example, mutagenesis of a 9 amino acid peptide using NNN triplets results in a DNA complexity of 4^{27} and a protein complexity of 20^9. The use of NN (G, C) as described above, represents a slightly improved dope. It should be remembered that since different codons may translate to the same amino acid, each protein sequence does not appear equiprobably in the mutant library. Arkin and Youvan (1992) described a set of mixtures e.g. NTG/C encodes the hydrophobic amino

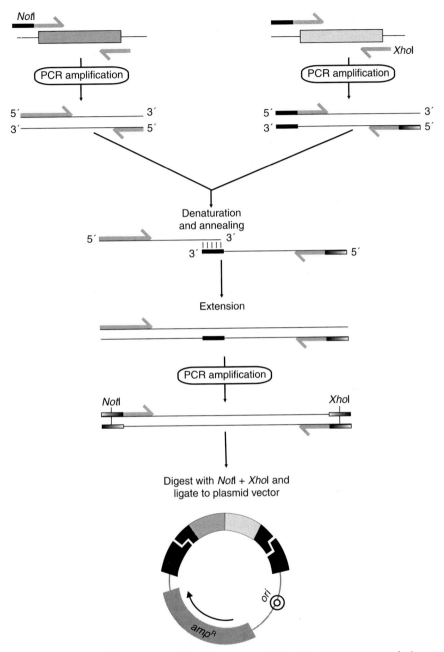

Fig. 19.11 PCR-mediated site-specific mutagenesis provides a convenient method to create gene fusions and nucleotide substitutions rapidly. Novel gene fusions can be created by performing an initial pair of PCR reactions that introduce complementary DNA sequences at the ends of two sequences that will be spliced together. The overlapping segment functions as a bridge to generate a new template for a second amplification reaction that uses the two most outside primers from the initial reactions. The net result is a spliced PCR product that can be directly inserted into a cloning vector through the addition of restriction enzyme recognition sequences to the primers. Concept of the figure taken from Miesfeld: *Applied Molecular Genetics*. Copyright © 1999 John Wiley and Sons. This material is used by permission of John Wiley and Sons, Inc.

Fig. 19.12 Mutagenesis by means of doped oligonucleotides. During synthesis of the upper strand of oligonucleotide, a mixture of all four nucleotides is used at the position indicated by the letter N. When the lower strand is synthesized inosine (I) is inserted at the position shown in figure. The double-stranded oligonucleotide is inserted into the relevant position of the vector. Reproduced from *Principles of Gene Manipulation*, R.W. Old and S.B. Primrose, by kind permission of Blackwell Science, Osney Mead, Oxford, U.K.

acids phenylalanine, isoleucine, leucine, methionine and valine. Other mixtures can be designed that encode primarily hydrophilic or aromatic amino acids.

Another method of generating mutants is by PCR. Mutagenic PCR is a random mutagenesis technique that exploits the elevated error rate of *Taq* polymerase in the presence of $MnCl_2$ and high $MgCl_2$. *Taq* DNA polymerase lacks a $3' \rightarrow 5'$ exonuclease activity and is therefore unable to proof-read copied sequences. This result in a high error frequency in PCR-amplified DNA, the average frequency being one mistake per 10,000 nucleotides per cycle. Mutagenic PCR is similar to random chemical mutagenesis with the added advantage that the chosen pair provides a convenient way to target the random base pair mutation to a defined segment of DNA.

Insertional mutagenesis

A highly efficient procedure for obtaining mutants in genes identified in sequencing programs takes advantage of the availability of large collections of plants mutagenized by an insertion element. In the plant system, the insertional element is either a T-DNA or a transposon. Insertional mutagenesis is known to occur naturally in many species through the excision and reintegration of endogenous transposable elements, which insert into a gene. The disrupted gene is not functionally redundant; that is, if other genes are unable to complement partially or completely the function of that gene, the progeny homozygous for the mutation exhibits a detectable altered phenotype. *In vitro* insertional mutagenesis involves using known DNA sequences as the insertion elements. In addition to providing a powerful approach to gene identification through their loss of or change in function, it has the added advantage of facilitating gene cloning since the insertion element acts as a tag of the mutant gene (Walbot, 1992). Random or directed insertion of DNA sequences into a gene mutates and tags the gene. Both T-DNA and transposon-mediated gene tagging have been used successfully in the identification and isolation of several plant genes.

The use of insertional mutagenesis, in principle, provides a more rapid way to clone a mutated gene. Because the sequence of inserted element is known, the gene in which it is inserted can be easily recovered using various cloning or PCR-based strategies.

Transposon-mediated insertional mutagenesis

All higher organisms possess multiple copies of several kinds of transposable elements, the DNA sequences capable of moving from one site to another within a genome through a process of excision and reintegration. Transposons were first discovered and described by Barbara McClintock in the late 1940s and early 1950s. Majority of the information on these mobile elements has been obtained from studies on their behavior in maize and snapdragon. These mobile elements can be transposed from one site to another either within the same replicon or to another replicon present in the cell. When inserted at a given locus, transposons disrupt the function of the gene at the site of their integration and cause large-scale rearrangements of adjacent DNA sequences. If the site of transposon integration is in, or in the vicinity of, a particular gene, the expression of that gene can be altered in many ways depending on the proximity of the element to the structural or regulatory region of the gene. This lead to transposon-mediated insertion mutagenesis and is one of the best means to induce a high rate, frequency, and types of *in vitro* mutagenesis. Transposable elements cause mutations in two principal ways.

1. *Transposition in a coding region of DNA*: In a natural state, transposable elements are present in the noncoding regions of the genome and are therefore silent and harmless. But they can insert into a coding region of a gene and disrupt its function. Because most transposable elements contain their coding regions, their transcription interferes with the transcription of the gene and gets terminated within the transposable element. Even if the transcription proceeds through the element, the phenotype is altered and a mutant results. Its coding region when compared to the nonmutated organism contains altered and incorrect sequences.

2. *Recombination:* Transposable elements induce *in vitro* mutagenesis by introducing *in vitro* recombinations. Transposable elements are present in multiple copies, often with two or more copies in the same chromosome. Whenever the two copies are either together or near to each other, they pair during synapsis, undergo crossing over, resulting in the deletion of the DNA region between these two elements. Alternatively, the downstream element can pair with the upstream element in the homologous chromosome, producing a chromatid that lacks the element in the region between the elements and one chromatid in which this region is duplicated.

Details on transposon structure and function and transposon as a gene tag are discussed in Chapter 20.

Advantages of transposon-based approaches are as follows:
1. They generate large populations of insertions.
2. They have the ability to use the propensity of many transposable elements to transpose to linked sites, which makes it possible to remobilize the element for insertion in the vicinity of the starting insertion site.

Limitations of transposon-based approaches are as follows:
1. Not many isolated and fully characterized transposons are available except those from maize and snapdragon. This limits the approach of transposon tagging to these plants. To overcome this restriction, transposon elements in heterologous systems are being used.
2. Transposons characterized so far all have a tendency to tag a gene located at some distance from the transposon-associated region in the parent. To overcome such a problem, a heterologous system is used. For this, nonautonomous elements such as *Ds* or *dSpm* are introduced at various locations of the plant genome. When this plant is crossed with a plant containing an active element such as *Ac* or *Spm*, this element activates the defective elements and induces new mutations. The selfed progeny yields new transposon-tagged mutations. Once the mutation of interest is identified, the autonomous element is outcrossed, thereby stabilizing the mutation. Through backcrossing, the defective element cosegregating with the phenotype can be identified and the tagged gene isolated. Another advantage of using heterologous

systems is that the species receiving a transposable element does not contain any element of the same family in its genome. This makes identification of a transposable element cosegregating with the mutant phenotype easy.

T-DNA mediated insertional mutagenensis

The soil-borne plant pathogen *Agrobacterium tumefaciens* is a natural genetic engineer of plants. Under the inductive effects of phenolics released by wounded dicotyledonous plant tissues to which the bacteria attach, the virulence region of the Ti plasmid is activated and endonucleases encoded by the *vir*D operon cleave the borders of the T-DNA region. The procedure of T-DNA transfer has been explained in Chapter 21. Transgene integration in plants is by illegitimate recombination, which occurs at a low frequency. T-DNA is transferred as a single-stranded molecule associated with the VirD2 and VirE proteins comprising the so-called T-complex. Breaks in chromosomal DNA are required for insertion of the T-DNA strand, and analyses of insertion sites indicate sequence similarities of only a few nucleotides between T-DNA and plant DNA. There are some rearrangements or deletions induced in the host DNA during this process. In some cases, the T-DNA molecules may insert in tandem arrays or at a single or multiple loci as inverted or direct repeats. The characteristic of *Agrobacterium*-mediated plant transformation is that the T-DNA can be used as an insertional mutagen and by extension, a gene tag. A large number of mutants have been obtained by T-DNA insertion. Versatility of this experimental system can be further exploited to tag DNA sequence sthat control gene expression. A number of plant species have been used to identify and clone of genes by T-DNA insertional mutagenesis, the principal requirements being ease of transformation and a diploid chromosome complement. But T-DNA tagging is rather inefficient in most plant species because of the high percentage of noncoding and highly repetitive DNA. For dicots, the main tool is T-DNA

insertion mutants with the most effort in *Arabidopsis*. This weedy crucifer offers several advantages as its genome being too small facilitates the screening of genomic libraries with cloned T-DNA flanking probes. *Arabidopsis* has only 10–14% highly repetitive DNA; the detailed integrated genetic/RFLP linkage map is available and genetic transformation is routine in several laboratories.

T-DNA tags provide dominant markers for genetic mapping and for mutations. To identify a gene by T-DNA tagging, a large population of transgenic plants is generated that contains T-DNA harboring selectable and screenable reporter genes. This population is screened for a phenotype of choice and its stability over a number of generations is ascertained. The segregation of the mutant phenotype is due to a single mutation or the mutation is consequence of a T-DNA insertion event. If the phenotype and T-DNA cosegregate 100% in a sample of atleast 100–200 progeny following both selfing and backcross, cloning can be done using inverse PCR, plasmid rescue of the screening of genomic libraries made from individual transgenic lines of interest. The proof that the cloned gene is allelic to the mutant gene of interest can be obtained by genetic complementation in which the wild type is used to transform the mutant and the particular mutant is rescued. Introduction of the wild-type allele is done by *Agrobacterium*-mediated transformation of a line homozygous for the mutation; self-fertilization of the complemented line results in segregation of the mutant and wild type alleles in a Mendelian fashion.

Advantages of T-DNA mediated muta genesis:

1. Results in fewer (1–2) insertions per line.
2. Insertions are stable.
3. Easy to maintain and does not show strong insertional biases.

Limitations of T-DNA mediated muta genesis:

1. It is more difficult to achieve saturation with T-DNA insertional mutagenesis.

2. Most of the transformation systems developed for *Arabidopsis* involve tissue culture, which often induces high frequencies of somaclonal variation. To overcome this problem, a nontissue culture transformation system based on infecting germinating seeds with *Agrobacterium* has been developed. In *Arabidopsis*, highly efficient plant transformation techniques have been instrumental in generating large populations of T-DNA insertion lines while minimizing the effect of somaclonal variation (Azpiroz-Leehan and Feldman, 1997).

The first published examples of targeted mutations in *Arabidopsis* show that careful analysis of mutants is often required to detect the deleterious effects of a given mutation (Gilliard *et al.*, 1998; Hirsch *et al.*, 1998).

A further extension of this strategy involves the use of an EST-like approach to systematically sequence the flanking regions of insertions. Using PCR-based techniques, it is possible to isolate and sequence a large number of insertion sites from pooled or individual lines. The insertion sites can be mapped by comparing the sequences with the genomic sequences. Databases of such flanking sequences will be established and can be searched for hits in any particular gene.

Further reading

Kaul, M.L.H. and Nirmala, C. 1999. Biotechnology: Miracle or mirage. *In vivo* and *in vitro* mutagenesis. In: *Breeding in Crop Plants: Mutations and in vitro Breeding*. Siddiqui, B.A. and Khan, S. (Eds.), Kalyani Publishers, India.

Miesfeld, R.L. 1999. *Applied Molecular Genetics*. Wiley-Liss, New York.

Walker, J.M. and Gaastra, W. (Eds.). 1987. *Techniques in Molecular Biology*. Croom Helm, Australia.

Old, R.W. and Primrose, S.B. 1995. *Principles of Gene Manipulation*. Blackwell Science, UK.

20

Transposon Genetic Elements and Gene Tagging

Transposons are mobile genetic elements. They are segments of DNA that are not autonomous and can insert at random into the genome, plasmids, or bacterial chromosomes independent of host cell recombination system. They have been described as introgenic parasites and have the ability to integrate at different sites within the genome and to move around the genome i.e. transpose. The process of integration and excision of transposable elements is called transposition. All transposable elements share two basic properties:

1. The ability to move from place to place in the genome–hence their designation as mobile DNAs or transposable elements.
2. The ability to amplify their copy number within the genome via this transposition, thereby providing a selectable function that can make them selfish or parasitic DNAs.

Barbara McClintock discovered transposons almost 50 years ago (McClintock, 1951). By insertion, transposons can abolish gene function, resulting in mutant phenotypes. Transposon-induced mutations are often unstable because on excision of the transposon, gene function is restored, and as a consequence, reversion of the phenotype can be recognized by variegated pattern. Most of the activities of a transposable element give rise to changes in gene and/or genome structure, often with accompanying alterations in gene activity. It was for this reason that McClintock originally named these entities

as "controlling elements."

Transposons contain two functional regions: (1) DNA repeat sequences at the termini of the transposon, which are required for genomic integration and excision and (2) an internal segment encoding a transposase enzyme, which is required for transposition.

Transposons have been identified in many organisms both prokaryotes and eukaryotes. Bacterial transposons were the first to be characterized at the molecular level. In the last decade, plant transposons were isolated and characterized. In maize, a variety of transposons have been identified, which are classified into families with several members. Transposons have also been identified in other plant species such as snapdragon, petunia, *Arabidopsis*, etc. In many plants with large and complex genomes, transposable elements make up over 50% of the nuclear DNA.

Transposable elements in bacteria

Genetic instabilities have been found in bacteria, and in many cases led to the identification of transposable elements. Bacterial transposons were the first to be studied at the molecular level, and have been categorized as follows.

IS elements

The simplest bacterial transposons are the insertion sequences or IS elements. Each type is given the prefix IS followed by a number that identifies the type reflecting the history of their isolation. A standard strain of *E. coli* may contain several (<10) copies of any one of the more common IS elements. The IS elements are autonomous units, each coding for the proteins needed to sponsor its own transposition. Each IS element is different in sequence, but there are some common features in organization. An IS element is compactly organized, less than 1500 nucleotide pairs long. Typically, it contains a single coding sequence with short,

identical or nearly identical sequences at both the ends. The gene codes for a protein called transposase, which is involved in promoting or regulating transposition. The terminal sequences at both the ends are always in inverted orientation with respect to each other, and therefore they are called terminal inverted repeats (TIRs), their length ranging from 9 to 40 nucleotide pairs (Fig. 20.1). The structure of a generic transposon before and after insertion at a target site is shown in Fig. 20. 2. When IS

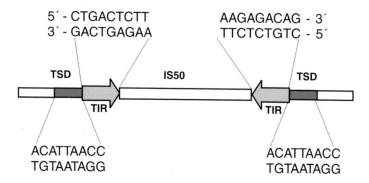

Fig. 20.1 Structure of an IS50 element. TIR – Terminal inverted repeat; TSD – Target site duplication. Terminal inverted repeats are imperfect because the fourth nucleotide pair from each end is different.

Fig. 20.2 Transposons with terminal inverted repeats generate direct repeats of flanking DNA at the target site. The ends of the transposon consist of inverted repeats of 9 base pairs (1 to 9).

elements insert into a chromosome or plasmid they create a duplication of the DNA sequence at the site of the insertion. One copy of the duplication is located at each side of the element. These short (3–12 bp) direct repeat sequences are therefore called target site duplications and are thought to arise from staggered breaks in double-stranded DNA (Table 20.1). Thus an IS element displays a characteristic structure wherein its ends are identified by inverted terminal repeats whereas the adjacent ends of flanking host DNA are identified by short direct repeats.

Composite transposons

Two homologous IS elements combine with other genes (generally those for drug resistance) to form a composite transposon, denoted by symbol *Tn*. Thus, a composite transposon has a central region carrying the drug marker(s) flanked on either side by "arms" that consist of the IS elements. The arms may be in either the same (direct repeat) or inverted orientation (Fig. 20.3).

Direct repeat

Inverted repeat

Fig. 20.3. Components of a composite transposon

The arms consists of IS modules, and each module has inverted repeats at the ends; thus, a composite transposon also ends in the same short inverted repeats.

Compound transposons are created when two IS elements insert near each other. The sequence between them can then be transposed by the joint action of the flanking elements. The gene present in the composite transposon is not necessary for transposition. Figure 20.4 shows three examples of composite transposons. In *Tn9*, the flanking IS elements are in direct orientation with respect to each other, whereas in *Tn5* and *Tn10*, the orientation is inverted. Each of these composite transposons carries a gene for antibiotic resistance – chloramphenicol in *Tn9*, kanamycin in *Tn5*, and tetracycline resistance in *Tn10* (Table 20.2). It has been observed that sometimes the flanking IS elements in a composite transposon are not quite identical. For example, in *Tn5*, the element on the left, IS50L, is incapable of stimulating transposition, but the element on the right, IS50R, is capable of causing transposition. This difference is due to a single nucleotide pair change that prevents IS50L from synthesizing a necessary transposition factor. However, IS50R synthesizes the factor called transposase by which the entire composite transposon can be mobilized. Detailed analysis has shown that IS50R produces another protein, a shortened version of the transposase, which inhibits transposition of an incoming transposon when the bacterial cell is already infected and carries

Table 20.1 Features of some IS elements

Name	Element size (bp)	TIR left/right (bp)	TSDs (bp)	Target site preference
IS1	768	20/23	9	AT rich terminal G or C
IS2	1327	32/41	5	Hot spots in P_2 genome
IS3	1258	29/40	3	?
IS4	1426	16/18	11–13	$AAAN_{20}TTT$
IS5	1195	16	4	Hot spots
IS10	1329	17/22	9	NGCTNAGCN
IS50R	1531	9	9	Hot spots
IS91	~1800	8/9	0	?

Note. TIRs – Inverted terminal repeats; TSD – Target site duplication ; bp – base pairs

Fig. 20.4 Genetic organization of composite transposons. *cam* – chloramphenicol; *kan* – kanamycin; *ble* – bleomycin; *str*–streptomycin; *tet* – tetracycline.

Table 20.2 Characteristic features of some composite transposons

Element	Length (bp)	Marker genes	Flanking IS elements	IS elements orientation
Tn903	3100	*kan*^{R}	IS903	Inverted
Tn5	5700	*kan*^{R}	IS50R	Inverted
			IS50L	
Tn9	2500	*cam*^{R}	IS1	Direct
Tn10	9300	*tet*^{R}	IS10R	Inverted
			IS10L	

a copy of *Tn5*. When the modules of a composite transposon are identical, either module can sponsor movement of the transposon (e.g. in *Tn9*). What is responsible for transposing a composite transposon instead of just the individual module (especially where both modules are functional)? A major force supporting the transposition of composite transposons is the selection for the marker(s) carried in the central region.

Complex transposons: the Tn3 family

Complex transposons do not contain IS elements. Elements in this group of transposons have inverted terminal repeats that are 38–40 nucleotide pairs long and produce target site duplications of 5 nucleotide pairs on insertion. They are larger than IS elements (typically 5000 nucleotide pairs long) and usually contain accessory genes as well as genes needed for transposition. The genetic organization of *Tn3*,

Fig. 20.5 Genetic organization of *Tn3* complex transposon.

the most thoroughly studied is shown in Fig. 20.5. Three genes, *tnp*A, *tnp*R and *bla*, encode for transposase, a resolvase/repressor, and beta lactamase, respectively. The beta lactamase confers resistance to ampicillin while the other two proteins play important role in transposition. The transposition stage involves the ends of the elements as in IS type elements.

Transposable elements in eukaryotes

Transposable elements are ubiquitous in the plant kingdom and share many common features, both structural and mechanistic, with mobile elements from other eukaryotes.

Classification

On the basis of structure. This includes the size and nature of the terminal repeats, which could be long or short and could be direct repeats (e.g. CCCAGAC and GGGTCTG) or inverted repeats (e.g. CCCAGAC and GTCTGGG). There are four types of transposons:
  i. With long terminal direct repeats (*Copia* in *Drosophila*, *Ty* in yeast, *IAP* in mice)
 ii. With long terminal inverted repeats (*FB, TE* in *Drososphila*)
iii. With short terminal inverted repeats (*P* and *I* in *Drosophila*, *Ac/Ds* in maize, *Tam*1 in snapdragon)
 iv. Without terminal repeats (*Alu* in mammals)
      On the basis of mechanism of transposition. According to this classification, transposable elements have been grouped into two major superfamilies.

  i. *Class I elements (RNA transposable elements):* These transpose by reverse transcription of an RNA intermediate (DNA–RNA–DNA). They are mobile because of their ability to make DNA copies of their RNA transcripts, and these DNA copies then integrate at new sites in the genome. Hence, these elements do not excise when they transpose; instead they make a copy that inserts elsewhere in the genome.
 ii. *Class II elements:* These transpose directly (DNA to DNA), existing as sequences of DNA coding for proteins that are able to manipulate DNA directly so as to propagate themselves within the genome.

Class I elements

Mobile elements that transpose via RNA intermediates through reverse transcription induce stable mutations. In most or all plant species, they comprise the greatest mass of transposable elements. Retroelements make up over 70% of the nuclear DNA in maize (SanMiguel and Bennetzen, 1998) and are equally or even more numerous in other plant species with large complex genomes. Three main transposable element families belonging to class I: (1) Retroviruses, (2) long terminal repeat (LTR) retrotransposons, and (3) non-LTR retrotransposons. Only the last two families have been found in plants.

Retroviruses: Retroviruses have genomes of single-stranded RNA that are replicated through a double-stranded DNA intermediate.

Retroposons/retrotransposons are genomic (duplex DNA) sequences that occasionally transpose within a genome but do not migrate between cells. Reverse transcription is the unifying mechanism for reproduction of retroviruses and perpetuation of retrotransposons. The cycle of each type of element is in principle similar, although retroviruses are usually regarded from the perspective of the free viral RNA form, wheras retroposons are regarded from the stance of the genomic (duplex DNA). Retroposons differ from retroviruses in not passing through an independent infectious form and resemble them in the mechanism used for transposition.

A typical retroviral sequence contains *gag*, *pol*, and *env* genes, which are translated into polyproteins, each of which is cleaved into smaller functional proteins (Fig. 20.6). The *gag* gene gives rise to the protein components of the nucleoprotein core of the virion and the *env* gene codes for components of the envelope of the particle, thus packing and generating virion. The *pol* gene codes for proteins concerned with nucleic acid synthesis and recombination. The *pol* gene codes for reverse transcriptase, which converts the genome into a complementary DNA strand. *Pol* gene also has a DNA polymerase activity, which enables it to synthesize a duplex

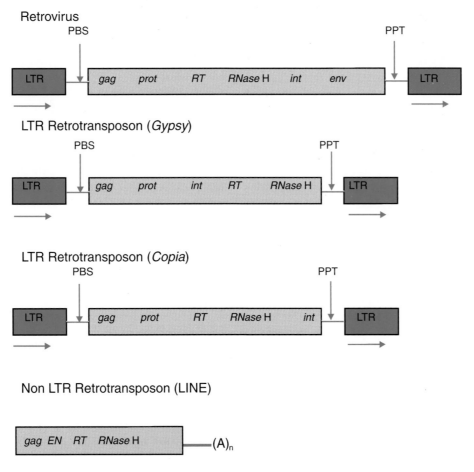

Fig. 20.6 Class I mobile elements. PBS is the primer binding site and PPT is polypurine tract, key sequence regions needed for replication/transposition of the LTR containing elements.

DNA from the single-stranded reverse transcript of the RNA. The enzyme has an RNase H activity, which can degrade the RNA part of the RNA–DNA hybrid. The DNA copy synthesized is longer than the RNA template; by switching template strands, reverse transcriptase (RT) copies the 3′ sequence of the RNA to the 5′ end of the DNA, and copies the 5′ sequence of the RNA to 3′ end of DNA. This generates the characteristic LTRs. Linear duplex DNA is inserted into the host genome by the enzyme integrase (int). Transcription of the integrated DNA from a promoter in the left LTR generates further copies of the RNA sequence. The integration event generates direct target repeats. An inserted provirus therefore has direct terminal repeats of the LTRs, flanked by short repeats of the target DNA. Mammalian and avian genomes have endogenous (inactive) proviruses with such structures.

LTR retrotransposons: The most numerous class of large retroelements in plants is the retrotransposons, which contain direct long terminal repeats (LTRs). These vary in size from several hundred bases to over 10 kb, with LTRs that are usually a few hundred bases to several thousand bases in length. LTR retrotransposons are found in all eukaryotes, either because they originated early in the eukaryotic lineage or because of their potential horizontal transfer. These resemble the integrated copies of retroviruses which are flanked by long terminal direct repeats and contain an internal domain encoding proteins analogous to the group specific antigen (*gag*) and the *pol* polyprotein (Fig. 20.6). LTRs carry the regulatory sequences required for transcription, which is the first step in the transposition. The *pol* polyprotein has conserved domains characteristic of reverse transcriptase, integrase, and endonuclease (rt-int-RNaseH). By sequence homology and the order of these domains, retrotransposons have been divided into two major subclasses, the *Ty 1-copia* group and the *gypsy* group (named after their first representatives observed in *Drosophila*). They differ in the position of integrase

within the encoded polyprotein. Both these groups are found in plants, but the *Ty 1-copia* group has been better characterized. Recent studies have shown that retrotransposons are a significant fraction of large plant genomes. Due to this character, most of the plant retrotransposons were identified as repetitive sequences or insertion elements in cloned DNA. Several *Ty 1-copia* group elements have been fully sequenced in the plant kingdom. Even algae contain LTR retrotransposons, indicating that such sequences are from a very ancient lineage.

LTR retrotransposons within individual plant species are always heterogeneous in sequence and are often highly repetitious, varying from about 20 copies per genome for the *Ta* elements of *Arabidopsis thaliana* to 13,000 for *Del 1* elements in *Lilium henryi*. The *Bis 1* family accounts for 5% of the wheat genome and one *Ty 1-copia* subgroup is present in more than one million copies in a bean species (*Vicia faba*). The total copy number of retrotransposons in the rice genome was estimated to be about 1000, and 32 families were isolated showing that retrotransposons are a major class of transposable elements in rice. Although these retrotransposons appear inactive during normal growth conditions, 5 out of 32 families were active under tissue culture conditions. Grandbastein *et al.* (1989) reported the isolation of the first tobacco (*N. tabacum*) transposable element *Tnt 1* which seems to be the most complete mobile retrotransposon characterized in higher plants. The retrotransposon have been isolated from different plant species and their characteristics are shown in Table 20.3.

Ty in yeast: The *Ty* elements comprise a family of dispersed repetitive DNA sequences that are found at different sites in different strains of yeast (*Saccharomyces cerevisiae*). *Ty* is an abbreviation for transposon yeast. The yeast carries about 35 copies of a *Ty* transposable element in its haploid genome. These transposons are about 5900 nucleotide pairs and are bounded at each end by a DNA segment called the δ sequence, which is approximately 340 base

Table 20.3 Characteristic features of some retrotransposons

Organism	Retrotransposon	Length of internal domain	LTR Left/Right (bp)	TSD (bp)
Yeast	*Ty 912*	5250 bp	334	5
Drosophila	*Copia*	4594 bp	276	5
Tobacco	*Tnt 1*	3984 bp	610	5
Arabidopsis	*Ta 1-3*	4190 bp	514	5
Rice	*Tos 3-1*	5200 bp	115	5
Barley	*Bare-1*	1255 bp	1829	4
Maize	*BS-1*	3203 bp	302	5
Wheat	*WIS 2-1A*	5114 bp	1755	5

Note. LTR – Long terminal repeat; TSD – Target site duplication; bp – base pair.

Fig. 20.7 Genetic organization of yeast *Ty* element.

pairs. Each δ sequence is oriented in the same direction, known as long terminal repeats or LTRs (Fig. 20.7). *Ty* elements are flanked by five base pair direct repeats created by the duplication of DNA at the site of the *Ty* insertion. These target site duplications do not have a standard sequence, but they tend to contain AT base pairs. This may indicate that *Ty* elements preferentially insert into AT rich regions of the genome. The genetic organization of the *Ty* elements resembles that of the eukaryotic retroviruses. These single-stranded RNA viruses synthesize DNA from their RNA after entering a cell. The DNA then inserts itself into a site in the genome, creating a target site duplication. The simplest proviruses possess three genes, *gag*, *pol* and *env*, which encode structural, catalytic, and membrane proteins, respectively. *Ty* elements have only two genes, *TyA* and *TyB*, which are analogous to the *gag* and *pol* genes of the retroviruses. Biochemical studies have shown that the product of these two genes can form virus-like particles inside yeast cells. However, it is not known whether these particles are genuinely infectious. But it has been shown that

transposition of *Ty* elements involves an RNA intermediate and hence *Ty* elements are called retrotransposons.

Non-LTR retrotransposons: These are also called nonviral retrotransposons. This class of transposons is identified by the absence of long terminal repeats, besides other features. They are targets for a transposition event by an enzyme system coded elsewhere. They do not code for proteins that have transposition functions. A significant part of the moderately repetitive DNA of genomes consists of retrotransposons. Two families account for most of this material: (1) long interspersed nuclear elements (LINEs) and (2) short interspersed nuclear elements (SINEs). LINEs are derived from transcripts of RNA polymerase II, whereas SINEs are derived from transcripts of RNA polymerase III.

Non-LTR transposons have been best characterized in mammals and *Drosophila*, but their list is growing steadily in plants. LINEs have the structure of an integrated DNA version of an mRNA (Fig. 20.6). A fully intact LINE encodes for gag proteins (involved in intracellular packaging of RNA transcript) and a polymerase (*pol*) function that includes the enzyme reverse transcriptase (RT). The *pol* functions have the ability to reverse-transcribe a LINE RNA into DNA, while an endonuclease (EN) also encoded by the element is probably associated with integration into the genome. LINE insertions are flanked by short direct duplications of target DNA, like those seen for all other mobile DNAs, usually created by the action of a transposase or integrase. LINEs terminate in an A-rich tract.

The mammalian genome contains 20–50,000 copies of LINEs, called L1. The typical member is ~6,500 bp long, terminating in an A-rich tract. The first plant non-LTR retrotransposons to be discovered were the *Cin4* elements in maize. *Cin4* elements range in size from 1 kb to 6.5 kb. The copy number of plant non-LTR retrotransposons can be very high: *Del2* in *Lilium speciosum* is present in 250,000 copies, occupying 4% of the genome. Members of this family have been discovered in tobacco also. Sequencing of LINEs suggests that there is homology between an open reading frame in L1 and reverse transcriptase. This suggests that L1 could have originated as a mobile gene coding for its own transposition. Since LINEs originate from RNA polymerase II transcripts, the genomic sequences are inactive because they lack the promoter that was upstream of the original start point for transcription. Because they possess the features of the mature transcript, they are called processed *pseudogenes.*

SINEs are highly abundant in animals but rare in most plant genomes. SINEs are usually only 100 bp to 300 bp in size and appear to be derived from reverse transcription of RNA polymerase III products. They encode no known peptides and must use *trans*-acting polymerase and integrase functions in order to transpose. SINEs are usually derivatives of tRNA or snRNA genes that have mutated to a structure that can be reverse transcribed and integrated. Because these RNA polymerase III transcribed genes carry a promoter specified within the RNA itself, a newly inserted element can usually be transcribed in any active part of the genome, thus creating a high potential for amplification. Both LINEs and SINEs lack introns because they are derived from mature mRNA, often have an integrated poly(A) tail at their 3′ end. These integrated RNA copies are seen as intronless pseudogenes.

The most prominent SINEs are the Alu family in humans and B1 family in mouse. In the human genome, a large part of the moderately repetitive DNA exists as sequences of ~300 bp interspersed with nonrepetitive DNA. The duplex DNA is cleaved by the restriction enzyme *Alu*I at a single site located 170 bp along the sequence. Cleaved sequences are members of a single family and are named as the Alu family after their origin of identification. There are ~300,000 members in the haploid genome.

Class II elements

These mobile elements transpose directly, i.e. DNA to DNA. These transposable elements are a mixed bunch. The three superstars of this class are the maize transposons *Ac-Ds*, *Spm/dspm* (*En/I*), and *Mu*. *Ac-Ds* elements were the first transposable elements of any organism to be studied systematically by Barbara McClintock (1951). Genetic studies on maize initiated in the 1940s identified changes in the genome during somatic cell division. These changes are brought about by controlling elements, recognized by their ability to move from one site to another. Originally identified by McClintock using purely genetic means, the controlling elements can now be recognized as transposable elements directly comparable to those in bacteria. The basic structure of the three major plant class II transposable element families is simple if one ignores the intron/exon structure (Fig. 20.8). All of them contain terminal inverted repeats and probably only one or two genes essential for transposition. A family of DNA transposable elements is characterized by their sharing the same terminal inverted repeat (TIR) sequences. Hence, all *Ac* and all *Ds* transposable elements have approximately the same 11-bp TIR, while

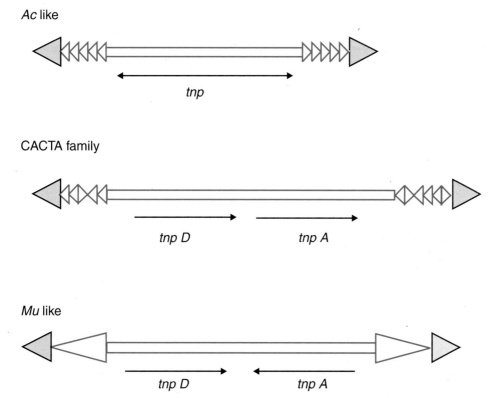

Fig. 20.8 Schematic representation of the structural features of class II plant transposons. Filled triangles are short terminal inverted repeats and unfilled triangles are short repeats, with the polarity shown by the direction of triangle. The exact number and polarity of the short repeats is constant for particular element types in the same family.

the 13-bp inverted repeat termini shared by all *Spm/dspm* elements are completely different from the *Ac/Ds* TIR. Within a family, one or more members will encode an enzyme called transposase, which recognizes the family's TIRs. All families come in two varieties, i.e. two-element system. The elements can be classified into two classes: (1) Autonomous elements, which have the ability to excise and transpose autonomously, rendering sites of their insertion mutable or *m* alleles. This element is analogous to bacterial transposons in its ability to mobilize (2) Nonautonomous elements, which are deleted defective elements that do not transpose autonomously but require helper functions from an autonomous element. These are stable and can be derived from autonomous element (often internal deletion derivatives) of the same family. They become unstable only when an autonomous member of the same family is present elsewhere in the genome. When complemented by an autonomous member of the same family, the nonautonomous element has the ability to transpose. Families of controlling elements are defined by the interactions between autonomous and nonautonomous elements. A family consists of a single type of autonomous element accompanied by many varieties of nonautonomous elements. A nonautonomous element is included in the family by its ability to be activated in *trans* by the autonomous elements. A nonautonomous element of one family cannot be cross-inactivated by the autonomous elements of any other family; therefore, the relationship between autonomous and nonautonomous elements of a family seems to be quite specific.

Ac-Ds elements: The activator (*Ac*) and dissociation (*Ds*) elements of *Zea mays* represent a relatively simple transposable element system (Fig. 20.9). The autonomous *Ac* element is 4565 bp long, which encodes a major transcript of 3.5 kb that contains a long GC-rich untranslated leader (652 nucleotides). Following splicing out of four introns from the primary transcript, a mature mRNA is obtained with 3312 nucleotides that is translated into a protein of 807 amino acids, and the protein is of 92 kDa. The *Ac* element is bounded by a direct repeat of 8 nucleotide pairs. This direct repeat is created when the element inserts into a site on a chromosome. Other repeat sequences are found within the element itself. The most conspicuous is 11-bp sequence at one end is repeated at the other end in the opposite orientation. These terminal inverted repeats (TIRs) are thought to play an important role in transposition. The product of *Ac*, termed as transposase, is sufficient for transposition of *Ac* in a growing number of transgenic heterologous plant species. The *cis* determinants for transposition of *Ac* are represented by about 200 bp at each end. Mutational alterations within these regions reduce or abolish excision of *Ac* elements. DNA-binding tests showed that the *Ac* transposase binds to the hexamer motif AAACGG that occurs in clusters within these regions (Kunze and Starlinger, 1989). The other *cis* determinant that is clearly defined and absolutely required for excision is 11-bp terminal inverted repeats, which seem to be recognized by *Ac* transposase. The asymmetry of the two *Ac* ends is important with respect to excision. *Ac* elements with identical ends are defective in excision.

Defective element of the *Ac* family, termed dissociation (*Ds*), form a rather heterogeneous group. Common to all *Ds* elements is the 11-bp TIR. Nonautosomal elements are derived from autonomous elements by deletions that inactivate the *trans*-acting transposase, but leave the sites intact (termini) on which the transposase acts. Because of deletions, *Ds* elements cannot produce transposase. However, transposase produced by *Ac* element is diffusable and *trans*-acting and causes transposition of *Ds* elements.

Two features of maize have helped to follow transposition events: (1) Controlling elements

Fig. 20.9 Structural organization of the members of *Ac/Ds* family of transposable elements in maize.

often insert near genes that have visible but nonlethal effects on the phenotype; and (2) maize displays clonal development and due to this occurrence and timing of a transposition event can be visualized by the phenotype. Mitotic descendents of a given cell remain in the same location and give rise to a sector of tissue with changed phenotype, which is called variegation.

Ac seems to be the prototype of a family of elements whose members are widely distributed in plants (Gierl *et al.*, 1992). Characteristics of the *Ac* transposable element family are shown in Table 20.4.

***En/Spm* transposable element family:** Another maize transposon family discovered by McClintock is the suppressor mutator (*Spm*) and the enhancer (*En*) family. In this family the autonomous element is called the *Spm* and the nonautonomous elements are called *dSpm* (d stands for deleted or defective) while *En* is autonomous and inhibitor (*I*) is nonautonomous element. The *Spm* and *En* autonomous elements are virtually identical. The 8287-bp long autonomous enhancer (*En*) or suppressor mutator element (*Spm*) encodes at least two functional products derived by alternative splicing from a precursor transcript. *En/Spm* has 13-bp terminal inverted repeats. These TIRs are *cis*-acting sequences that are absolutely required for excision. When they insert into a chromosome, they create a nucleotide pair target site duplication. The *dSpm* elements are smaller than the *Spm* elements because part of their DNA sequence has been deleted. The deleted *dSpm* elements are unable to stimulate their own

transposition. Two functional products – TNPA, a 67 kDa protein, and TNPD, a 131 kDa protein – are encoded by the two genes. TNPA is about 100 times more abundant than TNPD, but both proteins are required for transposition of *En/Spm*. Similar to *Ac* transposon, the asymmetry at the two ends is also required for *En/Spm*. A possible autoregulatory role for TNPA has been suggested, because the innermost TNPA binding motif at the 5' end overlaps with the "TATA" box of the single *En/Spm* promoter (Gierl *et al.*, 1988). It has been speculated from a comparison with *En/Spm* related elements from other plant species that the TNPD interacts with 13-bp TIRs and may accomplish endonucleolytic cleavage at the elements ends. *En/Spm* belongs to the so-called CACTA family of elements. These elements produce a 3-bp target site duplication on insertion and share nearly identical 13-bp TIRs (Table 20.5).

Unclassified elements: In addition to the elements described previously, a number of less well-characterized mobile elements are known. Small transposable elements in plants, the miniature inverted repeat transposable elements (MITEs), have a structure indicating that they are likely to be DNA transposable elements. The MITEs are unusual in having no identified autonomous elements. The existence of undiscovered autonomous elements that encode MITE-specific transposases is likely, although it is also possible that these tiny elements utilize a *trans*-acting transposition that is itself not encoded on a mobile DNA. Such an activity might be specified by a standard host gene involved in some other cellular process (e.g. DNA replica-

Table 20.4 Characteristics of *Ac* transposable element family

Element	TIR	TSD	Size	Species
Ac	CAGGGATGAAAT	8	4565 bp	*Zea mays*
Bg	CAGGG	8	4869 bp	*Zea mays*
Tam 3	TAAAGATGTGAA	8	3629 bp	*Antirrhimum majus*
Tpc 1	TAGGGTGTAAA	8	927 bp	*Pisum crispum*
Tps-r	TAGGGGTGGCAA	8	8000 bp	*Pisum sativum*
Tst 1	CAGGGGCGTAT	8	736 bp	*Solanum tuberosum*

Note. TIR – Terminal inverted repeat; TSD – Target site duplication; bp – base pair.

Table 20.5 Characteristics of *En/Spm* or CACTA transposable element family

Elements	TIR	TSD (bp)	Size	Species
En/Spm	CACTACAAGAAAA	3	8.287 kb	*Zea mays*
MB 11	CACTACCGGAATT	3	9 kb	*Zea mays*
Tam 1	CACTACAACAAAA	3	15.164 kb	*Zea mays*
Tam 2	CACTACAACAAAA	3	5.187 kb	*Zea mays*
Tam 4	CACTACAAAAAAA	3	4.329 kb	*Zea mays*
Tam 7	CACTACAACAAAA	3	7 kb	*Zea mays*
Tam 8	CACTACAACAAAA	3	3 kb	*Zea mays*
Tam 9	CACTACAACAAAA	3	5.5 kb	*Zea mays*
Tgm	CACTATAAGAAAA	3	1.6–1.2 kb	*Glycine max*
Pis 1	CACTACGCCAAAA	3	2.5 kb	*Pisum sativum*

Note. TIR – Terminal inverted repeat; TSD – Target site duplication; kb - kilo base

tion, recombination or repair), although there are no obvious candidates at this time. For example, Tourist–Stowaway family elements are found in high copy number in maize. These Tourist–Stowaway sequences are present in a wide variety of flowering plants. These elements are small (about 100–350 bp) and heterogeneous in sequence. The structures of these elements are similar to several different types of transposable elements, but identical to none.

Transposon Gene Tagging

Isolation of plant genes by conventional techniques requires knowledge of a gene product. In routine procedures for gene cloning, genomic or expression libraries are screened with probes made of mRNA or protein respectively. Also, based on the cDNA or protein sequence, oligonucleotide primers can be designed to apply a PCR approach. However, for genes with unknown products, alternative strategies that have a genetic basis are required. Recently, gene isolation techniques have been successfully applied, starting with the genomic map position of the gene. These map-based cloning approaches, however, do not involve mutant phenotypes that may hamper the identification of genes. To obtain mutants that allow simultaneous isolation of the corresponding genes, strategies have been developed using transposons. By insertional mutagenesis, mutant phenotypes are obtained and the genes responsible for this are molecularly tagged and hence can be cloned.

It has been demonstrated that *Ac* and *Ds*, like other transposons, can be used to isolate plant genes. The procedure of gene tagging by transposons is known as transposon tagging. This has been shown to be generally applicable and is one of the most useful methods to identify and isolate specific gene loci of an organism. The advantage of transposon cloning over most other gene isolation methods is that except for a mutant phenotype, no other information on the gene product or its function is required. In principle, any gene can be mutated by a transposon and if its inactivation can be recognized on the basis of a phenotype, it can be cloned using the transposon as a probe. Only in plant species with a well-defined endogenous transposon system, such as maize and snapdragon, was transposon tagging previously possible; in other plant species lacking characterized transposons, transposon tagging could not be applied. However, lately it has been demonstrated that the *Ac* element from maize can carry out transposition successfully in tobacco. Since then, *Ac* and *Ds* have been introduced in other heterologous hosts. There are two major features to the method of transposon tagging:

1. Integration of a transposon into a genetic locus often leads to a loss of function whereas excision from that locus may result in partial or full restoration of its original tasks. 2. Excision and integration of a transposon result in a change in the physical size of the donor site as well as

that of the new insertion site. As a consequence, the result of transposition becomes detectable as restriction fragment length polymorphisms (RFLPs) when the transposon is used as a molecular probe.

Isolation of genes through transposon tagging involves insertion of a transposon within a gene (Fig. 20.10). This can induce a mutation that can be detected phenotypically. A DNA fragment containing the transposon and gene sequences can then be isolated using the transposon sequences as a probe. Then sequences flanking the transposon can be used to screen a library made from nonmutant plants for isolation of an intact gene. The major steps are

1. isolation of the transposable element,
2. gene tagging,
3. characterization of the transposon-induced mutation,
4. isolation of the mutated gene by using

transposon as a probe, and
5. isolation of the intact gene by using sequences flanking the transposon as a probe.

Isolation of the transposable element

Many transposons are isolated by trapping them in genes that have been previously isolated (rev. Balcells et al., 1991). For example, the waxy (*Wx*) locus encodes a glycosyl transferase, which is required for amylose production in the kernel. Kernels that harbor insertions of the maize transposon activator (*Ac*) at the *Wx* locus were identified because they contained sectors of tissue with and without amylose. The first step in obtaining *Ac* from this material was the isolation of a cDNA clone for the *Wx* gene made from mRNA expressed in nonmutant plants. Once the nonmutant copy of the gene was cloned, it was straightforward to use this as a hybridization probe to isolate the mutant copy, and therefore the *Ac* element (Fedoroff *et al.*,

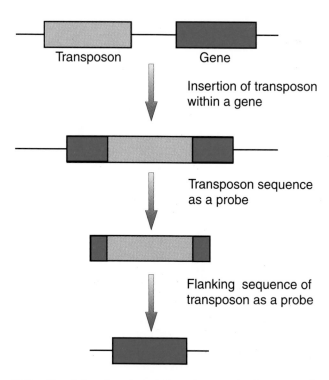

Fig. 20.10 The major steps in isolation of a gene by transposon tagging.

1983). Other transposable elements were isolated in maize by similar methods. Once the transposable elements were isolated, they could be used to clone genes previously tagged by insertion of one of these elements.

Gene tagging

The insertion of a transposon causes genetic instability, which has to be identified in a genetic screening procedure. Subsequently, the gene can be isolated by cloning the DNA sequences flanking the transposable insertion. This technique, called transposon tagging, has been widely used to isolate genes. The phenomenon of genetic instability caused by insertion of transposons is illustrated by an example of coloration of flowers (Fig. 20.11). As shown in the figure, the inserted transposon inactivates the color gene, and the cells are not able to synthesize the pigment during petal development (shown as white

hexagonal cells). Due to any excision events in mitotically active cells, the cell autonomous function of the color gene may be restored (shown as gray hexagonal cells). Clonally derived daughter cells of such somatic reversions form pigmented sectors, thus displaying a variegated color system of the petals (left flower). Excisions occurring at an early stage result in large sectors, whereas later-stage excisions result in the development of small sectors of color in the petal. When excisions occur in cells or their ancestors that contribute to the germ cells (germinal excision), selfing may lead to the appearance of wild-type colored plants in the progeny. In fact, an excision event early in the development of a plant may result in a fully revertant and wild-type colored flower (right flower). Assuming that no further transposition occurs, RFLP analysis using the colored gene as a probe will reveal a 1:1 ratio between the wild

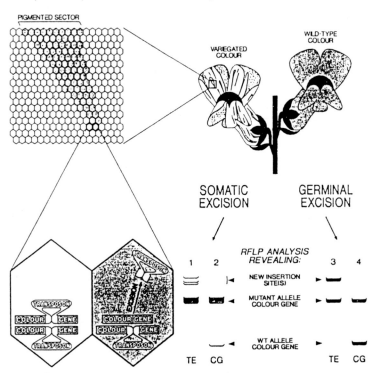

Fig. 20.11 The phenomenon of genetic instability when an active transposon resides at a locus involved in coloration of flowers. Reproduced from Lonning, W.E. and Huijser, P., *Gene tagging by endogenous transposons, Plant Molecular Biology Manual* **K1:** 1-15, Eds.: W.E. Lonning and P. Huijser, © 1994 Kluwer Academic Publishers. Reproduced by kind permission of Kluwer Academic Publishers, Dordrecht, Netherlands.

type (revertant) allele and the transposon carrying the mutant allele (Lane 4). From selfing of such flowers, a Mendelian 3:1 segregation of wild type to variegated colored plants in the progeny is expected. However, the same situation occurs with respect to the transposon carrying the allele, and upon excision, the new transposition insertion site (Lane 3 now using the transposon as a probe). If the new insertion has caused a recessive mutation, a quarter of the progeny of the selfed revertant flower will also reveal the new mutant phenotype. This is the basis of the nontargeted gene tagging approach. The large number of revertants needed for this approach can be obtained from the progeny of selfed variegated flowers.

In contrast to fully revertant flowers, RFLP analysis performed on the genomic DNA ex- tracted from variegated flowers reveals a strong disturbance in the distribution of different alleles. Generally, although dependent on the activity of the transposon, majority of the cells will be homozygous for the transposon carrying the mutant allele, resulting in only a weak band representing the wild type (revertant) allele (Lane 2, when color gene used as a probe). In addition, the revertant sectors of variegated flower repre- sent independent excision events, and thus multiple new insertion sites can be expected (Lane 1, when transposon was used as a probe). In fact, any site might become tagged, although, due to late excision events, many of these new insertion sites will remain undetectable in a Southern blot analysis. However, many cells of a variegated flower can be genetically screened for the presence of a new insertion at an already

Fig. 20.12 The basis for a nontargeted gene tagging approach. In the mutant plants, the newly tagged (and recessive) allele is always homozygous. Homo- and heterozygosity are distinguished by the thickness of bands. FS – flanking sequence of tagged locus probe; T – transposon; TE – transposable element (tagging transposon) specific sequence probe; WT – wild type. Reproduced from Lonning, W.E. and Huijser, P., *Gene tagging by endogenous transposons, Plant Molecular Biology Manual* **K1:** 1-15, Eds.: W.E. Lonning and P. Huijser, © 1994 Kluwer Academic Publishers. Reproduced by kind permission of Kluwer Academic Publishers, Dordrecht, Netherlands.

known locus. For that purpose the variegated flower is fertilized with pollen from a plant homozygous for the respective mutant allele. Any new mutation at the locus should become visible in the progeny of otherwise wild-type plants. This strategy is followed in the targeted gene tagging approach.

Two different approaches to isolate genetic loci with the help of transposons are based on the following observations: (1) *Targeted*, with the goal to recover a specific gene and (2) *Nontargeted*, not aiming for a specific gene but any phenotype that is correlated with transposon insertion. These two approaches require different experimental set-ups involving different crosses (Lonnig and Huijser, 1994).

Nontargeted or random gene tagging: This is designed to identify the reinsertion of an active transposon at any new position in the genome where it causes a detectable mutation (Fig. 20.12). The basis for a nontargeted gene tagging approach is the relatively rare germinal excision events. In the first step, a plant population is generated that contains active transposable elements. Many individuals of this population will contain a transposable element at a new chromosomal location. If this new position is within a gene, the transposon will probably inactivate it. However, as the insertion will almost certainly be heterozygous in this generation, no mutant phenotype will be detected. These plants are therefore allowed to self-fertilize and the subsequent generation screened for mutants.

This approach is where it is supposed that the newly tagged mutant allele is fully recessive and, furthermore, that the transposon donor site and the insertion site are unlinked. In such a case, the revertant is self-fertilized so that F_2 progenies inherit the transposition events. It means the progeny will segregate for plants with a wild-type and the (new) mutant phenotype (represented in the figure as a flower with altered morphology) in an approximate ratio of 3:1. For RFLP analysis, genomic DNA is extracted from single plants. The numbered lanes show the possible hybridization patterns and their ex-

pected ratios when genomic DNA of wild-type or of mutant plants is probed with the transposon or with sequences flanking the transposon at its new insertion site.

The frequency of insertion events is generally correlated with that of excision events. Thus one can increase the probability of finding such new mutations by selecting for germinal revertants from lines carrying an active transposon in an already characterized locus. In such a program one never knows beforehand which mutants will appear in the next generation. Therefore, this approach is preferred when the aim is to mark any new locus in a genetically controlled process involving many genes.

This method is advantageous over targeted tagging in that although one may be screening for a mutation in a particular gene, other interesting mutations can be detected. Strategies of this type are used if instead of a particular gene, a group of genes that belong, for example, to a particular physiological pathway or to a developmental program are to be isolated. Such strategies have been used successfully in snapdragon, maize, and petunia.

Targeted gene tagging: This strategy is directed toward tagging a specific gene of interest, and for this stable mutant alleles are available. Wild-type plants containing active transposable elements are crossed with plants that are homozygous recessive for the gene of interest (stable mutation). Majority of the F_1 progeny will exhibit the wild-type phenotype, since these plants are heterozygous for this gene. In exceptional cases, however, a transposable element in the wild-type parent will have been inserted into the gene and be transmitted to the gametes. These rare events are then directly uncovered in the F_1 generation by individuals that exhibit the mutant phenotype. The new mutants are selfed and revertants looked for in the next generation. Transposons are probably involved if revertants appear. The frequency of mutations caused by transposable elements may vary because of the nature of the active element system used and the position of

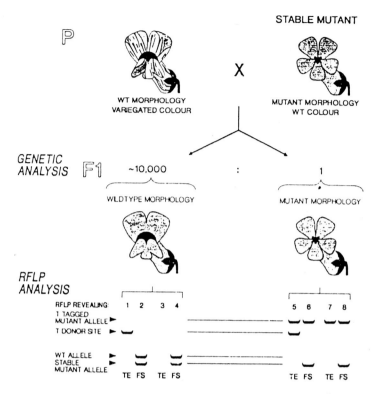

Fig. 20.13 The basis for a targeted gene tagging approach. With RFLP analysis using the transposon (TE, odd numbered lanes) or its flanking sequences as its new insertion site (FS, even numbered lanes) as probes, the newly tagged mutant allele should always be detectable in the mutant (Lanes 5–8) but be absent from the wild type plants (Lanes 1–4). FS – flanking sequence of tagged locus probe; T – transposon; TE – transposable element (tagging transposon) specific sequence probe; WT – wild type. Reproduced from Lonning, W.E. and Huijser, P., *Gene tagging by endogenous transposons, Plant Molecular Biology Manual* **K1:** 1-15, Eds.: W.E. Lonning and P. Huijser, © 1994 Kluwer Academic Publishers. Reproduced by kind permission of Kluwer Academic Publishers, Dordrecht, Netherlands.

the element in chromosome. The procedure is outlined in Fig. 20.13. It starts with the generation of phenotypically wild type plants heterozygous for the gene of interest by crossing homozygous stable recessive mutant plants with a wild-type line carrying an active transposon in its genome. In Fig. 20.13, the unstable line is represented as a variegated flower (should be taken as a female parent) and is crossed with plants homozygous for a stable recessive mutant as flower with altered morphology. New (recessive) mutant alleles that might have arisen in the active transposon carrying line (variegated color) are thus brought together with the stable mutant allele. This rare situation should alter the otherwise wild-type appearance of F_1 plants. An

F_1 population of at least 10,000 plants is needed, but to improve the probability of getting alleles with transposon insertions, this number should be raised to 30,000. Among these plants, new unstable alleles of the gene might appear. New mutants are selfed and revertants are looked for in the next generation. This method has been used successfully to isolate mutations of the *deficiens* locus of snapdragon (Sommer *et al.*, 1990) as well as transposon-induced mutations in maize (Table 20.5).

Problems of targeted gene tagging

1. Insertion within the gene of interest is a rare event. In the above-cited example of the *deficiens* mutant, out of 45,000 plants only

17 mutants could be isolated. If a mutation in the desired gene does not arise, it is unlikely that any other transposon-induced mutation will be apparent.

2. There is difficulty in cross-fertilizing some plant species and success depends on obtaining large number of hybrid progeny, which is difficult in plants such as *Arabidopsis*.

Characterization of a transposon insertion mutation involves testing which type of transposon and which copy of that particular type has caused the mutation.

First, it is necessary to prove that the mutation is due to the insertion of a transposon. The instability of transposon insertion mutation can be used to achieve this. Most mutations caused by the insertion of a transposon should produce sectors of revertant tissue due to somatic transposition events and should produce nonmutant progeny as a consequence of germinal transposition.

Second, species such as *Zea mays* and *Antirrhinum majus* contain several different active transposable element systems and these elements occur in more than one copy per genome. In any case, cosegregation of transposon copy with the mutant phenotype has to be demonstrated. This is done by analyzing DNA from the progenitor, mutant, and revertant plants in genomic Southern experiments. The success of gene tagging therefore increases with the extent to which the different potentially mobile transposable elements of a given organism are molecularly characterized.

Isolation of mutant gene: Once the transposable elements have been isolated, they can be used to clone genes that have been tagged by the insertion of one of these elements. For example, the Bronze (*Bz*) locus encodes an enzyme required for anthocyanin synthesis in kernels and *Ac*-induced mutations of the *Bz* locus were identified genetically. This locus could then be isolated relatively easy by using the *Ac* element sequences to screen a library made from mutant plant DNA.

Isolation of intact gene: A nonmutant copy

of the gene is later identified by using DNA flanking *Ac* to screen a library made from nonmutant plants.

Transposon tagging in heterologous species

A major limitation in transposon tagging is the low frequency with which a specific gene may be targeted. Conceivably, there are two aspects of the process, which could be optimized to increase the frequency of transposon inserting within a gene of interest: (1) the transposon frequency could be increased; (2) the proportion of transposition events which lead to the transposon inserting within the gene of interest could be raised.

For the use in heterologous species, *Ac* and *Ds* components are designed separately and are then used to transform heterologous species (Fig. 20.14). These are also known as transposon traps. It has two components. The first component carries an immobile transposase source, i.e. which expresses transposase (Ac protein) but terminal sequences essential for transposition have been deleted. This type of element is referred to as *Ac* stable and can be modified further. For example, the low level of expression of *Ac* mRNA in maize and tobacco suggests that this is a major factor limiting transposition frequency. To overcome this, the *Ac* promoter is removed and replaced with plant promoters, which are capable of much higher transcription rates. The second component is similar to the *Ds* element of maize; it comprises an element that possesses both termini but does not encode the Ac protein and can, therefore, only move in the presence of *Ac* stable elements. To follow transposition of this internally deleted element, it is inserted within genes whose activity can be followed phenotypically. For example, *Ds* insertion within genes required for resistance to antibiotics results in individuals in which transposition can be followed with the appropriate drug. For many plant species, the most appropriate antibiotic for this purpose is streptomycin. Thus internal deletion derivative of *Ac* is inserted

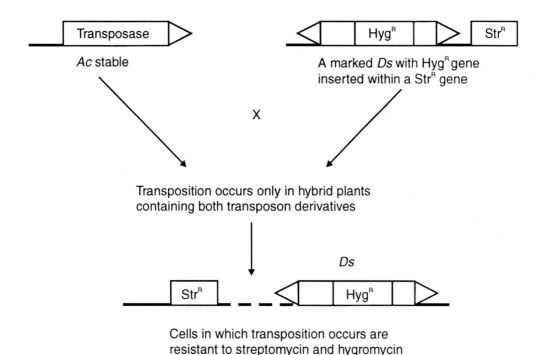

Fig. 20.14 A diagram illustrating two component transposon system based on *in vitro* engineered derivatives of the maize transposon *Ac*. *Ac* stable component comprises the *Ac* transposase gene expressed from a strong plant promoter but cannot transpose as one terminus has been removed. The second derivative consists of *Ds* element marked by the insertion of a hygromycin (Hyg) resistance gene. The marked *Ds* is inserted between the promoter and coding sequence of streptomycin (Str) resistance gene, which inactivates the streptomycin resistance gene. Plants containing these constructs are crossed. Transposition of *Ds* occurs in the hybrid plants. In the progeny this can be detected as green clones of green streptomycin-resistant cells on a white background. In the subsequent genertion fully streptomycin-resistant individuals occur which are derived from pollen or egg cells in which *Ds* has been excised from the streptomycin resistance gene. If these seedlings are also resistant to hygromycin then the *Ds* must be present at a new location in the plant genome.

within a streptomycin resistance gene *in vitro*. The insertion inactivates the gene, which is then introduced into plants.

Progeny of these plants are grown and crossed to plants containing *Ac* stable elements. The hybrid seeds, which have inherited both *Ac* derivatives, are then germinated on a medium containing streptomycin. If no transposition occurs, the streptomycin gene remains inactive and the seedlings appear white because streptomycin prevents normal chloroplast development. However, if the transposon excises from streptomycin-resistance gene, the activity of the

gene is restored, enabling those cells harboring an excision event to develop normal chloroplasts. This results in clones of green cells on the cotyledons of germinating seedlings. Excision of the transposon also occurs in cell lineages destined to form gametes, and in this case fully streptomycin-resistant seedlings might appear in subsequent generations. These seedlings are fully green when grown on a streptomycin-containing medium, and might contain the transposon at a new location in the plant genome. However, seedlings in which excision has occurred do not necessarily still contain the transposon. This difficulty can be overcome in an engineered transposon tagging system by inserting a second antibiotic-resistance gene (e.g. hygromycin) within the transposon. Thus the second component, a *Ds* element, is marked by the insertion of a hygromycin-resistance gene (*Hyg*^R). The marked *Ds* is inserted between the promoter and coding sequence of a streptomycin-resistance gene. This inactivates the gene. The *Ds* can only transpose when the same cell has *Ac* stable element. A plant containing one of these components is cross-pollinated with a plant containing the second derivative and transposition of *Ds* occurs in the progeny. In this way excision and reinsertion is selected for with a combination of streptomycin and hygromycin antibiotics. This can be detected as clones of green streptomycin-resistant cells on a white background. In the subsequent generation, fully streptomycin-resistant individuals are obtained, which are derived from pollen or egg cells in which *Ds* has excised from the streptomycin-resistance gene. If the seedlings are resistant to hygromycin and streptomycin then the *Ds* must be present or got inserted at a new location in the plant genome.

Transposon trap elements (Fig. 20.15) can be designed to isolate either the gene itself (gene trap) or its control elements (enhancer trap). Gene traps are based on the proper integration of a mobile element carrying a reporter gene that lacks its own promoter sequences into an actively expressed genetic locus. When the trap element integrates into an exon, in the proper reading frame and orientation, the reporter gene is expressed. To ensure that the trap works, appropriate splicing signals are provided immediately upstream of the reporter gene sequence if gene trap elements get inserted into an intron. The enhancer trap construct carries a minimal weak promoter upstream of the reporter gene. When this element gets inserted in the vicinity of an actively transcribed region, the reporter gene is activated by the neighboring regulatory sequences. The technique of enhancer trapping allows the expression pattern of a large number of genes to be visualized and screened rapidly.

Improving targeting of transposition events: How can one increase the frequency with which transposons insert within the gene of interest? In maize, majority of transposition events result in the transposon moving relatively short genetic distances. A study found that new positions in the genome of *Ac* elements which excised from the *P* gene of anthocyanin were approximately 65% and were shown to be geneti-

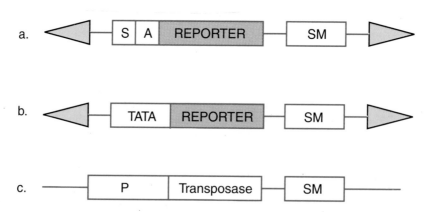

Fig. 20.15 Transposable element constructs for trapping genetic regions. **a.** Gene trap construct; **b.** enhancer trap construct; **c.** immobilized transposase construct without the terminal inverted repeats. The arrows represent the conserved terminal repeat sequences required for transposition. SM – selectable marker; S – conserved splicing signal; A – splice acceptor; TATA – minimal promoter sequence; P – strong promoter.

cally linked to *P* gene itself (Balcells *et al.*, 1991). This suggests that in maize *Ac* transposes preferentially to genetically linked sites. It was later shown that plants in which *Ac* had transposed from the *P* gene produced a relatively high proportion of progeny in which *Ac* reinserted within the gene. This was interpreted as subsequent transposition from the new locations resulting in reinsertion of transposon within the *P* gene. These experiments suggest that the most effective way to tag a specific gene with a transposon would be to start with a transposon closely linked genetically to the gene. This assumption is supported by recent experiments showing that *Ac* also transposes to linked sites in tobacco and this approach has now been incorporated into most gene tagging experiments with *Ac*.

Isolation of genes

Transposon tagging is now being adopted in a wide range of plant species to isolate genes involved in developmental and metabolic processes, which have been very difficult to approach by traditional biochemical methods. Genes controlling resistance to several plant diseases, floral initiation, and development as

well as synthesis and detection of hormonal signals are among the various types of genes tagged (Tables 20.6 and 20.7). The *deficiens* gene, which is necessary for normal development of petals and stamens (Sommer *et al.*, 1990) and *floricaula* gene which seems to govern the developmental switch from shoot to floral meristems (Coen *et al.*, 1990) have been isolated. Transposon tagging has been used to isolate genes necessary for resistance to fungal pathogen (Jones *et al.*, 1994; Izawa *et al.*, 1997).

Advantages of transposon tagging

1. Generation of large number of transposon insertions by simply crossing a small number of transgenic parental lines.
2. Visual selection based on reporter gene expression.

Significance of transposon tagging

Transposable elements and mutations

Transposable elements are responsible for mutations in a wide variety of organisms. Transposons have a number of effects on the genomes. DNA sequence alterations caused by transposons can

Table 20.6 Genes isolated using transposon tagging in plants

Gene	Gene function	Species	Transposon
A1	Enzyme required for anthocyanin biosynthesis	Maize	*Mu; En/Spm*
A2	Enzyme required for anthocyanin biosynthesis	Maize	*En*
Bz2	Anthocyanin pathway	Maize	*Ds; Mu*
C1	Regulatory gene	Maize	*En/Spm*
C2	Chalcone synthase	Maize	*En/Spm*
hcf-106	Chloroplast development	Maize	*Mu*
Kn1 (knotted)	Regulatory gene	Maize	*Ds*
Opaque 2	Regulatory gene	Maize	*Ac; En/Spm*
R	Regulatory gene	Maize	*Ac*
P	Regulatory gene	Maize	*Ac*
d1 and d5	GA biosynthesis	Maize	*Ac-Ds*
Rp1	Resistance gene	Maize	*Mu; Ac-Ds*
Deficiens	Regulatory gene	A. majus	*Tam7*
Delila	Regulatory gene	A. majus	*Tam2*
Floricaula	Regulatory gene	A. majus	*Tam3*
Globosa	Regulatory gene	A. majus	*Tam3; Tam9*
Olive	Chloroplast development	A. majus	*Tam3*
An9	Anthocyanin biosynthesis	Petunia	*dTph-4*
An2	MyB domain protein	Petunia	*dTph-1*

Table 20.7 Genes isolated using transposon tagging in heterologous species

Gene	Gene function	Species	Transposon
ALB-3	Accumulation of anthocyanin	Arabidopsis	En/I
Ms-2	Chloroplast differentiation	Arabidopsis	Ac-Ds
N	Male sterility	Arabidopsis	En/I
cf-4A	TMV resistance	Tobacco	Ac
Dem	Avr-4 response	Tomato	Ac-Ds
Cf-9	Organization of shoot apical tissues of developing embryos	Tomato	Ds
Pi-z^t	Resistance gene	Tomato	Ds
Ei-Ef-1	Blast resistance	Rice	Ds
Ei-Ef-1	Flowering time gene	Rice	Ds

range from one or a few base pairs to large chromosomal rearrangements via deletion, inversion, and duplication. Transposons affect genes directly by inserting into them, leading to subtle or major effects on gene expression. Many insertions even create a null phenotype because insertion into exons results in altered proteins. Transposable element insertions into introns can also have a variety of phenotypic effects, depending on the creation or destruction of splice sites, damage to transcriptional control regions, or transcriptional interference.

Insertion of class II elements into 5′ transcriptional control regions can also lead to mutations that depend upon the presence or absence of active elements in the plants. Insertion of transposable element into 3′ untranslated regions can create hybrid 3′ sequences and new polyadenylation sites; for instance, *Mu* and LTR transposons have contributed polyadenylation signals to some genes (Ortiz and Strommer, 1990).

Loss of mobile elements by excision often results in mutations. Excision is virtually always imprecise at the nucleotide level, leaving behind a few extra base pairs (a footprint) to mark the elements passage through the region.

The natural ability of transposable elements to cause mutations has been exploited to stimulate the transposition of a particular family of elements, thereby increasing the mutation rate.

Transposons and gene expression

The sequences introduced by insertion or altered by excision have the ability to impose new temporal and spatial patterns of gene expression. This can be mediated by alterations to transcriptional regulatory control regions or post-transcriptional RNA processing. Transposons are therefore not only capable of producing mutations but can also generate new regulatory units. This suggests that transposable elements have played and are playing a major role in plant gene evolution.

Transposons in vector development

Transposable elements are useful in the genetic transformation of higher organisms. Perhaps the most sophisticated system has been developed by using the P elements of *Drosophila* for construction of cloning vectors.

Members of P elements vary in size. The largest elements are 2907 bp long. These transposons terminate in 31-bp long terminal inverted repeats and are flanked by 8-bp target site duplication. These complete elements are autonomously mobile because they carry a gene that encodes a transposase protein. Other P elements are structurally incomplete because they lack the ability to produce transposase, but they possess terminal and subterminal sequences needed for transposition. Cells of higher organisms can be transformed if DNA fragments are inserted into the transposable elements. A system has been developed by using the P elements of *Drosophila* wherein a nonautonomous element serves as the transformation vector and a complete element serves as the source of transposase that is needed to insert the vector into the chromosome of a *Drosophila* cell.

The nonautonomous element carried a fragment of DNA with a rosy gene and a marker gene to identify transgenes within a P element and was inserted in a plasmid. The second plasmid carried the P element for the expression of transposase, but without the capability of its own integration. The two plasmids were coinjected into the embryos of a strain of *Drosophila melanogaster* homozygous deficient for the rosy gene. It was found that 50% of the progeny did carry *rosy*[+] progeny, indicating successful transfer. *Ac* and *Ds* elements can also be used in a manner similar that for *Drosophila*.

Further reading

Balcells, L., Swinburne, J., and Coupland, G. 1991. Transposons as tools for the isolation of gene. *Trends Biotech.*, **9**: 31-37.

Bennetzen, J.L. 2000. Transposable element contributions to plant gene and genome evolution. *Plant Mol. Biol.*, **42**: 251-269.

Gierl, A., Saedler, H. and Peterson, P.A. 1989. Maize transposable elements. *Annu. Rev. Genet.*, **23**: 71-85.

Flavell, A.J., Pearce, S.R., and Kumar, A. 1994. Plant transposable elements and the genome. *Curr. Opin. Genet. Develop.*, **4**: 838-844.

Lewin, B. 1995. *Genes V.* Oxford University Press, Oxford, UK.

Lonnig, W.E. and Huijser, P. 1994. Gene tagging by endogenous transposons. *Plant Mol. Biol. Manual* **K1**: 1-15.

Snustad, D.P., Simmons, M.J. and Jenkins, J.B. 1997. *Principles of Genetics.* John Wiley & Sons, New York.

21

Gene Isolation

Molecular manipulation of specific genes is a very popular area of research in biotechnology. Since no crop is perfectly suited to human needs, plant breeders are routinely concerned with improving traits such as tolerance to disease or environmental stress, yield, quality, and plant architecture. Several recent technological developments promise to facilitate greatly in the identification and isolation of genes.

General strategies for cloning genes from plants

All techniques for gene isolation exploit one or more of the following characteristics that define genes:

i. They have a defined primary structure (sequence).
ii. They encode an RNA with a particular expression pattern.
iii. They occupy a particular location within the genome.
iv. Most genes have a function.

Some techniques may permit the isolation of genes from any plant, whereas others are only applicable to one or a few plant species. Consequently, genes isolated from one flowering plant can generally be used to isolate the corresponding genes from other plants by heterologous hybridization. Therefore, genes isolated from noncrop plants, such as *Arabidopsis*, may be useful in manipulating crop plants.

Isolation of genes coding for specific proteins

The most widely used method for isolation of genes based on their function involved protein purification and complementation of mutant phenotypes. Isolation of genes for specific proteins became possible only after the discovery of reverse transcriptase enzyme because this enzyme can be used for the synthesis of cDNA from mRNA. This cDNA can then be used for isolating the corresponding gene from genomic DNA. Thus, techniques should be available for the isolation of specific mRNA. For this purpose, antibodies are produced against a specific protein for which the gene is to be isolated. The different steps involved are shown in Fig. 21.1. cDNA may be cloned in expression vectors such as λgt11, which can accept gene insertions into the β galactosidase gene. The chimeric vector will then produce a hybrid protein, which can be identified using antibodies. Plaques belonging to the chimeric vector carrying the desired gene can also be identified by the reaction of an antibody attached to a radioactive protein (see immunological detection procedure in Chapter 17).

Isolation of genes that are tissue-specific in function

Genes that are expressed in specific tissues are much easier to isolate. For example, genes for

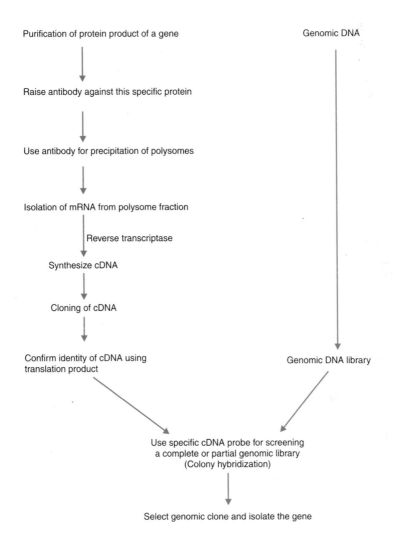

Purification of protein product of a gene

↓

Raise antibody against this specific protein

↓

Use antibody for precipitation of polysomes

↓

Isolation of mRNA from polysome fraction

| Reverse transcriptase
↓

Synthesize cDNA

↓

Cloning of cDNA

↓

Confirm identity of cDNA using translation product

Genomic DNA

↓

Genomic DNA library

Use specific cDNA probe for screening a complete or partial genomic library (Colony hybridization)

↓

Select genomic clone and isolate the gene

Fig. 21.1 Isolation of a gene for a specific protein.

storage proteins are expressed only in developing seeds. Such genes can be isolated because mRNA extracted from these specific tissues will either exclusively belong to the gene of interest or be rich in this specific mRNA. Other mRNA molecules in minor quantities can be eliminated since these can be identified through their isolation from tissues where this gene is silent. The strategy is outlined in Fig. 21.2.

Specific molecular probes (DNA or RNA), if available, can be used for isolation of specific genes. These probes may be available from another species for the same gene or may be artificially synthesized using a part of the amino acid sequence of the protein (see probes in Chapter 17). The probes obtained from one species and used for another species are called heterologous probes. These are very effective in identifying gene clones during colony hybridization/ plaque hybridization/ Southern blots. Heterologous probes are generally used with cDNA library and not with genomic library.

Synthetic oligonucleotide probes can also be used if protein sequencing is known for 5–15 consecutive amino acids, and this information can be used for the synthesis of a synthetic oligonucleotide probe. These oligo probes can be used for screening of cDNA or genomic libraries for isolation of gene.

Cloning methods based on DNA insertions

Several techniques exploit the fact that certain kinds of mutations lead to relatively large alterations of chromosomal structure, which can be used to isolate the corresponding genes. In some cases when the phenotypic effect is known but the gene product is not known or not identified, genes can be isolated. This phenomenon is known as reverse genetics, when one starts from the phenotype and then traces it back to the gene without knowing the gene product. The methods used for isolating these genes are different from those used for genes coding for known proteins. Two approaches involves either transposon or T-DNA from the Ti plasmid of *Agrobacterium tumefaciens* are frequently employed. In both these cases, transposons or T-DNA insert into or near the gene of interest, thereby causing a mutation that marks the gene. The chromosomal region surrounding this molecular tag of transposon or T-DNA can then be isolated by probing a genomic library of the tagged line with the tag (transposon or T-DNA).

Transposon and transposon tagging has been discussed in detail in Chapter 20. Transposon tagging was earlier confined to species such as snapdragon and maize, where well-characterized endogenous transposable elements were available. To make transposon tagging feasible in other plant species, various derivatives of maize *Ac* element have been developed and introduced by transformation into several plant species where they have been shown to transpose. Several genes have been cloned by transposon tagging.

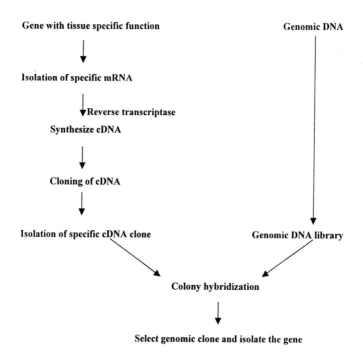

Fig. 21.2 Isolation of a gene for a tissue-specific function.

T-DNA tagging exploits the fact that during the production of transgenic plants by *A. tumefaciens* mediated transformation, one or more copies of the T-DNA are inserted at random locations into the host genome. The procedure of T-DNA transfer and integration of DNA is explained in Chapter 22. For *Arabidopsis*, a method for large-scale transformation of intact plants has been developed. When this transformation procedure is used, 35–40% of the mutations generated are tagged by a T-DNA insert. The major limitation of this approach is the relatively large amount of effort required to produce and propagate considerable numbers of intact transformed plants.

Subtractive cloning

In this technique, either the total genomic DNA or mRNA of the target cell line is used for isolating a gene. Total genomic DNA from a line containing a deletion mutant is hybridized in excess to highly fragmented DNA from wild type plants. Fragments that do not hybridize to DNA from the mutant correspond to the deleted region. Cloning by subtractive hybridization from total genomic DNA is shown in Fig. 21.3. Total chromosomal DNA from a wild type plant is cut into small segments with a restriction endonuclease such as *Sau*3A. Total chromosomal DNA from a plant containing a deletion mutation is randomly sheared and then biotinylated. The two DNA samples are then denatured and the nonbiotinylated DNA (wild) is hybridized to a large excess of the biotinylated DNA (mutant) in solution. The DNA is then applied to a column containing avidin-coated beads that bind biotinylated DNA molecules. Most of the DNA from the wild type plant hybridizes to the excess biotinylated DNA from the mutant plant and therefore is bound to the column. However, DNA from the wild type plant (DNA fragment B) that corresponds to the DNA fragment missing from the mutant plant cannot hybridize to a biotinylated strand and so will be enriched in the column eluate. The eluated DNA is then rehybridized to an excess of biotinylated mutant DNA and the column enrichment process is repeated. After several cycles of enrichment, the resulting DNA is amplified using PCR and then cloned. The clones are tested by hybridization to Southern blots of the mutant and wild type to ensure that they hybridize to the wild type but not to the mutant DNA.

The utility of the method was demonstrated by Sun *et al.* (1992) by isolating gene (*ga*1) involved in gibberellin biosynthesis in *Arabidopsis*. Salt stress induced genes have been isolated from wheatgrass by using this approach (Gulick and Dvorak, 1990).

When mRNA is used for isolation of genes, the mRNA of the target cell line is used as a substrate to prepare a set of cDNA molecules corresponding to all the expressed genes. To remove sequences that are not specific for the target cell, the cDNA preparation is exhaustively hybridized with the mRNA of the other closely related cell. This step removes all the sequences from the cDNA preparation that are common to the two cell types. Therefore, the specificity of the technique will depend on the closeness of the relationship between the two cells. After discarding all the cDNA sequences that hybridize with the other mRNA, those that are left (<5%, if technique works well) are hybridized with mRNA from the target cell to confirm that they represent coding sequences. These clones should contain sequences specific to the mRNA population of the target cell and they can be characterized. The different steps involved in subtractive hybridization for isolation of specific mRNA are outlined in Fig. 21.4.

The limitations of subtractive cloning are as follows:

1. It is practicable only for plants with small genomes.
2. Methods have not been developed for reliably producing mutagenized plant populations with small deletions.

Map-based cloning

Most important eukaryotic genes, for example, disease resistance genes, are known by their

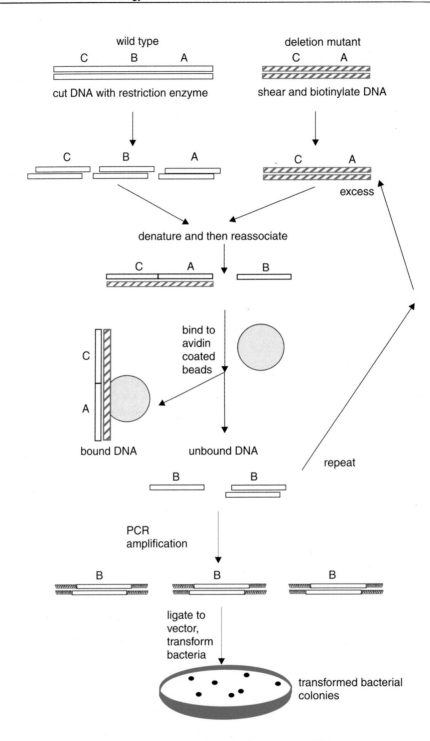

Fig. 21.3 A strategy for cloning by subtractive hybridization for genomic DNA.

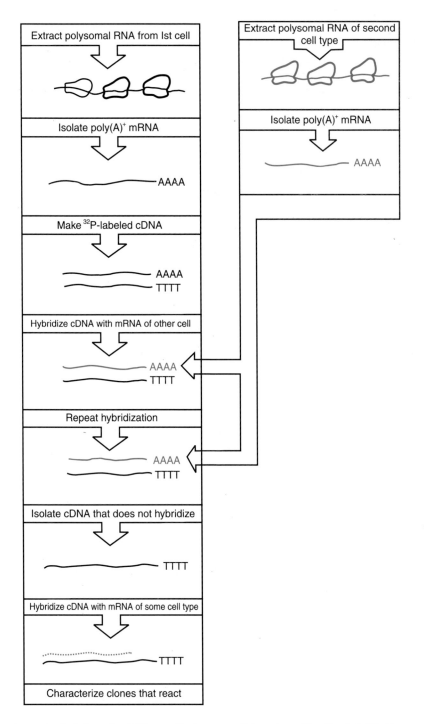

Fig. 21.4 A scheme for isolation of mRNA specific to one cell type by subtractive hybridization.

phenotype and not by their biochemical effects. Traditional cloning strategies have been ineffective with these genes. Researchers have therefore sought new approaches to gene cloning that are not based on the prior knowledge of the biochemical effect of a gene, and map-based cloning is one such technique. The concept behind map based or positional cloning is to find a DNA marker (DNA sequence) linked to a gene of interest (target gene) and then "walk" to the gene via overlapping clones. It has certain requirements: (i) The individuals within the population have genetically based differences in the trait of interest; (ii) The genes responsible for this difference can be mapped to a chromosomal position(s) adjacent to segments of DNA that have already been cloned (e.g. RFLP or

microsatellite markers); (iii) The availability of comprehensive genomic libraries of relatively large DNA fragments, typically in vectors like YAC, BAC, PAC, etc.; and (iv) The availability of DNA probes that are closely linked to the gene of interest (ideally less than a few hundred kb apart. The following steps are involved in map-based cloning (Fig. 21.5).

Identification of DNA markers linked to target gene and construction of genetic maps

The first step in map-based cloning is that large number of cloned DNA sequences are mapped in terms of their genetic location (e.g. RFLP mapping). Earlier, only RFLP marker was used, but now various other molecular marker tech-

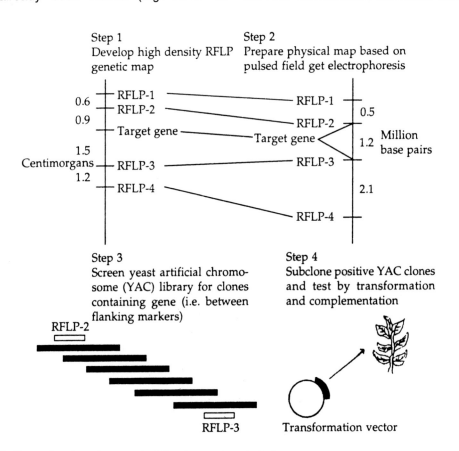

Fig. 21.5 The major steps for gene isolation in map-based cloning.

niques are also used for genetic location of a DNA sequence. Those DNA sequences that are tightly linked to the gene of interest are identified. The procedure is outlined in Fig. 21.6 for identification of DNA markers linked to target gene. For example, when RFLP analysis is conducted on three regions of chromosomes from two strains of the same species which is indicated by straight lines in Fig. 21.6. Restriction enzyme recognition sequences on these regions are indicated by vertical lines. In strain 2, presence of a mutation eliminates a restriction endonuclease site as shown by asterisk in the figure. The loss of this recognition site can be visualized. Genomic DNA from both the strains is isolated, cut with restriction endonuclease, and size fractionated by agarose gel electrophoresis. The number of bands would be very large and it would be impossible to distinguish individual

bands. Therefore, for visualization, DNA is transferred from the gel to a nitrocellulose filter and probed with a small piece of genomic DNA in a Southern blot experiment (explained in Chapter 14). The probe DNA will hybridize to two DNA restriction fragments from strain 1 but, due to the loss of a recognition site from strain 2, it will hybridize to only one larger DNA restriction fragment. Likewise, RAPDs can also be used to mark the target gene. Assuming the same three regions of chromosomes from two strains of the same species indicated by straight lines in Fig. 21.6. Binding sites for a particular oligonucleotide primer on the chromosomes are indicated by arrows, with the direction of arrow showing the orientation in which the primer binds at that site. When PCR is performed, DNA will be amplified between the pairs of oligonucleotide primers that bind close together and with opposite orientation

Fig. 21.6 A procedure for identification of molecular markers linked to target gene using: a – RFLP; b – RAPD.

with respect to each other. The chromosomal regions that will be amplified are indicated by the numbers showing the distance in kilobase pairs between the relevant pairs of binding sites. The asterisk indicates the mutation that has eliminated an oligonucleotide primer binding site from the strain 2. The loss of this binding site leads to the production of one less PCR product from strain 2 as compared to strain 1. The loss of this PCR product can be visualized by size fractionating the PCR products on a gel.

Rice is the most extensively analyzed species by RFLPs. The polymorphism detected by RFLP markers is very reliable as it involves the recognition by specific restriction enzymes and hybridization with a specific probe. Genetic maps for rice with more than 2300 DNA markers, tomato with more than 1400 markers distributed over 12 chromosomes, *Arabidopsis* with 500 markers distributed over five chromosomes, and likewise for barley, wheat, maize, etc. have been developed. In many cases, disease resistance genes have also been mapped relative to RFLP markers. One of the major problems with RFLPs as starting points for map-based cloning is the difficulty in specifically saturating chromosomal regions near gene of interest with sufficient numbers of RFLP markers. This is due to the large amount of DNA in the genome of most plants. Plant species have typically more than 1 billion base pairs (10^9) of nuclear DNA: rice – 400 mb; tomato – 1 billion bases; maize – 3 billion bases; and wheat genome – more than 16 billion base pairs. DNA clones used for generating an RFLP map are, by their very nature, randomly distributed throughout the genome. This means that many DNA clones must generally be mapped before one is found that is very tightly linked to a target gene. This becomes apparent by noting that in theory over 1500 randomly distributed DNA clones would have to be analyzed in order to be 95% certain of finding at least one clone within 1 cM of a target gene in a typical plant genome of total size of 1000 cM. Construction of such high-resolution genetic maps provides information primarily about the orientation of DNA markers relative to one another, and the target gene and is essential before physical mapping, chromosome walking and gene cloning can begin.

Construction of high-resolution physical maps

Once molecular markers (RFLPs, RAPDs, microsatellites) tightly linked to a gene of interest are identified, the next step in map-based cloning is physical mapping. Although segregation analysis yields vital information on the orientation and genetic distance between RFLPs in terms of recombination frequency (cM), it is still essential to know the actual physical distance in terms of nucleotides between RFLPs that flank a target gene. Moreover, RFLPs that appear to be very close to a gene in terms of genetic distance may still be very far away in terms of physical distance. It is clearly essential to know the magnitude of the physical distance between RFLPs before initiating the next step of chromosome walking and map-based cloning.

Until recently, detailed characterization of the large DNA molecules necessary for map-based cloning was impossible. Routine agarose gel electrophoresis only separates DNA molecules up to 50 kb, whereas map-based cloning generally involves distance up to 1 megabase (Mb) or more. Now, DNA molecules up to 10 Mb can be separated by pulsed field gel electrophoresis (PFGE). Recent developments in PFGE, which is capable of separating DNA molecules upto 10 million bp in length, has made long-range physical mapping practical. PFGE includes several types of related electrophoretic systems such as CHEF, field inversion gel electrophoresis (FIGE), which are discussed in Chapter 14. In each of these systems, DNA molecules are separated not only on the basis of migration through a gel matrix, but also on how long it takes for DNA molecules to reorient themselves in an electric field whose orientation changes periodically. Larger DNA molecules tend to take longer to change directions, and, consequently, travel more slowly.

The average relationship between the genetic and physical distance of a genome is calculated from the genome size and length of the genetic map. However, actual values for any specific location vary widely from the average. In tomato, 1 cM averages 900 kb, but in the *Tm2a* region of chromosome 9, 1 cM is 4–16 Mb. This physical distance is currently well beyond the capabilities of map-based cloning. In wheat, where the kb:cM ratio is much larger than in tomato, 1 cM is approximately equal to 1 Mb in the region of the α-amylase gene on Chromosome 6.

Chromosome walking

After a physical map of the genome near a target gene is constructed, this region can then be cloned by chromosome walking. Walking along the chromosome is a term used to describe an approach that allows the isolation of gene sequences whose function is quite unknown. It is a technique that has been in use for more than a decade to identify genomic clones contiguous to a cloned DNA starting point. The principle is that a cloned genomic fragment is used as a starting point for the walk. This should be as close as possible to the target point. The starting point may be RFLP sequence. Two examples are given to illustrate chromosome walking.

Figure 21.7 shows that gene A is to be cloned, which has been identified genetically but no probe is available. Sequence of a nearby gene B is available in cloned fragment 1. The genomic library is then screened with the chosen clone as probe to identify other clones containing DNA with which it reacts and which represent clones overlapping with it. The overlap can be to the right or left. This step will identify overlapping clones 2, 3, and 4. Clone 3 can in turn be used as a probe to identify clone 5. Repetition of this walking step along the chromosome is done. A large-scale map of the walk can be made by assembling contiguous clones.

In the second example, molecular markers such as RFLPs have been identified and are used as a starting point for chromosome walking. Two RFLPs lying closest to the gene of interest, yet flanking the gene on either side, are chosen. Chromosome walking begins by identifying genomic clones that overlap the initial RFLPs, which is accomplished by probing a complete genomic library with radiolabeled nucleic acid probes synthesized from the initial RFLPs. The newly identified genomic clones are then isolated and the ends of these clones become starting points for the next step in the chromosome walk. In Fig. 21.8, a mutation in a gene has been determined by genetic mapping to lie between two particular RFLP markers. DNA from RFLP-X is used to probe a genomic DNA library cloned in YAC vector in order to identify clones that contains an overlapping DNA insert. DNA from the end of this clone (YAC-A) is then isolated by inverse PCR and is used to screen the genomic DNA library to identify clones containing additional overlapping DNA fragments (YAC-B1 and YAC-B2). Each cycle of isolating the ends of the clone and then using this end to identify clones

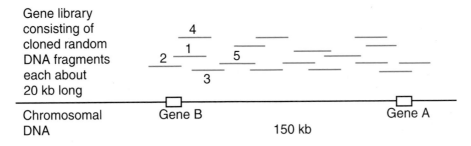

Fig. 21.7 Chromosome walking to clone DNA sequences.

Fig. 21.8 Gene isolation by map-based cloning using chromosome walking.

containing overlapping DNA fragments represents one step in chromosome walking. The process is continued until clones spanning the gene of interest are identified. In this case, since RFLP-Y is known from genetic mapping to lie on the far side of the gene of interest from RFLP-X, steps are taken until a clone (YAC-C2) that contains an insert that overlaps the DNA contained in RFLP-Y is identified. In practice, the walking process will usually be initiated from both RFLPs simultaneously.

When chromosome walking is performed in plants and animals, a problem of repeated DNA sequences must be realized. These may occur dispersed at several places in the genome and could disrupt the orderly progress of the walk. For this reason, the probe used for stepping one genomic clone to the next must be a unique sequence clone. For chromosome walking, improved vectors and strategies have been developed. Two λ phage derivatives, λDASH and λFIX, have been designed by Stratagene Cloning Systems.

For map-based cloning where positional location is done and the distances to be traversed may be as great as 1 Mb or more, cloning vectors in the form of yeast artificial chromosomes (YACs) and bacterial artificial chromosomes (BACs) are used. YACs and BACs can have fragments of 400–600 kb and 100–150 kb, respectively. YAC libraries have been constructed for *Arabidopsis*, tomato, maize, rice, etc. Although a fewer number of clones are needed to make a contig for a specific genomic region by using YACs, chimerism in the inserted DNA fragment can be a problem. BAC carries rather small DNA fragments compared to that by YACs, but is less prone to chimerism. Using these libraries and DNA markers for overlapping contigs, physical maps are being developed. The most advanced map developed in plants is for *Arabidopsis*. In rice, about 1200 RFLP and STS markers were used for screening a YAC library containing about 7000 clones with the average insert size of 350 kb. As a result, about 2200 independent YACs have been identified and ordered. Both *Arabidopsis* and rice genomes have been sequenced.

Testing of gene by transformation and complementation

The last step in map-based cloning is pinpointing the target gene among the overlapping DNA clones identified during chromosome walking. Whenever feasible, this can be accomplished by transformation of recessive individuals with subclones derived from chromosome walk. Then, the subclone containing the target gene can be identified in transformed individuals expressing the appropriate phenotype, assuming that introduced copy of the gene is expressed. If transformation is not feasible or transforming sequences are not expressed, other strategies are required.

Difficulties in map-based cloning

Although simple in concept, map-based cloning is generally very difficult in practice due to the following reasons:

1. The enormous size of eukaryotic genome: The genomes of higher plants are typically more than 1 billion base pairs, and in crops such as wheat and oats, they are more than 16 billion base pairs in length. This means that even when 1000 RFLP markers have been placed on a linkage map, the average physical distance between RFLPs will be 1 Mb in most plants and over 16 Mb in wheat and oats. Such distances are formidable even with the best chromosome walking technique.

2. Relationship between genetic distance, as measured by recombination frequency, and physical distance, as measured in number of nucleotides, is not uniform throughout eukaryotic genomes. Thus, even RFLP markers that appear to be close to a gene of interest in terms of recombination frequency may still, infact, be far away in terms of actual number of nucleotides.

3. Most eukaryotic genomes contain significant amounts of repetitive DNA, which makes it extremely difficult to perform chromosome walking. Dispersed, repetitive DNA sequences can cross-hybridize with the overlapping

stretch of DNA required to identify the next step of a chromosome walk and when this happens, further progress in chromosome walking along the genome may be almost impossible.

Chromosome jumping

Chromosome walking is simple in principle, but walking on a large-scale is too demanding. For large distances, walking is usually combined with chromosome jumping, which overcomes the difficulties caused by repeated, unclonable, or unstable genomic segments. Chromosome jumping has been described by Collins and Weissman (1984) and Poustka and Lehrach (1986). This technique depends on circularization of very large genomic DNA fragments, followed by cloning DNA from the region covering the closure site of these circles, thus bringing together DNA sequences that were originally located far apart in the genome. These cloned DNAs from the closure sites make up a jumping library, which greatly speeds up the process of long-range chromosome walking. Chromosome jumping utilizes type II restriction endonucleases that cut DNA sequences infrequently. Such enzymes have octanucleotide or longer recognition sequences, for example, *Not* I (GCGGCCGC), *Pac* I (TTAATTAA) and *Sfi* I (GGCCNNNNNGGCC). Some restriction endonucleases have shorter recognition sequences, but have the property of being uncommon sequences in mammalian DNA, for example, *Nru* I (TCGCGA) and *Bss* HII (GCGCGC). Their target sequences include two CG dinucleotides, which is rare in mammalian DNA. Additional specificities in target sites may be created by combining the specificities of certain methylases with that of restriction endonucleases. For example, *Dpn* I cuts DNA at the sequence G-^{m}A-T-C to produce flush ends. This enzyme requires the methylation of adenines in both strands of DNA for cleavage to occur.

A strategy for creating a jumping library is shown in Fig. 21.9. Poustka *et al.* (1987) used *Not* I restriction enzyme to digest genomic DNA, which cuts rarely. Partial digestion of genomic

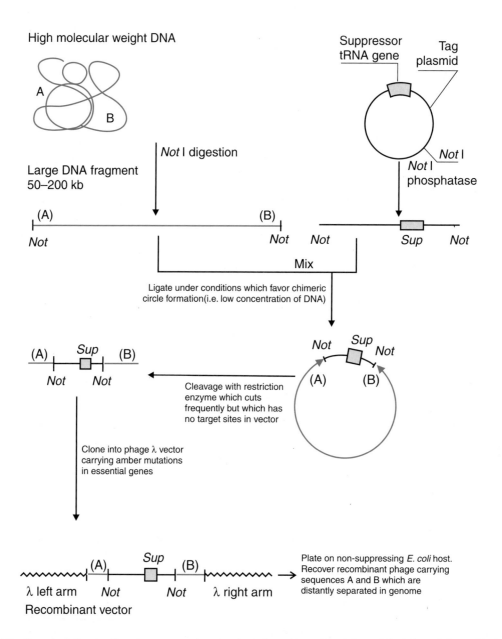

Fig. 21.9 Gene isolation by the chromosome jumping technique. Reproduced from *Principles of Gene Manipulation, R.W. Old and S.B. Primrose, by kind permission of Blackwell Science, Osney Mead, Oxford, U.K.*

DNA would result fragments of 50–200 kb. As shown in the figure, X and Y genes are far apart. The vector is also digested with *Not*I enzyme, and the fragments are ligated in the vector molecules, which favors chimeric circle formation. This chimeric vector is then cleaved with a restriction endonuclease that cuts frequently but has no target sites in the vector. Thus, it will bring together two marker genes (*X* and *Y*) near, which will be separated only by small portion of vector DNA. This technique has been applied to a jump of 100 kb in the cystic fibrosis locus in humans.

Chromosome landing approach

In map-based cloning, the central step is chromosome walking, which is a very time consuming process. When map-based cloning was first proposed for use in species with large genomes, molecular linkage mapping was in its infancy and molecular maps were not available for most plant species. It could take months, or even years, to find a single molecular marker linked to a target gene. Recent developments both in technology and analytical procedures led Tanksley *et al.* (1995) to propose the chromosome landing approach. In this technique, efforts are directed toward selectively identifying molecular markers that are so close physically to the target gene that little or no chromosome walking is required. These tightly linked markers are used to screen a genomic library to directly isolate a single clone containing the gene of interest. Molecular linkage maps have now been constructed for more than 30 species, and work is underway to construct maps for many more species by using all the new molecular techniques that have been already described and also by many new techniques. The availability of high-density molecular linkage maps provides a ready starting point to search for molecular markers tightly linked to a target gene. Additional help can be sought from comparative genetic linkage maps of species that have evolved from a common ancestral genome and thus share portions of their linkage maps that are conserved in both gene content and order. Extensive comparative maps have been constructed for a number of plant taxa, including species in the grass family (rice, maize, sorghum, barley, and wheat), the night-shade family (tomato, potato and pepper), and the mustard family (*Arabidopsis*, cabbage, cauliflower and rape).

Although high-density maps and comparative maps can provide some of the linked markers needed for chromosome landing in plants, it is clear that additional methods are needed by which new markers can be selectively isolated from specific regions of a genome. The probability of identifying one or more markers within a specified physical distance of a target gene decreases as one gets closer to the target. Ideally one would like to identify a marker at a physical distance from a target that is less than the average insert size of the genomic library from which one expects to isolate the gene. For example, for a YAC library with an average insert size of 200 kb, it would be desirable to identify at least one marker within a 400-kb window containing the target (200 kb on each side). For species with large genome of 1 billion base pairs, the probability would be $4 \times 10^2/10^6 = 4 \times 10^{-4}$. To have a 90% chance of identifying at least one marker within a 400-kb interval, ca. 6000 markers would need to be sampled. It is not likely that many plant species will have molecular linkage maps of this density in the foreseeable future.

Several strategies allowing one to screen a large number of random, unmapped molecular markers in a relatively short time and selecting just those few markers that reside in the vicinity of the target gene have been developed. These high-volume marker technologies that have shown efficacy are RFLP, STS (sequence tagged sites), EST (expressed sequence tag), SSCP (single strand conformation polymorphism), microsatellites, minisatellites, RAPD, SCAR (sequence characterized amplified regions), CAPs (cleaved amplified polymorphic sequences), AFLP, etc. These methods rely on two principles: (i) generate hundreds or even thousands of potentially polymorphic DNA segments and visualize rapidly from single preparations of DNA; and (ii) use of genetic stocks to identify among these thousands of DNA fragments, those few that are derived from a region adjacent to the targeted gene.

In the past few years, thousands of loci scattered throughout the genome have been assayed in just weeks or months by using one or more of these high volume markers technologies. The next problem is to determine which of the amplified loci lie near the targeted gene. Two strategies have proved effective: (i) nearly isogenic lines and (ii) bulk segregant analysis. These are described in Chapter 22.

High-volume marker techniques such as those based on RAPDs, RFLPs, and AFLPs can be used to compare the genetic profiles of pairs of NIL or pairs of bulks to search for DNA markers near the target gene.

In segregating populations of a modest size, a target gene can usually be placed within an interval of 5–10 cM with a high degree of certainty. The markers defining this interval can then be used to screen a large segregating population and identify individuals derived from one or more gametes containing a crossover in the given interval. Only such individuals are useful in orienting other markers closer to the

target gene. Once identified, these individuals can be analyzed in relation to all molecular markers within the region to identify those closest to the target. This flanking marker approach was recently used to refine the map position of the *Cf-2* gene, which encodes resistance to *Cladosporium fulvum* in tomato to a 40-kb region between two DNA markers. Through the chromosome landing approach, a tomato gene *Pto*, which confers resistance to the bacterial pathogen *Pseudomonas syringae* has been isolated (Martin *et al.*, 1993).

Further reading

Gibson, S. and Sommerville, C. 1993. Isolating plant genes: A review. *Trends in Biotech.*, **2**: 306-312.

Gupta, P.K. 1996. *Elements of Biotechnology*. Rastogi Publishers, Meerut.

Miesfeld, R.L. 1999. *Applied Molecular Genetics*. Wiley-Liss, New York.

Old, R.W. and Primrose, S.B. 1995. *Principles of Gene Manipulation*. Blackwell Science, UK.

22

Molecular Markers and Marker-assisted Selection

The science of plant genetics traces back to Mendel´s classical studies on garden peas. Since then, researchers have been identifying, cataloging, and mapping single gene markers in many species of higher plants. In the early part of the twentieth century, scientists discovered that Mendelian factors controlling inheritance, which we now call genes, were organized in linear order on cytogenetically defined structures called chromosomes. Clues to the location of a gene can come from comparing the inheritance of a mutant gene with the inheritance of marker(s) of known chromosomal location. Coinheritance or genetic linkage of a gene of interest (e.g. resistance gene) and a marker suggests that they are physically close together on the chromosome. Linkage is a familiar concept in genetics and dates back to early studies on *Drosophila*. It was shown that combination of genes tended to be inherited as groups, linked together because they are close to each other on the same chromosome. In fact the first chromosome map was produced by Sturtevant with segregation data derived from studies on *Drosophila* (Crow and Dove, 1988). The markers on the first genetic map were phenotypic traits scored by visual observation of morphological characteristics of the flies. A marker must be polymorphic, that is, it must exists in different forms so that chromosome carrying the mutant gene can be distinguished from the chromosome with normal gene by the form of marker it also carries, This polymorphism in the marker can be detected at three levels: phenotype (morphological), differences in proteins (biochemical), or differences in the nucleotide sequence of DNA (molecular).

MORPHOLOGICAL MARKERS

Morphological markers generally correspond to the qualitative traits that can be scored visually. They have been found in nature or as the result of mutagenesis experiments. Morphological markers are usually dominant or recessive. Until recently, however, the genetic markers used to develop maps in plants have been those affecting morphological characters, including genes for dwarfism, albinism, and altered leaf morphology.

Scientists have long theorized about the use of genetic maps and markers to speed up the process of plant and animal breeding. In 1923, Sax proposed identification and selection of minor genes of interest by linkage with major genes, which could be scored more easily. This idea has resurfaced many times, and many workers have extended the concept. Unfortunately, there are constraints on the use of morphological markers. (i) They cause such large effects on phenotype that they are undesirable in breeding programs. (ii) They mask the effects of linked minor gene(s), making it nearly impossible to identify desirable linkages for selection. (iii) They are highly influenced by the environment. The morphology/phenotype is the result of genetic constitution and its interaction with environment. Due to varying levels of G×E interaction, it is not appropriate to compare the morphological data of varieties that have been collected across different years and/or locations.

BIOCHEMICAL MARKERS

In recent years, isozymes have been used very successfully as biochemical markers in certain aspects of plant breeding and genetics as nearly neutral genetic markers. Biochemical markers are proteins produced by gene expression. These proteins can be isolated and identified by electrophoresis and staining. Isozymes, the different molecular forms of the same enzyme that catalyze the same reaction, are proteins. They are revealed on electrophoregrams through a colored reaction associated with enzymatic activity. They are the products of the various alleles of one or several genes. Monomer and dimer isozymes are most often used because analysis of their segregation is easier. Isozymes are generally codominant. It should be noted that in most cases the polymorphism of isozyme markers is rather poor within a cultivated species. This explains why most of the known cases originate from interspecific crosses. As a result, even with the use of isozymes as genetic markers, the full potential of genetic mapping in plant breeding has yet to be realized.

MOLECULAR MARKERS

A molecular marker is a DNA sequence that is readily detected and whose inheritance can easily be monitored. The use of molecular markers is based on naturally occurring DNA polymorphism, which forms the basis for designing strategies to exploit for applied purposes. A marker must be polymorphic; that is, it must exist in different forms so that chromosome carrying the mutant gene can be distinguished from the chromosome with the normal gene by a marker

it also carries. Genetic polymorphism is defined as the simultaneous occurrence of a trait in the same population of two or more discontinuous variants or genotypes. The first such DNA markers to be utilized were fragments produced by restriction digestion – the restriction fragment length polymorphism (RFLP) based genetic markers. Consequently, several marker systems have been developed. One of the main aspects of this technology is that linkage between molecular markers and traits of interest can be detected in a single cross. With morphological and biochemical markers, a separate cross is required to test linkage with each new marker. A molecular marker should have some desirable properties.

i. It must be polymorphic as it is the polymorphism that is measured for genetic diversity studies.
ii. Codominant inheritance: The different forms of a marker should be detectable in diploid organisms to allow discrimination of homo- and heterozygotes.
iii. A marker should be evenly and frequently distributed throughout the genome.
iv. It should be easy, fast, and cheap to detect.
v. It should be reproducible.
vi. High exchange of data between laboratories.

Unfortunately, no single molecular marker meets all these requirements. A wide range of molecular techniques is available that detect polymorphism at the DNA level. These have been grouped into the following categories based on the basic strategy, and some major ones are described.

Non-PCR-based approaches: Restriction fragment length polymorphism (RFLP).

PCR-based techniques: Random amplified polymorphic DNA (RAPD), microsatellites or simple sequence repeat polymorphism (SSRP), amplified fragment length polymorphism (AFLP), arbitrarily primed PCR (AP-PCR), etc.

Targeted PCR and sequencing: Sequence tagged sites (STS), sequence characterized amplified region (SCARs), sequnece tagged microsatellites (STMs), cleaved amplified polymorphic sequences (CAPS), etc.

NON-PCR-BASED APPROACHES

Restriction fragment length polymorphism (RFLP)

RFLP was the first technology that enabled the detection of polymorphism at the DNA sequence level. Genetic information, which makes up the genes of higher plants, is stored in the DNA sequences. Variation in this DNA sequence is the basis for the genetic diversity within a species. In most cases, DNA sequence variation is phenotypically neutral because the majority of mutations altering the order of nucleotides are only maintained during evolution if they are silent.

In this method, DNA is digested with restriction enzyme(s), which cuts the DNA at specific sequences, electrophoresed, blotted on a membrane, and probed with a labeled clone. Polymorphism in the hybridization pattern is revealed and attributed to sequence difference between individuals. The DNA sequence variation detected by this method was termed restriction fragment length polymorphism (Botstein *et al.*, 1980). Grodzicker first described this phenomenon for mutant strains of adeno-virus. Botstein, White, Skolnick, and Davis realized that RFLPs were not necessarily associated with specific genes but were distributed throughout the genome, and they made a detailed proposal for using RFLPs to map all the genes in the human genome. By definition, RFLP variation is environmentally independent. Variation in one DNA fragment obtained with a specific enzyme is treated as one RFLP. Plants are able to replicate their DNA with high accuracy and rapidity, but many mechanisms causing changes in the DNA are operative. Simple base pair changes or large-scale changes as a result of inversions, translocations, deletions or transpositions may occur. This will result in loss or gain of a recognition site and in turn lead to restriction fragments of different lengths. Genomic restriction fragments of different lengths between genotypes can be detected on Southern blots and by a suitable probe. The majority of single

nucleotide changes are innocuous, although they may alter a restriction enzyme site so that an enzyme no longer cuts DNA at that site. For example, the restriction endonuclease *Eco*RI cuts DNA at a specific sequence

$$\downarrow$$
...GAATTC...   ...G   AATTC...
...CTTAAG...   ...CTTAA   ....G...
$$\uparrow$$

When one or more nucleotides in the endonuclease recognition sequence are altered, *Eco*RI will fail to cut the DNA strands at the altered site and a longer DNA fragment is produced. When a given restriction enzyme site is present in the DNA molecule of one chromosome but absent from the DNA molecule of the other homologous chromosome, a shorter fragment is produced from the chromosome with the site, and a longer fragment from the other chromosome. It is now possible to distinguish the two chromosomes in such an individual on the basis of this restriction fragment length polymorphism. The individual is heterozygous for this RFLP and, in the jargon of the geneticist, is said to be informative at that locus. Because the RFLP is inherited just like a gene, one can follow the individual chromosomes as they pass from generation to generation by tracing the inheritance of the marker fragments.

Procedure

RFLP analysis comprises the following basic steps:
i. DNA isolation,
ii. Cutting DNA into smaller fragments using restriction enzyme(s),
iii. Separation of DNA fragments by gel electrophoresis,
iv. Transferring DNA fragments to a nylon or nitrocellulose membrane filter,
v. Visualization of specific DNA fragments using labeled probes, and
vi. Analysis of results,
Genomic DNA of an organism is isolated and digested with a particular restriction endonuclease

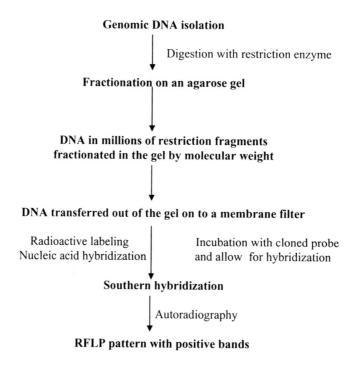

Fig. 22.1. An outline of conventional RFLP analysis.

(Fig. 22.1). It will cut the DNA into a large number of fragments. Since DNA has a large number of phosphate groups, which are negatively charged at neutral pH, it will migrate toward the anode in an electric field. In a porous medium of agarose or polyacrylamide, DNA fragments will migrate toward the anode at a rate that is proportional to their molecular weight. Agarose gel electrophoresis has already been explained in Chapter 14.

Relatively small DNAs of chloroplasts will usually produce about 40 discrete restriction fragments when digested with *Eco*RI and can be separated on agarose gels. These are stained with ethidium bromide and the pattern can be seen under UV light. But digestion of nuclear DNA yields millions of DNA fragments in a continuous range of sizes, and the gel appears to run as a smear. However, individual restriction fragments are still well resolved in the gel and RFLPs are still present between DNAs from different organisms.

Since RFLPs of nuclear DNA cannot be directly visualized, the usual procedure is to use small pieces of chromosomal DNA as probes to detect individual restriction fragments. Using the high specificity of DNA–DNA hybridization, such probes can detect individual restriction fragments in the complex mixture of fragments of nuclear DNA present in a restriction digest. To use this technique, a set of chromosomal DNA fragments are prepared as probes. For preparation of probes, DNA is isolated from the species of interest, digested with a restriction enzyme to generate relatively small fragments, usually of 0.5–2.0 kb. Individual restriction fragments are ligated into a bacterial plasmid and the plasmid is transformed into a bacterial cell. By growing these transformed bacteria, one obtains a large supply of a single plant DNA restriction fragments, which are suitable for use as a hybridization probe. Such a set of probes is called a library. Southern blots are prepared from the digested

DNA as explained earlier and then probed with one of the cloned probes from the library. For more details, read the Southern hybridization detection technique in Chapter 14.

The potential of RFLPs as diagnostic markers became evident from studies of the human globin genes, where a direct correlation between the sickle cell mutation carried by a specific β-globin allele and the presence of certain RFLP fragments was evident. The molecular basis of sickle cell anemia is now precisely known. This is not the case for most heritable traits, particularly for those showing quantitative inheritance. For such traits, genetically closely linked markers that do not influence the phenotypic expression of the trait can be identified and they can serve as diagnostic tools. When a genetic linkage map saturated with marker loci is available, the genome of a species can be systematically scanned for markers cosegregating with a specific trait of interest. Because of their abundance, RFLPs are the first class of genetic markers allowing the construction of highly saturated linkage maps. This was first suggested for the human genome by Botstein *et al.* (1980). The potential of this approach in plant breeding was recognized few years later, and its feasibility was experimentally tested in maize and tomato. Since then, an impressive amount of RFLP data has been accumulated in several plant species including important crops. Extensive RFLP linkage maps have been constructed and several genes contributing to agronomically relevant traits have been mapped by RFLP markers.

Construction of RFLP genetic maps

In order to discuss the utility of RFLP markers in genetic mapping, it is useful to think of genetic mapping in a slightly different fashion. Consider a genetic cross in a diploid organism in which two homozygous parents are crossed to produce an F_1, and the F_1 is then selfed to produce an F_2 population. It is clear that the F_1 hybrid from two such parents will have one set of chromosomes from each parent and that the two members of a homologous chromosome pair will be different

from one another to the degree to which the parents differ in DNA base sequence.

When the F_1 plant undergoes meiosis to produce gametes, its chromosomes will undergo recombination by crossing over. This recombination process is the basis of conventional genetic mapping. The chromosome segments on the same (homologous) chromosome will undergo recombination according to a function that depends on their physical distances. Chromosome segments that are close together (closely linked) will undergo recombination less frequently than segments that are further apart. Thus, genetic distance or map distance is defined as a function of the recombination that occurs during gamete formation. Previously, the only way to follow chromosome segments and observe their recombination during gamete formation was to observe the phenotype caused by the action of genes. By observing a phenotype such as flower color, plant height, or insect resistance, the behavior of chromosome segments has been utilized to construct genetic maps of several plant species. However, the method is cumbersome and extremely time consuming.

Fortunately, RFLP markers provide a way to directly follow chromosome segments during recombination and greatly simplify the construction of genetic maps. Instead of looking at the phenotype caused by the presence of a gene on a chromosome segment, one looks directly at RFLP markers on the segments itself. When one considers the inheritance pattern of RFLP markers, it is seen that they segregate exactly in the same way as do conventional gene markers and follow strict Mendelian rules. This means that the segregation pattern of RFLP markers can be analyzed by conventional Mendelian methods and that maps of RFLP markers can be constructed in the same fashion as maps of conventional markers (Fig. 22.2).

To construct RFLP map, a fairly well defined series of steps outlined below is followed. For several crop species such as rice, tomato, barley, and maize, RFLP maps are already

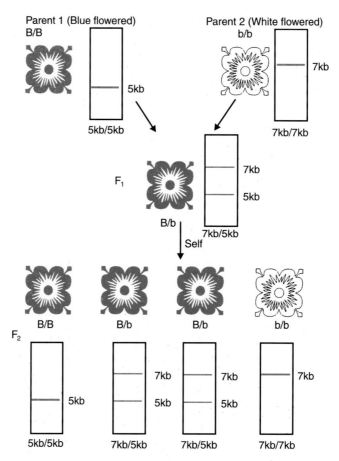

Fig. 22.2 A diagrammatic representation of inheritance of RFLP marker as compared to inheritance of conventional single gene marker controlling flower color.

available and one can simply use them without having to construct a new one.

Select the parent plants. As parent plants in a cross for RFLP map construction, one wants to select plants that are genetically divergent enough to exhibit sufficient RFLPs but not so far distant as to cause sterility of progeny. It would be advantageous also to select a cross in which some desirable agronomic traits are segregating, but this has not always been possible. Since one does not know *a priori* how much RFLP variation will be present in a crop species, it is necessary to make a survey using a random selection of cloned probes. Such a survey should include cultivars or plant introductions thought to represent

the spectrum of variability present in the taxon. Related wild species that can be crossed to the cultivar should also be included. One should utilize any available information to select the plants to be surveyed.

After selecting a range of plants to be surveyed, DNA is isolated and individual plants of each of the accessions are digested with restriction enzymes and screened for polymorphisms by the usual method. Each accession can then be stored for RFLP alleles present, and two accessions showing a usable amount of polymorphism can be selected for crossing.

Produce a mapping population. The

selected parent plants are crossed to produce an F_1 plant(s). In the casoduce an F_1 plant(s). In the case, wheat, or barley, all F_1 plants should be alike and heterozygous for all RFLPs exhibited between the parent plants. A mapping population can be derived by selfing the F_1 to produce an F_2, which is then scored for segregation of the RFLPs or by backcrossing the F_1 to one of the parents and observing the segregation in the first backcross generation. It is better to use an F_2 population if this is possible, as more information can be gained from this than from a backcross population of comparable size. A mapping population of about 50 F_2 or backcross plants is sufficient for a fairly detailed map. With plants in which plant breeding studies are being actively carried out, suitable mapping populations might already be available as remnant seed from previous crosses. If this is the case, it may be possible to make a map without having to make any crosses. This can greatly speed up the process.

Recombination frequencies can also be estimated from doubled haploids (DH) derived from the pollen of F_1 plants. In maize, recombinant inbred (RI) lines have also been produced from a number of crosses through continuous selfing or sib mating in F_2 plants. The crossing scheme is outlined in Fig. 22.3.

Scoring of RFLPs in the mapping population. Once the mapping population is obtained, DNA is isolated from each individual plant in the population. It is important to realize that the chromosome of each plant in the mapping population contains a unique array of parental chromosome segments. To derive a RFLP map it is necessary to determine which parent's chromosome segments would be destroyed by recombination if the F_2 or backcross plants were allowed to reproduce sexually. Every effort should be made to keep the individual plants in the mapping population alive so that repeated DNA extractions can be made, and a large amount of DNA can be accumulated. Mapping proceeds by sequentially scoring RFLPs in the individual plants of the mapping population

by the following series of procedures:

- *Screening for polymorphism:* Probes are selected sequentially from the library and tested against the parents to determine which restriction enzymes will detect a polymorphism between the parents. The number of enzymes required will depend on the amount of variation present in the parent plants. Very polymorphic plants, such as inbred lines of maize, will usually require only one or two enzymes to detect polymorphism with most probes. In rice, 11 enzymes were used and about 75% of the probes were polymorphic with these.

- *Scoring:* When a polymorphic probe/enzyme combination is detected, it can then be scored in the mapping population. To accomplish this, a series of agarose gels must be prepared and filters prepared from them for DNA hybridization. These filters are generally called F_2 survey or mapping filters. There must be one filter set for each restriction enzyme used in the mapping project, and each filter set must contain digested DNA from each individual plant in the mapping population. Thus, one set of mapping filters would contain DNA from each of the F_2 plants digested with *Eco*RI, and another set would contain DNA from each of the plants digested with *Pst*I. If a polymorphism between the parent plants has been detected with a given probe/enzyme combination, then that probe will be scored on the corresponding mapping filter set. For example, if probe No. 36 is found to be polymorphic between the parents using *Pst*I, then it could be scored on the *Pst*I mapping filter set. In an F_2 mapping population scored for a single RFLP, only three types of plants can be present: two homozygous (one similar to each of the parents) and one heterozygous. Thus, each plant in the mapping population can be scored as a heterozygote or as one of the two possible types of homozygotes when tested with each library probe.

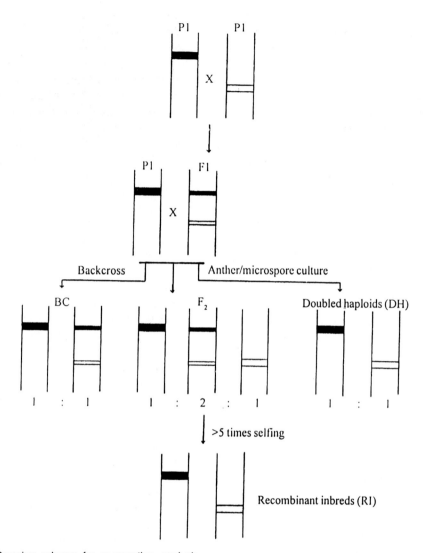

Fig. 22.3. Crossing scheme for segregation analysis.

- *Linkage analysis:* Data accumulated from scoring the mapping population sequentially with probes from the library is used to construct the linkage map. Linkage analysis is based on the degree to which probes tend to cosegregate. The first probe scored in a mapping study will, of course, provide no information on linkage, but beginning with the second probe one can determine whether linkage is indicated. If the second probe is linked to the first one, they will tend to cosegregate (F_2 plants heterozygous for the first probe will also tend to be heterozygous for the second probe; plants homozygous for the allele found in one of the parents for the first probe will also tend to be homozygous for the second probe). If no linkage is indicated, distribution of homozygotes and the heterozygotes for the first two probes will tend to be random. Simple statistical tests, such as a Chi-square analysis, will test for randomness of segregation and hence linkage. The first few probes, because they are randomly selected, are unlikely to be linked and hence to show cosegregation. As one sequentially adds probes, however, linkages will be detected. At first there will be many more linkage groups than there are chromosomes, but the two numbers will tend to converge as more markers are added. As each probe is screened, it is tested for possible linkage to all the other markers mapped before. Thus, when one is testing probe No.100, one can test it against the 99 other probes that have been scored before to see if it is linked to any of them. This is essentially like making a 100-point cross. It is evident that a great amount of data is rapidly accumulated in an RFLP mapping project, and a computer is required for efficient data storage and analysis. When a new probe is scored, if linkage with one or more markers is indicated by the data, the map distance is determined by computer analysis using an algorithm such as maximum likelihood. Recombination data can be converted to map distances using any of the commonly used mapping functions.

Uses of RFLP

1. It permits direct identification of a genotype or cultivar in any tissue at any developmental stage in an environment independent manner.
2. RFLPs are codominant markers, enabling heterozygotes to be distinguished from homozygotes (Fig. 22.4).

3. It has a discriminating power that can be at the species/population (single locus probes) or individual level (multi locus probes).
4. The method is simple as no sequence-specific information is required.
5. Indirect selection using qualitative traits: The greatest potential use of RFLP is its role as indirect selection criteria. It is useful in conjunction with analysis of conventional markers. RFLP maps can be used to supplement regular plant breeding protocols when one wants to select for a conventional gene, but directly selecting for that gene would be expensive, difficult, or time consuming. With indirect selection, one does not directly select for the gene of interest, but rather for one more closely linked RFLP markers. If the RFLP markers are indeed closely linked, they will remain associated with the gene of interest during segregation. This allows one to select for the RFLP marker with confidence that the conventional gene will also be present, since only relatively rare recombination events would separate the two.

Indirect selection for an RFLP marker has advantages in several plant breeding scenarios. One such situation might involve selection for a recessive gene. Suppose one was trying to introgress a recessive gene into a cultivar by using a backcross breeding program. Backcross protocols involve alternate backcross and selection phases. It is necessary to select progeny bearing the desired gene at several points in the backcross cycle. If a gene is dominant it can be directly selected, but recessive genes will

A	B	F1	F2			BC$_A$		DH
▬	▬ ▬	▬				▬ ▬		▬
	▬ ▬		¼	½	¼	½	½	½	½

Fig. 22.4 Segregation patterns and theoretical ratios with a RFLP codominant marker. A and B are the two parents; BC$_A$: back cross with parent A; DH: double haploids.

not be expressed in any of the backcross plants, and it would be necessary to carry out progeny testing (probably by selfing the backcross plants to test for the presence of the recessive gene). A recessive allele could be indirectly selected, however, by selecting for a linked RFLP marker. No progeny testing would be necessary, and the process could be greatly speeded up.

6. *Tagging of monogenic traits with RFLP markers*: In inbreeding species, the availability of nearly isogenic lines (NILs) facilitates detection of RFLP markers linked to monogenic disease-resistance genes. NILs are obtained by crossing a donor parent carrying the resistance gene with a susceptible recipient parent, followed by repeated backcrosses to the recipient parent under continuous selection for the resistance trait. The donor genome is therefore progressively diluted in the introgressed lines until, in the ideal situation, only a short chromosomal segment is retained around the resistance gene. When the parents and the NILs derived from them are compared with molecular probes, only those that are linked to the resistance locus are expected to be polymorphic in the NILs. The linkage and the genetic distance between RFLP markers and the resistance locus are then verified by segregation analysis in normal mapping populations. This technique has been successfully employed for tagging genes of economic value in several crops, and some of the examples are given in Table 22.1.

7. *Indirect selection using quantitative trait loci (QTL)*: Several characters of plant species, among which are traits of agronomic importance, are inherited quantitatively. This type of genetic variation is due to multiple factors acting collectively on the expression of a trait. These loci have been designated the acronym QTL (for quantitative trait loci). Each of the individual genes of such a polygenic system contributes a small positive or negative effect to the trait of interest. Clear dominance is not exhibited and the phenotype has a large environmental component. All these characteristics conspire to make quantitative traits very difficult to analyze.

Table 22.1 Some examples of molecular markers associated with resistance traits in crop plants

Host species	Pathogen	Resistance gene	Marker
Rice	*Pyricularia oryzae*	*Pi*-2(t), *Pi*-4(t)	RFLP
		Pi-10(t)	RAPD
	Xanthomonas oryzae	*Xa*2l	RAPD
		*Xa*3, *Xa*4, *Xa*10	RFLP & RAPD
	Orseolia oryzae	*Gm*2, *Gm*4t	RAPD
Wheat	*Puccinia recondita*	*Lr*9, *Lr* 24	RFLP & RAPD
	Erysiphe graminis	*Pm*1, *Pm*2, *Pm*3	RFLP
	Hessian fly	*H*21	RAPD
Maize	Leaf blight	*rhm*	RFLP
Barley	Stem rust	*Rpg* 1	RFLP
	Barley yellow mosaic	*ym*4	RFLP
	Rhyncosporium secalis	—	RFLP
	and barley mild mosaic virus	—	
	Erysiphe graminis	—	RFLP
Brassica napus	*Leptosphaeria maculans*	—	RFLP
Pea	*Erysiphe polygoni*	*er*	RFLP
Mungbean	Bruchid *Callosobruchus*	—	RFLP
Tomato	*Fusarium oxysporum*	*I*2	RFLP
	Cladosporium fulvum	Cf_2/Cf_5	RFLP
Potato	Cyst nematode	*H*1	RFLP

Conventional Mendelian methods of analysis, which are suitable for single gene traits, cannot be applied to the analysis of quantitative traits, and one is forced to utilize biometrical methods and extensive testing over different years and in different environments in an effort to advance toward the desired state. QTLs are inherited in the same way as the genes of major effects. They segregate, recombine, and exhibit the linkage theory of association of marker loci with QTLs.

Fundamental advances in this area of plant breeding seemed unlikely before the advent of RFLP mapping techniques. As one can use RFLP markers to simultaneously follow the segregation of all chromosome segments during a cross, the basic idea is to look for correlation between the quantitative trait of interest and specific chromosome segments marked by RFLPs. If a correlation exists, then the chromosome segment must be involved in the quantitative trait (one or more genes determining the trait must be on that chromosome segment). The difficult part in the procedure is establishing a correlation between the trait and specific chromosome segments. The RFLP markers are easily scored, but the quantitative trait must be scored in the conventional fashion.

RFLP-QTL linkage relationship can be studied in self-pollinated species and among inbred lines in cross-pollinated species. Parents that differ for a number of RFLPs as well as mean values of the quantitative traits are first selected. These parents are crossed and the F_2 population is obtained. In the F_2 population, a chromosomal segment representing linkage between a RFLP marker and a QTL will be present in the background of random genetic variation due to independent assortment and recombination. By growing samples of F_2 population in different environments, one may also study the linkage under different environments. In these cases if mean values of a particular quantitative trait are determined in two groups of plants representing alternative alleles of RFLP, a significant difference in means (of two groups) for quantitative traits will indicate a RFLP-QTL linkage. Individual marker alleles can be assigned breeding values according to the realized effect of the QTL to which they are linked. The realized effect of the QTL is a function of how large an effect the QTL has, and how tightly it is linked to the marker. The selection can then be practiced simultaneously for a number of markers that will have the effect of selecting for QTLs with a positive effect on the quantitative trait. Lander and Botstein (1989) have elaborated a method for mapping QTLs using RFLP maps, which is aimed at identifying promising crosses for QTL mapping, exploiting the full power of complete linkage maps by the approach of interval mapping of QTLs, and decreasing the number of progenies to be genotyped. Interval mapping assesses the effects of each genome segment, located between a pair of marker loci, rather than the effect of a QTL associated with a single RFLP.

When the estimated phenotypic effect b of a single allele substitution at a putative QTL is known, maximum likelihood equations (MLEs) can be calculated, which maximize the probability that the observed data will have occurred. These MLEs are compared to MLEs obtained under the assumption that no QTL is linked ($b = 0$), and the evidence that a QTL is existing is indicated by the LOD score (log of the score of odds for the presence of a linkage against the odds of not having this linkage, e.g. if the score is 100:1 in favor of linkage). The LOD score is the ratio of the probability that the data have arisen assuming the presence of a QTL to the probability that no QTL is present. If the LOD score exceeds a predetermined threshold T, a QTL is accepted as being present.

Linkage between RFLP and QTL can also be used to predict combining ability. As

RFLPs that are positively correlated for important traits can be identified, inbred lines with RFLPs that should give high general combining ability can be selected.

Problems

1. Conventional RFLP analysis requires relatively large amount of highly pure DNA.
2. A constant good supply of probes that can reliably detect variation are needed.
3. It is laborious and expensive to identify suitable marker/restriction enzyme combinations from genomic or cDNA libraries where no suitable single-locus probes are known to exist.
4. RFLPs are time consuming as they are not amenable to automation.
5. RFLP work is carried out using radioactively labeled probes and therefore requires expertise in autoradiography.

PCR-BASED TECHNIQUES

Random amplified polymorphic DNA (RAPD) markers

RAPD analysis is a PCR-based molecular marker technique. Here, single short oligonucleotide primers are arbitrarily selected to amplify a set of DNA segments distributed randomly throughout the genome. A number of closely related techniques based on this principle were developed almost at the same time and are collectively referred to as multiple arbitrary amplicon profiling (MAAP). These techniques are:

RAPD	Randomly amplified polymorphic DNA
DAF	DNA amplifying fingerprinting
AP-PCR	Arbitrarily primed polymerase chain reaction

All these techniques refer to DNA amplification using single random primers and share the same principle with some differences in the experimental details. Of these, RAPD was the first to become available and is most commonly and frequently used. Williams *et al.* (1990) showed that the differences as polymorphisms in the pattern of bands amplified from genetically distinct individuals behaved as Mendelian genetic markers. Welsh and McClelland (1990) showed that the pattern of amplified bands so obtained could be used for genomic fingerprinting.

RAPD amplification is performed in conditions resembling those of PCR, using genomic DNA from the species of interest and a single short oligonucleotide (usually a 10-base primer). The DNA amplification product is generated from a region that is flanked by a part of 10-bp priming sites in the appropriate orientation. Genomic DNA from two different individuals often produces different amplification patterns [randomly amplified polymorphic DNAs (RAPDs)]. A particular fragment generated for one individual but not for other represents DNA polymorphism and can be used as a genetic marker. Using different combinations of nucleotides, many random oligonucleotide primers have been designed and are commercially available. Such primers can be synthesized in an oligonucleotide synthesizing facility based on sequences chosen at random. No separate information is required from the plant to be studied. The choice of single primers (or RAPD primers) to use is done operationally. Since each random primer will anneal to a different region of the DNA, theoretically many different loci can be analyzed.

The PCR reaction typically requires cycling among three temperatures, the first to denature the template DNA strands, the second to anneal the primers, and the third to extend at temperatures optimal for *Taq* polymerase. This cycle is usually repeated 25 to 45 times (Fig. 22.5).

It is usually sufficient to heat the reaction mixture at 94°C for 30 to 60s. However, the initial denaturation step for the plant DNA should be for at least 3 min. Insufficient denaturation is a common problem leading to the failure of the PCR reaction.

The temperature for the annealing step will depend to some extent on base composition and the length of primers. Generally, an annealing

DNA isolation

↓ *Taq* DNA polymerase,
  primers, dNTPs
  and buffer

Keep the tubes in PCR thermocycler

↓ Denature DNA
  (94°C, 1 min)

DNA strands separated

↓ Annealing of primers
  (36°C, 2 min)

Primers annealed to template DNA strands

↓ DNA synthesis
  (72°C, 1.5 min)

Complementary strand synthesis

↓ 35 to 45 cycles

Amplified products separated by gel electrophoresis

Fig. 22.5. An outline of RAPD analysis.

temperature of 35–40°C is used in PCR reactions for RAPD analysis. Annealing may require only a few seconds, but this will depend on a variety of factors such as interference from secondary structure and primary concentration. A time in the range of 30 s to 1 min is usually sufficient.

The extension temperature depends on length and concentration of the target sequence, and in general 72°C temperature is used. As a thumb rule, 1 min per kilo base is probably ample. After cycling, an extended extension period of 5–10 min is included to ensure that all annealed templates are fully polymerized.

Most genomes from bacteria to higher plants

to human beings contain enough perfect or imperfect binding sites for a short primer of arbitrary sequence. These markers cover the entire genome, that is, single to repeated sequences. Binding of primer to the complementary sequences allows the amplification of several bands. A DNA sequence difference between individuals in a primer-binding site may result in the failure of the primer to bind, and hence in the absence of a particular band among amplification products. The reaction products are conveniently analyzed on agarose gels, stained with ethidium bromide, and seen under UV light.

Presence of a RAPD band corresponds to a dominant allele against absence of band that corresponds to a recessive allele. Thus, heterozygous and homozygous dominant individuals cannot be differentiated with RAPD markers (Fig. 22.6). The segregating F_2 population may therefore be scored as follows:

Band present   AA (homozygote)
                         Aa (heterozygote)
Band absent   aa (recessive homozygote)

In F_2, these two classes would be expected to segregate in a 3:1 ratio. This causes a loss of information relative to RFLP markers which show codominance. Backcross, recombinant inbred, and the doubled haploid populations do not suffer this loss of mapping information, because the complete information available from a backcross can be obtained from scoring the presence or absence of a polymorphic marker. Keeping in mind this fact, it is easy to select the populations best suited for the construction of genetic maps with RAPD markers

A comparison of the properties of RAPDs in relation to other markers is given in Table 22.2.

A	B	F1	F2		BC$_A$	BC$_B$	DH	
—	—	—			—	—	—	
			¾	¼	1	½	½	½	½

Fig. 22.6 Segregation patterns and theoretical ratios with a RAPD dominant marker. A and B are the two parents; BC$_A$: backcross with parent A; BC$_B$: backcross with parent B; DH: double haploids.

Table 22.2 A comparison of properties of restriction fragment length polymorphisms and randomly amplified polymorphic DNAs

Characteristic	RAPD	RFLP	AFLP	Sequence-tagged microsatellites
Principle	DNA amplification	Restriction digestion	DNÀ amplification	DNA amplification
Detection	DNA staining	Southern blotting	DNA staining	DNA staining
DNA required-quantity	Crude	Relatively pure	Relatively pure	Crude
DNA required-quantity	Low	High	Medium	Low
Primer requirement	Yes (random primer)	None	Yes (selective primer)	Yes (selective primer)
Probe requirement	None	Set of specific probes	None	None
Use of radioisotopes	No	Yes	Yes-no	Yes-no
Part of genome surveyed	Whole genome	Generally low copy coding region	Whole	Whole
Dominant/codominat	Dominant	Codominant	Dominant (codominant)	Codominant
Polymorphism	Medium	Medium	Medium	High
Automation	Yes	No	Yes	Yes
Reliability	Intermediate	High	High	High
Recurring cost	Low	High	Medium	Low

The following characteristics of RAPD markers make them relevant for marker-assisted selection.

Advantages

1. Need for a small amount of DNA (15–25 ng) makes it possible to work with populations which are inaccessible for RFLP analysis.
2. It involves nonradioactive assays.
3. It needs a simple experimental set-up requiring only a thermocycler and an agarose assembly.
4. It does not require species-specific probe libraries; thus, work can be conducted on a large variety of species where such probe libraries are not available.
5. It provides a quick and efficient screening for DNA sequence-based polymorphism at many loci.
6. It does not involve blotting or hybridization steps.

Application

RAPD markers have been successfully used for the following applications:

1. *Construction of genetic maps:* Their potential as genetic markers for map construction has been exploited and maps consisting largely or exclusively of RAPD markers have been developed in *Arabidopsis*, pine, *Helianthus*, etc.
2. *Mapping of traits:* The use of RFLPs for selection of markers linked to monogenic and polygenic traits has been discussed. Likewise, RAPD markers may also be used for indirect selection in segregating populations during plant breeding programs. This technique has been employed for tagging genes of economic value, examples of which are given in Table 22.1.
3. Analysis of the genetic structure of populations.

4. Fingerprinting of individuals.

5. Targeting markers to specific regions of the genome.

6. Identification of somatic hybrids (see Chapter 9).

7. A useful system for evaluation and characterization of genetic resources. Some success has already been achieved in characterization of collections from tree genera *Gliricidia* and *Thenobroma.*

Limitations

1. RAPD polymorphisms are inherited as dominant-recessive characters. This causes a loss of information relative to markers which show codominance.

2. RAPD primers are relatively short, a mismatch of even a single nucleotide can often prevent the primer from annealing; hence, there is loss of band.

3. The production of nonparental bands in the offspring of known pedigrees warrants its use with caution and extreme care.

4. RAPD is sensitive to changes in PCR conditions, resulting in changes to some of the amplified fragments.

DNA amplification fingerprinting (DAF)

A single arbitrary primer, as short as 5 bases, is used to amplify DNA using PCR. DAF requires careful optimization of parameters; however, it is extremely amenable to automation and fluorescent tagging of primers for early and easy determination of amplified products. DAF differs from RAPD in following respects:

i. Higher primer concentration are used in DAF.

ii. Shorter primers of 5–8 nucleotides are used in DAF.

iii. Two temperature cycles in DAF as compared to three in RAPD.

iv. DAF usually produces a very complex banding pattern.

v. Products of DAF are analyzed on acrylamide gels and detected by silver staining.

Arbitrarily primed polymerase chain reaction (AP-PCR)

This is a special case of RAPD wherein discreet amplification patterns are generated by employing single primers of 10–50 bases in length in PCR of genomic DNA. In the first two cycles, annealing is under nonstringent conditions. AP-PCR differs from RAPD in following respects:

i. In AP-PCR the amplification is in three parts, with its own stringency and concentration of constituents.

ii. In AP-PCR higher primer concentrations are used in the first PCR cycles.

iii. Primers of variable length, and often designed for other purposes, are arbitrarily chosen for use (M13 universal sequence primer).

iv. Products of AP-PCR are mostly analyzed on acrylamide gels and detected by autoradiography.

Amplified fragment length polymorphism (AFLP)

This is a highly sensitive method for detecting polymorphism throughout the genome, and it is becoming increasingly popular. It is essentially a combination of RFLP and RAPD methods, and it is applicable universally and is highly reproducible. It is based on PCR amplification of genomic restriction fragments generated by specific restriction enzymes and oligonucleotide adapters of few nucleotide bases (Vos *et al.,* 1995). It is a novel DNA fingerprinting technique. DNA fingerprinting involves the display of a set of DNA fragments from a specific DNA sample. Fingerprints are produced without prior sequence knowledge, using a limited set of genetic primers. AFLP technique uses stringent reaction conditions for primer annealing and combines the reliability of RFLP technique with the power of PCR technique.

In principle, AFLP involves the following steps:

1. DNA is cut with restriction enzymes (generally by two enzymes), and double-stranded (ds) oligonucleotide adapters are ligated to the ends of the DNA fragments.

*Eco*RI adapter

*Mse*I adapter

5' – CTCGTAGACTGCGTACC
          CATCTGACGCATGGTTAA – 5'

5' - GACGATGAGTCCTGAG
          TACTCAGGACTCAT – 5'

AFLP primers consist of three parts, a core sequence, an enzyme specific sequence (ENZ) and a selective extension (EXT). This is illistrated below :

	CORE	**ENZ**	**EXT**
*Eco*RI	5' - GACTGCGTACC	AATTC	NNN --3'
*Mse*I	5' - GATGAGTCCTGAG	TAA	NNN --3'

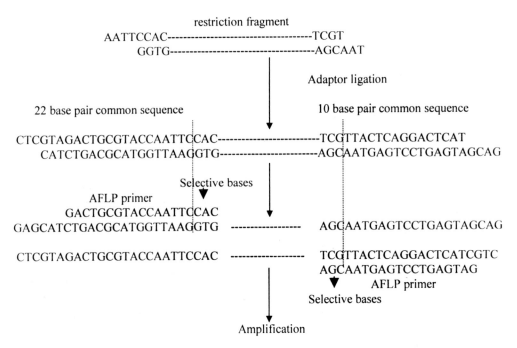

restriction fragment

AATTCCAC-----------------------------------TCGT
  GGTG----------------------------------AGCAAT

Adaptor ligation

22 base pair common sequence   10 base pair common sequence

CTCGTAGACTGCGTACCAATTCCAC-------------------------TCGTTACTCAGGACTCAT
CATCTGACGCATGGTTAAGGTG-----------------------AGCAATGAGTCCTGAGTAGCAG

Selective bases

AFLP primer
          GACTGCGTACCAATTCCAC
GAGCATCTGACGCATGGTTAAGGTG ------------------ AGCAATGAGTCCTGAGTAGCAG

CTCGTAGACTGCGTACCAATTCCAC ------------------ TCGTTACTCAGGACTCATCGTC
                                         AGCAATGAGTCCTGAGTAG
                                         AFLP primer
Selective bases

Amplification

Fig. 22.7. Schematic representation of the AFLP technique.

2. Selective amplification of sets of restriction fragments is usually carried out with ^{32}P-labeled primers designed according to the sequence of adaptors plus 1–3 additional nucleotides. Only fragments containing the restriction site sequence plus the additional nucleotides will be amplified.

3. *Gel analysis of the amplified fragments.* The amplification products are separated on highly resolving sequencing gels and visual-ized using autoradiography. Fluorescent or silver staining techniques can be used to visualize the products in cases where radiolabeled nucleotides are not used in the PCR.

A schematic representation of the technique is shown in Fig. 22.7. The restriction fragments for amplification are generated by two restriction enzymes, a rare cutter and a frequent cutter. The figure shows that fragments are generated from

*EcoR*I and *Mse*I. Oligonucleotides are generally used as adapters and primers. AFLP adapters consist of a core sequence and an enzyme specific sequence.

The restriction fragments with their 5′ protruding ends are obtained and then adapters are ligated. The sequence of the adapters and the adjacent restriction site serves as primer-binding sites for subsequent amplification of the restriction fragments. Selective nucleotides are included at the 3′ ends of PCR primers, which therefore can only prime DNA synthesis from a subset of the restriction sites. Only those restriction fragments in which nucleotides flanking the restriction site match the selective nucleotides will be amplified.

The AFLP procedure results in predominant amplification of restriction fragments having a rare cutter sequence on one end and a frequent cutter sequence on the other end.

The method allows specific coamplification of a high number of restriction fragments. Typically 50–100 restriction fragments are amplified and detected on denaturing polyacrylamide gels.

This method generates a large number of restriction fragment bands, thereby facilitating the detection of polymorphism. Choosing the different base number and composition of nucleotides in the adapters can control the number of DNA fragments that are amplified. Although not many maps using AFLP markers have been developed so far, this method is now widely used for developing polymorphic markers. AFLP is a technique that detects genomic restriction fragments and resembles RFLP in this respect, with the major difference that PCR amplification instead of Southern blotting is used for detection of restriction fragments. In general, there is a linear correlation between the number of amplified fragments and genome size. Most AFLP fragments correspond to unique positions on the genome and hence can be exploited as landmarks in genetic and physical maps, each fragment being characterized by its size and its primers required for amplification. A comparison of the properties of AFLPs in relation to other markers has been shown in Table 22.2.

Advantages

1. This technique is extremely sensitive.
2. It has high reproducibility, rendering it superior to RAPD.
3. It has widescale applicability, proving extremely proficient in revealing diversity.
4. It discriminates heterozygotes from homozygotes when a gel scanner is used.
5. It permits detection of restriction fragments in any background or complexity, including pooled DNA samples and cloned DNA segments.
6. It is not only a simple fingerprinting technique, but can also be used for mapping.

Disadvantages

1. It is highly expensive and requires more DNA than is needed in RAPD (1 mg per reaction).
2. It is technically more demanding than RAPDs, as it requires experience of sequencing gels.
3. AFLPs are expensive to generate as silver staining, fluorescent dye, or radioactivity detect the bands.

Simple sequence repeats (microsatellites)

The term *microsatellites* was coined by Litt and Lutty (1989). Simple sequence repeats (SSRs), also known as microsatellites, are present in the genomes of all eukaryotes. These are ideal DNA markers for genetic mapping and population studies because of their abundance. These are tandemly arranged repeats of mono-, di-, tri-, tetra- and penta-nucleotides with different lengths of repeat motifs (e.g. A, T, AT, GA, AGG, AAAC, etc.). These repeats are widely distributed throughout plant and animal genomes that display high levels of genetic variation based on differences in the number of tandemly repeating units at a locus (Table 22.3). These SSR length polymorphisms at individual loci are detected by PCR, using locus-specific flanking region primers where the sequence is known (discussed in the following section under sequence tagged microsatellites). Thus, STMs require precise DNA sequence information for each marker locus from which a pair of identifying flanking markers are designed. This is impractical for many plant and animal species that are not well-characterized genetic systems. Nevertheless, high levels of polymorphism inherent to SSRs has prompted an alternative approach: the detection of SSR-derived polymorphisms without the need of sequencing. The approaches described below all involve PCR amplification fingerprinting with oligonucleotide primers corresponding directly to simple sequence repeat target sites. Some of these SSR-based methods have been collectively termed *microsatellite-primed PCR (MP-PCR)* and shown in Fig. 22.8.

Table 22.3 List of microsatellites or simple sequence repeats in plants

Species	Repeats
Arabidopsis	AT,AG,GA,CT,TC,CA
Barley	AT,CT,TG,TCT,CTT,ATTT
Maize	CT,CA,AC,AG,GA,AG
Rice	CA,GT,GA,AT,GGT
Sorghum	AG,AC,AAC,AAG
Wheat	GA,GT,CT,CA,AT,GT
Soybean	AT,CT,TA.ATT,AAT,TAT,TAA
Sunflower	CA
Tomato	GT,GA,ATT,GATA
Grape vine	GA,GT,AG
Citrus	ATT,TAT,TAG
Yam	CT,AT,TA,CA

Microsatellite-directed PCR: unanchored primers: This approach is also known as SPAR (single primer amplification reaction). Oligonucleotides composed wholly of defined, short tandem repeat sequences and representing a variety of different microsatellite types have been used as generic primers in PCR-amplification-based application. Most often, a SSR oligonucleotide is used in a single primer amplification reaction (Fig. 22.8). The microsatellite repeat primed reaction is similar to the RAPD reaction in that exponential amplification will occur from these single primer reactions only

when the particular repeat used as a primer is represented in multiple copies, which are closely spaced (<2–3 kb) and inversely oriented in the template DNA. These reactions are multiplexed, multiple products are simultaneously coamplified, and therefore mutiple loci can be detected from a genome using a single PCR reaction. Apart from the SSR sequence used as a primer, there is no other DNA sequence information required to design and perform these amplifications. The banding pattern after amplification can be resolved on low-resolution agarose gels, using ethidium bromide staining. The same set of primers can be used across species and even kingdoms, and therefore constitutes a general purpose primer set. It has been reported that MP-PCR reactions can be more specific than RAPD reactions as the longer SSR-based primers (15-mers to 20-mers) enable higher stringency amplifications. This may result in fewer problems with reproducibility, a complaint frequently leveled against the low stringency RAPD assays with shorter primers.

This approach is simple to perform and requires little upfront locus-specific or species-specific information. The unanchored SSR primers appears to behave during PCR much like a RAPD primer with considerable dependence on specific amplification conditions. SPAR appears to be subject to irreproducibility effects and requires careful optimization of the conditions for specific primer genome combinations to ensure that heterogeneity of banding patterns observed is the result of true genetic polymorphism at fixed loci.

Microsatellite-directed PCR: anchored primers: This method is known variously as anchored microsatellite primed PCR (AMP-PCR), inter-microsatellite PCR, and inter SSR amplification (ISA or ISSR). In this approach, radiolabeled di- or trinucleotide repeats (SSR primers) are modified by the addition of either 3' or 5' anchor sequences of 2–4 nucleotides composed of nonrepeat bases (Fig. 22.8; Zietkiewicz et al., 1994). The anchored primers can be $(CA)_8RG$, $(CA)_8RY$, $(CA)_7RTCY$ (where Y = pyrimidine; R = purine). One of the first examples of the use of anchored simple

Fig. 22.8. Schematic representation of different microsatellite primed PCR methods. RE - restriction enzyme; Ad-synthetic adaptor; 10-mer-arbitrary sequence decamer.

sequence repeat primers was for first-strand cDNA synthesis with oligo-d(T) primers carrying one or two non-T bases at the 3' end. The anchor in the primer serves to fix the annealing of the primer to a single position at each target site on the template, such that every new polymerization event initiates at the same target position. Thus, there is little or no chance for primer slippage on the template. It has been reasoned that since dinucleotide repeats are most abundant in eukaryotic genomes than other types of microsatellites and at least $(CA)_n : (GT)_n$ repeats appear to be ubiquitously represented in genomes, a small defined set of primers representing these repeats can be developed that would produce complex fingerprints.

Random amplified microsatellite polymorphisms (RAMP): This approach is based on the random distribution of nucleotide sequences immediately flanking a simple sequence repeat (Wu *et al.*, 1994). Amplification is performed between a 5'-anchored mono-, di-, or trinucleotide repeat primer and a random sequence decamer (RAPD primer) to generate coamplified products capable of carrying codominant length polymorphism that may be present at the SSR target site (Fig. 22.8). The RAPD primer binding site serves as an arbitrary endpoint for the SSR-based amplification product, and, therefore, the products obtained are not as restricted by the relative genomic position of SSRs in the genome as they are with ISA. Individual combinations of different SSR-based primers with different decamers allows for the generation of nearly limitless numbers of unique amplification products, all of which are bordered at one end by a microsatellite repeat. The SSR primer is ^{32}P labeled and the amplification products are resolved on polyacrylamide gels. The RAMP method was shown to produce polymorphic coamplification products in *Arabidopsis*. Both dominant and codominat polymorphisms were detectable.

Selective amplification of microsatellite polymorphic loci (SAMPL): It is an SSR-based modification of AFLP approach, termed selective amplification of microsatellite polymorphic loci (Vogel and Morgante, 1995). AFLP is a PCR-based method for detecting differences in a selected set of restriction fragments and involves the following steps: (1) restriction digestion of total genomic DNA with two endonucleases (one a rare cutter and the other a frequent cutter); (2) ligation of adapters of known sequences to the two ends of each fragment, and (3) selective amplification of a set of anonymous restriction fragments by using primers complementary to the sequences, including adapter sequence + restriction site sequence + selective nucleotides at the 3′ end. Amplification may involve two steps: (1) preamplification, using primers with fewer (0 or 1) selective oligonucleotides and (2) amplification, using primers with three selective nucleotides. In the SAMPL approach, amplification is performed with one labeled SSR primer (anchored through the use of compound repeats) and one unlabeled adapter primer (Fig. 22.8). The SAMPL method uses the same adapter modified restriction fragment templates as those for AFLP, but the amplification is performed using a ^{32}P- or ^{33}P-labeled SSR primer combined with an unlabeled adaptor primer. In SAMPL, during second amplification, one AFLP primer with three selective nucleotides is used in combination with a SAMPL primer that is at least in part complementary to microsatellite sequences. Generally, the SAMPL primer (18–20 nucleotides long) is based on the sequences of two different adjacent SSRs and an associated intervening sequence, known to be found in compound repeats.

Many different combinations of restriction endonucleases, SSR primers, and selective adapter primers are possible with SAMPL, allowing for the detection of nearly a limitless number of SSR-based polymorphisms in the genome. Since SSRs are ubiquitously present in eukaryote genomes, complex fingerprints have been reported in both plant and animal genomes. Both dominant and codominant polymorphisms have been detected in plant species.

It is not possible to state definitely which of the methods would be most suitable for every application. Unanchored SSR primer approach requires a heavy note of caution, since tri-tetra- and pentanucleotide-based oligonucleotides have been shown to exhibit a great deal of nonspecificity, and hence the bands generated may not be SSR based and may not be reproducible. Inter-repeat amplifications with anchored primers are well suited for fingerprinting and germplasm discrimination, but the bands produced are always limited to SSRs within amplifiable distances of one another in the genome. For map construction and trait marking, either RAMP or SAMPL involving amplification between a single SSR and an arbitrary fixed site will allow for detection of nearly limitless numbers of SSR-based markers. SAMPL may be well suited for detecting large amounts of diverse SSR-based codominant polymorphism, although the assay itself is more technically demanding than any other method.

TARGETED PCR AND SEQUENCING

An approach opposite to the arbitrarily amplicon profiling procedure is to design primers to target specific regions of the genome. This approach is a targeted PCR and sequencing. The targeted product can be compared to the corresponding product from another individual on an agarose gel, but the resolution achieved will only detect differences in length of fragment resulting from many base pair changes. To resolve all the possible sequence differences, sequencing the entire fragment becomes necessary. A *sequence tagged site (STS)* is the general term given to a marker that is defined by its primer sequences.

Sequence tagged sites

STS is a short unique sequence (60–1000 bp) that can be amplified by PCR, which identifies a known location on a chromosome (Olson *et al.*, 1989). Specific PCR markers that match the nucleotide sequence of the ends of a DNA fragment can be derived from primers, for example, an RFLP probe or an expressed sequence tag. To date, all STSs that have been used in mapping projects have been derived from well-characterized probes or sequences. STSs are the physical DNA landmarks and PCR is the experimental method used to detect them. STS maps simply represent the relative order and spacing of STSs within a region of DNA. Using this technique, tedious hybridization procedures involved in RFLP analysis can be overcome. STSs have been extensively used for physical mapping of genome. Examples of STSs are

    Sequence tagged microsatellites (STMs)
    Sequence characterized amplified regions (SCARs).
    Cleaved amplified polymorphic sequence (CAPS).

Sequence tagged microsatellites (STMs)

The term *microsatellites* was coined by Litt and Lutty (1989). Simple sequence repeats, also known as microsatellites, are present in the genomes of all eukaryotes. These are ideal DNA markers for genetic mapping and population studies because of their abundance. These are tandemly arranged repeats of mono-, di-, tri- and tetranucleotides with different lengths of repeat motifs (e.g. A, T, AT, GA, AGG, AAAC, etc.). A motif is A, AT, AGG, etc. and repeat number is denoted by *n*. Thus a repeat $(AT)_9$ means AT nucleotides are tandemly arranged one after another nine times. In a genome of a particular species when this repeat is identified in a gene, which constitues a microsatellite, the gene is sequenced with its flanking sequences to design primers for amplification of microsatellites. The regions flanking the microsatellite are generally conserved among genotypes of the same species. PCR primers to the flanking regions are used to amplify the SSR-containing DNA fragment. Length polymorphism is created when PCR products from different individuals vary in length as a result of variation in the number of repeat units in the SSR. Genebank sequence data have also been used for designing primers for amplification of microsatellites. Thus, SSR polymorphism (SSRP) reflects polymorphism based on the number of repeat units in a defined region of the genome.

Poly G and poly A are the simplest of microsatellites. It now appears that any short nucleotide motif can be found as a microsatellite in the eukaryotic genome. The number and composition of microsatellite repeats differ in plants and animals. The frequency of repeats longer than 20 bp has been estimated to occur every 33 kb in plants, whereas in mammals it is found to occur every 6 kb. In humans, AC or TC are very common repeats; in plants, AT is more common, followed by AG or TC. In general, plants have about ten times less SSRs than do humans. Microsatellites have been identified in several plant species such as *Arabidopsis thaliana*, *Brassica napus*, rice, maize, wheat, barley, etc. Some of the SSRs in plants are listed in Table 22.3 (rev Mohan *et al.*, 1997).

Procedure

A specific microsatellite contained within a stretch of DNA can be amplified by PCR using flanking primer sequences, and then analyzed on Metaphor agarose or polyacrylamide gels. The gels are stained with ethidium bromide and seen under UV light. The principle underlying the detection of SSR length polymorphism among parents and their F_1 hybrid is illustrated in Fig. 22.9. The variation in length of the PCR product is a function of the number of SSR units.

This is a relatively new technique and is specifically useful in inbreeding crops such as wheat and barley, which are characterized by low levels of RFLP variation. A comparison of the properties of STMs in relation to other markers has been shown in Table 22.2.

Fig. 22.9. The principle involved in the detection of SSR length polymorphism.

Advantages

1. This technique is very simple and easy to use.
2. Assay is PCR-based. It is sufficient to merely separate the amplification products by electrophoresis to observe the results. This reduces the time required considerably to obtain a result comparable to the methods that are based on Southern blotting.
3. The use of radioisotopes can be avoided because the size polymorphism between alleles is frequently large enough to be seen in agarose gels.
4. Microsatellites segregate as codominant markers.
5. Polymorphism generated is highly reproducible.
6. They are perfectly suited for use in map-based cloning because identification of marker anchored clones is conveniently carried out by PCR.

Disadvantages

The major disadvantage of microsatellites is the cost of establishing polymorphic primer sites and the investment in synthesizing the oligo nucleotides.

Sequence characterized amplified regions (SCARs)

A SCAR is a genomic DNA fragment at a single genetically defined locus that is identified by PCR amplification using a pair of specific oligonucleotide primers. Williams *et al.* (1991) converted RFLP markers into SCARs by sequencing two ends of genomic DNA clones and designing oligonucleotide primers based on the end sequences. These primers were used directly on genomic DNA in a PCR reaction to amplify the polymorphic region. If no amplified fragment length polymorphism is noticed, then the PCR fragments can be subjected to restriction digestion to detect RFLPs within the amplified fragment. SCARs are inherited in a codominant fashion in contrast to RAPDs, which are inherited in a dominant manner. Paran and Michelmore (1993) converted RAPD markers into SCARs. Amplified RAPD products are cloned and sequenced. The sequence of primers derived from the termini of a band is identified as a RAPD marker. Two 24-base oligonucleotide primers corresponding to the ends of the fragment (the 5′ ten bases are the same as the original 10-mer used in the RAPD reaction and 14 internal bases from the end) have been synthesized. These primers with their increased specificity generally amplify a single highly repeatable band, not the 5–10 bands for the progenitor 10 base primers. SCARs are similar to STSs, but do not involve DNA hybridization for detection and can therefore contain repeated DNA sequences.

SCARs have several advantages over RAPD markers. RAPDs show dominant nature, amplification of multiple loci, and are sensitive to reaction conditions. The mapping efficiency of RAPD markers in F_2 populations is decreased by their dominant nature. The conversion of dominant RAPDs to codominant SCARs increases the amount of information per F_2 individual. As the annealing conditions for SCARs are more stringent than for RAPDs, SCAR primers detect only one locus. Also, the use of longer oligonucleotide primers for SCARs allows a more reproducible assay than the one obtained with the short primers

used for RAPD analysis. SCARs can readily be applied to commercial breeding programs as they do not require the use of radioactive isotopes.

Cleaved amplified polymorphic sequences (CAPS)

This is also sometimes known as PCR-RFLP. In this technique, PCR-amplified DNA fragments are digested with a restriction enzyme to reveal restriction site polymorphisms. The digested restriction fragments are subjected to agarose gel electrophoresis followed by staining with ethidium bromide for visualization of bands. Konieczny and Asubel (1993) first adapted the CAPS procedure for genetic mapping by developing a set of CAPS markers for use with *Arabidopsis*. They designed 18 sets of primers, which amplified products ranging in size from 0.316 to 1.728 kb from Columbia and Landsberg ecotypes. These amplified products were digested with a panel of restriction endonucleases to identify restriction enzymes that generated ecotype specific patterns.

Advantages

1. The sizes of the cleaved and uncleaved amplification products can be adjusted arbitrarily by the appropriate placement of the PCR primers.
2. CAPS markers can be readily assayed using standard agarose gel electrophoresis.
3. CAPS markers are codominant.
4. A small quantity of DNA is required to assay CAPS markers.
5. CAPS mapping procedure is technically simple but the results are robust because an amplification product is always obtained.
6. CAPS markers can be assayed relatively quickly.

Advantages of targeted PCR and sequencing approaches

1. The fragment in which polymorphisms are studied is of known identity, therefore avoiding ambiguities of analyzing RAPD and AFLP bands or random probes.

2. Data based on DNA sequences or restriction site mapping provide both the means of classifying individuals into different classes and assessing relationships among the classes.

3. For population studies, the use of an organellar sequence in complementation with a nuclear sequence can provide useful data with respect to mechanism of differentiation, gene flow, and dispersal.

4. The development of genealogical-based analytical methods, coupled with studies of DNA sequence variation within and among populations, is likely to reveal immense information on demographic processes.

Disadvantages

1. Unless the frequency of variants is high enough to be easily detected by PCR-RFLP or other sensitive gel assays, sequencing is required, which in turn necessitates investment of adequate resources and experienced researchers.

2. Although the quality of data is high because the approach is often resource intensive, the coverage of the genome is highly restricted, usually to only one sequence and, therefore, to one point of comparison.

3. There is difficulty in identifying sequences that are reliably variable enough at the population level in any system under study.

FINGERPRINTING

The term *DNA fingerprinting* was coined for multilocus DNA profiles detected by using multilocus hypervariable probes, which simultaneously score genotypes at several loci without recognizing the allelic relationships of the individual DNA fragments (Jeffreys *et al.*, 1985). DNA typing methods have been widely used in criminal investigation and for determining the pedigree and parentage of an individual. The technique of classical DNA fingerprinting is a derivative of RFLP analysis, but differs from the latter by the type of hybridization probe applied to reveal genetic polymorphisms. For a typical multilocus DNA fingerprint, probes are used which create complex banding patterns by recognizing multiple genome loci simultaneously.

For characterization of genome (fingerprinting) a variety of molecular marker techniques are available, which are either hybridization based or PCR based. The markers are RFLP, RAPD, AP-PCR, DAF, microsatellites, AFLP, etc. These molecular marker techniques are used as such or have been modified to reveal polymorphism. It has been reported that techniques that give high information content (estimated as either the polymorphic information content or genetic diversity index or expected heterozygosity) should be used.

RFLP approach has been widely used in fingerprinting of individuals. The Southern blot restriction fragment length polymorphism method of DNA typing and population databases are available, although PCR-based databases are also coming up for DNA typing.

The AFLP approach is widely used for developing polymorphic markers. The high frequency of identifiable AFLPs coupled with high reproducibility makes this technology an attractive tool for detecting polymorphism and fingerprinting. This technique is extremely useful in detection of polymorphisms between closely related genotypes. AFLP analysis depicts unique fingerprints regardless of the origin and complexity of the genome.

A hybridization-based approach is oligonucleotide fingerprinting. It reveals multiple genome loci simultaneously. Each of these loci is characterized by more or less regular arrays of tandemly repeated DNA motif that occur in different numbers at different loci. Two categories of such multilocus probes are mainly used. The first category comprises cloned DNA fragments or synthetic oligonucleotides, which are complementary to minisatellites (tandem repeats of a basic motif about 10 to 60 bp long). The second category of probes is short, synthetic oligonucleotides, which are complementary to microsatellites (tandem repeats of one to five bp).

RFLP analysis with microsatellite-complementary oligonucleotides is also referred to as oligonucleotide fingerprinting or hybridization-based microsatellite fingerprinting. Both minisatellite and microsatellite probes have been applied to RFLP fingerprinting of numerous plant and animal species. High levels of polymorphism between related genotypes are often observed, and the technique is being used for genome analysis, paternity testing, genotype identification, and population genetics. The procedure for oligonucleotide fingerprinting of plants is described below.

1. Isolation of genomic DNA suitable for RFLP analysis.
2. Complete digestion of genomic DNA with an appropriate restriction enzyme.
3. Electrophoretic separation of restriction fragments in agarose gels.
4. Denaturation and immobilization of the separated DNA fragments within the gel or blotting the DNA fragments on a membrane.
5. Hybridization of the dried gel or the membrane to radioactive or non-radioactive labeled microsatellite complementary oligonucleotide probes.
6. Detecion of hybridizing fragments (i.e. fingerprints) by autoradiography or by chemiluminescence.
7. Evaluation of banding patterns and documentation by photography.

This technique of in-gel hybridization offers seveal advantages over Southern hybridization and has been increasingly utilized to reveal hypervariable target regions in a variety of plant materials. The quality, intensity, and number of fragments that hybridize with synthetic oligonucleotides vary considerably from probe to probe and from species to species. Since SSRs are abundant and uniformily distributed in the genome, a large number of hybridized fragments, differing in size (few hundred base pairs to as many as 8–10 kb), is generally obtained. Therefore, they may be used as multilocus probes for identification and characterizaton of varieties, cultivars, and breeding lines. Multilocus probes have the advantage of revealing polymorphism at many loci simultaneously, so that a small collection of probes is enough to cover a representative part of the genome.

Microsatellite-complementary oligonucleotides are also useful for banding patterns when hybridized to PCR products generated by arbitrary primers. This novel strategy combines arbitrarily or microsatellite-primed PCR with microsatellite hybridization. In this technique, genomic DNA is amplified with either a single arbitrary 10-mer primer (as in RAPD analysis) or a microsatellite complementary 15-mer or 10-mer primer. PCR products are then electrophoresed, blotted, and hybridized to a ^{32}P or digoxigenin-labeled mono-, di-, tri-, or tetranucleotide SSR probes [e.g. $(CA)_8$, $(GA)_8$, $(GTG)_5$, $(GCGA)_4$]. Subsequent autoradiography reveals reproducible probe-dependent fingerprints, which are completely different from the ethidium bromide staining patterns and polymorphic at an intraspecific level. This provides for speed of the assay along with high sensitivity, thereby detecting a high level of polymorphism. This method is termed *RAMPO (random amplified microsatellite polymorphism)*, *RAHM (random amplified hybridization micosatellite)* or *RAMS (randomly amplified microsatellites)*.

Evaluation of fingerprints for oligonucleotide fingerprinting: A fingerprinting pattern obtained from a particular DNA sample is rarely informative on its own. Patterns originating from different samples have to be compared to each other, individual bands are assigned to particular positions, and different lanes are screened for matching bands. This multilocus DNA profiles are usually translated into discreet characters, such as "0" for absence and "1" for presence of band at a particular position. These "0/1 matrices" can then be exploited for different types of studies. Since bands are generally inherited in a Mendelian fashion, an application of RFLP fingerprinting is the estimation of relatedness between parents and offspring. For paternity testing, the fingerprint patterns of mother, father, and offspring are compared. All offspring bands must either be derived from mother or father. A large number of nonparental bands will exclude paternity. For the evaluation of higher order relationships, a similarity index (*F*) can be calculated from band sharing data of each of the fingerprints according to the formula:

$$F = \frac{2m[ab]}{m[a] + m[b]}$$

where *m*[*a*] and *m*[*b*] represent the total number of bands present in lanes [*a*] and [*b*] respectively, and *m*[*ab*] the number of bands which both lanes share. *F* can acquire any value between 0 and 1, where 0 means no bands in common and 1 means patterns are identical. A distance matrix can be prepared from the similarity values, which can be further transformed into a dendogram.

The application of RAPDs and the related modified marker AP-PCR in variability analysis and individual specific genotyping has largely been carried out. A modification of RAPD is DAF, which has been shown to generate reliable fingerprints, because many more bands are resolved due to short primers. Thus, DAF is useful for detecting polymorphisms even between organisms that are closely related. ISSRs or ISA (inter-SSR amplification), which has been explained, is a suitable marker for fingerprinting.

For fingerprint evaluation obtained by PCR approaches, quantification of pairwise similarities or differences between fingerprints can be done using different algorithms. Most commonly, a similarity index is calculated as a depictor of band sharing. For example, Jacard's similarity coefficient (*J*) takes into consideration only those matches between bands that are present (Jacard, 1908), whereas Nei and Li's coefficient (*N*) measures the proportion of bands shared as the result of being inherited from a common ancestor and represents the proportion of bands present and shared in both samples divided by the average of the proportion of bands present in each sample (Nei and Li, 1979). These similarity coefficients are calculated as follows:

$$J = \frac{sp}{sp + (1-s)} \qquad N = \frac{2sp}{2sp + (1-s)}$$

where *p* is the proportion of shared bands present

in both samples and *s* is the proportion of the total number of bands that are shared either present or absent in both samples. The *N* coefficient has been recommended for analysis of multilocus fingerprinting.

These DNA fingerprints can also be used for genetic mapping. Bands that are polymorphic btween two parents of a cross can be analyzed for linkage to each other as well as to other types of markers. A variety of computer programs are available that calculate recombination values and map distances and give a most likely order of markers within linkage groups and thus create genetic maps (e.g. LINKAGE-1, MAPMAKER, GMENDEL, JOINMAP).

MARKER-ASSISTED SELECTION (MAS)

In this technique, linkages are sought between DNA markers and agronomically important traits such as resistance to pathogens, insects and nematodes, tolerance to abiotic stresses, quality parameters, and quantitative traits. Instead of selecting for a trait, the breeder can select for a marker that can be detected very easily in the selection scheme. The essential requirements for marker-assisted selection in a plant breeding program are as follows:

1. Marker(s) should cosegregate or be closely linked (1 cM or less is probably sufficient for MAS) with the desired trait.
2. An efficient means of screening large populations for the molecular marker(s) should be available. At present, this means relatively easy analysis based on PCR technology.
3. The screening technique should have high reproducibility across laboratories.
4. It should be economical to use and be user friendly.

Several strategies have been developed that allow screening a large number of random, unmapped molecular markers in a relatively short time and to select just those few markers that reside in the vicinity of the target gene. These high-volume marker technologies that have shown efficacy are RAPD, AFLP, RFLP, microsatellites, etc. These methods rely on two principles: (i) generate hundreds or even thousands of potentially polymorphic DNA segments and rapidly visualize from single preparations of DNA; and (ii) use genetic stocks to identify among these thousands of DNA fragments those few that are derived from a region adjacent to the targeted gene.

In the past few years, by using one or more of these high-volume markers technologies, thousands of loci scattered throughout the genome have been assayed in a matter of weeks or months. The next problem is to determine which of the amplified loci lie near the targeted gene. Two strategies have proved effective.

1. *Nearly isogenic line (NIL) strategy:* Breeders have developed NIL genetic stocks and have been maintaining these inbred lines that differ at the targeted locus. Nearly isogenic lines are created when a donor line (P1) is crossed to a recipient line (P2). The resulting F_1 hybrid is then backcrossed to the P2 recipient to produce the backcross 1 generation (BC$_1$). From BC$_1$, a single individual containing the dominant allele of the target gene from P1 is selected. Selection for the target gene is normally made on the basis of phenotype. This BC$_1$ individual is again backcrossed to P2, and the cycle of backcross selection is repeated for a number of generations. In the BC$_7$ generation, most, if not all, of the genome will be derived from

Fig. 22.10 Pattern of markers based on RAPDs/AFLPs can be used to compare the genetic profiles of NILs and bulks.

P2, except for a small chromosomal segment containing the selected dominant allele, which is derived from P1. Lines homozygous for the target gene can be selected from the BC_7F_2 generation. The homozygous BC_7F_2 line is said to be nearly isogenic with the recipient parent, P2. The banding pattern by high-volume marker techniques is shown Fig. 22.10.

2. *Bulk segregant analysis (BSA):* This method is more generally applicable, and relies on the use of segregating populations (Michelmore *et al.*, 1991). It requires the generation of populations of bulked segregants (bulks). When P1 and P2 are hybridized, the F_2 generation derived from the cross will segregate for alleles from both parents at all loci throughout the genome. If the F_2 population is divided into two pools of contrasting individuals on the basis of screening at a single target locus, these two pools (Bulk 1 and Bulk 2) will differ in their allelic content only at loci contained in the chromosomal region close to the target gene. Bulk 1 individuals selected for recessive phenotype will contain only P2 alleles near the target, whereas Bulk 2 plants selected for dominant phenotype will contain alleles from both P1 and P2 (dominant homozygote and heterozygote). Bulk 1 and Bulk 2 will contain alleles from both P1 and P2 at loci unlinked to the target.

High-volume marker techniques, such as those based on RAPDs, RFLPs, and AFLPs, can be used to compare the genetic profiles of pairs of NIL or pairs of bulks to search for DNA markers near the target gene. DNA fragments indicated by arrows that are observed in the NIL and P1 populations, but not in P2, or that are unique to either Bulk 1 or Bulk 2, should be derived from a locus tightly linked to the target gene.

Marker-assisted selection has become a promising and potent approach for integrating biotechnology with conventional and traditional breeding. The interest of plant breeders in molecular markers revolves around certain basic issues:

1. *Resistance breeding:* At present, breeding for disease and pest resistance is conducted on the segregating populations derived from crosses of susceptible cultivars with resistant donors. These populaions are then selected either under natural disease or pest hot spots or under artificially created conditions. Although these procedures have given excellent results, they are time consuming. Besides, there are always susceptible plants that escape attack. Screening of plants with several different pathogens and their pathotypes or pests and their biotypes simultaneously or even sequentially is difficult, if not impossible. Availability of tightly linked genetic markers for resistance genes will help in identifying plants carrying these genes simultaneously without subjecting them to pathogen or insect attack in early generations. The breeder would require a low amount of DNA from each individual plant to be tested without destroying the plant, and see the presence or absence of the product of PCR reaction (marker band) on the gel. Only materials in the advanced generations would be required to be tested in disease and insect nurseries. Thus, with MAS, it is now possible for the breeder to conduct many rounds of selection in a year without depending on the natural occurrence of the pest or pathogen as well. Some illustrative examples of molecular markers with resistance genes are given in Table 22.1.

2. *Pyramiding of major/minor genes into cultivars for development of durable resistance/multiple resistance:* Pathogens and insects are known to overcome resistance provided by single genes. Single-gene resistances are fragile and often broken down easily. Therefore, breeders intend to accumulate several major and minor resistance genes into one cultivar to achieve durable reistance. Durability of resistance has been increased by developing multilines and by pyramiding of resistance genes. MAS for resistance genes can be useful in these approaches. Based on

host pathogen or host insect interaction alone it is often not possible to discriminate the presence of additional resistance (R) gene(s). With MAS, new R gene segregation can be followed even in the presence of the existing R gene(s), and hence R genes from diverse sources can be incorporated in a single genotype for durable resistance. Pyramiding of bacterial blight resistance genes Xa1, Xa3, Xa4, xa5, and Xa10 in different combinations using molecular markers has been reported in rice (Yoshimura *et al.*, 1995).

3. *Improvement of qualitative characters:* RFLP markers have been linked to the linolenic acid content Fan locus in soybean (Brummer *et al.*, 1995). Not only this, RAPD markers that control somatic embryogenesis in alfalfa have been identified (Yu and Pauls, 1993).

4. *Molecular markers for hybrid vigor:* Hybrids in crops such as maize, sorghum, rice pearl millet, cotton, and several vegetable crops have contributed greatly toward increasing the yield potential of these crops. The use of molecular markers has been investigated in maize and rice. Using molecular markers on a set of diallel crosses among eight elite parental lines widely used in Chinese hybrid rice production program, high correlation was found between specific heterozygosity and midparent heterosis. Photoperiod sensitive genic male sterility (PSGMS) genes in rice, designated pms-1 and pms-2, were located on chromosomes 7 and 2, respectively (Zhang *et al.*, 1994).

5. The MAS is especially useful for the traits that are arduous and/or expensive to evaluate, such as male fertility restoring genes for cytoplasmic male sterility in which the presence of the fertility restorer genes (Frs) in the breeding lines cannot easily be detected by conventional breeding techniques as they involve careful and extensive evaluation and analysis of innumerable segregants. The multiple marker alleles at the RG-532 locus closely linked to Fr_3 locus offer a possibility of marker-assisted screen-

ing for more effective sterility. Zhang *et al.* (1997) have identified RAPD markers linked to a restorer gene for WA gene-cytoplasmic male sterility. In fact, the marker-assisted identification of the effective alleles and the use of MAS in breeding can facilitate commercial hybrid rice production. In sugarbeet, selected polymorphic RAPD fragments from PCR fingerprint patterns of mitochondrial DNA from two nearly isogenic cytoplasmic male sterile and male fertile lines were detected, cloned, and sequenced by Lorenz *et al.* (1997).

6. *Molecular markers and abiotic resistance:* Using molecular markers, biochemists and physiologists have identified specific traits beneficial in improving drought responses such as osmotic adjustments, water use efficiency, and efficient root system. All these add to the yield improvement in crop plants. In rice and maize, QTLs for root traits have been identified and are being used to breed high yielding drought-resistant rice and maize genotypes. In many crops, RAPD markers are population specific. MAS has helped to improve yield performance under drought in beans, soybean, and peas.

PROTOCOL

22.1. FOR RAPD ANALYSIS

1. Isolate the genomic DNA.
2. Add the reagents in a 0.2 ml microcentrifuge tube placed on ice

Stock	Final	25 µl
Taq DNA polymerase buffer (10×)	1×	2.5 µl
dNTPs (10 mM each)	0.2 mM each	0.5 µl each
Random primer (5 pmoles)	0.2 µM	1.5 µl
DNA (10-50 ng)	10–50 ng	1 µl
Taq DNA polymerase (5 units/µl)	1 U	0.2 µl

Sterile water to make final volume to 25 µl.

3. Prepare cocktail of components. Keep one tube with only cocktail and water (without DNA) always as a control.
4. Mix well and spin briefly.
5. Overlay the reaction with a 25 ml of mineral oil to prevent evaporation of reagents at high temperatures (in those PCR models where hot lid is not provided).
6. Place the tubes in thermocycler block and set the machine for following cycling conditions.

Step 1	94°C, 2 min
Step 2	(25–30 cycles)
	94°C, 1 min
	40°C, 1 min 30 sec
	72°C, 2 min
Step 3	72°C, 10 min
Step 4	4°C, end

After amplification, add 2.5 µl of tracking dye in the sample and store at 4°C.
7. Prepare agarose gel of 1.4 to 1.8%.
8. Load the samples carefully in the wells.
9. Add suitable molecular weight, e.g. 1-kb ladder in one of the lanes.
10. Perform electrophoresis, stain the gel with ethidium bromide and photograph it.

Further reading

Caetano-Anolles, G. and Gresshoff, P.M. (Eds.). 1997. *DNA Markers: Protocols, Applications and Overview.* Wiley-VCH, New York.

Gebhardt, C. and Salamini, F. 1992. Restriction fragment length polymorphism analysis of plant genomes and its application to plant breeding. *Int. Rev. Cytol.,* **135**: 201-235.

Kochert, G. 1990. Introduction to RFLP mapping and plant breeding applications. The Rockfeller Foundation-International Program on Rice Biotechnology, pp. 1-14.

Mohan, M., Nair, S., Bhagwat, A., Krishna, T.G., Yano, M., Bhatia, C.R. and Sasaki, T. 1997. Genome mapping, molecular markers and marker-assisted selection in crop plants. *Mol. Breed.,* **3**: 87-103.

Vos, P., Hogers, R., Bleeker, M., Reijans, M., van de Lee T., Hornes, M., Fritjers, A., Pot, J., Peleman, J. Kuiper, M., and Zabeau, M. 1995. AFLP: a new technique for DNA fingerprinting. *Nucl. Acids Res.,* **23**: 4407-4414.

Welsh, J. and McClelland, M. 1990. Fingerprinting genomes using PCR with arbitrary primers. *Nucl. Acids Res.* **8**: 7213-7218.

Williams, J.G.K., Kubelik, A.R., Livak, K.J., Rafalski, J.A. and Tingey, S.V. 1990. DNA polymorphisms amplified by arbitrary primers are useful as genetic markers. *Nucl. Acids Res.,* **18**: 6531-6535.

23

Gene Transfer in Plants

Gene transfer or DNA uptake refers to the process that moves a specific piece of DNA (usually a foreign gene ligated to a bacterial plasmid) into cells. In plant breeding, techniques involving gene transfer through sexual and vegetative propagation are well established, the aim being to introduce genetic diversity into plant populations, to select superior plants carrying genes for desired traits and to maintain the range of plant varieties. The application of these conventional or classical techniques has produced significant achievements in the yield improvement of major food crops. However, this takes a long time. For example, it usually takes 6–8 years to produce a new variety of wheat or rice by sexual propagation. With the rapid development of genetic engineering techniques based on the knowledge of gene structure and function, plant breeding has been dramatically broadened.

The directed desirable gene transfer from one organism to another and the subsequent stable integration and expression of a foreign gene into the genome is referred as genetic transformation. The transferred gene is known as *transgene,* and the organisms that develop after a successful gene transfer are known as *transgenics.* Transgenic plants are the plants that carry the stably integrated foreign genes. These plants may also be called transformed plants.

In vitro gene transfer is the laboratory technique of transferring refined desirable genes across taxonomic boundaries into plant and animals from other plants, animals and microbes, or even to introduce artificial, synthetic or chimeric genes into plants. The development of gene transfer techniques in plants has been very rapid. The techniques presently available rely on natural plant vectors as well as vectorless systems, which include directed physical and chemical methods for delivering foreign DNA into plant cells. Some terms and expressions that will be used in the gene transfer methods have been explained before the actual discussion on gene transfer methods.

TRANSIENT AND STABLE GENE EXPRESSION

One of the main considerations for any gene transfer technique is the potential of the recipient cell to express the introduced gene. The transferred DNA may be expressed for only a short period of time following the DNA transfer process and this is called *transient expression.* Only a small fraction of the DNA introduced into the cell by direct gene transfer methods becomes stably integrated into the chromosome of the cell. The DNA introduced into majority of the cells is lost with time and cell division. Fortunately, this transient DNA is expressed in the cell and forms the basis of extremely useful transient assays. Transient assays are commonly used for the analysis of gene expression and rapid monitoring of gene transfer. Different variables associated with gene transfer can be optimized using transient expression.

Stable transformation occurs when DNA is integrated into the plant nuclear or plastid genomes, expression occurs in regenerated plants and is inherited in subsequent generations. For stable transformation, the developmental potential of the transformed cells is very important. Ideally, the transformed cells should be capable of regeneration into fertile plants. In general, the smaller the number of cells required for regeneration, the better the culture system is for gene transfer. Therefore, a small cell cluster size gives a larger ratio of transformed to nontransformed cells.

MARKER GENES

Monitoring and detection of plant transformation systems in order to know whether the DNA has been successfully transferred into recipient cells is done with the help of a set of genes. The marker gene(s) is/are introduced into the plasmid along with the target gene. Marker genes are categorized into two types:
  i) Reporter or scoreable genes
  ii) Selectable marker genes

The advantages of using these genes lie in their easy assay, which requires no DNA extraction, electrophoresis, or autoradiography. Some genes are used both as selectable markers and reporter genes, whereas some are used either as selectable markers or as reporter genes.

Reporter genes

A number of screenable or scoreable or reporter genes are available that show immediate expression in the cells/tissue. A reporter gene is a test gene whose expression results in quantifiable phenotype. A reporter system is useful in the analysis of plant gene expression and standardization of parameters for successful gene transfer in a particular technique. An assay for a reporter gene is usually carried out at the level of protein quantification. The principle of

using reporter genes in studying molecular processes in a living cell means that in the natural gene, a synthetic modification is introduced (or the protein coding sequence is deleted and replaced by another gene) in order to either simplify the detection of the gene product or to distinguish it from similar or identical genes in the genome. The use of reporter genes requires a method of gene transfer, either transient or stable. An ideal reporter should have the following features:
  i. Detection with high sensitivity,
  ii. Low endogenous background,
  iii. There should be a quantitative assay,
  iv. Assay should be nondestructive, and
  v. Assay should require a minimal amount of effort and expense.

The most commonly used reporter genes are described below and also summarized in Table 23.1.

Opine synthase

The octopine or nopaline (the two most common opines) synthase genes (*ocs, nos*) are present in the T-DNA of Ti (tumor inducing) or Ri (root inducing) plasmids of *Agrobacterium*. The presence of opines in any plant material clearly indicates the transformed status of the plant cells. Furthermore, opines can be detected in some cases before the appearance of any visual symptoms related to the transformed status of the plant material. The opine synthase genes in principle constitute good reporter genes since no natural equivalent to their gene products has been found in plant cells. The value of opines as screenable marker resides in the simplicity of protocols for opine detection, as enzymes are stable and the enzymatic assay is inexpensive and easy to perform.

There are basically two ways to assess the presence of opines in a transformed plant tissue. The first one aims at detecting the enzymatic activities responsible for the synthesis of opines. This technique involves a simple protein extraction from the plant sample. The extract is then incubated with the appropriate opines precursors, arginine, pyruvate and NADH for *ocs* and arginine, ketoglutaric acid and NADH for *nos,* and the reaction products are separated by paper chromatography, whereas the second technique aims at directly detecting the presence of opines in the plant tissue. This involves extraction of the plant metabolite from the tissue, the separation of these compounds (opines) by paper electrophoresis, and the detection of opines by appropriate staining treatments.

Chloramphenicol acetyl transferase

The gene for chloramphenicol acetyl transferase (*cat*) is one of the most commonly used reporter genes in eukaryotic organisms. Two different *cat* genes have been identified, but the most commonly used reporter was found in the transposable element *Tn*9. The structural gene codes for the enzyme chloramphenicol acetyl transferase (CAT), which carries out acetylation of chloramphenicol. The enzymatic assay for CAT is very sensitive and relatively simple, using acetyl CoA and (^{14}C) chloramphenicol to convert chloramphenicol into its acetylated derivatives. This is done by incubating the labeled chloramphenicol with a crude extract of plant tissue in a suitable buffer. The presence of acetyl chloramphenicol is detected through autoradiography. The *cat* gene was the first bacterial gene to be introduced into plant cells (Herrera-Estrella *et al.*, 1983).

β Glucuronidase (GUS)

During the last few years the bacterial gene *uid*A, encoding β-glucuronidase (GUS), has become the most frequently used reporter gene for the analysis of plant gene expression. *uid*A encodes a soluble β-glucuronidase enzyme, a homotetramer of approximately 68 kDa mass, requiring a pH optimum of 7–8. β-glucuronidase breaks down glucuronide substrate to give a colored reaction, so that its presence can be detected *in situ*. Its wide acceptance has mainly resulted from the availability of a highly sensitive nonradioactive assay using fluorogenic substrate 4-MU Gluc and of a histochemical assay using X-Gluc, that allows a quantitative analysis of cell- and tissue-specific expression. The major advantage of this reporter is that it does not require DNA extraction, electrophoresis, or autoradiography.

Bacterial luciferase (Lux F2)

Bacterial luciferases have originated from *Vibrio harveyi* and *V. fischerii*. In these organisms, luciferase is expressed as a heterodimeric flavin monooxygenase that is responsible for catalyzing a light-emitting reaction. The *V. harveyi* luciferase (Lux) is encoded by the *lux*A and *lux*B cistrons that are part of a longer operon. The *lux* genes in plant cells can be expressed as separate subunits or as a fused polypeptide. The subunits are correctly assembled in plant cells and the

Table 23.1 Reporter genes used as scoreable markers and their assays

Reporter gene	Substrate and assay	Identification
Octopine synthase (*ocs*)	Arginine + pyruvate + NADH	Electrophoresis, chromatography
Nopaline synthase (*nos*)	Arginine + ketoglutaric acid + NADH	Electrophoresis, chromatography
Chloramphenicol acetyl transferase (*cat*)	^{14}C Chloramphenicol + acetyl CoA: TLC separation	Autoradiography
β-Glucuronidase (*gus*)	Glucuronides (X-Gluc, 4-MUGluc, PNPG, NAG, REG)	Fluorescence detection, colorimetric, histochemical
Bacterial luciferase (*Lux F2*)	Decanal and FMNH$_2$	Bioluminiscence
Firefly luciferase	ATP + O$_2$ + luciferin	Bioluminiscence
Green fluorescent protein	No substrate required; UV light irradiation	Fluorescence detection
Anthocyanin regulatory genes	No substrate required	Visual

fused protein has been designated as Lux F2. The activity of luciferase is normally measured as the initial maximum light intensity upon mixing the enzyme with its aldehyde and reduced flavin mononucleotide substrates ($FMNH_2$) in the presence of oxygen. Light is measured as photons or quanta per second (qu/s), and if calibrated against a known light standard, the absolute number of enzyme molecules in the reaction can be calculated. Light emission can be monitored visually, photographically, or electronically. A simple low cost luminometer can detect *ca.* 23-5 qu/s, corresponding to about 10^6 luciferase molecules.

The firefly luciferase (Luc)

Luciferase isolated from the North American firefly *Photinus pyralis* catalyzes the oxidative decarboxylation of luciferin, a 6-hydroxy-benzothiazole, to oxyluciferin in the presence of ATP, oxygen, and magnesium ions. Luminescense in terms of light production can be recorded in a luminometer.

Green fluorescent protein (GFP)

The gene for green fluorescent protein (GFP) has been obtained from jellyfish *Aequorea victoria*. GFP is a small protein of 238 amino acids. This owes its visible absorbance and fluorescence to a β-hydroxy benzylidine imidazolinone chromophore. The formation of green fluorescence can be detected under fluorescent microscope.

Anthocyanins

Anthocyanins are prominent red and purple pigments of flowering plants. These anthocyanins have been used as visible marker for transformation. At least 10 genes that encode regulatory or structural proteins of the biosynthetic pathway control anthocyanin pigmentation within the plant. The regulatory genes C_1 and B/R co-ordinately regulate the activity of structural genes. Chimeric gene constructs have been prepared with C_1 and B/R regulatory genes, which when introduced into the plant cells produce purple red coloration. These red spots can be

visually or microscopically identified. These regulatory genes have been used as a visible marker in maize and wheat transformation. This reporter has advantages that a tissue is not sacrificed, transient expression can be seen in the position of target, and no addition of substrates is required.

Selectable markers

Selection of transformed cells is a key factor in developing successful methods for genetic transformation. This is done by certain selectable marker genes that are present in the vector along with the gene of interest. Selectable markers are an integral part of plant transformation strategies. Selectable marker genes enable the transformed cells to survive on media containing toxic levels of selection agent whereas nontransformed cells get killed. A large number of such selectable marker genes have become available, *viz.* antibiotic, antimetabolite and herbicide resistance genes, hormone biosynthetic genes, and genes conferring resistance to toxic levels of amino acids or amino acid analogs. The usefulness of a particular resistance marker depends on (i) the characteristics of selection agent, (ii) the resistance gene, and (iii) the plant material. The selection agent must be toxic to plant cells, though not so toxic that products from dying, nontransformed cells kill adjacent, transformed cells. Different selectable marker genes and their substrates have been summarized in Table 23.2.

Antibiotic resistance markers

Neomycin phosphotransferase II (NPT II)

NPT II or aminoglycoside 3′-phosphotransferase II (APH (3′)-II) enzyme is the most widely used marker. The enzyme is encoded by the *npt*II gene, which has been derived from the Tn5 transposon. It inactivates a number of aminoglycoside antibiotics such as kanamycin, neomycin, geneticin (G418), and paromycin. Kanamycin is the most commonly used selective agent, with concentrations ranging from 50 to 500 mg/l.

Table 23.2 Selectable marker genes used for gene transfer

Selectable marker gene	Substrates used for selection
Neomycin phosphotransferase (*npt*II)	G418, kanamycin, neomycin, paromycin
Hygromycin phosphotransferase (*hpt*)	Hygromycin B
Gentamycin acetyl transferase	Gentamycin
Streptomycin phosphotransferase	Streptomycin
Dihydrofolate reductase (*dhfr*)	Methotrexate
Phosphinothricin acetyltransferase (*bar*)	L-phosphinothricin (PPT), bialaphos
5-enolpyruvyl shikimate -3-phosphate (EPSP) synthase (*aro*A)	Glyphosate (roundup)
Acetolactate synthase mutant form (*als*)	Sulphonylurea, imidazolinones
Bromoxynil nitrilase (*bxn*)	Bromoxynil

Resistance to kanamycin in putative transformants and their progeny can be checked by applying kanamycin solution locally or sprayed on soil grown plants to select for kanamycin-resistant progeny.

Hygromycin phosphotransferase

The hygromycin phosphotransferase gene (*hpt*) was originally derived from *E. coli*. Hygromycin B is an aminocyclitol antibiotic that interferes with protein synthesis, is more toxic than kanamycin, and kills sensitive cells faster. Hygromycin resistance can also be checked by a nondestructive callus induction test and segregation of *hpt* gene in the progeny of transgenic plants can be scored by seed germination assay.

Gentamycin acetyltransferase

Three genes encoding aminoglycoside-3-*N*-acetyltransferases (AAC (3)) have been used as selectable markers in combination with gentamycin antibiotic selection. The genes are AAC (3)-I, AAC (3)-III, and AAC (3)-IV.

Streptomycin and spectinomycin resistance

Two dominant genes, streptomycin phosphotransferase (*spt*) gene from Tn5 and the aminoglycoside-3´-adenyltransferase gene (*aad* A) conferring resistance to both streptomycin and spectinomycin, have been described for plant transformation. Streptomycin and spectinomycin

resistance markers differ from other markers in that they allow differentiation by color. Sensitive plant cells bleach but do not die, whereas resistant cells remain green under appropriate conditions.

Antimetabolite marker

Methotrexate-insensitive dihydrofolate reductase (dhfr)

Methotrexate is an antimetabolite that inhibits the enzyme dihydrofolate reductase (DHFR) and thus interferes with DNA synthesis. A mutant mouse *dhfr* gene encoding an enzyme with very low affinity to methotrexate has been isolated. This *dhfr* gene fused to CaMV promoter yields a methotrexate resistance marker, which is used for plant transformation.

Herbicide resistance markers

Bar gene

Bialaphos, phosphinothricin (PPT), and glufosinate ammonium are nonselective herbicides that inhibit the enzyme glutamine synthase (GS). GS is necessary for the assimilation of NH_3 and in the regulation of nitrogen metabolism in plants. Inhibition of GS results in the accumulation of NH_3, which causes death of plant cells. The *bar* gene codes for the enzyme phosphinothricin acetyl transferase (PAT), which converts PPT/bialaphos into a nonherbicidal acetylated form by transferring the acetyl group from acetyl CoA on the free amino group of PPT and confers resistance to the cell. This gene has been cloned from two different strains of *Streptomyces hygroscopicus* and *S. viridochromogenes*.

Enolpyruvylshikimate phosphate (EPSP) synthase (aro A) gene

Glyphosate (N-phosphonomethyl-glycine) is a broad-spectrum herbicide, which inhibits the essential photosynthesis process. It works by competitive inhibition of the enzyme EPSP synthase (5-enolpyruvyl-shikimate 3-phosphate). This enzyme catalyzes the synthesis of 5-enolpyruvylshikimate 3-phosphate from phosphoenolpyruvate and shikimate 3-phosphate, an essential step in the synthesis of phenylalanine, tyrosine, and tryptophane. In plants, much of the EPSP synthase activity is encoded by nuclear genes, but enzyme is localized in chloroplasts. Mutants of the bacterium *Salmonella typhimurium* resistant to glyphosate have been obtained. These map at the *aro* A locus, which encodes EPSP synthase.

Acetolactate synthase (als) gene

The herbicidal action of sulphonylureas and imidazolinones is based on their ability to noncompetitively inhibit the enzyme acetolactate synthase (ALS) in the pathway leading to the production of branched chain amino acids, *viz.* valine, leucine, and isoleucine. Mutant genes have been isolated from *E. coli* and yeast. Mutation to resistance leads to loss of herbicide binding to ALS. This mutant gene has been cloned and used as a selectable marker in transformation experiments.

Bromoxynil nitrilase (bxn)

Bromoxynil, a nitrilic herbicide is an inhibitor of photosystem II. Bromoxynil is used as a sole nitrogen source by *Klebsiella ozaene*. The bromoxynil nitrilase gene has been isolated, which can convert bromoxynil herbicide into 3,5-dibromo-4-hydroxybenzoic acid. This gene has been successfully used as a selectable marker to select for transformed plants.

CHIMERIC GENE VECTORS

The nuclear plant gene consists of different regions, each involved in different functions of transcription and translation of mRNA. Starting with the 5′ end, there is a promoter region that is involved in the initiation of transcription, together with enhancer/silencer regions that confer regulation of expression, a transcriptional start or cap site, and the so-called CAAT and TATA boxes, which help in binding RNA polymerase. One or more untranslated or intron regions are present within the transcribed region. The end of the translation region is determined by a stop codon and followed by a terminator at the 3′ end with polyadenylation signal.

In the majority of cases, whether direct vectorless delivery systems or natural vectors are used for transfer of marker genes, reporter genes and the genes of interest are modified prior to the transfer by different methods. These are placed under the control of different promoter, terminator, and enhancer sequences. These are called chimeric or transgene constructs, since they consist of components derived from different origins (Fig. 23.1). Plants are usually transformed with relatively simple constructs in which the gene of interest is coupled with an appropriate promoter, 5′ leader and a 3′ terminator sequences to ensure efficient transcription, stability, and translation of mRNA. The promoter can be of

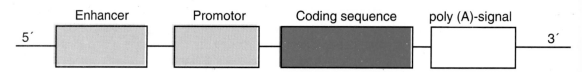

Fig. 23.1 Sequence components of a chimeric gene construct.

plant, viral, or bacterial origin. Some promoters confer constitutive expression, whereas others may be selected to permit tissue-specific expression or environmentally inducible expression. The promoters of bacterial origin are *lac, trp,* and *tac,* and phage promoters are T3, T7, and SP6. The cauliflower mosaic virus (CaMV) 35S RNA promoter is often used because it is a plant virus promoter, and plant viruses are dependent on transcription and translation factors of plants. It directs high levels of expression in most tissues. Others such as maize ubiquitin I promoter, *rbc*S ribulose bisphosphate carboxylase small subunit, *Adh*1 alcohol dehydrogenase, *nos* nopaline synthase, and the rice actin promoter/ intron sequence are often preferred for expression in monocots.

The control of gene expression plays an important part in plant development and it is obvious that consideration has to be given to the regulation of genes introduced artificially. As the DNA constructs become more sophisticated, they will contain complex chimeric nucleotide sequences that are a combination of enhancer and silencer sequences, transcription promoters and terminators, protein-coding open reading frames possibly with organelle targeting or signal sequences, selectable markers, reporter genes and vector sequences. The DNA constructs can be introduced as naked DNA or as a plasmid, or can be cloned into a specialized vector for use with *Agrobacterium.* Thus, it is the chimeric gene constructs that are used for expression of transferred genes.

GENE TRANSFER METHODS

Various gene transfer techniques are grouped in two main categories:
1. Vector-mediated gene transfer and
2. Vectorless or direct DNA transfer.

Vector-mediated gene transfer

The term *plant gene vector* applies to potential vectors both for the transfer of genetic information between plants and the transfer of genetic information from other organisms (bacteria, fungi and animals) to plants. Plant gene vectors being exploited for transfer of genes are plasmids of *Agrobacterium,* viruses, and transposable elements.

Agrobacterium-mediated transformation

The *Agrobacterium* system was historically the first successful plant transformation system, marking the breakthrough in plant genetic engineering in 1983. The breakthrough in gene manipulation in plants came by characterizing and exploiting plasmids carried by the bacterial plant pathogens *Agrobacterium tumefaciens* and *A. rhizogenes.* These provide natural gene transfer, gene expression, and selection systems. In recent times, *A. tumefaciens* has been treated as nature's most effective plant genetic engineer.

Classification of Agrobacteria

Bacteria of the genus *Agrobacterium* are well known as gene vectors for plant cells. *Agrobacterium* are gram-negative rods that belong to the bacterial family Rhizobiaceae. These are often found near soil level at the junction of plant stem and root. The old system of classification according to their phytopathogenic characteristics is as follows:

A. tumefaciens: Induces crown gall disease
A. rhizogenes: Induces hairy root disease
A. radiobacter: An avirulent strain

Recently, agrobacteria have been reclassified according to their growth patterns. In this way, three different groups (biotypes) were found. Strains belonging to biotypes 1 and 2 are used in plant genetic engineering experiments.

Biotype 1: These strains can be distinguished from other bacteria by the production of ketolactose from lactose. Well-known *A. tumefaciens* strains such as Ach5, A6, B6, C58, T37, Bo542, and 23955 belong to biotype 1. They can proliferate at temperatures up to 37°C, but they are usually grown in the laboratory at 28 or 29°C because certain Ti plasmids are somewhat unstable at higher temperatures. Classic *E. coli* media such as LB (Luria Bertini), NB (nutrient broth), and TY (tryptone yeast extract) can be used for growth, but incubation in a defined minimal media such as MM or in a medium such as YMB (yeast maltase broth) also permits growth.

Biotype 2: These are identified by their ability to grow on erythritol as a carbon source is a phenotype by which they can be distinguished from other agrobacteria. Laboratory strains of *A. rhizogenes* include the biotype 2 strains NCPPB 1855, ATCC 23834, and A4. Biotype 2 strains proliferate poorly at temperatures higher than 30°C and are therefore grown at 28 or 29°C. They do not grow well in the classic *E. coli* media, but can instead be cultured on media such as TY + Ca, YMB, or in a defined minimal medium (MM).

Tumor-inducing principle and the Ti plasmid

Smith and Townsend showed in 1907 that a bacterium is the causative agent of crown gall tumors. Experiments of Braun and coworkers showed that the continued presence of viable bacteria is not required for tumor maintenance (Braun and Stonier, 1958). The bacteria do not penetrate into the plant cells that are converted into tumor cells. Rather, bacteria penetrate into the intercellular spaces and into injured cells and attach themselves only to the wall of healthy plant cells. Attention thus focussed on the identification of a putative "tumor inducing principle." Zaenen et al. (1974) first noted that virulent strains of *A. tumefaciens* harbor large plasmids and the virulence trait is plasmid borne. Virulence is lost when the bacteria are cured of the plasmid.

A. tumefaciens induces tumors called crown galls, whereas *A. rhizogenes* causes hairy root disease. Large plasmids in these agrobacteria are called tumor-inducing plasmids (Ti plasmids) and root inducing plasmids (Ri plasmids), respectively, which confer on their hosts the pathogenic capacity. Both sorts of diseases result from transfer and functional integration of a particular set of the Ti or Ri plasmid into plant chromosomes. It was later revealed that only a small part of the Ti plasmid, the T-DNA or transferred DNA, is transferred to and integrated into the host plant nuclear genome (Chilton et al., 1977).

Crown gall disease caused by *A. tumefaciens* is characterized by unlimited plant cell proliferation (gall formation). Naturally transformed crown gall tumor tissue differs from nontransformed tissue in its ability to grow autonomously on synthetic media in the absence of hormones, i.e. auxins and cytokinins. The plant cells in the tumor acquire two new properties:

1. They show phytohormone independent growth.

2. They contain one or more of unusual amino acid derivatives known as opines. Opine synthesis is a unique characteristic of tumor cells; normal plant tissues do not usually produce these compounds. The ability to produce these opines was specified by the transferred bacterial DNA and not by the host plant genome. They do not appear to have any known useful function in the plant cells and are not directly responsible for the tumorous state of the cells. However, opines are useful to the bacterium since they serve as a source of carbon and nitrogen. In addition, some of the opines also induce conjugation between bacteria. Opines distinguish agrobacteria from other bacteria in the rhizosphere. Agrobacteria are able to utilize opines, thus providing a competitive advantage in the crown gall rhizosphere to the inciting agrobacteria over other soil organisms. This phenomenon has been termed genetic colonization.

It was earlier observed that there were two families of bacteria and the crown galls they established differed by producing two different amino acid derivatives, subsequently called octopine and nopaline. The octopine family has carboxyl-ethyl derivatives of arginine whereas nopaline family has related compounds that are dicarboxypropyl derivatives of arginine. For the synthesis of octopine and nopaline, the corresponding enzyme octopine synthase (*ocs*) and nopaline synthase (*nos*) are coded by T-DNA. Now, Ti plasmids are defined as octopine, nopaline, succinamopine, agropine, or leucinopine, depending on the type of opine produced in the tumors.

Organization of Ti plasmid

The Ti plasmids of *Agrobacterium* are large circular DNA molecules, up to 200 kb in length with molecular weights of about 1.2×10^8 (3 to 8% of the *Agrobacterium* chromosome). They exist in the bacterial cells as independent replicating genetic units. Ti plasmids have major regions for virulence, origin of replication, conjugation, oncogenicity, and catabolism of opines. Genetic organization of these regions has been shown in octopine Ti plasmid DNA (Fig. 23.2).

Studies on *A. tumefaciens*-mediated gene transfer system have revealed that the following elements are required for T-DNA transfer to plants.

T-DNA: T-DNA is the transferred DNA or transforming DNA of Ti plasmid. The length and number of the T-DNA region vary in Ti and Ri plasmids. The nopaline type Ti plasmid has a single 23-kb T-DNA region (e.g. pTiC58), whereas the octopine type Ti plasmid (pTiAch5) has two closely adjoining regions of 13-kb left-hand piece

(T_L) and 8 kb right-hand piece $(T_R$ DNA). The T-DNA carries genes for opine synthesis (*nos* or *ocs*) and phytohormone biosynthesis (Fig. 23.3). An oncogenic (*onc*) region contains rooty locus (root inhibition, *roi* or tumor morphology root, *tmr*) and results in tumors developing abnormal roots and a shooty locus (shoot inhibition, *shi* or tumor morphology shoot, *tms*) results in tumors with abnormal shoot tissue. Mutations of a further morphogenetic locus, *tml*, produce larger than normal tumors. Two genes comprise the *shi* or *tms* shooty locus and one gene *tmr* represents the rooty locus. These *tms*1 and *tms*2 are responsible for the synthesis of the phytohormone indole 3-acetic acid from tryptophan. Rooty locus *tmr* gene encodes isopentyl transferase, which catalyzes the formation of isopentyl adenosine 5′ monophosphate (cytokinin) (Fig. 23.4). The phytohormones in turn alter the developmental program, leading to the formation of crown gall.

An opine synthesis *os* region is responsible for the synthesis of unusual amino acid or sugar derivatives, which are collectively given the name opines. Opine synthesis is controlled by enzyme octopine synthase and nopaline synthase at loci called *ocs* and *nos* on octopine and nopaline plasmids, respectively.

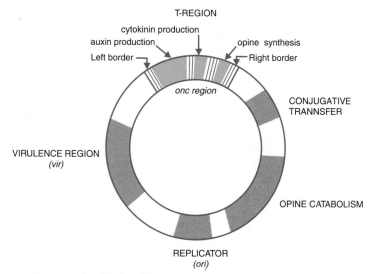

Fig. 23.2 Genetic map of an octopine Ti plasmid.

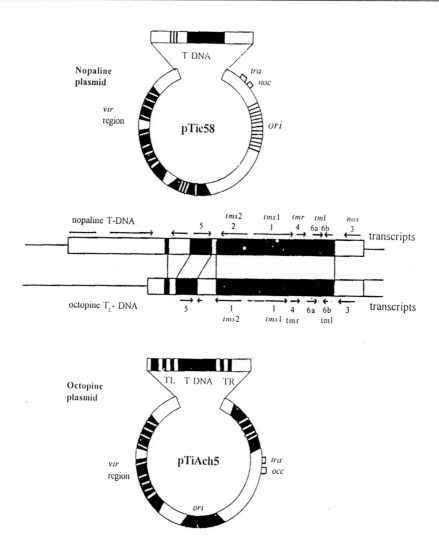

Fig. 23.3 Genetic organization of nopaline (pTiC58) and octopine (pTiAch5) plasmid DNAs. Tumor morphology is controlled by expression of genes 1 and 2, gene 4 is responsible for cytokinin biosynthesis and *tml* locus specifies size. *nos* and *ocs* control tumor opine synthesis.

The third component of T-DNA region on all Ti and Ri plasmids is a 25-bp highly conserved directly repeated DNA sequence that flanks and delimits the T-DNA. These sequences act as borders and are referred to as left and right borders. These borders play an important role in the transfer process. In fact, any DNA sequence that is located between these borders is transferred to the plant, and this feature has been exploited in genetic manipulation experiments. The nucleotide sequence of the border repeats is similar in nopaline and octopine Ti plasmids. There are no genes within the T-DNA of naturally occurring Ti plasmids that are necessary for T-DNA transfer. The naturally occurring T-DNA encoded "oncogenes" are necessary for tumorigenesis and opine production but not necessary for T-DNA transfer. Thus, these

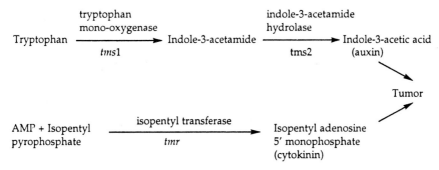

Fig. 23.4 Pathway of auxin and cytokinin biosynthesis in *Agrobacterium* induced tumor tissue formation.

oncogenes can be deleted, thereby disarming the T-DNA. Such disarmed Ti plasmids are capable of transferring T-DNA to plants but will not cause tumors. In these disarmed plasmids, novel foreign genes can be inserted between the T-DNA borders.

Virulence genes: The second gene system necessary for T-DNA transfer consists of the virulence (*vir*) genes encoded by the Ti plasmid in a region outside the T-DNA. The 40-kb *vir* region of the octopine Ti plasmid embraces 24 genes involved in virulence. These genes are present in eight operons called *vir* A–*vir* H, which are coregulated and thus form a regulon. Most of the operons except *vir* A, *vir* F, and *vir* G contain several genes, i.e. they are polycistronic. With the exception of the *vir* A and *vir* G genes, the *vir* operons are not transcribed during normal vegetative growth. The genes for virulence are silent till they are induced by certain plant factors. *vir* A and *vir* G make up a two-component regulatory system important for transcriptional activation of the remaining *vir* genes and to some extent their own increased activation.

The remaining *vir* genes are involved in the processing of the T-DNA from the Ti plasmid, and in T-DNA transfer from the bacterium to the plant cell. Certain Vir proteins may also be involved in the targeting of the T-DNA to the plant nucleus, and perhaps in T-DNA integration into plant DNA. Thus, genetic manipulation that stimulates *vir* gene induction (or increase Vir protein activity) may result in increased T-DNA transfer to the plant.

Chromosomal genes: The third component necessary for T-DNA transfer consists of a number of genes encoded by the *A. tumefaciens* chromosome. Some of these genes *viz.* chvA, chvB, and pscA (*exo*C) present on the *A. tumefaciens* chromosome are essential for T-DNA transfer. Genes chvA and chvB code for exopolysaccharide production, which is important for attachment of the bacterium to the plant cell. Genes at these two loci are constitutively expressed nonregulated genes. The gene chvE encodes a glucose/galactose transporter, and is important for binding specific sugars that act as *vir* gene coinducers. Other chromosomal genes, including chvD, ilv, miaA, and att, contribute to virulence in more minor and in some cases yet undefined ways.

Opine catabolism is controlled by loci noc and occ for nopaline and octopine catabolism, respectively. It lies on the right limb of the molecule near to locus tra specifying conjugative transfer of the Ti plasmid to avirulent *Agrobacterium* strains. These noc and occ genes do not form a part of T-DNA.

T-DNA transfer

T-DNA transfer begins with the introduction of bacteria into a plant wound. Wounding is a necessary event in the process and may, at least in part, be required for the synthesis by the plant certain compounds that induce the expression of the *vir* genes. Two of the most active substances identified are acetosyringone and β-hydroxy-acetosyringone. The *vir* A product is a periplasmic membrane protein that senses specific phenolic compounds (such as acetosyringone and related molecules) synthesized by wounded plant cells. It is most likely that the Vir A protein also interacts with another protein encoded by *chv*E gene present on bacterial chromosome and is important for the binding of sugar coinducers. The presence of these phenolic and sugar inducers results in the autophosphorylation of Vir A, which then transfers the phosphate group to the Vir G protein, thus activating Vir G. The activated Vir G protein acts as a transcriptional activator of the other *vir* genes (most likely by binding to specific DNA sequences, the "*vir* box" that precedes the *vir* genes). An early event in the T-DNA transfer is the nicking of Ti plasmid (Fig. 23.5). Two of the proteins coded by *vir*D, Vir D_1 and D_2, provide an endonuclease that initiates the transfer process by nicking T-DNA at a specific site. Nicking takes place between the third and fourth base of the bottom strand of each 25-bp repeat. A nick at the right border 25-bp repeat provides a priming end and synthesis of new DNA strand begins from this nick in the $5' \rightarrow 3'$ direction. A new strand displaces the old T-DNA single strand. Recent evidence indicates that T-strand retains the Vir D_2 protein covalently attached to the 5' terminus. The presence of Vir D_2 makes the 5' end of the T-strand less vulnerable to an attack by exonuclease. Besides, the Vir D_2 protein may also act as a pilot to direct the T-strand to the nucleus of the transformed cell, since it contains a nuclear targeting sequence.

Outside the T-DNA, but immediately adjacent to the right border, is another short sequence called overdrive, which greatly stimulates the transfer process. Overdrive functions like an enhancer. Vir C1 and Vir C2 may act at the overdrive sequence. Vir E2 is a 69-kDa protein that binds single-stranded T-DNA and is involved in transporting T-DNA into plant cell nucleus. The remaining Vir proteins are not involved in regulation of expression or T-strand expression; only those encoded by the *vir* B operon are essential for virulence. Most of the proteins predicted for the *vir* B operon are located in the

membrane and these proteins may form a conjugal pore or pilus structure in the membrane through which the T-DNA is delivered into the plant cell. The *vir* B_{11} gene has an ATP-binding site and the protein has an ATPase activity. It may therefore be involved in delivering energy required for T-DNA transfer. The *vir*H operon consists of two genes and it has a role in the detoxification of certain plant compounds that might otherwise adversely affect the growth of *Agrobacterium*. Enhanced tumorigenicity was observed for bacteria having the *vir* H genes as compared to those lacking these in certain hosts. *vir* F plays a role in T-DNA delivery rather than symptom formation. Presence of *vir* F gene in octopine Ti strains makes them vastly superior in transferring DNA.

The transfer of T-DNA closely resembles the events involved in bacterial conjugation when the *E. coli* chromosome is transferred from one cell to another in single-stranded form. A difference is that the transfer of T-DNA is usually limited by the boundary of the left repeat, whereas transfer of bacterial DNA is indefinite. It is still not very clear how the transferred DNA is integrated into the plant genome. At some stage the newly generated single strand must be converted into duplex DNA.

Agrobacterium vectors

The *Agrobacterium* Ti plasmid system exploits the elements of the *Agrobacterium* transformation mechanism. But the following properties of Ti plasmids do not allow their direct use: (i) large size; (ii) tumor induction property and (iii) absence of unique restriction enzyme sites. *Agrobacterium* plasmids are disarmed by deleting naturally occurring T-DNA encoded oncogenes and replacing them with foreign genes of interest. Besides, restriction enzyme sites are also added into the plasmids.

Cointegrate vectors

Vectors that recombine via DNA homology into a resident Ti plasmid are often referred to as integrative or cointegrate vectors. In a cointegrate

Fig. 23.5 Schematic summary of the events leading to the transformation by *Agrobacterium;* **a**–activation of the *vir* genes via VirA and VirG; **b**–key events in the induction of *vir* genes and T-DNA transfer, **c**–key events in the transfer of T-DNA to plant cell. OM: outer membrane; IM: inner membrane; AS: acetosyringone.

vector, the disarmed Ti vector is covalently linked to donor vector with gene of interest-T-DNA border sequences present in a *A. tumefaciens* strain to act as one unit. In a cointegrate vector, much of the wild type T-DNA (especially the hormone biosynthetic genes and not the border regions) is replaced with a segment of DNA common to many *E. coli* cloning vectors. Homology between the *E. coli* plasmid-based segment of the modified T-DNA and the donor vector with identical sequences provides a site for recombination to occur, which is catalyzed by native *Agrobacterium rec* functions, resulting in the formation of a hybrid or cointegrate Ti plasmid. Because most cointegrates result from a single recombination event, the ensuing product places the gene of interest along with the entire donor vector with its marker genes between the native T-DNA borders of the disarmed Ti plasmid. The formation of the cointegrate vector, using the disarmed vector pGV3850, is illustrated in Fig. 23.6. This vector has pBR322 DNA in the T-DNA region, which provides a region of homology with most other cloning vectors. In this example, donor vector pLGVneo1103 carries a plant selectable marker *nos-npt*-II hybrid gene and a bacterial selectable marker, the *kan* gene. Recombination between the homologous pBR322 DNA in the two plasmids produces the cointegrate.

The standard cointegrate vector contains (i) convenient sites for insertion of the gene of interest, (ii) antibiotic selectable marker gene or genes active in both *E. coli* and *A. tumefaciens*, (iii) a plant functional selectable marker gene, and (iv) *E. coli* functional origin of replication that does not operate in *Agrobacterium*. Research has not demonstrated any specific advantage to retaining the T-DNA and *vir* functions on a single replicon.

Binary vectors

The binary vector system consists of two autonomously replicating plasmids within *A. tumefaciens*: a shuttle (more commonly referred to as a binary) vector that contains gene of interest between the T-DNA borders and a helper Ti plasmid that provides the *vir* gene products to facilitate transfer into plant cells. Disarmed helper Ti plasmids have been engineered by removing the oncogenic genes while still providing the necessary *vir* gene products required for transferring the T-DNA to the host plant cell.

The standard components of a binary vector are (i) multiple cloning site, (ii) a broad host range origin of replication functional in both *E. coli* and *A. tumefaciens* (e.g. RK2), (iii) selectable markers for both bacteria and plants, and (iv) T-DNA border sequences (although only the right border is absolutely essential). A typical binary vector system is shown in Fig. 23.7. The helper Ti plasmid pAL4404 is a derivative of the octopine plasmid pTiAch5 that has the complete T-DNA region deleted but contains *vir* genes. The mini T-DNA donor vector pBin19 contains a truncated T-DNA from the nopaline plasmid pTiT37 comprising the left and right border sequences and a *nos-npt* II hybrid gene as a selectable marker. A *lacZ* polylinker is inserted in the mini T-DNA to provide unique cloning sites and a color test for insertion. The linker carries the galactosidase that produces blue colonies on IPTG/X-GAL indicator plates and white colonies if it is disrupted by an insertion of foreign gene, i.e. pBin 19.

There are a number of advantages associated with using a binary system as compared to using a cointegrate vector:

1. Binary vectors do not need *in vivo* recombination whereas cointegrate vectors require a recombinational event for stable maintenance within the target *A. tumefaciens* strain.
2. Binary vectors require only that an intact plasmid vector be introduced into the target bacterium, making the process of bacterial transformation both more efficient and quicker (2–3 days versus 4–7 days).
3. Binary vector systems with plant ready genes in agrobacteria are easily and efficiently obtained.
4. In binary system, the binary plasmids exist as separate replicons; thus copy number is not strictly tied to that of Ti plasmid. Because of this, in most cases, confirmation of the transformation event is accomplished via agrobacteria mini preps as compared to Southern hybridization used for cointegrate plasmids.

Most of the recently developed plant transformation vectors are binary, largely due to

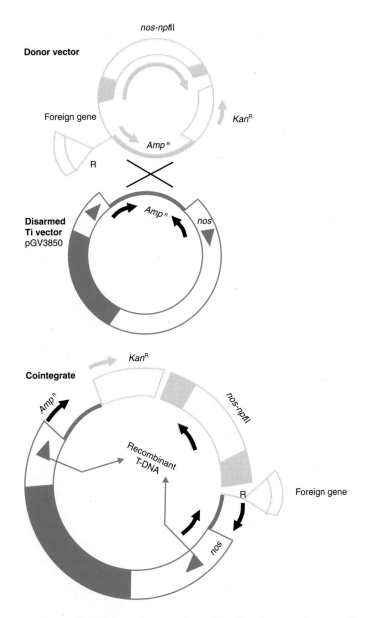

Fig. 23.6 Producing a recombinant T-DNA by cointegrate formation. The donor vector recombines with the disarmed vector by homologous recombination produce the cointegrate.

the ease of both *in vivo* and *in vitro* DNA manipulation and their higher transformation efficiencies, allowing the use of direct *Agrobacterium* transformation techniques.

Transformation Techniques Using *Agrobacterium*

Agrobacterium-mediated plant transformation techniques are now extensively utilized to

Fig. 23.7 A typical binary Ti vector system.

generate transgenic plants. The important requirements for *Agrobacterium*-mediated gene transfer in higher plants are as follows:

i. The plant explants must produce acetosyringone or other related compounds for activation of *vir* genes or *Agrobacterium* may be preinduced with synthetic acetosyringone.

ii. The induced agrobacteria should have access to cells that are competent for transformation.

iii. Transformation competent cells/tissues should be able to regenerate into whole plants.

A general scheme for transformation using *A. tumefaciens* as a vector is shown in Fig. 23.8.

The explant used for inoculation or cocultivation with *Agrobacterium* carrying the desired cointegrate or binary vectors include cells, protoplasts, callus, tissue slices, whole organs or sections, etc. There are three general approaches for inoculation of *Agrobacterium*.

i) *Infection of wounded plants*: In vivo inoculation of seedlings or well-established plants that have been propagated *in vitro* under aseptic conditions is the traditional way to obtain *Agrobacterium* transformed cells. Seedlings are decapitated and the freshly cut surface wound is inoculated with an overnight culture of *Agrobacterium*. The tumor produced is excised and grown as a callus culture. The transforming calluses are picked off and regenerated.

ii) *Cocultivation*: Protoplasts are isolated and during the cell wall reformation stage, these are incubated for 24–40 h in a suspension of *Agrobacterium* at about 100 bacteria per protoplast. Transformation occurs during the subsequent few days of cocultivation and exposure to selective agent.

iii) *Leaf disc method*: In practice, this procedure can be performed with any tissue explant that provides a good source for initiation of whole plant differentiation. In this regard, newly emerged cotyledons have provided useful material. In the disc, or explant method, a tissue segment is excised and tissue is allowed to incubate in *Agrobacterium* suspension for few hours to three days and then cultured on a media for bacterial growth to take place. Tissue explants are then moved into a media containing a bacteriostatic agent that eliminates the bacteria. Then explants are moved to a media designed for the selection of trans-

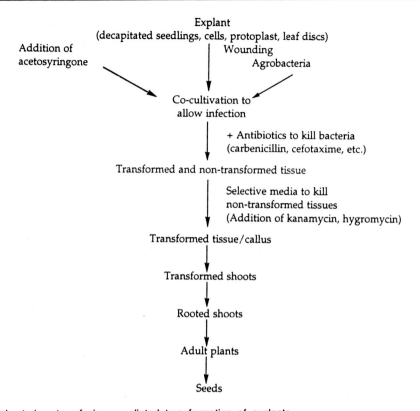

Explant
(decapitated seedlings, cells, protoplast, leaf discs)

Addition of
acetosyringone

Wounding

Agrobacteria

Co-cultivation to
allow infection

+ Antibiotics to kill bacteria
(carbenicillin, cefotaxime, etc.)

Transformed and non-transformed tissue

Selective media to kill
non-transformed tissues
(Addition of kanamycin, hygromycin)

Transformed tissue/callus

Transformed shoots

Rooted shoots

Adult plants

Seeds

Fig. 23.8 *Agrobacterium tumefaciens* mediated transformation of explants.

formed plant cells, and therefore, contains the appropriate antibiotics. The disc method is technically simple and produces transgenic plants very rapidly and is therefore the favored procedure at present.

Procedure

A particular *Agrobacterium* strain is first grown in solid media plates (e.g. YEP medium) containing required antibiotics as per bacterial selectable markers. *Agrobacterium* suspension for cocultivation/inoculation is prepared by picking a single colony from media plate and inoculating in 5–10 ml of liquid medium (YEP) containing appropriate antibiotics. If *Agrobacterium* is to be preinduced by acetosyringone (e.g. in monocots), then it should be added in the liquid medium. After 2 days of growth in dark at 28°C, the OD of the culture is measured at 600 nm. The suspension is diluted or concentrated to bring the value at 0.1

OD for cocultivation. Explants are then inoculated/ cocultivated with *Agrobacterium* suspension for few hours to few days. Explants are then blot-dried on a sterile filter paper and transferred to nutrient culture medium of explant without bacteriostatic agent for 3 days. This allows the *Agrobacterium* to grow, multiply, and transfer the T-DNA. The explants are then transferred to media containing bacteriostatic agent such as cefotaxine and carbenicillin to kill the bacteria. These explants are then transferred to a selection medium containing appropriate selection agent (antibiotics, herbicides, etc.) depending on the plant selectable marker genes employed for transformation. Resistant/tolerant tissues are continuously grown and selected on selection agent followed by regeneration. These transgenic plants are then tested for stable integration and expression of genes by PCR method or by Southern hybridization. These are T_0 transgenic plants and seeds obtained from these plants are T_0 seeds. Plants obtained from these T_0 seeds are T_1 plants.

Agrobacterium-mediated virus infection – Agroinfection

The delivery of viral or viroidal sequences to plants using bacterium as a route is referred to as "agroinfection" or "agroinoculation." It has now been shown that plant virus genomes that are cloned into Ti plasmids can be introduced into plants from *Agrobacterium* and they will then initiate a typical infection. This *Agrobacterium*-mediated virus infection has been termed "agroinfection." Successful transfer of a single viral genome can be detected as a result of massive amplification during viral replication.

Grimsley *et al.* (1987) constructed a vector that contained the cloned double-stranded copy of the maize streak virus (MSV) genome, a gemini virus. This virus has a single-stranded DNA genome that replicates in the plant nucleus via DNA intermediates that are autonomous. The naked viral DNA and cloned DNA are not infectious and are transmitted in nature by insects. A head-to-tail dimer of the MSV genome was inserted into a mini-T-DNA vector between border sequences. The plasmid was mobilized into *A. tumefaciens* in a triparental mating and the host agrobacteria were inoculated into leaf wounds of maize seedlings. The maize plants developed all the symptoms of MSV infection. This provides a sensitive assay for T-DNA transfer to monocots.

Advantages

Agrobacterium-mediated transformation has the following advantages
1. It is a natural means of transfer and is therefore perceived as a more acceptable technique to those who feel natural is the best.
2. *Agrobacterium* is capable of infecting intact plant cells, tissues, and organs. As a result, tissue culture limitations are much less of a problem. Transformed plants can be regenerated more rapidly.

3. *Agrobacterium* is capable of transferring large fragments of DNA very efficiently without substantial rearrangements.
4. Integration of T-DNA is a relatively precise process.
5. The stability of gene transferred is excellent.

Disadvantages

Although the introduction of new traits into plants via the *Agrobacterium* system is now a common practice, there are still shortcomings in the system.
1. It has the limitation of host range; some important food crops cannot be infected with *Agrobacterium*. Although lately, much progress has been achieved to overcome this limitation by the development of highly virulent strains of *Agrobacterium*.
2. Sometimes, cells in a tissue that are able to regenerate are difficult to transform. It might be that embryogenic cells are in deep layers to be reached by *Agrobacterium*, or simply are not targets for T-DNA transfer.

Viruses-mediated gene transfer

Vectors based on viruses are desirable because of the high efficiency of gene transfer that can be obtained by infection and the amplification of transferred genes that occurs via viral genome replication. Also, many viral infections are systemic so that gene can be introduced into all the cells in a plant. Viruses provide natural examples of genetic engineering since viral infections of a cell result in the addition of new genetic material which is expressed in the host. Additional genetic material incorporated in the genome of a plant virus might be replicated and expressed in the plant cell along with the other viral genes. Once the biological characteristics of a virus have been suitably selected and manipulated, new genes must be incorporated into the virus in such a way that they are expressed in the plant. Vectors for transferring genes into plants are based on DNA or RNA molecules that naturally express their genetic information in plant cells. The replicating genomes of plant viruses are nonintegrative vectors as compared to those vectors based on the T-DNA of *A. tumefaciens*, which are integrative gene vectors.

Nonintegrative vectors as plant virus vectors do not integrate into the host genome; rather, they spread systemically within a plant and accumulate to high copy numbers in their respective target cells. In most cases, viral genomes have been modified to accommodate the insertion of foreign sequences, which are transferred, multiplied, and expressed in a plant as part of the recombinant virus genome. Apart from their pathogenicity, recombinant viruses are designed to mimic their wild type counterparts in all other respects. A viral vector should possess the following characteristics:

1. It should have a broad host range, virulence, ease of mechanical transmission, and rate of seed transmission.
2. It should have the potential to carry additional genetic information since there are strict packaging limitations. Thus viruses whose capsid are filamentous or rod-shaped, or viruses that possess a multipartite genome or a helper or satellite component, offer the potential for carrying extra nucleic acid.
3. Virus suitability as a vector depends on the fact that genetic material must be able to be manipulated and be infectious.

At present three groups of viruses are being studied for vector development and transformation.

Caulimoviruses

Among the plant viruses type, the virus of the caulimovirus group, cauliflower mosaic virus (CaMV), is cited as the most likely potential vector for introducing foreign genes into the plants. This is mainly because caulimoviruses are unique among plant viruses in having a genome made of double-stranded DNA, which of course lends itself more readily to the manipulations. They were the first plant viruses to be manipulated by the use of recombinant DNA technology. The caulimoviruses group

consists of 6–19 viruses, each of which has a limited host range. The commonest are the carnations etched virus (CERV), cauliflower mosaic virus (CaMV), dahlia mosaic virus (DaMV), mirabilis mosaic virus, and strawberry vein banding virus. The best-known member CaMV infects many members of Cruciferae and *Datura stramonium*. The virus is naturally transmitted by aphids but can very easily be transmitted mechanically. It has been found that virion DNA alone or cloned CaMV DNA is infectious when simply rubbed on the surface of susceptible leaves. Virus particles can be found at a high copy number (up to 10^6 virions per cell) and spread rapidly in 3–4 weeks in a systemic manner throughout the entire plant. Virus accumulates in the cytoplasm as inclusion bodies, which consist of a protein matrix with embedded virus particles. The virus particle is spherical, isometric, about 50 nm in diameter, and may be isolated from the inclusion body using urea and nonionic detergents. The DNA exists in linear, open circular, and twisted or knotted forms; however, none of the circular forms is covalently closed due to the presence of site-specific single stranded breaks (Fig. 23.9).

The genome of CaMV consists of a relaxed circular molecule of 8 kb. DNA sequence analysis revealed that six major and two minor open reading frames (ORF) are present on one coding strand in a very tightly packed arrangement. The caulimoviruses as well as badnaviruses replicate via reverse transcription of a more than genome length RNA. The mechanism of replication is as follows: The infecting CAMV DNA enters the plant nucleus, where the single-stranded overlaps are digested and gaps ligated to give a supercoiled minichromosome. The function of this minichromosome is to act as a template for plant nuclear RNA polymerase II. The transcript thus formed is transported to the cytoplasm where it is either translated or replicated by reverse transcription. The RNA transcript is then copied into the minus-strand DNA. Synthesis of the plus strand DNA starts at two primer binding sites near gaps 2 and 3. From gap 2, synthesis proceeds to the 5′ end of the minus strand DNA, whereas synthesis from gap 3 continues to gap 2. This DNA molecule gets packaged into virus particles or re-enters the nucleus and undergoes another round of transcription and/or translation/ replication.

Two genomic regions, ORF II that codes the insect transmission factor and ORF VII with unknown function, can be dispensed with and replaced by the gene of interest. There is one report on the insertion of a foreign gene into the plant describing the use of CaMV. Brisson *et al.* (1984) achieved the expression of foreign protein encoded by the bacterial dihydrofolate reductase gene in turnip plant cells by a CaMV vector.

Uses as a vector

1. Naked DNA is infective, being able to enter plant cells directly if rubbed on to a leaf with a mild abrasive.

2. As a DNA virus whose genome is known to be packaged in nucleosomes and transcribed by RNA polymerase II, it is more suited for exploitation as an experimental tool than any other plant virus.

Problems

1. The genome is so tightly packed with the coding regions that there is little room for insertion of foreign DNA. Most deletions of any significant size destroy virus infectivity. The theoretical packaging capacity is of 1000 nucleotides, but attempts to propagate and stably express genes larger than 500 bp have not been successful.

2. CaMV-derived vectors are restricted to members of Cruciferae that can be infected by viral DNA. Recently, some mutant strains of CaMV infecting species of Solanaceae have also been described.

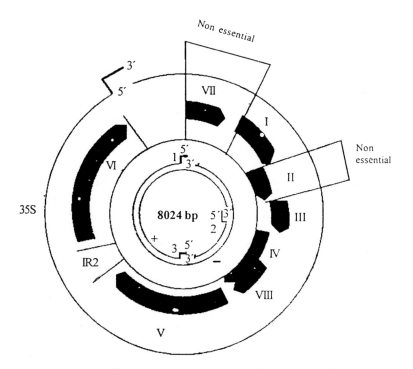

Fig. 23.9 A functional map of the cauliflower mosaic virus genome. The coding regions are shown as dark boxes. The different reading frames are indicated by the position of the boxes. The thin line in the centre is the DNA and outside is 35S transcript. Concept of the figure was taken from *Applied Molecular Genetics,* Roger L. Miesfeld, by kind permission of Johny Wiley & Sons, Inc., 605 Third Avenue, New York, USA.

3. CaMV DNA has multiple cleavage sites for most of the commonly used restriction endonucleases. This would limit the usefulness of wild isolates of CaMV.
4. Infectivity with CaMV is another problem because once established it becomes systemic, spreading throughout the whole plant. This lack of inheritance through the germline may be advantageous in that the CaMV DNA and any inserted gene sequence would be highly amplified in the host plant cells, potentially permitting the expression of large quantities of the foreign gene product. However, it appears that to propagate CaMV and to allow its movement throughout the vasculature of the plant, the DNA must be encapsidated, and this would impose serious constraints on the size of foreign DNA that can be inserted into the viral genome.

Gemini viruses

The gemini viruses replicate and cause diseases in a wide variety of plant species, including many of agricultural importance. These viruses have some properties that make them ideal vectors for the expression of foreign genes in plants. The potential of gemini viruses as gene cloning vectors for plants stems from work on several plant diseases now recognized as being caused by these agents. Both the curly top virus (CTV) and maize streak virus (MSV) cause important diseases.

The gemini viruses are characterized on the basis of their unique virion morphology and possession of single-stranded (ss) DNA. The gemini virus group takes its name from the unusual twin icosahedral capsid structure of its members. These have a small capsid size, 18–20 nm × 30 nm, and geminate (paired particles) morphology, which makes them different from all other classes of viruses. They have covalently closed circular (ccc) topography of the ssDNA, which is in the molecular weight range of $7 \times 10^5 - 9 \times 10^5$. All gemini viruses have a single major coat protein subunit in the range $2.7–3.4 \times 10^4$ daltons.

The genome of geminiviruses consists of either one or two circular, single-stranded DNA molecules. The single-stranded viral DNA, 2.6–3.0 kb long, is converted into a double-stranded replicative form in the nucleus of plant cells. Many copies of the replicative form of a geminivirus genome accumulate inside the nuclei of infected cells. There is no evidence to date for a reverse transcription step in geminivirus replication.

The bean golden mosaic virus (BGMV) DNA was found to be 2510 nucleotides long, and if this was the complete genome it would be less than half the length of any other known autonomously replicating plant virus. By comparing the ssDNA of the virus particles with the viral dsDNA found in infected plants, it was found that the nucleotide sequence had twice the complexity expected on the physical size of the viral DNA. This indicates that BGMV DNA is heterogeneous, the virus having a divided genome consisting of two DNA molecules of approximately the same size but different genetic content.

The genomic ssDNA replicates in the nucleus through double-stranded DNA (dsDNA) intermediates, most likely by a rolling circle mechanism. Despite an overall similarity in capsid and genome structure, the geminivirus group is diverse and can be divided into at least three subgroups.

The genomes of all group members contain two elements that must be part of any functional geminivirus vector system, although these elements need not be physically present on the same molecule. These elements are an origin of replication and an essential viral replication protein.

Most geminivirus vectors described to date are relatively primitive and simply permit the investigator to replace the coat protein coding sequence with a reporter or other gene of interest. Geminivirus expression vectors may be used to deliver, amplify, and express foreign genes in several different systems of protoplasts and cultured cells, leaf discs, and plants.

The cereal geminivirus, wheat dwarf virus, is under development as a vector for introducing genes into cereals. It is capable of accommodating and replicating gene inserts up to about 3 kb in length. Three bacterial genes inserted into the wheat dwarf virus genome have been successfully replicated and expressed after transfer into cultured cells of *Triticum monococcum* and *Zea mays.*

Uses as a cloning vector ·

1. These viruses contain single-stranded DNA that appears to replicate via a double-stranded intermediate and thus makes *in vivo* manipulation in bacterial plasmids more convenient.
2. An attractive feature is the ability of bipartite geminiviruses to contain a deletion or a replacement of virus coat protein sequences by foreign genes without interfering with the replication of the virus genome.

Problems

1. These are not readily transferred by mechanical means from plant to plant, but transmitted in nature by insects in a persistent manner.
2. The small particle size may present packaging problems for modified DNA molecules and any useful genetic modifications will have to solve the problems of a vector, which in its natural state causes severe disease in susceptible plants.

RNA viruses

RNA viruses have advantages and disadvantages, which make them the ideal choice for certain applications as vectors. Their assets include ease of use, perhaps the highest levels of gene expression, and the vector's ability to infect and spread in differentiated tissue.

There are two basic types of single-stranded RNA viruses. The monopartite viruses have undivided genomes containing all the genetic information and are usually fairly large, for example, tobacco mosaic virus (TMV). The multipartite viruses, as the name suggests, have their genome divided among small RNAs either in the same particle or separate particles; for example, the brome mosaic virus (BMV) contains four RNAs divided between three separate particles. The RNA components of multipartite genomes are small and appear to be able to self-replicate in plants. With some members of the group, the genes encoding the coat protein may be dispensable, as their loss does not affect viral DNA multiplication.

The second group, subgenomic RNAs (e.g. RNA IV of BMV), is unlikely to find application as cloning vectors as it is unable to self-replicate in infected plants.

The third group, satellite RNAs, has perhaps the greatest potential as it is totally dispensable to the virus. Satellite RNAs vary in size from 270 bases (tobacco ringspot virus satellite) to 1.5 kb (tomato black ring virus satellite). These satellites appear to share little homology with the viral genomic RNAs, templates for their own replication, and utilize the machinery for replication set up by the virus. They are not required for virus replication, but are capable of altering the pathogenicity of the viral infection. These satellite RNAs have a number of other unusual properties, including the ability to code for proteins and stability in the plant in the absence of other viral components.

Plant viruses have only few genes, but these are expressed to high levels by a variety of means. High gene product yields and activities are associated with RNA viral vectors. Plant viral genomes contain a gene for movement through the plant, a gene for the protein making up the protective coat, and a gene for replication. Coat proteins are the most abundant viral gene product. One example of a gene replacement construction is the replacement of RNA viral vector coat protein gene of BMV by the chloramphenicol acetyl transferase gene (French *et al.*, 1986). Expression of cDNA in barley protoplasts suggests that RNA virus may be a useful vector for gene manipulation.

A second common type of vector construction is the insertion of a foreign gene into an intact viral genome. An example is the TMV vector TB2 that was the first plant RNA viral vector able to spread systemically in whole plants (Donson *et al.*, 1991). In TB2, the foreign gene is inserted at the 3´ end of the movement protein, which is the normal position for the TMV coat protein ORF (Fig. 23.10). Thus, foreign gene expression is driven by the native coat protein subgenomic promoter (sgp) and a sgp from related virus ORSV (odontoglossum ringspot tobamovirus) drives the expression of ORF of coat protein.

Most RNA viral vectors carrying large or small inserts replicate stably in protoplasts and/or inoculated leaves. Thus, experiments can be carried out in either protoplasts or differentiated tissue, using the present generation of plant viral vectors.

A variety of DNA and RNA viral vectors can be used to express genes in protoplasts, but the choice of vector is limited for use in inoculated leaves of whole plants. The other disadvantages are limitation of insert size and the induction of symptoms in the host.

Vectorless or direct DNA transfer

Direct gene transfer has proved to be a simple and effective technique for the introduction of foreign DNA into plant genomes. It has been further subdivided into three categories:
1. Physical gene transfer methods;
2. Chemical gene transfer methods; and
3. DNA imbibition by cell, tissue, and organs.

Physical gene transfer methods

Species- and genotype-independent transformation methods wherein no natural vector is involved but which are based on the direct delivery of naked DNA to the plant cells have been grouped under this category. This is also referred to as DNA-mediated gene transfer (DMGT). A general scheme for the production of transgenic plants using various direct DNA delivery methods is shown in Fig. 23.11. The various methods are described in detail.

Electroporation

Electroporation is the process where electrical impulses of high field strength are used to reversibly permeabilize cell membranes to facilitate uptake of large molecules, including DNA. The electroporation method is based on Neumann *et al.* (1982) for animal cells. It uses a relatively high initial field strength (1–1.5 kV) with a low capacitance and therefore a short decay time. Other methods with low initial field strength and long decay time have been described. In this procedure, a sample of protoplasts is pulsed with high/low voltage pulses in the chamber of an electroporator. The chamber is cylindrical in form with a distance of 1 cm between parallel steel electrodes. The pulse is applied by discharge of a capacitor across the cell. It has been reported that by using linear DNA rather than circular DNA, a field strength of 1.25 kV/cm and employing polyethylene glycol (PEG) can increase protoplast

Insert : [DHFR, *npt*II, α trichosanthin]
rep   : viral replicase

sgp-t & sgp-o: sgp from TMV and ORSV (Odontoglossum ringspot tobamovirus)
CP : viral coat protein;   MP : viral movement protein

Fig 23.10 Genomic map of TMV : TB2 plant RNA viral vector.

transformation frequency. PEG is believed to assist the association of the DNA with the membrane.

Electroporation has been used for a long time for transient and integrative transformation of protoplasts. Only recently, conditions under which DNA molecules can be delivered into intact plant cells of sugarbeet and rice that are still surrounded by a cell wall have been standardized. Further, transformability of intact plant cells or tissues depends on pretreatment of the cells or tissues to be transformed, either by mechanical wounding or by treating the cells or tissues with hypertonic or enzyme-containing (e.g. 0.3% macerozyme) solutions.

The range of tissues that can be transformed by electroporation seems to be narrower. For tissues that are susceptible to DNA uptake by electroporation, this method is convenient, simple, fast, has low cell toxicity, and it is inexpensive to obtain transient and stable transformation in differentiated tissues. The disadvantage of the technique is the difficulty in regenerating plants from protoplasts.

Particle bombardment/microprojectile/ biolistics

The technique of particle bombardment, also known as biolistics, microprojectile bombardment, particle acceleration, etc., has been shown to be the most versatile and effective way for the creation of many transgenic organisms, including microorganisms, mammalian cells, and plant species. The procedure in which high-velocity microprojectiles were used to deliver nucleic acids into living cells was described by Klein *et al.* (1987) and Sanford *et al.* (1987).

The basic system that has received attention employs PDS 1000 (gun-powder-driven device) or the PDS-1000/He (helium driven particle gun). The DNA bearing tungsten or gold particles (1– 3 μm in diameter), referred to as microprojectiles, is carried by a macroprojectile or macrocarrier and is accelerated into living plant cells. The DNA-bearing particles (microprojectiles) are placed on the leading surface of the macrocarrier and released from the macrocarrier upon impact with a stopping plate or screen. The stopping plate is designed to halt the forward motion of macroprojectile while permitting the passage of the microprojectiles. In this procedure, when helium gas is released from the tank, a disc known as rupture disc blocks its entry to the chamber. These discs are available with various strengths to resist the pressure of gas, which varies from 500 to 1700 p.s.i. When the disc ruptures, compressed helium gas is suddenly released, which accelerates a thin plastic sheet carrying microprojectiles into a metal screen. Upon impact with a stopping plate or screen, the macroprojectile movement is stopped, but this permits the passage of microprojectiles through the mesh screen. The microprojectiles then travel through a partial vacuum until they reach the target tissue. The partial vacuum is used to reduce the aerodynamic drag on the microprojectiles and decrease the force of the shock wave created when the macrocarrier impacts the stopping plate. See Fig. 23.12 for a schematic diagram of the PDS-1000/He.

The use of particle bombardment requires careful consideration of a number of parameters. These can be classified into three general categories:

Physical parameters

Nature, chemical, and physical properties of the metal particles utilized to carry the foreign DNA: Particles should be of high mass in order to possess adequate momentum to penetrate into the appropriate tissue. A variety of metals such as tungsten, gold, platinum, palladium, rhodium, iridium, and possibly other second- and third-row transition metals can be used. Size range of the particles is *ca.* 1 μm. Tungsten particles of diameter 1.2 μm and gold particles of diameter 1.0–1.6 μm have been most commonly used. Metals should be chemically inert to prevent adverse reactions with the DNA or cell components.

Nature, preparation and binding of DNA on to the particles: The nature of DNA, that is, as single versus double stranded, may be important under some conditions, but this may not be a significant variable in specific cases. In the process of coating metal particles with DNA, certain additives such as spermidine and calcium chloride appear to be useful.

Target tissue: It is very important to target the appropriate cells that are capable of cell division and are competent for transformation. It is apparent that different tissues have different requirements, thus extensive studies need to be performed in order to ascertain the origin of regenerating tissue in a particular transformation study. Penetration of microprojectile (DNA coated to inert metal of gold or tungsten) is one of the most important variables, and the ability to tune a system to achieve particle delivery to specific layers may result in success or failure in recovering transgenic plants from a given tissue. Generally the cells near the center of the target are injured and thus cannot proliferate.

Environmental parameters: These include variables such as temperature, photoperiod and humidity of donor plants, explants, and bombarded tissues. These parameters affect the physiology of the tissues, influence receptiveness of target

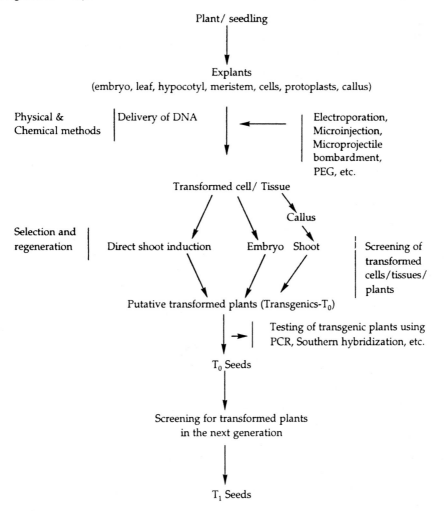

Fig. 23.11 A general scheme for the production of transgenic plants using direct gene delivery methods.

tissue to foreign DNA delivery, and also affect its susceptibility to damage and injury that may adversely affect the outcome of the transformation process. Some explants may require a healing period after bombardment under special regimes of light, temperature, and humidity.

Biological parameters: Choice and nature of explants, and pre- and postbombardment culture conditions determine whether experiments utilizing particle bombardments are successful. The explants derived from plants that are under stress or infected with bacteria or fungi, over- or under-watered will be inferior material for bombardment. Osmotic pre- and post-treatment of explant with mannitol has been shown to be important in transformation. Experiments performed with synchronized cultured cells indicate that the transformation frequencies may be also influenced by cell cycle stage.

Advantages

Several advantages make this technique a method of choice for engineering crop species.

1. It is clean and safe.
2. *Transformation of organized tissue:* The ability to engineer organized and potentially regenerable tissue permits introduction of foreign genes into elite germplasm.
3. *Universal delivery system:* Transient gene expression has been demonstrated in numerous tissues representing many different species.
4. *Transformation of recalcitrant species:* Engineering of important agronomic crops such as rice, maize, wheat, cotton, soybean, etc. has been restricted to a few noncommercial varieties when conventional methods are used. Particle bombardment technology allowed recovery of transgenic plants from many commercial cultivars.
5. *Study of basic plant development processes:* It is possible to study developmental processes and also clarify the origin of germline in regenerated plants by utilizing chromogenic markers.

Disadvantages

1. In plants, gene transfer leads to nonhomologous integration into the chromosome, and is characterized by multiple copies and some degree of rearrangement.
2. The emergence of chimaeral plants.
3. Lack of control over the velocity of bombardment, which often leads to substantial damage to the target cells.

Macroinjection

It is the injection of DNA solution (5–10 µl) by micropipettes into the developing floral side shoots (tillers) of plants. Within the floral tillers are archesporial cells that give rise to pollen in the developing sac by two meiotic cell divisions. It was suggested that such cells might also be able to take up large molecules such as DNA. Thus a plasmid encoding kanamycin resistance and under the control of a promoter was injected into the tillers of rye plants. It resulted two plants showing resistance to kanamycin (De la Pena *et al.*, 1987). However, these experiments could not be repeated.

Microinjection

Microinjection is the direct mechanical introduction of DNA under microscopical control into a specific target. A target can be a defined cell within a multicellular structure such as embryo, ovule, and meristem or protoplasts, cells, or a defined compartment of a single cell. As a direct physical method, microinjection is able to penetrate intact cell walls. It is host-range independent and does not necessarily require a protoplast regeneration system. This method has been proposed for the transfer of cell organelles and for the manipulation of isolated chromosomes. When cells or protoplasts are used as targets in this technique, glass micropipettes of 0.5–10.0 µm diameter tips are used for transfer of micromolecules into the cytoplasm or the nucleus of a recipient cell or protoplast. Recipient cells can be immobilized using methods such as agarose embedding, agar embedding, poly-lysine treated glass surfaces, and suction holding

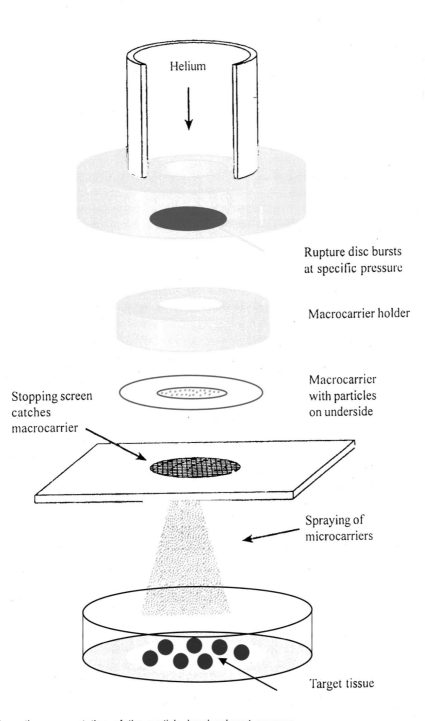

Helium

Rupture disc bursts
at specific pressure

Macrocarrier holder

Stopping screen
catches
macrocarrier

Macrocarrier
with particles
on underside

Spraying of
microcarriers

Target tissue

Fig. 23.12 Schematic representation of the particle bombardment process.

pipettes (Fig. 23.13). Once injection has been achieved, the injected cell must be cultured properly to ensure its continued growth and development. With this method the operator has the ability to manoeuver the cell and thus more accurately target the nucleus. Transgenic chimaeras have actually been obtained in tobacco and *Brassica napus* by this approach. The main disadvantage of this technique is the production of chimaeric plants with only a part of the plant transformed. The microinjection process is slow, expensive, and requires highly skilled and experienced personnel.

Liposome-mediated transformation

Liposomes are artificial lipid vesicles surrounded by a synthetic membrane of phospholipids, which have been used in mammalian tissue culture to deliver drugs, proteins, etc. into the cells. These can be induced to fuse with protoplasts using PEG and have therefore been used for gene transfer. In this method, DNA enters the protoplasts due to endocytosis of liposomes, and this involves the following steps:

i) Adhesion of liposomes to the protoplast surface,
ii) Fusion of liposomes at the site of adhesion, and
iii) Release of plasmids inside the cell.

Uptake of liposomes containing DNA molecules has been demonstrated in protoplasts of tobacco and carrot (Fig. 23.14). Liposome-mediated transformation showed higher efficiency when it was used in conjunction with PEG. Liposomes offer many advantages for protoplast transformation, including (i) protection of nucleic acids from nuclease digestion prior to introduction into cellular environment, (ii) reduced levels of cell toxicity, (iii) applicability to a wide variety of cells types, (iv) high degree of reproducibility, and (v) stability and storage of nucleic acids once encapsulated.

Silicon carbide fiber-mediated transformation

Silicon carbide fibers (SCF) average 0.6 μm in diameter and 10–80 μm in length. It has been demonstrated that these fibers have the capability to deliver DNA into plant cells. The method involves vortexing a mixture of plasmid DNA encoding a selectable or screenable marker gene, silicon carbide fibers, and the explant tissue in an eppendorf tube. Silicon carbide fibers have great intrinsic hardness with sharp cutting edges. DNA delivery in this system is presumably due to cell wall penetration by DNA-coated silicon carbide fibers during vortexing of SCF with explant and DNA adhering to fibers might enter the cells. It is possible that SCF function as numerous needles facilitating DNA delivery into the cells. During the mixing process, DNA penetrates the cell to become stably integrated into the nuclear genome. Advantages of the silicon carbide fiber-mediated transformation method over other procedures include the ability to transform walled cells thus avoiding protoplast isolation, relative ease of the procedure, and very low equipment costs. The disadvantage is that silicon carbide has some carcinogenic properties.

The feasibility of this technique has been demonstrated in fertile transgenic plants of maize and tobacco.

Ultrasound-mediated DNA transformation

Ultrasound is used described for stimulating uptake of foreign DNA by plant protoplasts and leaf segments of tobacco. The procedure involves immersion of explant (leaves/protoplasts) in sonication buffer containing plasmid DNA and sonication with an ultrasonic pulse generator at an acoustic intensity of 0.5 w/cm^2 for 30 min. Samples are rinsed in a buffer solution and then cultured for growth and differentiation. This technique has the advantages of being simple, inexpensive, and multifunctional. One can use standardized conditions and there is no requirement for tissue culture expertise. Little success has been achieved by this technique.

DNA transfer via pollen

Pollen has been suggested as a vector for gene transfer by various workers. It has been reported that introduction of DNA into gametes followed by

Magnification of
microinjection

OIL SUCTION

Pressure Oil

DNA

HOLDING PIPETTE   CELL   INJECTION PIPETTE

RETURN INJECTED CELLS TO PETRI DISH

Fig. 23.13 Microinjection by holding pipette method.

fertilization and zygotic embryogenesis will result in gene transfer. This kind of approach would be simpler, faster, and cheaper than the *in vitro* methods. Since ovules are difficult to isolate and the injection of DNA into embryo sac *in situ* seemed to be too tedious and unpredictable, it was logical to favor the use of pollen as a vector for DNA. The use of DNA-treated pollen as a vector for pollinating fertile plants of maize has been studied. There are various reports indicating that DNA can be either taken up by germinating pollen and integrated into sperm nuclei or reach the zygote via pollen tube pathway. The main problems are the presence of cell wall and action of nucleases on the DNA. *Agrobacterium* has also been used to transfer the T-DNA via pollen as a vector to overcome the nuclease action on DNA. Both these approaches have been tried and some success has been achieved.

Chemical gene transfer methods

This involves plasma membrane destabilizing and/or precipitating agents. Protoplasts are mainly used being incubated with DNA in buffers containing PEG, poly L-ornithine, polyvinyl alcohol, or divalent ions. The chemical transformation techniques work for a broad spectrum of plants.

PEG-mediated gene transfer

The first conclusive demonstration of uptake and integration of isolated Ti plasmid DNA into plant protoplasts was reported in *Petunia* and tobacco in the presence of poly L-ornithine or polyethylene glycol (PEG). The presence of Ti DNA in the plant genome was demonstrated both by the phenotype of hormone auxotrophic growth, the production of the expected opine and by Southern blot analysis of DNA from the transformants.

In a general procedure, protoplasts are isolated and a particular concentration of protoplast suspension is taken in a tube followed by addition of plasmid DNA (donor and carrier). To this 40% PEG 4000 (w/v) dissolved in mannitol and calcium nitrate solution is slowly added because of high viscosity, and this mixture is incubated for few minutes (*ca.* 5 min). As per the requirements of the experiment, transient or stable transformation studies are conducted.

Among the most important parameters that affect the efficiency of PEG-mediated gene transfer are the concentration of magnesium or calcium ions in the incubation mixture, and the presence of carrier DNA. The linearized double-stranded plasmid DNA molecules are more

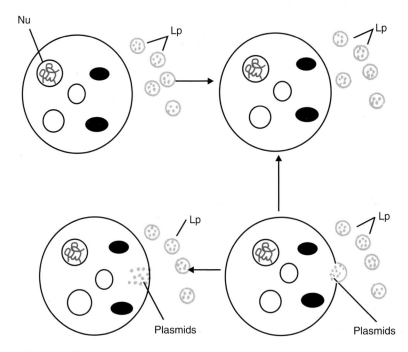

Fig. 23.14 Fusion of plasmid-filled liposomes with protoplast. Nu: nucleus; Lp: liposomes.

efficiently expressed and integrated into the genome than supercoiled forms. Integration of foreign DNA into the nuclear genome occurs predominantly at random sites.

The main application of the technique, apart from analyzing the transformation process itself, is in introducing foreign genes to plant cells. This can be accomplished by constructing a molecule containing a selectable marker and the gene of interest, or more easily by simply mixing DNA of the gene of interest with the selectable marker plasmid in a molar ratio of approximately 3:1 to 10:1, transforming, selecting for the marker, and analyzing transformants for the presence of the second gene. The method can also be applied with DNA from different sources. Manipulation of nucleic acids prior to transformation is also possible and there are no host range limitations.

The advantage of the method is that the form of the DNA applied to the protoplasts is controlled entirely by the experimenter and not by an intermediate biological vector. The main disadvantage is that the system requires protoplasts and a functional system for regeneration of these protoplasts to calluses and whole plants. It is therefore not applicable to many plant systems. In addition, the relatively random way in which the DNA is integrated into the genome means that, for the introduction of nonselectable genes, a thorough characterization of the transformants by Southern blot analysis is necessary to confirm the nature of the integration event.

Calcium phosphate coprecipitation

In this method, DNA is mixed with calcium chloride solution and isotonic phosphate buffer to form a DNA–CaPO$_4$ precipitate. The precipitate is allowed to react with actively dividing cells for several hours, washed, and then incubated in fresh culture medium. Giving them a physiological shock with DMSO can increase the efficiency of transformation to a certain extent. Relative success depends on high DNA concentration and its apparent protection in the precipitate.

The polycation DMSO technique

It involves the use of a polycation, polybrene, to increase the adsorption of DNA to the surface followed by a brief treatment with 25–30% DMSO to increase membrane permeability and enhance the uptake. The major advantage of polybrene is that it is less toxic than other polycations and a high transformation efficiency requires very small quantities of plasmid DNA to be used.

DEAE dextran procedure

Transformation of cells with DNA complexed to the high-molecular-weight polymer diethyl amino ethyl (DEAE) dextran is used to obtain efficient transient expression. The efficiency increases to 80% when DMSO shock is given. But this technique does not produce stable transformants.

Advantages

Direct gene transfer for cell transformation circumvents the host range limitation. It is possible to introduce DNA sequences directly into cells and organs by biolistic or electroporation approaches. Cereals, the world's most important crops, have been transformed by direct gene transfer methods.

Disadvantages

i. Direct gene transfer usually depends on a culture system that allows regeneration of mature plants from protoplasts/cells/organs, which is not usually available for every desired plant species. The problem may be overcome by further improvement of plant tissue culture methods.

ii. These methods show unpredictable pattern of foreign DNA integration. During their passage into the nucleus, the DNA is subjected to nucleolytic cleavage resulting in truncation, recombination, and rearrangement or silencing.

DNA imbibition by cells, tissues, embryos and seeds

Incubation of cells, tissues, and organs with DNA for transformation has met with little or no success and hence has not resulted in any proven case of integrative transformation, though reporter gene has been expressed from imbibed dry seed and embryos. In all these cases, the plant cell walls not only work as efficient barriers but are also efficient traps for DNA molecules.

STATUS AND EXPRESSION OF TRANSFERRED GENES

DNA is introduced into plant cells and these cells are then grown *in vitro* to regenerate plants. Selection and growth of plant cells on selective media provide initial phenotypic evidence for transformation. This includes resistance to antibiotics, herbicides, etc. Several studies have analyzed the inheritance and expression of transferred DNA. The location of the genes and temporal and spatial aspects of their expression may also be important. In general, when transgenic plants are produced by a direct DNA transformation system, more screening has to be done to select lines that contain a limited number of integrated copies. In most cases it will not be possible to separate these copies by crossing because they will often be tightly linked. The transgenic plants so obtained are self-fertilized. DNA introduced by methods other than *Agrobacterium*-mediated transformation integrates into the genome by as yet unknown mechanisms. Generally the DNA integrates at a single locus, but it usually consists of rearranged multimers of the donor DNA. Molecular evidences (e.g. PCR analysis, Southern hybridization) are essential to confirm integration of transferred genes. Seeds are analyzed in the T_1 (transgenic first) generation for segregation data. The simple pattern of inheritance (3:1) indicates a single site of insertion and this can be confirmed by Southern blotting. The copy number is usually low, between one and five copies per cell. The DNA is transmitted through meiosis as a simple Mendelian trait. Occasionally, however, more complex patterns of integration are seen.

In order to apply gene technology successfully to modern agriculture, it is essential to understand transgene expression. Sometimes when a transgene is introduced into an organism it may not show its expression. This is known as gene silencing.

Gene silencing

Partial or complete inactivation of gene(s) is known as gene silencing. Presence of multiple copies of a transgene in a plant nucleus can also lead to silencing of some or all copies of the gene(s) including the endogenous gene having homology with the transgene. It has also been established that in all such cases of transgene silencing, loss of expression of the transgenes is not due to the loss of these genes but due to their inactivation. Before considering the real causes of gene silencing it is must to know how a foreign DNA sequence is recognized by cell?

Detection of intrusive DNA: The ability to recognize "self" from "non-self" is a characteristic feature that occurs at cellular level. The observations from genetically engineered plants suggest that an ability to recognize "self" from "nonself" exists at the nucleic acid level. The possible mechanisms by which intrusive DNA is detected by cell are (i) Foreign sequence has different base sequence composition from that of endogenous chromosomal environment; (ii) Gene transfer via direct DNA delivery method has no control on copy number. These copies may form independent domains in the cell and work independently, with the result that they get recognized by genome scanning machinery; (iii) Every cell has its own modification and restriction system that shows xenophobic effect, which does not allow the cell to contaminate its chromosomal environment by any foreign sequence.

Genomes are made of isochores that are very long stretches of DNA with high compositional homogeneity. On this basis, a concept of gene space has been given. According to this concept, if a GC-rich transgene is integrated in to a GC-rich gene space (GC isochore) or an AT-rich transgene is integrated into the AT-rich gene space (AT isochore), it is normally transcribed.

But if the GC-rich transgene gets integrated into the AT-rich gene space or vice versa, it is inactivated, as there is no compositional homogeneity with neighboring sequence. A powerful initial response by the cell for the presence of foreign DNA may be fragmentation of invasive DNA via the action of cytoplasmic nucleases.

Causes of gene silencing: Silencing of a gene can be complete or partial. It may occur at transcriptional or translational level. After the entry of transgene into a receptive genome, there is an increase in methylation pattern. This is a major cause of gene silencing, leading to inactivation of gene(s). There are various causes of gene silencing.

1. *DNA methylation:* The inactivation of transgene is often accompanied by an increase in DNA methylation, and inactivation very frequently correlates with number of copies of integrated transgenes. When foreign DNA is introduced in plants, the modification and restriction system of cell causes methylation of foreign sequence to make it inactivated. DNA methylation is imposed on cytosine residues within symmetrical target sequences. In plants an inverse correlation between gene transcription and cytosine methylation has been observed for certain controlling elements as methylated DNA prevents transcription by directly hindering trans-acting factors and basal transcription machinery accessibility. The presence of methylated cytosine residue in the opposite strand provides the information to methylate C-residue in newly synthesized complementary strand with the help of the enzyme methyl transferase and S-adenosyl methionine as a cofactor. On the basis of DNA methylation, gene silencing may be (i) transcriptional silencing, which is linked to methylation of promoter region; (ii) post-transcriptional silencing, which involves inactivation of the coding sequence by methylation.

Induction of *de novo* methylation of transgenes involves different modes:

i) *DNA methylation via DNA–DNA pairing:* When multiple transgene copies integrate as a concatamer and DNA coiling occurs, then these copies come in front of each other just like homologous sequences. Due to their mutual suppressing effect, there is increase in methylation pattern that leads to inactivation of genes.

ii) *Transgene recognition:* The transgene gets integrated properly, but due to older age of transgene(s) or certain environmental stresses the transgene get hypermethylated and inactivated.

iii) *Insertion into hypermethylated genomic regions:* When a transgene gets integrated into a hypermethylated region, then there may be spreading of hypermethylation pattern that leads to inactivation of transgene.

2. *Homology-dependent gene silencing:* Not only did homologous sequences affect the stability of transgene expression but also the activity of endogenous genes could be altered after insertion of transgene into genome. This homology-dependent inactivation may involve different modes:

i) *Inactivation of homologous transgenes:* When transgenic plants were retransformed with constructs that are partly homologous with the integrated transgene. In the presence of second construct, the primary transgene becomes inactivated and hypermethylated within the promoter region. Silencing is influenced by length of homology. Minimum length of homology reported between two homologous sequences for inactivation is 90 bp. It is also influenced by position of interacting sequences (linked copies are more efficiently silenced).

ii) *Paramutation:* It is interaction of homologous plant alleles that leads to heritable epigenetic effects. Thus, combination of two homologous alleles that differed in their state of methylation results in paramutation phenomenon.

iii) *Cosuppression:* Expression of endogenous genes can be inhibited by the introduction of a homologous sense construct that is capable of transcribing mRNA of same strandedness as transcript of the gene. In co-suppression, there is suppression of both the transgene and homologous resident gene or inactivation of either of two. This phenomenon was first described in Petunia when a chalcone synthase gene (*chs*) involved in floral pigment biosynthesis was introduced in a deep violet flowering line of *P. hybrida*. Forty-two percent of the transgenic plants had white or variegated flowers. It was shown that cosuppression was due to a post-transcriptional event, and was independent of promoter, which was necessary for many different transgenes, and a single transgene can cause cosuppression of two or more endogenous genes.

3. *Suppression by antisense genes:* It is post-transcriptional inhibition of gene expression. Introduction of antisense and sense transgenes into plants has been widely used to generate mutant phenotype. Since 1980, a large number of transgenic plants containing antisense and sense genes have been generated. One of the first successful applications of antisense gene approach was the inhibition of fruit softening by antisense polygalacturonase gene, which lead to first commercial transgenic tomato "Flavr Savr" variety.

However, it is of major concern that antisense RNA may block RNA production along different pathways. Antisense gene often leads to reduced levels of target mRNA and can potentially form a double-stranded structure with complementary mRNAs. Antisense RNA may interfere with the following processes:

i. Antisense transcripts affect the target gene directly in the nucleus, thereby preventing synthesis of mRNA, i.e. transcription.

ii. Antisense RNA may block processing of mRNAs by masking the sequences recognized by splicing and the polyadenylation apparatus. Antisense RNA may disturb the normal transport of mRNAs out of the nucleus by forming a hybrid with their target mRNAs and disturbs the regular flow of transcripts.

iii. Many antisense RNAs complementary to ribosome binding site have been shown to inhibit translation initiation.

iv. Antisense RNA prevents the accumulation of target mRNA. Sense and antisense RNAs form a double RNA intermediate that is rapidly degraded by ds RNA specific ribonucleases.

4. *Position effect.* Whenever a transgene gets integrated into an improper region (hypermethylated, heterochromatic, telomeric, compositionally different genomic region), due to suppressing effect of the adjacent region or environment, it gets inactivated.

5. *Increased copy number.* A correlation between number of integrated copies and the frequency of inactivation is well documented for copies arranged in *cis* position. A reduction of copy number inside a locus was shown to increase gene expression or decrease the suppressing effect. The multiplicity does not need to include the whole gene. It may be either duplication of promoter or addition of a truncated transgene-coding region that can cause decreased expression. Increased copy number may be due to direct gene transfer methods as there is no control on copy number.

Strategies for avoiding gene silencing: To obtain stable expression and inheritance of transgenes in genetically transformed plants, the following criteria should be considered.

i. Gene silencing is frequently observed upon integration of complex inserts. Thus, it is better to opt for integration of single inserts of transferred gene without duplication in the form of tandem or inverted repeats and consisting of single unique elements. This can be achieved by using the vector-mediated gene transfer method, which generally introduces one copy (rarely two copies) in the cell.

ii. Irrespective of exact mechanism, homology at DNA or RNA levels seems to stimulate silencing events. If homology can be avoided or length and degree of homology should be controlled by interrupting perfect homology with mismatch or intron sequences.

iii. Integration of single-copy, unmethylated sequences of plant genome may increase the probability for continuous stable expression as the structure of integrated DNA itself, the environment of insert, may influence the stability of gene expression.

iv. Gene silencing occasionally becomes evident only after transmission of the gene to next generation. In some cases transgene expression decreased progressively over subsequent generations. Therefore, continuous monitoring of expression levels in progeny of even well-established transgenic lines might be a precaution against unexpected epigenetic modifications.

It is also necessary to arrange for the gene product to appear in the correct subcellular location. Also, it may be acceptable to have nuclear-encoded genes that are involved in functions such as detoxification or resistance switched on all the time in some situations. Sometimes, however, it is desirable to ensure the expression of a foreign gene at a particular developmental stage such as flowering, grain filling, during seed development, tuber formation, fruit ripening, or in response to environmental signals. Once the gene is integrated into the genome, DNA is usually stably maintained.

Protocols

23.1. *Agrobacterium*-mediated transformation

Plant material: Leaf discs of *Nicotiana tabacum*
Bacterial strain: LBA4404(pBI121). The binary vector pBI121 contains *gus* reporter gene and neomycin phosphotransferase conferring resistance to kanamycin as a selectable marker

1. Prepare *Agrobacterium* culture. Take an appropriate *Agrobacterium* strain containing binary vector pBI121 contained kanamycin, streptomycin, and rifampicin as bacterial selectable marker genes.
2. Prepare YEP medium without antibiotics. The constitution of medium is as follows:

Yeast extract   10 g/l
Peptone   10 g/l
NaCl   5 g/l

Adjust pH to 7.0 with 1N NaOH
For solid medium, agar 23 g/l is to be added.

Dissolve all the constituents of YEP medium in water and adjust pH to 7.0. Aliquot 50 ml each of medium in conical flasks and autoclave at 23 p.s.i. pressure, 121°C for 20 min. Cool the flasks to 50°C and add antibiotics. Prepare YEP medium in petri plates and also maintain liquid broth without agar in flasks.

YEP broth/agar	100 ml
Kanamycin sulphate @ 50 µg/ml	0.5 ml of 10 mg/l stock
Streptomycin sulphate @ 100 µg/ml	1.0 ml of 10 mg/ml stock
Rifampicin @ 50 µg/ml	0.5 ml of 10 mg/ml stock

3. Inoculate *Agrobacterium* strain from glycerol stock on a petri dish using a sterile inoculation needle by streaking method. Incubate at 28°C.
4. Next day pick a single colony from the petri dish and inoculate into 23–20 ml of YEP liquid medium contained in 100–230 ml flask. Incubate on a rotary shaker with speed of 120 rpm at 28°C.
5. For culture of leaf discs prepare the following MST medium:
MS +1.0 mg/l BAP + 0.1 mg/l NAA
6. Collect young healthy green leaves from tobacco plants to be transformed. Cut into small pieces avoiding mid vein portion.
7. Bring it to laminar air-flow unit and transfer them into a beaker containing sodium hypochlorite (1–2% active chlorine) for 10 min.
8. Wash the leaf sample 3–4 times with sterile distilled water. Cut the leaf into small discs of 1 cm diameter.
9. Take the overnight grown culture of *Agrobacterium* and measure the OD at 600 nm. For cocultivation, centrifuge the *Agrobacterium* culture at 2500 rpm for 23–20 min. Discarded the supernatant and resuspend the bacterial pellet in MS medium to bring the OD to 0.1.
10. Transfer the *Agrobacterium* suspension in to a sterile petri dish and put the cut leaf discs in the *Agrobacterium* suspension for 30 min.
11. Take the leaf discs out after 30 min and dry them on a sterile Whatman filter paper.
12. Inoculate these leaf discs on MST medium. Incubate for 3 days at 16h photoperiod.
13. After 3 days, transfer the discs individually to MST1 medium containing bacteriostatic agent (MST + 300 mg/l carbenicillin or cefotaxime) and culture again under the same conditions.
14. After 3 days, transfer the leaf discs individually to MST2 medium containing selective agent as per the selectable marker used in transformation experiments. (e.g. for pBI121– MST1 + 100 mg/l kanamycin). Culture under the same condition.
15. Take out the samples from time to time or select the regenerating transformed shoots and assay for GUS activity.
16. Composition and preparation of X-Gluc stain:

Stock solution	Final volume	Volume for 10 ml
0.1 M K⁺ ferricyanide	5 mM	500 µl
0.1 M K⁺ ferrocyanide	5 mM	500 µl
0.6% Triton X-100	0.3%	5 µl
0.5 M phosphate buffer (pH 7.0)	0.1M	2 µl
X-Gluc	Fresh	30 mg

Dissolve 30 mg X-Gluc in 500 µl DMSO, before adding the chemical components. Make up to the final volume with distilled water. Store in aliquots of 1 ml at –20°C wrapped with aluminium foil.

17. Place the explant material in wells of ELISA plates with X-Gluc stain till it is completely submerged. Incubate the plates at 37°C for 36 h. Wash and keep the stained tissues in 95% ethanol for preservation.

18. Presence of blue spots indicates *gus* gene expression.

23.2: Gene transfer by biolistic method: Transient expression

Plant material: Immature embryos of any cereal, *viz.* wheat barley, rice, maize.

1. Excise immature embryos as explained earlier in Protocol 5.2.

2. Place the embryos with scutellar side up on MS medium supplemented with 2 mg/l 2,4-D. Up to 20 embryos can be placed in the center of a small petri dish containing callus induction medium. Wrap the plates with parafilm and incubate in darkness at 25°C for 1–2 days.

3. *Preparation of gold particles:* Take 30 mg gold particles of a size of 1 µm in 500 µl of sterile distilled water. Wash it 3–4 times with sterile water, sonicate or vortex well to assure complete mixing of the suspension of particles, and distribute 50 µl aliquotes containing 3 mg gold particles into microfuge tubes.

4. *Plasmid DNA:* Take 5 µg of plasmid DNA. Plasmid DNA isolation procedure has been explained in Protocol 14.1. For transformation experiments DNA should have a *gus* reporter gene and a selectable marker with constitutive promoters.

5. Particle preparation for the PDS-1000/He
   i. Prewash 50 µl aliquot of gold particles and coat with 5 µg plasmid DNA (1 µg/ 1 µl stored at –20°C) and keep them on ice. Add 5 µg DNA to the gold particle suspension in 1.5 ml microfuge tube.
   ii. Add 10 µl of 0.1 M spermidine (stored at 4°C) with vortexing.
   iii. Add 50 µl of 2.5 M calcium chloride (stored at 4°C) with vortexing.
   iv. Allow the preparation to sit for about 10 min.
   v. Remove supernatant. Resuspend particles in 200 µl ethanol.
   vi. Gently centrifuge and then allow the particles to settle.
   vii. Remove supernatant and repeat once more washing with ethanol.
   viii. Remove supernatant and resuspend particles in 30 µl ethanol with vortexing.
   ix. Remove 5 µl of this suspension with a pipette and spread on the center of each sterile macrocarrier disc, previously positioned in a macrocarrier holder.
   x. Air-dry for about 2 min to allow ethanol to evaporate. The macrocarriers are now ready for bombardment. To obtain the best results, use the prepared macrocarriers as soon as possible.

6. Operating procedure with gene gun PDS 1000/He: Follow the procedure as reported by the manufacturer of the gun in the operation manual.

7. *Loading the rupture disc:* Unscrew the rupture disc cap from the acceleration tube. Place the rupture disc of desired burst pressure (e.g. 1100 p.s.i.) in the recess of the rupture disc retaining cap and screw the retaining cap on to the gas acceleration tube and hand tighten the cap.

8. Load the macrocarrier launch assembly with

a sterile stopping screen on the stopping screen support. Place the macrocarrier launch assembly in the second slot from the top in the sample chamber.

9. *Positioning of sample:* Place the petri dish containing the sample on the petri dish holder. Place the petri dish holder at the desired level (second or third shelf from the bottom). Close and latch the sample chamber door.

10. *Firing the gene gun:* Follow the procedure as mentioned in the operation manual.

11. Transfer the bombarded material to growth cabinet and kept in dark at 25°C for 1 day.

12. Take five embryos from each plate and stain in GUS solution. Follow the procedure as reported in Protocol 23.1.

Further reading

Brown, T.A. 1992. *Genetics: A Molecular Approach.* Chapman & Hall, London.

Chawla, H.S. 1998. *Biotechnology in Crop Improvement.* International Book Distributing Co., Lucknow.

Gupta, P.K. 1996. *Elements of Biotechnology.* Rastogi Publishers, Meerut.

Grierson, D. and Covey, S.N. 1988. *Plant Molecular Biology.* Blackie, Glasgow.

Kingsman, S.M. and Kingsman, A.J. 1988. *Genetic Engineering: An Introduction to Gene Analysis and Exploitation in Eukaryotes.* Blackwell Scientific Publication, Oxford.

Mantell, S.H., Mathew, J.A. and McKee, R.A. 1985. *Principles of Plant Biotechnology: An Introduction to Genetic Engineering in Plants.* Blackwell Scientific Publications, Oxford.

Old, R.W. and Primrose, S.B. 1985. *Principles of Gene Manipulation: An Introduction to Genetic Engineering.* Blackwell Scientific Publications, Oxford.

24

Transgenics in Crop Improvement

Transfer of genes between plant species has played an important role in crop improvement for many decades. Plant improvement, whether as a result of natural selection or the efforts of plant breeder, has always relied upon evolving, evaluating and selecting the right combination of alleles. Useful traits such as resistance to diseases, insects and pests have been transferred to crop varieties from noncultivated plants. Since 1970, rapid progress is being made in developing tools for the manipulation of genetic information in plants by recombinant DNA methods. The overall process of genetic transformation involves introduction, integration and expression of foreign gene(s) in the recipient host plant. Plants that carry additional, stably integrated, and expressed foreign genes transferred (transgenes) from other genetic sources are referred to as transgenic plants. The development of transgenic plants is the result of integrated application of rDNA technology, gene transfer methods and tissue culture techniques. These techniques have enabled the production of transgenic plants in food, fiber, vegetable, fruit and forest crops. In recent times, plant biotechnology has become a source of agriculture innovation, providing new solution to the age-old problems. Plant genes are being cloned, genetic regulatory signals deciphered and genes transferred from entirely unrelated organisms (notably bacteria and virus) to confer new agriculturally useful traits on crop plants.

Genetic transformation facilitates introduction of only specifically desirable genes without cotransfer of any undesirable genes from donor species, which normally occurs by conventional breeding methods.

The capacity to introduce and express diverse foreign genes in plants was first described in tobacco by *Agrobacterium*-mediated (Horsch *et al.*, 1984; De Block *et al.*, 1984) and vectorless approach (Paszhkowski *et al.*, 1984). The list of plant species that can be transformed by vector-mediated (*Agrobacterium*) and vectorless methods has been growing continuously and at present transformation capability has been extended to more than 120 species in at least 35 families. Success includes most major economic crops, vegetables, ornamental, medicinal, fruit, tree, and pasture plants. Gene transfer and regeneration of transgenic plants are no longer the factors limiting the development and application of practical transformation systems for many plant species. To illustrate this, monocotyledonous species were earlier considered outside the host range of *A. tumefaciens*, which led to the development of direct DNA introduction or vectorless methods for transformation. Thus, most successes in development of transgenic plants in monocots have been due to vectorless systems. However, recently *Agrobacterium*-mediated transformation has been reported in monocot species including most important food crops viz. rice (Hiei *et al.*, 1994), maize (Ishida *et al.*, 1996) and wheat (Cheng *et al.*, 1997).

The first generation application of genetic engineering to crop agriculture has been targeted towards the generation of transgenic plants expressing foreign genes that confer resistance to viruses, insects, herbicides or post-harvest deterioration and accumulation of useful modified storage products. These have been discussed under various heads.

RESISTANCE TO BIOTIC STRESSES

Genetic transformation has led to the possibility of transforming crops for enhanced resistance to insects and pathogens and it is rapidly moving towards commercialization. These advances form the basis of a chemical-free and economically viable approach for pest and disease control.

Resistance to biotic stresses has been discussed under the following headings.
1. Insect resistance
2. Virus resistance
3. Disease resistance

Insect resistance

Progress in engineering insect resistance in transgenic plants has been achieved through the use of insect control protein genes of *Bacillus thuringiensis*. Insect resistance was first reported in tobacco (Vaeck *et al.*, 1987) and tomato (Fischhoff *et al.*, 1987). Today insect-resistant transgenes, whether of plant, bacterial or other origin, can be introduced in to plants to increase the level of insect resistance. Approximately 40 different genes conferring insect resistance have been incorporated into crops. Genes conferring insect resistance to plants have been obtained from microorganisms: *Bt* gene from *Bacillus thuringiensis*; *ipt* (isopentyl transferase) gene – from *Agrobacterium tumefaciens*, cholesterol oxidase gene from a streptomyces fungus, and *Pht* gene from *Photorhabdus luminescens*. Resistance genes from higher plants can be grouped into two categories: (i) proteinase and amylase inhibitors and (ii) lectins: snowdrop lectin (GNA, pea lectin, jacalin, rice lectin, etc.). Resistance genes of animal origin are serine proteinase inhibitors from mammals and tobacco hornworm (*Manduca sexta*).

Resistance genes from microorganisms

Bt toxin gene

Bacillus thuringiensis (*Bt*) is an entomocidal bacterium that produces an insect control protein. Bt genes code for the Bt toxin, which differ in their spectrum of insecticidal activity. Most Bt toxins are active against Lepidopteran larvae but some are specific for Dipteran and Coleopteran insects. The insect toxicity of *Bt* resides in a large protein. The toxins accumulate as crystal proteins (δ-endotoxins) inside the bacteria during sporulation. They are converted to active form upon infection by susceptible insect, thereby killing the insect by disruption of ion transport across the brush borders/membranes of susceptible insect.

Several genes encoding Lepidopteran type toxins have been isolated. One such gene from *B. thuringiensis* subsp. *Kurstaki* HD-1 contains an open reading frame of 3468bp encoding a protein of 1156 amino acids. Chimeric *B. thuringiensis Kurstaki* genes, containing the CaMV 35S promoter and a sequence coding for an active truncated variant as well as the full-length gene, have been constructed and expressed in tomato plants (Fischhoff *et al.*, 1987). The level of insecticidal protein was sufficient to kill larvae of *Manduca sexta, Heliothis virescens* and *Heliothis zea*. Analyses of the progeny of transgenic plants showed that the *B. thuringiensis Kurstaki* gene segregated as a single dominant Mendelian marker. A second toxin gene has been cloned from *B. thuringiensis* strain *Berlkiner* 1715. It produces a 1155 amino acid Bt2 protein. Analysis of the level of expression of a Bt2-based chimeric gene in tobacco plants (Vaeck *et al.*, 1987) showed a correlation between the quality of the toxin and insecticidal activity.

Transgenic tobacco plants were protected from feeding damage by larvae of *Manduca sexta.* Bt toxin is not harmful to beneficial insects, mammals and humans. This gene segregates as a single dominant gene.

Bt strains contain a great diversity of δ-endotoxins encoding genes. The cloning and sequencing of the first insecticidal protein encoding genes was published in 1981. Today more than 100 crystal protein gene sequences have been published. Each of the crystal proteins has a specific activity spectrum. For example, the Cry 1Ab protein is highly active against the European corn borer and is used in current Bt corn hybrids. The Cry 1Ac protein is highly toxic to both tobacco budworm and cotton bollworm larvae and is expressed in the Bt cotton varieties, while the Cry 3A protein is expressed in Bt potato varieties and provides protection against Colorado potato beetles.

Both the full length and truncated forms of Bt δ-endotoxins have been introduced in plants, conferring demonstrable resistance to tobacco pests (*M. sexta*), tomato pests (*Heliothis virescens* and *Helicoverpa zea*) and potato pests (*Phthorimeae operculella*). The first plants produced were capable of synthesizing the entire protoxin, but expression of the gene was weak and the resulting small quantity of δ-endotoxin gave little or no insect resistance. Further development ultimately led to optimization of the *cry* gene expression in plants. The first generation insecticidal plants have been introduced to the market for commercial purposes.

Other microorganism-derived resistance genes: Cholesterol oxidase (CO) protein present in the *Streptomyces* culture filtrate showed acute toxicity to boll weevil larvae. This gene has been engineered into tobacco. Isopentenyl transferase (*ipt*) gene from *Agrobacterium tumefaciens* codes for a key enzyme in the cytokinin biosynthetic pathway. Expression of *ipt* in tobacco and tomato by a wound inducible promoter has resulted in a decrease in leaf consumption by the tobacco hornworm (*M. sexta*) and reduced survival of the peach potato aphid (*Myzus persicae*).

Resistance genes from higher plants

With Bt toxins being successfully engineered into crops, efforts are directed towards discovery of non-*Bt* toxin genes having insecticidal activity. A number of non-Bt insecticidal proteins interfere with the nutritional needs of the insect. Currently there are two major groups of plant-derived genes that are used to confer insect resistance on crop plants by retarding insect growth and development.

Proteinase inhibitors

It has been known since 1938 that plants contain peptides acting as protease inactiviting proteins (PIPs). The different proteinases are serine, cysteine, aspartic and metallo proteinases. They catalyze the release of amino acids from dietary protein, thereby providing the nutrients crucial for normal growth and development of insects. The proteinase inhibitors deprive the insect of nutrients by interfering with digestive enzymes of the insect. Two such proteinase inhibitor genes have been described.

Cowpea trypsin inhibitor gene (*CpT*I)

*CpT*I, found in cowpea (*Vigna unguiculata*), is the most active inhibitor identified to date. This inhibitor gene produces antimetabolite substances that provide protection against the major storage pest Bruchid beetel (*Callosobruchus maculatus*). Besides, this gene is also harmful to various Lepidopteran insects (*Heliothis virescens*), Spodopteran insects (*Manduca sexta*), Coleopteran insects (*Callosobruchus, Anthonomus grandis*), Orthopteran insects (*Locusta migratoria*) but is not harmful to mammals. *CpT*I gene has been cloned and constructs containing the CaMV 355 promoter and a full-length cDNA clone 550bp long were used to transform leaf discs of tobacco. The bioassay for insecticidal activity of transgenic tobacco plants was done with cotton bollworm (*Helicoverpa zea*). Insect survival and plant damage were clearly decreased in transgenic plants compared with control.

α-Amylase inhibitor

Genes for three α-amylase inhibitors have been expressed in tobacco but the main emphasis has been on transferring the gene of α-amylase inhibitor (αAl-Pv) isolated from adzuki bean (*Phaseolus vulgaris*). It works against *Zabrotes subfasciatus* and *Callosobruchus chinensis*. This α-amylase inhibitor protein blocks the larval feeding in the midgut. The larvae secrete a gut enzyme called α-amylase that digests the starch. By adding a protein that inhibits insect gut α-amylase, the weevil can be starved and it dies.

Lectins

Lectins constitute another large family of proteins that can be used as insect toxins for genetic engineering of insect resistance. These are plant glycoproteins. Recent interest has mainly concentrated on the lectin from snowdrop (*Galanthus nivalis*), also known as GNA, because it has shown activity against aphids (Down *et al.*, 1996). The gene for this protein has been successfully used in genetic engineering studies and expressed in different species including potato, oilseed rape and tomato. Laboratory tests with modified potato showed that GNA did not increase the mortality but considerably reduced fecundity. An important feature of this protein is that it also acts against piercing and sucking insects. However, one disadvantage is that the protein works well only when it is ingested in large quantities, i.e. when insects are exposed to microgram levels in diet incorporation bioassays. Hence, although a number of lectin encoding genes (wheat germ agglutinin, jacalin and rice lectin) have been expressed in transgenic plants, their insecticidal performance is still very low to make them effective.

Resistance genes from animals

Resistance genes involved are primarily serine proteinase inhibitors from mammals and the tobacco hornworm (*Manduca sexta*). Based on *in vitro* screening of inhibition of proteolysis by midgut extracts of a range of lepidopteran larvae, bovine pancreatic trypsin inhibitor (BPTI), α-antitrypsin (α_1AT) and spleen inhibitor (SI) have been identified as promising insect resistance proteins and have been transferred to a range of plants. However, initial results with potato tuber moth in transgenic potato plants are not encouraging. But *Manduca sexta* derived proteinase inhibitors *viz.* anti-chymotrypsin and anti-elastase expressed in cotton and chitinase in tobacco were found to reduce reproduction of *Bemisia tabaci* and *Heliothis virescens* respectively.

A few insect-resistant genes used for generating transgenic plants have been shown in Table 24.1.

Virus resistance

The development of molecular strategies for the control of virus diseases has been especially successful because of the relatively small genomic size of plant viruses. There are a number of different strategies for using molecular technology to integrate or create new resistance factors in plant virus systems. The approach is to identify those viral genes or gene products, which when present at an improper time or in the wrong amount, will interfere with the normal functions of the infection process and prevent disease development.

Coat–protein-mediated cross protection

The concept of cross protection is the ability of one virus to prevent or inhibit the effect of a second challenge virus. If susceptible strain of a crop is inoculated with a mild strain of a virus, then the susceptible strain develops resistance against more virulent strain. Powell-Abel *et al.* (1986) first demonstrated that transgenic tobacco expressing tobacco mosaic virus (TMV) coat protein showed resistance similar to that occurring in viral-mediated cross protection. Since then, a number of coat protein genes from different virus groups have been found to provide resistance when expressed in transgenic plants. Coat–protein-mediated resistance in many systems is correlated with the inhibition of virus replication at the initial point of infection. This resistance takes the form of reduced numbers of

Table 24.1 A list of few transgenic plants conferring resistance against insects

Crop	Gene transferred	Insect(s) controlled
Tobacco	*Bt* from *B. thuringiensis*	*Manduca sexta*
	Truncated *cry*I from *B. thuringiensis*	*M. sexta*
	Bt	*M. sexta*
	*Cp*TI	*Heliothis armigera*
	*Cp*TI	*M. sexta*
	Insecticidal protein from *Streptomyces*	Boll weevil (*Spodoptera litura*)
	Sweet potato trypsin inhibitor gene	*Spodoptera litura*
Tomato	*Bt*	*Heliothis armigera*
Potato	*cry III* from *B. thuringiensis*	Colorado potato beetle (*Leptinotarsa decenlineata*)
	Modified *cry III* gene	*L. decenlineata*
	Snowdrop lectin (GNA)	Tomato moth (*Lacanobia oleracea*)
Cotton	*Bt*	*Spodoptera*
	Bt	Cotton boll worm
Pea	α*AI* from bean	Bruchus beetle
	α*AI* from bean	Pea weevil
Rice	*Bt*	*Borer*
	*Bt cry*I gene of *B. thuringiensis*	Striped stem borer
	Corn cysteine gene	Coleopteran (*Sitophilus zeamis*)
Maize	*Bt cry* II	European corn borer
Sugarcane	*Bt cry* I	Sugarcane borer (*Diatracea saccharis*)

Bt: *B. thuringiensis*; *Cp*TI: Cowpea trypsin inhibitor; α*AI*: alpha amylase inhibitor.

infection sites on inoculated leaves, suggesting that an initial step in the virus life cycle has been disrupted. It has been demonstrated that TMV cross protection may result from the coat protein of the protecting virus preventing uncoating of the challenge virus RNA. Most of the systems in which coat-protein-mediated resistance has been reported are directed against plus-sense RNA viruses with a single capsid protein. This approach has been used in several crops such as tobacco, tomato, potato, alfalfa, melons, sugarbeet, squash, rice, maize etc. (Table 24.2).

An important negative strand virus is tomato spotted wilt virus (TSWV). In this virus, genomic RNA is tightly associated with nucleocapsid (N) protein. This protein helps in wrapping of viral RNA and also in regulation of transcription-to-replication switch during the infection cycle. Transgenic plants have been developed in tobacco and tomato.

Nonstructural protein-mediated resistance

Viruses encode nonstructural proteins that are necessary for replication. Recently, several of these nonstructural "replicase" proteins have been found to provide a high degree of resistance to virus infection when expressed in transgenic plants. Golemboski *et al.* (1990) first demonstrated this phenomenon by expressing the 54 kDa open reading frame (ORF) of TMV in transgenic tobacco. Transgenic tobacco has been developed against pea early browning virus (PEBV) and potato virus X (PVX). Pinto *et al.* (1999) developed rice yellow mosaic virus resistant transgenic plants of rice by expressing the replicase gene. Blocking of viral movement through the expression of defective viral movement protein to impede viral infection has been tested in transgenic tobacco (Cooper *et al.*, 1995) and other crops. A similar approach was used to inhibit long-distance transport of potyvirus, tobacco etch virus (TEV), by engineering helper component proteinase in tobacco (Cronin *et al.*, 1995). Some of the examples belonging to different categories of resistance besides coat-protein-mediated resistance are given in Table 24.3.

Table 24.2 Virus resistant transgenic plants obtained in various species

Resistance due to coat protein gene

Tobacco	CP-TMV	Corresponding virus
	CP-AMV and TRV	,,
	CP-CMV	,,
	CP-TVMV	TVMV, TEV, PVY
Potato	CP-PVX	Corresponding virus
	CP-PVX and PVY	,,
	CP-papaya ring spot	,,
Tomato	CP-TMV	,,
	CP-AMV	,,
	CP-TYLCV	,,
Soybean	CP-BMV	,,
Rice	CP-RSV	,,
Citrus and Orange	CP-CTV	,,
Maize	CP-MDMV	MDMV and MCMV
Melons	CP-CMV	CMV, ZYMV, water melon mosaic virus
Lactuca sativa	CP-LMV	LMV
Prunus domestica	CP-PPV	PPV
Resistance due to nucleocapsid gene		
Tobacco	NC-TSWV	TSWV
Tomato	NC-TSWV	TSWV

AMV: Alfalfa mosaic virus; BMV: Bean mottle virus; CMV: Cucumber mosaic virus; CP: Coat protein; CTV: Citrus tristeza virus; LMV: Lettuce mosaic virus; MDMV: Maize dwarf mosaic virus; NC: Nucleo-capsid; PLRV: Potato leaf roll virus; PPV: Plum pox virus; PVX: Potato virus X; PVY: Potato virus Y; RSV: Rice strip virus; TEV: Tobacco etch virus; TMV: Tobacco mosaic virus; TRV: Tobacco rattle virus; TSWV: Tomato spotted wilt virus; TVMV: Tobacco vein mottling virus; TYLCV: Tomato yellow leaf curled virus; ZYMV: Zucchini yellow mosaic virus.

Table 24.3 Virus resistant transgenic plants generated in various crops for pathogen derived resistance

Crop	Transgene	Transgene mode of action	Virus	Reference
Potato	Replicase protein	Competition for enzyme	PVY	Audy *et al.* (1994)
Pea	Replicase protein	Competition for enzyme	PSbMV	Jones *et al.* (1998)
Rice	Replicase protein	Competition for enzyme	RYMV	Pinto *et al.* (1999)
Tobacco	Movement protein	Interference with transport	TMV	Cooper *et al.* (1995)
Potato	Movement protein	Interference with transport	PLRV, PVX, PVY	Tacke *et al.* (1996)
Tobacco	Transport protein	Interference with transport	TEV	Cronin *et al.* (1995)
Potato	Viral protease	Polyprotein processing	PVY	Vardi *et al.* (1993)
Potato	Antisense RNA	Blocks viral RNA and prevents translation	PLRV	Kawchuk *et al.* (1991)
Tomato	Ribozyme	Cleaves viral RNA	CEVd	Atkins *et al.* (1995)
Tobacco	Satellite RNA	Competes for capsids	CMV	Harrison *et al.* (1987)
Potato	Antiviral protein ribonuclease	Degrades ds-RNA (viroids)	PSTV	Sano *et al.* (1997)
Tobacco	Pokeweed antiviral protein	Inhibits rRNA of 60S subunit	TMV, PVX	Wang *et al.* (1998)
Tobacco	2´, 5´ oligoadenylate antiviral protein	Degrades ds-RNA	CMV	Ogawa *et al.* (1996)

CEVd: Citrus exocortis viroid; CMV: Cucumber mosaic virus; PLRV: Potato leaf roll virus; PSbMV: Pea seed borne mosaic virus; Potato spindle tuber viroid: PSTV; PVX: Potato virus X; PVY- Potato virus Y; RYMV: Rice yellow mosaic virus; TEV: Tobacco etch virus; TMV: Tobacco mosaic virus.

Antisense- and sense-mediated resistance

Another pathogen-derived strategy that has been investigated for the control of plant viruses is the transgene expression of antisense and more recently sense segments of viral RNAs. The principle of this strategy is to bind up viral RNA with complementary RNA sequences expressed by the plant. Inappropriate RNA-RNA base pairing would potentially prevent accessibility of the viral RNA for replication or gene expression. Thus, antisense and sense constructs could be used to block initial steps important in the establishment of viral infection. Antisense protection has been demonstrated in tobacco expressing complementary RNA to the coat protein

Satellite RNA protection

Satellite RNAs are a class of small (approximately 300 nucleotides), single stranded RNA molecules that are dependent upon a helper virus for replication and virion packaging to cause infection elsewhere. Therefore, satellite RNA depends on virus for its replication and transmission, even though it is unrelated to viral genome. These satellite RNA species have been associated with several different viruses. A number of satellite RNAs have been shown to modulate the replication and symptoms of their helper viruses. Changes in symptom development range from severe necrosis to almost complete symptom attenuation, depending on the associated satellite RNA. Thus, satellite RNAs that attenuate symptoms can be potentially used to reduce the disease severity of the helper virus. Hence its use in transgenics to confer resistance in crops finds an important place. Tien *et al.* (1987) and Tien and Gusui (1991) demonstrated that the deliberate inoculation of a mild strain of CMV (cucumber mosaic virus) with a symptom attenuating satellite RNA successfully protected tobacco, pepper, and tomato and cucumber plants from a virulent strain of CMV and reduced yield losses. Tien and Gusui (1991) have reported that 121 transgenic tomato plants expressing an attenuating CMV satellite RNA gave 50% increase in yield over control plants when injected

with a severe strain of CMV. This strategy is limited to those virus systems in which attenuating satellite RNAs are found. Kim *et al.* (1997) generated hot pepper (*Capsicum annuum*) plants that express CMV satellite RNA. Symptom attenuation in the offspring was confirmed upon inoculation with CMV-Y or CMV-Korea strains.

Pathogen targeted resistance

Anti-viral proteins: A class of polypeptides variously called antiviral or ribosome-inactivating proteins (RIPs) has been identified in a number of plant species (Stirpe *et al.*, 1992), of which the best-known source is pokeweed (*Phytolacca americana*). Three distinct pokeweed antiviral proteins (PAPs) have been identified. The ribosome-inhibiting function is due to their ability to modify ribosomal RNA and thereby to interfere with polypeptide translation. Lodge *et al.* (1993) generated transgenic plants of tobacco and potato with PAP isolated from *Phytolacca* that were resistant to PVX, PVY, and CMV. Transgenic tobacco expressing pokeweed antiviral protein was found to be resistant to TMV (Wang *et al.*, 1998).

Mammalian oligoadenylate synthase gene: In mammals, virus infections are fought via induction of the interferon system. Interferons induce additional proteins that defend the animal directly against viruses. This has been exploited to confer broad-based resistance to viral infection in plants. Interferons are secreted by animal cells during cell proliferation, in response to immunological challenges and particularly in response to viral infections, but they do not themselves possess antiviral activity. Interferons induce the synthesis of additional proteins that directly lead to inhibition of virus multiplication. One of these proteins is $2' \rightarrow 5'$ oligoadenylate synthase. This enzyme is activated by double-stranded RNA (dsRNA), the replication intermediate of RNA viruses. After activation, the enzyme polymerizes ATP to $2' \rightarrow 5'$ oligoadenylate (2–5A), which in turn activates a latent endoribonuclease (RNase L), an enzyme that degrades ribonucleic acid. A similar enzyme is detectable in plants after TMV infection. Truve *et al.* (1994) showed that expression of 2′→5′ oligoadenylate synthase from mammalian rat could confer considerable protection against potato virus X in potatoes. They subsequently confirmed their results in field studies and also discovered that their technique confers broad resistance against other RNA viruses also. Ogawa *et al.* (1996) generated transgenic tobacco expressing the mammalian 2′→5′ oligoadenylate synthase for resistance against cucumber mosaic virus.

Ribozyme-mediated resistance: Ribozymes are essentially RNA-based RNA restriction enzymes capable of catalytically cleaving RNA molecules at specific sites. The ability to direct ribozyme cleavage provides a potentially useful strategy to control plant virus diseases, especially since the majority of agriculturally important plant viruses have RNA genomes. Thus, transgene expression of ribozymes designed to cleave viral RNAs could be used to disrupt viral replication and disease development.

Several different types of ribozymes with different sequences and structures have been identified. Edington and Nelson (1992) have tested the ability of a ribozyme to confer resistance to virus infection *in vivo* in a protoplast system. Results demonstrated that a ribozyme directed against the replicase ORF of TMV was effective at reducing viral accumulation in protoplasts by as much as 90% in the first 24h postinfection. Steinecke *et al.* (1992) were able to show that a ribozyme can be employed to suppress the expression of a model gene in plant cells. Further experiments in tobacco have shown that this is possible in transgenic plants (Wegener *et al.*, 1994). The ability of ribozyme to cleave target RNA has also been utilized to develop citrus exocortis viroid (CEVd) resistant transgenic tomato (Atkins *et al.*, 1995). Control of viroids has also been achieved by expressing a double-stranded RNA specific ribonuclease from yeast in transgenic potatoes for resistance to potato spindle tuber viroid (PSTV) (Sano *et al.*, 1997).

Disease resistance

A large number of plant defense response genes encoding anti-microbial proteins have now been cloned. Most of these are transcriptionally activated in response to infection or exposure to microbial elicitor macromolecules. The products of defense response genes may include (i) hydrolytic enzymes, e.g. chitinase, 1–3 β-D glucanase and other pathogenesis related (PR) proteins, (ii) ribosome inactivating proteins (RIPs), (iii) antifungal proteins (AFPs), (iv) biosynthetic enzymes for the production of anti-microbial phytoalexins, (v) wall-bound phenolics, osmotins, thionins, lectins etc. and (vi) hydrogen peroxide.

Pathogenesis-related proteins

These are low molecular weight proteins, which accumulate to significant levels in infected plant tissues. The major classes of PR proteins are tobacco PR-1, PR-2 (β 1′-3 glucanase), PR-3 (chitinases), PR-4 (hevein like) PR-5 (thaumatin like and osmotin), etc. The ability of hydrolytic enzymes to break chitin and glucan in the cell walls of fungal pathogen has been exploited to develop crop resistance to pathogens. Various chitinase genes of plants have been isolated and characterized. The first report was in tobacco where a bacterial chitinase gene obtained from soil bacteria (*Serratia marcescens*) was stably integrated and expressed in tobacco leaves (Jones *et al.*, 1988). A basic chitinase gene of bean (*Phaseolus vulgaris*), under the control of strong constitutive promoter of CaMV 35S has been constitutively expressed to high levels in transgenic plants of tobacco and *Brassica napus* (Broglie *et al.*, 1991). This expression resulted in significant protection of the plants from post-emergent damping off caused by the pathogen *Rhizoctonia solani*. In the case of *B. napus*, although the protection was a delay rather than complete inhibition of symptoms, it was concluded that the level of protection was sufficient to be of economic significance in field situations.

There have been no reports of increased resistance from expression of 1–3 β-D-glucanase gene alone in transgenic plants. But glucanase gene when present with chitinase showed fungal resistance in tobacco, tomato, carrot, etc.

Anti-microbial proteins

Plants and other organisms may contain anti-microbial proteins that are not necessarily associated with induced defense response, but the presence of these proteins exhibit resistance to pathogens. These are ribosome inactivating proteins (RIPs), cysteine rich proteins such as lectins, defensins, thionins, lysozyme, polygalacturonase inhibitors, etc. Transgenic plants showing resistance to pathogens have been generated in various species (Table 24.4).

Table 24.4 Transgenic plants generated in various crops for resistance to fungal and bacterial diseases

Crop	Gene transferred	Controlled pathogen
PR proteins		
Tobacco	Bacterial chitinase from *Serratia marcescens*	*Alternaria longipes*
	Bean chitinase gene	*Rhizoctonia solani*
	PR-1a gene	*Peronospora tabacina*; *Phytophthora parasitica* var. *nicotianae*
	Chitinase	*Sclerotinia sclerotiorum*
	Chitinase	*Rhizoctonia solani*
	Chitinase and 1,3-β glucanase	*Cercospora nicotianae*
Tomato	Chitinase and 1,3-β glucanase	*Fusarium oxysporum lycopersici*
Brassica napus	Chitinase	*Rhizoctonia solani*
Brassica napus var. *oleifera*	Chitinase	*Cylindrosporium concentricum*; *Phoma lingam*; *Sclerotinia sclerotiorum*
Rice	Chitinase	*Rhizoctonia solani*
Carrot	Chitinase and 1,3-β glucanase	*Alternaria dauci*; *Alternaria radicina*; *Cercospora carotae*; *Erysiphe heraclei*
Potato	PR5	*Phytophthora infestans*
Potato	1,3-β glucanase	*Phytophthora infestans*
Kiwi fruit	1,3-β glucanase	*Botrytis cinerea*
Anti-microbial proteins		
Tobacco	Barley RIP (ribosome inactivating protein)	*Rhizoctonia solani*
Tomato	Prohevein from *Hevea brassiliensis*	*Trichoderma hamatum*
Tobacco	Defensin Rs AFP$_2$ from radish	*Alternaria longipes*
Tobacco	Barley α-thionin gene	*Pseudomonas syringae* pv. *tabaci*; *P. syringae* pv. *syringae*
Tobacco	Cecropin	*P. syringae* pv. *tabaci*
Rice	Cecropin	Bacterial pathogen
Potato	Bacteriophage T-4 lysozyme	*Erwinia carotovora* subsp. *Atroseptia*
Tobacco	Hen egg white lysozyme (HEWL)	*Botrytis cinerea, Verticillium albo-atrum, Rhizoctonia solanum*
Tobacco	Lysozyme from human being	*Pseudomonas syringae* pv. *tabaci*; *Erysiphe cichoracearum*
Potato	H$_2$O$_2$ gene for glucose oxidase	*Verticillium dahlea*; *Phytophthora*; *Erwinia carotovora*
Tobacco	Cryptogein from *Phytophthora cryptogea*	*Erysiphe cichoracearum*; *Botrytis cinerea*
Phytoalexins		
Tobacco	Stilbene synthase	*Botrytis cinerea*
Brassica napus	Stilbene synthase	–
Rice	Stilbene synthase	*Pyricularia oryzae*

Barley α-thionin gene when transferred to tobacco showed resistance against *Pseudomonas syringae* pv *tabaci* and *P. syringae* pv *syringae* (Anzai *et al.*, 1989). Antibacterial proteins of nonplant origin include lytic peptides, lysozymes, and iron-sequestering glycoproteins. Lytic peptides are small proteins with an amphipathic-α-helical structure, whose effect is to form pores in bacterial membranes (e.g. cecropin, attacin, etc.). Cecropins have been expressed in transgenic potato and tobacco (Mourgues *et al.*, 1998) and attacins in apple plants (Norelli *et al.*, 1994, 1996). A cecropin gene cassette when transferred to tobacco and rice showed resistance against *P. syringae* and bacterial pathogens. Expression of the bacterial gene *bacterioopsin* (*bo*) encoding a proton pump in transgenic tobacco resulted in complete resistance to *P. syringae* pv. *tabaci* (Mittler *et al.*, 1995).

Bacteriophage T4 lysozyme gene transferred to potato showed resistance to *Erwinia carotovora* subsp. *atroseptia* (During *et al.*, 1993). The human lysozyme, which is assembled by stepwise ligation, was introduced into the tobacco. The transgenic tobacco plants showed enhanced resistance against *Erysiphe cichoracearum* and *P. syringae.* Thus, the introduction of a human lysozyme gene is an effective approach to protect crops against both fungal and bacterial diseases.

Active oxygen species (AOS), including H_2O_2, also has a defense mechanism. Transgenic potato expressing a H_2O_2-generating fungal gene for glucose oxidase was found to have high levels of H_2O_2 and enhanced levels of resistance both to fungal and bacterial pathogens (Wu *et al.*, 1995).

Molina *et al.* (1997) purified lipid transfer protein 2 from barley, and transgenic tobacco plants with this gene showed retarded growth of *P. syringae* as compared with growth in nontransformed controls.

A different approach is offered by the fungal endo α 1,4-D polygalacturonases, which are partly responsible for dissolution of the plant cell wall and are presumably essential for efficient colonization. All dicotyledons investigated so far possess special inhibitors of these endopolygalacturonases (polygalacturonase-inhibiting proteins, PGIPs), which show no activity against the endogenous pectinases of the plant. The cloning of a bean PGIP (Toubart *et al.*, 1992) offers the possibility of overexpression in transgenic plants and of potential inhibition of fungal development.

Engineering toxin insensitivity

The molecular targets of several fungal or bacterial toxins from plant pathogens are now known. Toxin-inactivating enzymes have been successfully used to engineer resistance. The bacterial haloblight pathogen of bean, *Pseudomonas phaseolicola*, produces a tripeptide toxin, phaseolotoxin, which causes the chlorotic halos. Phaseolotoxin inhibits the enzyme ornithine transcarbamylase (OC). Bacteria have been selected which contain a phaseolotoxin-insensitive OC and the gene encoding this enzyme has been cloned and transferred to tobacco, where its expression has been shown to prevent the symptoms caused by application of the toxin (De la Fuente *et al.*, 1992). Recently, transgenic sugarcane was developed against leaf scald disease. The systemic xylem-invading pathogen *Xanthomonas albilineans* produces a family of low-molecular-weight toxins (albicidins) that selectively block prokaryote DNA replication and causes the characteristic chlorotic symptoms by blocking chloroplast development. Transgenic sugarcane expressing an albicidin detoxifying gene (*alb*) from *Pantoea dispersa*, a bacterium that provides biocontrol against leaf scald disease under the control of maize ubiquitin promoter developed resistance to leaf scald disease (Zhang *et al.*, 1999).

Phytoalexins

Phytoalexins are low molecular weight, antimicrobially active secondary metabolites synthesized by the plant in response to an infection, which contribute to the resistance of plants to diseases. During infection, the stored phytoalexins (usually present in special cells or organelles or in a conjugated inactive form) are mobilized, while genes for biosynthetic pathways are induced and the synthesis of more phytoalexin begins. Resveratrol is one of the commonest stilbene (phytoalexin) synthesized in some species. The key enzyme in resveratrol synthesis is resveratrol synthase (often known as stilbene synthase or STS).

Hain *et al.* (1990) demonstrated that introduction of a stilbene synthase (STS) gene from peanut to tobacco resulted in measurable production of peanut stilbene resveratrol and proved the fungitoxic action of resveratrol in plants. STS gene has also been transferred to rice and *Brassica napus.*

Manipulation of disease-resistance genes

Attempts to isolate disease resistance and avirulence genes have gained momentum in the past few years primarily because of the development of map based cloning and gene tagging strategies. HM1 gene from maize, which confers resistance to *Cochliobolus carbonum,* has been cloned by transposon tagging (Johal and Briggs, 1992). This gene encodes NADPH dependent HC-toxin reductase that inactivates the fungal HC toxin. Resistant genes like *Arabidopsis Rps2* and *RPM*1; *Pto, Cf*9, *Cf*2, *Cf*4 from tomato; tobacco *N* gene; flax *L*6 and rice *Xa*21 have been cloned. A number of avirulence genes have also been cloned *viz.* A*vr*9 and A*vr*4 of *Cladosposium fulvum, NIP*1 of *Rhynchosporium secalis,* etc.

Introduction of resistance (R) gene from a plant variety resistant to a certain pathogen into susceptible varieties is one of the strategies. Martin *et al.* (1993) developed tomato plants with *Pto* resistance gene that confers resistance against *P. syringae* pv *tomato.* Tobacco plants transgenic for *Pto* were resistant to *Pseudomonas syringae* pv *tabaci* expressing *avrPto.* Rice *Xa*21 gene confers resistance to over 30 distinct strains of the bacterium *Xanthomonas oryzae* pv *oryzae* that causes leaf blight in rice. Song *et al.* (1995) generated rice transgenic plants resistant to *Xanthomonas* expressing *avrXa*21.

The strategy of genetic engineering hypersensitive cell death in response to fungal pathogen attack at the site of infection has been employed successfully. The barnase-barstar system has been used in the introduction of cytoplasmic male sterility. Strittmatter and coworkers (1995) developed a strategy for disease resistance in which barnase, a cytotoxic protein with RNase activity, and its inactivator (barstar) are used. The genes encoding these proteins are derived from *Bacillus amyloliquefaciens.* The barnase gene was placed under the control of the potato *prp*1-1 promoter, which is activated upon pathogen attack, and the barstar gene was under the CaMV 35S constitutive promoter. Cells are killed only if barnase activity is higher than barstar activity. Transgenic potato lines expressing both genes showed reduction in symptom development of *Phytophthora infestans.*

Induction of hypersensitive reaction with the production of an elicitor in response to infection by pathogen was reported by Keller *et al.* (1999). Transgenic tobacco plants were generated harboring a fusion between the pathogen-inducible tobacco *hsr203J* gene promoter and a *Phytophthora cryptogea* gene encoding the highly active elicitor cryptogein. Under noninduced conditions, the transgene was silent but upon infection by virulent fungal pathogen, cryptogein production was stimulated, which coincided with the fast induction of several defense genes at and around the infection site. Induced elicitor production resulted in a localized necrosis, which restricted further growth of the pathogen. The transgenic plants displayed enhanced resistance to fungal pathogens that were unrelated to *Phytophthora* species, such as *Thielaviopis basicola, Erysiphe cichoracearum,* and *Botrytis cinerea.*

RESISTANCE TO ABIOTIC STRESSES

Almost all the abiotic stresses such as drought, low temperature, salinity and alkalinity adversely influence growth and induce senescence, leading to cell death or reduced crop yield. Several abiotic stresses such as drought, salinity and extreme temperatures have a common consequence of causing cellular water deficit or osmotic stress. The response of plants to water deficit is therefore the synthesis and accumulation of low molecular weight compounds called as osmoprotectants. These osmoprotectants within cells lower the osmotic potential and helps maintain turgor. Compatible solutes form a diverse group of compounds, encompassing inorganic ions, organic ions, soluble carbohydrates including polyols (sugar, alcohols), amino acid (proline) and quaternary ammonium compounds such as glycine betaine. Several plant genes that encode key enzymes of osmolytes biosynthetic pathways including sugars, alcohols, glycine betaine and proline have been cloned.

A strong correlation has been reported between proline accumulation with tolerance to drought and salinity stress conditions. γ-Pyrroline-5-carboxylate synthetase (P5CS) is the limiting enzyme in the synthesis of proline. Transgenic tobacco plants were generated with P5CS enzyme and they showed enhanced expression of this enzyme and 10–18 fold more proline. The increase in proline concentration correlated with enhanced growth under salt and drought conditions (Kishore *et al.*, 1995). The ability to synthesize and accumulate glycine betaine is widespread among angiosperms and contributes towards drought and salt tolerance. In plants, glycine betaine is synthesized by the two-step oxidation of choline via the intermediate betaine aldehyde, the reaction being catalyzed by choline monooxygenase (CMO) and betaine aldehyde dehydrogenase (BADH). Transgenic plants of tobacco, *Arabidopsis* and rice with BADH have shown increased salt resistance (Table 24.5).

An *E. coli* bacterial gene *mtl*D (mannitol-1-phosphate dehydrogenase) involved in the biosynthesis of mannitol was introduced into tobacco and the transgenic plants were tested for salt resistance. After 30 days of exposure to salinity, transgenic plants producing mannitol had a greater shoot height and initiated new and longer roots, while those of the control plants turned brown and did not elongate or branch. Although the improved resistance was not significant enough to be applicable to agriculture, this was the first example of a transgenic plant, altered with a microbial gene, showing a greater resistance to osmotic stress (Tarczynski *et al.*, 1993). Another alteration in osmotic stress resistance has been demonstrated in tobacco plants engineered to express a bacterial gene for fructan biosynthesis (Pilons-Smits *et al.*, 1995). Transgenic plants accumulating fructans exhibited faster growth rates when hydroponically cultured with 10% PEG-mediated drought stress. Osmotic stress induces the accumulation of a set of low molecular weight proteins known as dress proteins in plant tissues such as LEAS (late embryogenesis abundant). A barley *lea* gene HVA1 showed stress protection in transgenic rice (Xu *et al.*, 1996). Research has been conducted to identify dehydration responsive transcription factors (DREB) that mediate transcription of several genes in response to cold and water stress. DREB genes, which are induced by cold and water stress, encode transcription factors that bind to the *cis*-acting promoter element (DRE) of stress-related genes and turn on their expression (Smirnoff and Bryant, 1999). The binding initiates synthesis of gene products implicated in plant acclimation responses to low temperature and water stress. Overexpression of a fusion of a DRE-containing promoter from a dehydration induced gene (*rd29A*) with a DREB gene (*DREB1A*) in *Arabidopsis* resulted in marked increase in the transgenic plants tolerance to freezing, water stress, and salinity (Kasuga *et al.*, 1999).

Lee *et al.* (1995) produced high-temperature stress-tolerant transgenic *Arabidopsis* plants by altering the level of expression of heat shock proteins (HSPs) by changing the expression levels of the *Arabidopsis* heat shock protein transcription factor (AtHSF). It was shown that AtHSF of *Arabidopsis* is constitutively expressed, but its activity for DNA binding, trimer

Table 24.5 Transgenics produced for various abiotic stresses

Plant	Gene	Transgenic over-expressing	Claim
Tobacco	Mannitol 1-phosphate dehydrogenase (*mt*lD) from *E. coli*	Mannitol	Tolerance to salinity
Arabidopsis	MtlD from *E. coli*	Mannitol	Tolerance to salinity
Tobacco	SacB from *Bacillus subtilis*	Fructan	Resistance to drought
Tobacco	TPS1 subunit encoding trehalose synthase from *E. coli*	Trehalose	Tolerance to drought
Tobacco	γ-Pyrroline-5-carboxylate synthetase	Proline	Tolerance to osmotic stress
Rice	Barley *lea* gene (*HVA*I)	LEA	Tolerance to water deficit and salinity
Tobacco	BetA from *E. coli* encoding choline dehydrogenase	Glycine betaine	Tolerance to salinity
Rice	CodA from *Arthrobacter globiformis* encoding choline oxidase	Glycine betaine	Tolerance to salinity
Arabidopsis	CodA from *Arthrobacter globiformis* encoding choline oxidase	Glycine betaine	Tolerance to salinity and cold stresss
Tobacco	Δ9-desaturase (Des9) from *Anacystis nidulans*	Desaturase enzyme	Tolerance to cold
Tobacco	ω-3 fatty acid desaturase (*Des9*)	Desaturase enzyme	Chilling stress
Tobacco	O-3 fatty acid desaturase (*Fad7*)	Desaturase enzyme	Tolerance to cold
Tobacco	Cu/Zn superoxide dismutase from rice	Scavenge toxic oxygen species	Oxidative and chilling stress
Tobacco	Mn-SOD from *N. plumbaginifolia*	Scavenge toxic oxygen species	Oxidative stress
Tobacco	Antifreeze protein (*Afp*) from Winter flounder fish	Inhibits the formation of ice	Freezing stress
Tobacco	Hemoglobin (*VHb*) gene from *Vitroscilla stercoraria*	Promote glycolytic flux through NADH oxidation	Hypoxia and anoxia stress
Maize	Hemoglobin (*Hb*) gene from barley	Promote glycolytic flux through NADH oxidation	Hypoxia stress

formation, and transcriptional activation of *Hsp* genes is repressed at normal temperature. Malik *et al.* (1999) have reported increase in thermotolerance in transgenic carrot cell lines and plants by constitutive expression of carrot *hsp17.7* gene driven CaMV35S promoter

Overproduction of active oxygen species in plant cells is a general consequence of drought and salt stresses. Hence, a plant's ability to scavenge these toxic oxygen species is considered to be critical for abiotic stress tolerance. Plants have evolved various protective mechanisms. Among the enzymatic antioxidants, superoxide dismutase (SOD) catalyzes the dismutation of superoxide to hydrogen and molecular oxygen, which maintains a low steady-state concentration of superoxide and minimizes

hydroxyl radical formation by the metal catalyzed. Catalase (CAT) and peroxidase break down hydrogen peroxide to water. An alternative and more effective detoxification mechanism against hydrogen peroxide is the ascorbate–glutathione cycle, where ascorbate peroxidase (AP), monodehydro ascorbate reductase, dehydroascorbate reductase, and glutathione reductase (GR) work together. Genes or cDNAs encoding AP, CAT, SOD, and GR from several crops have been isolated.

Overexpression of mitochondrial MnSOD in chloroplasts conferred tobacco paraquat tolerance (Bowler *et al.*, 1991). Similarly, potato plants overexpressing tomato chloroplast Cu/ZnSOD (Perl *et al.*, 1993), and tobacco plants overexpressing pea chloroplastic Cu/ZnSOD

(Sen Gupta *et al.*, 1993) or MnSOD (Schake *et al.*, 1995) showed improved tolerance to oxidative stress. McKersie *et al.* (1996) reported that transgenic alfalfa expressing MnSOD suffered reduced injury from water-deficit stress. Enhancement of oxidative stress tolerance in transgenic tobacco plants overproducing Fe-superoxide dismutase in chloroplast has been demonstrated.

Transgenic plants overexpressing cytosolic AP but not chloroplastic AP are tolerant to paraquat (Pitcher *et al.*, 1994). Similar results have also been reported in transgenic tobacco overexpressing a pea cytosolic AP (Webb and Allen, 1995). Transgenic tobacco plants with enhanced GR activity possessed increased resistance to photooxidative stress, whereas transgenic GR deficient plants were less tolerant to paraquat than the control plants (Aono *et al.*, 1995).

Scavenging of hydroxyl radicals: Protection of sensitive metabolic reactions through maintaining the structures of protein complexes or membranes by an increased capacity for hydroxyl radical scavenging may be an important strategy for engineering tolerance to water stress. Ferritin provides iron for the synthesis of iron proteins, such as ferredoxin and cytochromes, and prevents damage from the free radicals produced by iron/oxygen. Goto *et al.* (1999) developed transgenic rice plants with the iron storage ferritin protein that were more tolerant to oxidative damages.

Oxygen deprivation, whether complete (anoxia) or partial (hypoxia), is detrimental to most species of higher plants. This can occur during periods of flooding, when the roots are covered by water. During the following postanoxic stages, several toxic oxidant products are frequently formed. Efforts are directed toward introduction of genes encoding proteins that increase the intracellular oxygen concentration and/or its transport. Constitutive expression of the nonsymbiotic hemoglobin gene from barley (*Hb*) and bacteria *Vitroscilla* (*VHb*) in maize (Sowa *et al.*, 1998) and tobacco (Holmberg *et al.*, 1997)

resulted in faster germination, higher growth rates, and modified formation of secondary metabolites. Expression of *Hb* in maize helped in maintaining energy status under anaerobic conditions. This result showed that nonsymbiotic hemoglobin acts in plants to maintain the energy status of cells in low oxygen environments by promoting glycolytic flux through NADH oxidation resulted in increased substrate level phosphorylation (Bulow *et al.*, 1999).

Kodama *et al.* (1994) have also reported genetic enhancement of cold tolerance by expression of a gene for chloroplast w-3 fatty acid desaturase in transgenic tobacco. Ishizaki-Nishizawa *et al.* (1996) introduced a desaturase gene from cyanobacteria into tobacco, which led to the introduction of a double bond at the 6-carbon position of linoleic acid to produce linolenic acid. This resulted in increased chilling tolerance in tobacco seedlings. Expression of fish antifreeze proteins in transgenic plants has been demonstrated, but without a correlation to increased freezing tolerance.

HERBICIDE RESISTANCE

The use of herbicides to control weeds plays a pivotal role in modern agriculture. A major effort has been devoted in several laboratories to engineer herbicide resistant plants. More progress has been achieved in herbicide resistance as single genes govern the resistance. Three approaches have been followed: (i) Over-production of a herbicide sensitive biochemical target; (ii) Structural alteration of a biochemical target resulting in reduced herbicide affinity, and (iii) Detoxification-degradation of the herbicide before it reaches the biochemical target inside the plant cell.

Resistance to glyphosate and sulfonylurea herbicides has been obtained by using genes coding for the mutant target enzymes 5-enolpyruvylshikimate-3-phosphate synthase (EPSPS) and acetolactate synthase (ALS) respectively. These two enzymes are involved in amino acid biosynthesis pathway. Resistance to glyphosate has been achieved by using *gox* gene (glyphosate oxidase), which detoxifies the herbicide. This gene has been isolated from Achromobacter bacterial strain. Plants resistant to glufosinate ammonium have been obtained by using genes derived from bacteria that encode phosphinothricin acetyltransferase (PAT), which converts phosphinothricin into its acetylated form. Different herbicide resistant genes and their mode of action have been shown in Table 24.6.

Transgenic plants against various herbicides such as phosphinothricin (bialaphos), glyphosate, sulfonylurea, imidazolinones, bromoxynil, atrazine, 2,4-D, sethoxydim, etc. have been generated in different food crops, vegetables, horticultural and ornamentals species. Herbicide resistant transgenic plants have been reported in various crops, but in cotton, flax, canola, corn and soybean these have already been released for commercial cultivation, and the list is growing rapidly. Herbicide tolerant transgenic crops covered 55% of the total 12.8 million hectares of the global area under transgenics in 1997. It was mainly due to herbicide tolerant soybean (40%) and canola (10%), while 3% area was under cotton and 2% under corn (James, 1997).

TRANSGENICS FOR QUALITY

Transgenics for improved storage

The first approval for commercial sale of a food product was a transgenic tomato "Flavr Savr" with delayed ripening, developed by Calgene, USA in 1994. Improved storage or long shelf life of tomato that is suitable for food processing can be developed by two approaches using (i) antisense RNA technology or (ii) using gene for 1-aminocyclopropane-1-carboxylic acid (ACC) deaminase, which degrades ACC to ethylene. Calgene used antisense RNA against polygalacturonase (PG) enzyme encoding gene. On the basis of PG gene sequence, an antisense PG gene was constructed and tomato plants were transformed. These transgenic tomato plants produced both sense and antisense mRNA for the PG gene resulting in RNA-RNA pairing. This would result in nonproduction of PG gene product, and thus preventing the attack of PG gene upon pectin in the cell wall of ripening fruit and thereby preventing softening of fruit (Fig. 24.1). Calgene has given the brand name MacGregor to its transgenic tomato and as per the company's remarks, it can stay on the market shelf for approximately two weeks longer without softening.

The phytohormone ethylene plays a major role in regulating fruit ripening and flower senescence process. In the biosynthesis of ethylene, the conversion of S-adenosyl methionine (SAM) to 1-aminocyclopropane-1-carboxylic acid (ACC) and the conversion of ACC to ethylene are catalyzed by ACC synthase (ACS) and ACC oxidase (ACO) respectively. The genes encoding ACS and ACO (*acs* and *aco*) have been

Table 24.6 Herbicide resistant genes and their action

Herbicide	Inhibition	Gene	Product/action
Bialaphos/Basta/ Glufosinate	Glutamine synthase	*bar*	Phosphinothricin $\xrightarrow{\text{PAT}}$ Acetyl phosphinothricin
Glyphosate	EPSPS	*aro*A	Mutated EPSPS resistant to glyphosate
		gox	Glyphosate ⟶ Amino ethyl phosphate (detoxification)
Sulfonylurea and imidazolinone	ALS	*Hra, C*3, *csr*1-1, *ahas*3r	Mutated ALS resistant to herbicide
Bromoxynil	Photosystem II	*bxn*	Bromoxynil ⟶ 3,5,dibromo 4-hydroxy benzoic acid

EPSPS: 5 enol pyruvyl shikimate 3 phosphate synthase;
ALS: Aceto lactate synthase; PAT: Phosphinothricin acetyltransferase.

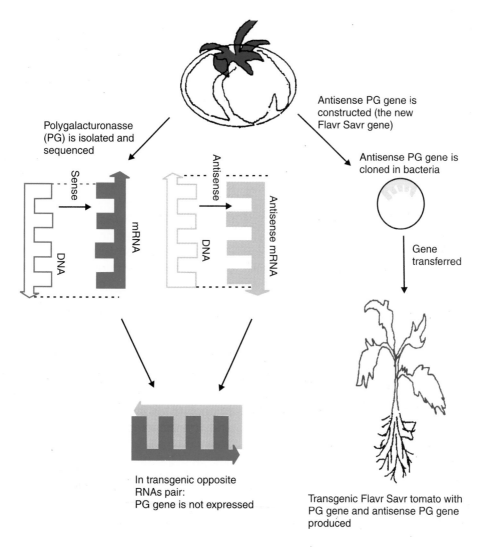

Polygalacturonasse (PG) is isolated and sequenced

Antisense PG gene is constructed (the new Flavr Savr gene)

Antisense PG gene is cloned in bacteria

Sense

Antisense

DNA

mRNA

DNA

Antisense mRNA

Gene transferred

In transgenic opposite RNAs pair: PG gene is not expressed

Transgenic Flavr Savr tomato with PG gene and antisense PG gene produced

Fig. 24.1 Production of transgenic Flavr Savr tomato using antisense RNA gene.

cloned from many species. Monsanto approached the problem of ethylene control by genetically engineering tomatoes to express a gene from *Pseudomonas* that encodes an enzyme capable of breaking down ACC into metabolites other than ethylene (Klee, 1993) which led to a similar delay in fruit ripening. Similarly, ethylene production has been reduced in tomato fruit by the overexpression of the gene for SAM hydrolase that had been isolated from bacteriophage T3.

Several companies such as Zeneca, DNA Plant Technology USA, Monsanto, Pioneer Hi-Bred and Limagrain's Vilmorin are developing transgenic and nontransgenic tomatoes for a longer shelf-life fresh market.

Longer life transgenic flowers

The post-harvest life of many flowers is determined by the onset of petal senescence. Petal senescence in carnation exhibit a characteristic "in rolling" behavior. This "in rolling" is also observed in response to exogenously supplied ethylene (Cornish, 1993). Using a cDNA clone for ACC oxidase (*aco*), Savin *et al.* (1996) produced transgenic carnations expressing antisense ACC oxidase and producing flowers with little detectable ethylene and a marked delay in petal senescence. This extended vase life of flowers by 200%. The only genetically engineered cut flower crop to be given approval for commercial release till 1997 is the longer life carnation containing an antisense *aco* gene, and its release has been granted in Australia. Approvals have been granted for field trials in the Netherlands, for glasshouse trials in USA. Approval has also been granted for the planted release of longer life carnation plants transformed with a sense *acs* gene in Australia (Tanaka *et al.*, 1998).

Transgenics for flower color and shape

As consumers of flowers are attracted to novel products, flower color has been the target for biotechnology. Anthocyanins are the major flower pigments in higher plants. A major target has been the development of blue pigmented flowers in species in which these colors do not naturally occur, such as rose or carnation. Delphinidin is the anthocyanin that normally leads to blue pigmentation. A research group from Florigene in Australia successfully identified and cloned a gene from *Petunia hybrida* encoding a flavonoid 3′ 5′-hydroxylase (F3′5′H), which is required for the biosynthesis of delphinidin (Holton *et al.*, 1993; Stevenson and Cornish, 1993). This provides important information necessary for producing transgenic plants with blue pigmentation in their flowers. Transgenic violet carnations have been successfully produced by the introduction of the petunia F3′5′H gene and its high expression in the petals. Transgenic violet carnations named Moondust™ are being marketed in Australia and Japan (Tanaka *et al.*, 1998).

Modification of intensity of flower color has also been achieved. White and pink flower varieties have been obtained by introducing

sense and antisense chalcone synthase (*chs*) transgenes into petunia, chrysanthemum, rose, carnation, gerbera, and torenia (Suzuki *et al.*, 1997). Flower color has been changed by the introduction of pathway-specific transcriptional activators that control the structural genes in anthocyanin biosynthesis. Expression of maize transcriptional factor "*R*" (*Lc* allele) under the control of the CaMV 35S promoter elevated the amount of anthocyanin in transgenic *Arabidopsis* and tobacco (Lloyd *et al.*, 1992). Constitutive expression of *Lc* gene in petunia resulted in change of leaf color to purple due to accumulation of anthocyanin (Bradley *et al.*, 1998). This approach may allow novel color patterns to be introduced into ornamental species by the controlled expression of anthocyanins and flavonols.

The size and shape of ornamental plants can be modified by altering the ratio of cytokinin:auxin in transgenic plants by expressing oncogenes from Ti and Ri plasmids from *Agrobacterium tumefaciens* and *A. rhizogenes*, respectively. Constitutive expression of the *rolC* gene encoding a cytokinin β glucosidase from the Ri plasmid into potato leads to a dwarf and bushy phenotype, transforming potato into an ornamental plant (Schmulling *et al.*, 1988). Similarly, introduction of this gene into roses have made them bushy (Tanaka *et al.*, 1998).

Transgenics for male sterility

Male sterility in plants is inherited both in a nuclear and cytoplasmic manner. Cytoplasmic male sterility (cms) is due to defects in the mitochondrial genome. Generally cms is associated with defective functioning of the tapetum of the anthers that supply nutrients to the developing pollen grains. The female components of these plants maintain fertility. Transgenics with male sterility and fertility restoration are produced in *Brassica napus* (Mariani *et al.*, 1990). In tobacco also, male sterility was introduced using mitochondrial mutated gene (Hernould *et al.*, 1992). This was achieved by introducing a ribonuclease gene. A hybrid gene with a ribonucleave coding sequence and a tapetum specific promoter was constructed. Mariani *et al.* (1990) formed a gene construct having an anther specific promoter (from *TA29* gene of tobacco) and bacterial coding sequence for a ribonuclease (barnase gene from *Bacillus amyloliquefaciens*). Barnase gene encoded a product ribonuclease enzyme, which was cytotoxic and killed only the tapetal cells, thus preventing pollen development and ultimately leading to male sterility. Using this gene construct (TA29-RNase), male sterile transgenics have been produced in tobacco, lettuce, cauliflower, cotton, tomato, corn, etc. But there was a limitation to this approach as the above-mentioned transgenic plants were always male sterile. Crossing with a restorer line could not restore male fertility in them. It was subsequently reported that male fertility in these transgenic plants could be restored by crossing them with a second set of transgenic plants which expressed a chimaeric ribonuclease inhibitor gene under the control of tapetum specific promoter. Mariani *et al.* (1992) used another gene construct barstar from *Bacillus amyloliquifaciens* with *TA29* promoter for development of transgenic male fertile plants in *B. napus*. Barstar gene product is an inhibitor of ribonuclease which forms a complex with ribonuclease and neutralizes the cytotoxic properties. When male sterile plants (barnase gene) were crossed to male fertile plants (barstar gene), the F_1 plants were fertile due to fertility restoration by suppression of cytotoxic ribonuclease activity in the anther by the formation of cell specific RNase/RNase inhibitor complex. This system was used for hybrid seed production.

Kriete *et al.* (1996) developed another strategy for inducing male sterility. In this system, a bacterial gene *argE* encoding N-acetyl ornithine deacetylase is used. This enzyme deacetylates the inactive herbicide (acetyl bialaphos) and makes it an active herbicide by deacetylation. The gene from *E. coli* has been introduced in tobacco under the control of a tapetum cell-specific promoter. Transgenic plants expressing *argE* gene when sprayed with acetylated (inactive) form of bialaphos showed the normal development of plants, but the anther tissue collapsed. Thus, the plant became male sterile. The advantage of this system is that one need not maintain the male sterile line because plants can be propagated sexually without any problem since *argE* does not affect the development of pollen. Moreover, since male sterility is induced by an inactive herbicide in normal plants, no genetic restorer of male sterility is required. This strategy can be applied to other plants where hybrid production is difficult.

Transgenics for terminator seed

The technology that terminates the viability/fertility of seeds after a given time is known as terminator technology and the gene involved is popularly known as terminator gene. The terminator gene technology has generated a lot of attention due to a patent (No. 5,723,765) on "Control of Plant Gene Expression" which was granted on March 3, 1998 by the United States Patent Office to United States Department of Agriculture and the Delta and Pine Land Co, USA. This patent was not for any new gene as is generally the case, but was for known genes and has been given for control of mechanisms to terminate the expression of desired traits either in the first generation of plant or only to the subsequent generations.

The terminator technology is based on the use of a suitable lethal gene which makes the second-generation seeds infertile. The seed company sells the first generation seeds which are fully developed, normal and fertile to produce healthy plants bearing seeds or fruits that can be used as food but will not germinate if planted. This will force the farmers to buy fresh seeds from the seed company to grow next season crop, since they cannot use the harvested seeds for next season crop.

The terminator technology employs three stretches of DNA (genes) which carry the necessary genetic information into the plants (Fig. 24.2).

1. *Terminator (lethal gene):* It could be any gene that produces a protein which is toxic to plants and does not allow the seeds to germinate. A lethal gene codes for ribosome inhibiting protein (RIP). Lethal gene encoding RIP interferes in the synthesis of all proteins in the plant cell, without being toxic to other organisms. Thus expression of RIP gene in the cells of embryo would prevent germination of the seeds. It is attached with a particular type of promoter one that is active only in the late stages of seed development. LEA (late embryogenesis abundance) promoter is ideal when the aim is to express the trait in the second generation of seed onwards. Such a promoter would be active only after the first generation plant has completed vegetative growth. To prevent the expression of lethal RIP gene in the first generation seed, a blocking sequence is placed in between the LEA promoter and the lethal gene. The blocking sequence itself is flanked by specific excision sequences (LOX sites). When the blocking sequence is excised out by site specific excision at flanking LOX sites, the lethal genes comes in direct contact with the promoter and thus show expression in all the subsequent generations during late embryogenesis stages.

2. *Recombinase gene:* The second gene construct consists of a gene coding for an enzyme called recombinase. This enzyme is able to recognize the excision (LOX) sequences and removes these sequences along with the blocking sequence from the first gene construct by a process of recombination. A preferred recombinase excision sequence system is a bacteriophage CRE/LOX system where the CRE protein (recombinase) performs site-specific recombination of DNA at LOX sites. This recombinase gene is placed behind a repressible promoter specific for a repressor encoded by the third gene. This promoter can be repressed and thus recombinase enzyme will not be produced if a particular protein is present. This site-specific recombination takes place during germination of first generation seed on sowing and thus removes the excision and blocking sequences from the first gene construct.

3. *Repressible gene:* A third gene produces a protein called repressor protein which represses the promoter of the recombinase gene in the second gene construct. The repressor protein itself becomes inactive when it binds to a specific chemical, i.e. tetracycline. The inactive repressor (i.e. repressor-tetracycline complex is not able to repress the promoter attached to recombinase gene, thus allowing the synthesis of recombinase enzyme.

Mechanism in pure line seed production

It is reported that plant cells will be genetically modified with these strips of DNA and the plants regenerated through tissue culture methods. During the first generation, i.e. when companies are producing the seed, the plants with these stretches of DNA will be normal. The blocking sequence is firmly present between the reporter and lethal gene. Seeds are therefore formed without any trouble.

When the first generation seeds mature, these seeds will be exposed to a certain chemical (tetracycline) and sold in the market to the farmers. The repressor protein is being produced by the third gene which in the presence of tetracycline becomes inactive and cannot bind on the repressible promoter site. Thus, recombinase gene will become active on the second strip of DNA. The recombinase promptly removes the excision and blocking sequences from the first gene construct. At this stage LEA promoter is in direct contact with the lethal gene. But the lethal gene is not expressed, because the promoter has been chosen to be active only at a particular stage of seed development i.e. late embryonic stage. As a result the seed germinates properly to produce healthy second-generation plants in the farmer's field.

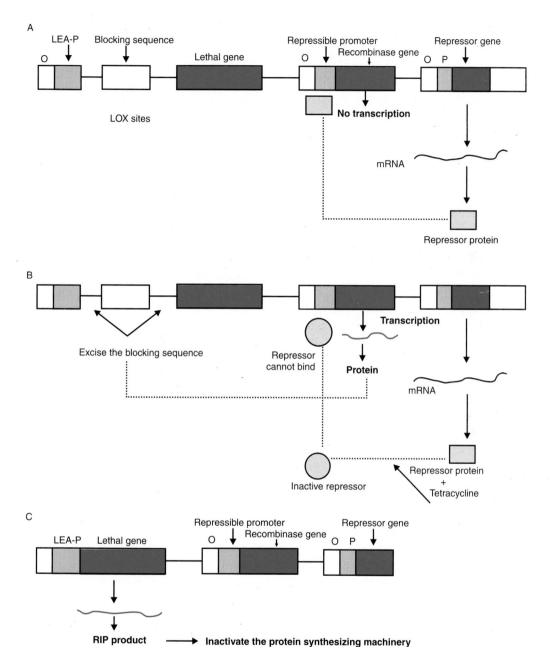

Fig. 24.2 A schematic diagram showing different gene sequences involved in terminator seed technology. **A**: Action of repressor gene on repressible promoter; **B**: Action of recombinase gene to excise blocking sequence; **C**: Synthesis of lethal gene product.

When the second generation plant starts producing seeds, in the late embryogenesis stage, the LEA promoter becomes active and produces a large amount of ribosome inactivating proteins, which in turn inactivate the protein synthesizing machinery of cells, i.e. ribosomes. This results in the production of infertile second generation seeds. These seeds can be used as food, but will not germinate if planted to grow as subsequent generation plants.

The technique is still under various stages of trial; it is most likely to appear in transgenic and hybrid seeds of cotton, tomato, potato, soybean etc.

Terminator technology for use in hybrid seed production

An alternative technique to using terminator gene is producing a pair of genetically engineered plants and then hybridizing them. One transgenic parent will contain the gene construct containing LEA promoter, excision sequences and the lethal gene. The second transgenic parent will have the gene construct containing the recombinase gene linked to a germination specific promoter. A cross between these two transgenic parent plants will produce hybrid seeds carrying genes from both the parents. When the hybrid seed is planted, although the recombinase gene will become active during germination and will remove the excision and blocking sequences from the first gene construct, yet it will produce healthy second-generation plants. The lethal gene will express only when the plant tries to produce the second generation seeds on the farmer's field, which will result in the production of sterile seeds. This technique is particularly suitable for use with plants that are being planted as a yearly crop from seeds.

TRANSGENIC PLANTS AS BIOREACTORS

Through genetic engineering techniques, compounds of commercial interest that were previously produced only from exotic sources of plants, animal and microbial origin are now produced in domesticated crops. Transgenic plants can act as living bioreactors for the inexpensive production of chemicals and pharmaceuticals, and this is also known as *molecular farming*. It has been possible to produce carbohydrates, fatty acids, polypeptides, vaccines, industrial enzymes and biodegradable plastics using the molecular farming approach (Table 24.7).

Carbohydrates

Starch is one of the components of plant cells. Tailor-made starches with reduced level of amylose or increased amount of starch have been development in potato. Starch precursors have also been rerouted into the biosynthetic pathways of other storage carbohydrates. Nonfructan storing tobacco and potato plants have been transformed with fructosyl transferase gene from *Bacillus subtilis* to accumulate fructan. Production of novel compounds has also been attempted through genetic engineering. Tobacco plants have been transformed with the mannitol-1-phosphate dehydrogenase gene (*mtl*D) from *E. coli.* These transgenics synthesized more mannitol and were also tolerant to high salinity (Tarczynski *et al.*, 1993). Similarly myo-inositol o-methyl transferase gene from *Mesembryanthemum crystallimum* (ice plant) has been incorporated into tobacco to produce pinnitol a myo-inositol derived cyclic sugar alcohol. Trehalose is an additive that improves processed and dried food by making them tastier. The plant biotechnological companies MOGEN (Leiden, The Netherlands) and Calgene (Davis, USA) have reported the synthesis of small quantities of trehalose in transgenic tobacco plants.

Sucrose, a disaccharide of linked glucose and fructose, is the most preferred sweetner. The key enzyme in sucrose synthesis is sucrose phosphate synthase (SPS). Overexpression of SPS holds promise for achieving the goal of increased sucrose synthesis in plants. Maize SPS has been overexpressed in tomato (Laporte

Table 24.7 The use of plants as bioreactors for the production of lipids, carbohydrates and proteins

Compound	Origin of gene(s)	Application	Plant species used
Lipids			
Medium chain fatty acids	California bay tree (*Umbellularia californica*) – Thioesterase	Food, detergent, industrial	Oilseed rape
Mono-unsaturated fatty acids	Rat–desaturase	Food	Tobacco
Poly-hydroxybutyric acid	*Alcaligenes eutrophus*	Biodegradable plastics	*Arabidopsis*, oilseed rape, soybean
Saturated fatty acids	*Brassica rapa*	Food, confectioneries	Oilseed rape
γ-Linolenic acid	Desaturase (*Synechocystis*); *Mortierella alpina*	For reducing coronary diseases	Tobacco; canola
Ricinoleic acid	Castor oleate hydroxylase	Nylon paints, varnishes, lubricants	*Arabidopsis*
Carbohydrates			
Amylose-free starch	*Solanum tuberosum* (GBSS)	Food, industrial	Potato
Cyclodextrins	*Klebsiella pneumoniae*-Cyclodextrin glucosyl transferase	Food, pharmaceutical	Potato
Fructans	*Bacillus subtilis*-Fructosyl transferase	Food, Industrial	Tobacco, potato
Increased amount of starch	*E. coli* (*glgC16*)	Food, industrial	Potato
Trehalose	*E. coli*	Food stabilizer	Tobacco
Sucrose	Maize sucrose phosphate synthase (SPS)	Sweetner	Tomato
Proteins			
Alpha-trichsantin	Chinese medicinal plant	Inhibition of HIV replication	*Nicotiana benthamiana*
Angiotensin converting enzyme inhibitor	Milk	Anti-hypersensitive effect	Tobacco, tomato
Antigens	Bacteria, viruses	Orally administered vaccines	Tobacco, tomato, potato, lettuce
Antigens	Pathogens	Subunit vaccine	Tobacco
Enkephalin	Human	Opiate activity	Oilseed rape, *Arabidopsis*
Hirudin	Synthetic	Thrombin inhibitor	Oilseed rape

Table adapted from Goddjin and Pen (1996).

et al., 1997). In transgenic plants, sucrose level increased with a concomitant reduction in starch.

Lipids

Monounsaturated fatty acid level in plants is desirable so that the nutritional value of oils is enhanced. When a rat desaturase gene was introduced into tobacco, level of palmitoleic acid (16:1) and oleic acid (18:1) levels increased (Grayburn *et al.*, 1992). For the manufacture of detergents and for specialized nutritional applications, thioesterase gene from California bay tree was incorporated into *Arabidopsis* for the accumulation of medium chain 12-carbon fatty acid as storage lipids (Voelker *et al.*, 1992). The first nonfood products of plant bioengineering in commercial production is a genetically engineered rape seed plant variety (Laurical) that was modified to produce lauric acid, a 12-carbon fatty acid used to make soaps and detergents

(Table 24.7). The researchers needed to introduce only one gene, which came from the California bay tree. That gene shut off fatty acid synthesis after 12 carbons rather than allowing the acids to grow to 18-carbon length normal for the plant, while having little effect on the productivity.

γ-Linolenic acid (GLA) is the first intermediate in the bioconversion of linoleic acid to archidonic acid. GLA is important in alleviating hypercholesteromia and many coronary heart diseases. It is not produced in oilseed crops. This conversion is catalyzed by the enzyme Δ^6 desaturase, and the gene encoding this enzyme has been cloned from cyanobacterium *Synechocystis*. Constitutive expression of this gene in transgenic tobacco resulted in the production of GLA (Reddy and Thomas, 1996). Recently, this desaturase gene has been cloned from filamentous fungi (*Mortierella alpina*) and borage transferred to canola and tobacco respectively (Knutzon *et al.*, 1998; Sayanova *et al.*, 1997).

The hydroxy fatty acid ricinoleic acid has industrial applications in the manufacture of nylon paints, varnishes, resins, lubricants, and cosmetics. Transgenic *Arabidopsis* expressing castor oleate hydroxylase gene accumulated ricinoleic acid up to 17% of the total lipid fraction. Similarly, an oleate 12-hydroxylase:desaturase gene has also been transferred from *Lesquerella fendleri* to *Arabidopsis* for production of hydroxylated fatty acid (Broun *et al.*, 1998).

Protein quality

Considerable progress has also been made in the improvement of nutritional quality of legume storage protein by expression of the gene responsible for a methionine-rich protein. The Brazil nut (*Bertholletia excelsa*) seed protein, 2S albumin of Brazil nut under the promoter of phaseolin gene was introduced into tobacco. The accumulation of 2S albumin resulted in a 30% increase in the methionine content in the transgenic tobacco seed (Altenbach *et al.*, 1989). Similarly, expression and accumulation of the Brazil nut 2S albumin protein in transgenic

Arabidopsis, rapeseed, soybean, French bean seeds, and potato have also been reported. Unfortunately, the Brazil nut 2S albumin was found to be allergenic (Nordlee *et al.*, 1996), which raised the safety concern of development of transgenic plants expressing this protein. Sunflower seed albumin, rich in methionine, is not an allergen and has been used for development of transgenic plants. The sunflower albumin gene under the control of the pea vicilin gene promoter has been transferred to lupin, and the methionine content of seed protein of transgenic plants increased by twofold. Rat feeding experiments demonstrated the superior nutritive value of transgenic lupin (Molvig *et al.*, 1997). Another seed storage protein gene (*Ama1*) isolated from Amaranthus is a good candidate for introduction into crop plants as the protein has well-balanced amino acid composition (Raina and Datta, 1992). This gene has been introduced into potato, and the level of accumulation of *Ama1* was found to be 0.3% of total soluble protein (Chakraborty *et al.*, 1998). The soybean glycinine gene has been introduced into rice to increase its protein content and to make it more digestible. In addition, some important amino acids, such as lysine, lacking in quantity in normal rice have been replinshed (Momma *et al.*, 1999). The gene encoding a human milk β-casein was expressed in transgenic potato under the control of an auxin-inducible promoter (Chong *et al.*, 1997). These findings open the way for reconstitution of human milk proteins in plant foods.

Vitamin and mineral

Vitamin A content in the diet plays a major role in preventing disorders such as color blindness. Rice, the staple food for billions of people, does not contain any provitamin A in the endosperm. Immature rice endosperm is capable of synthesizing the early intermediate geranyl geranyl diphosphate (GGPP), which can be used to produce the uncolored carotene phytoene by expressing the enzyme phytoene synthase in rice endosperm. The synthesis of β-carotene requires three additional plant enzymes: phytoene desaturase and ξ-carotene desaturase, each catalyzing the introduction of two double bonds, and lycopene β-cyclase, encoded by the *lcy* gene. Ye *et al.* (2000) used the *Agrobacterium*-mediated transformation method in rice to introduce the entire β-carotene biosynthetic pathway into the rice endosperm. Plant phytoene synthase (*phy*) gene originated from daffodil (*Narcissus pseudonarcissus*), bacterial phytoene desaturase (*crtl*) originating from *Erwinia uredovora* placed under the control of endosperm-specific glutelin (*Gt1*) and the constitutive CaMV35S promoter respectively. The phytoene synthase cDNA contained a 5'-sequence coding for a functional transit peptide, and the *crtl* gene contained the transit peptide (*tp*) sequence of the pea Rubisco small subunit. These two direct the formation of lycopene in the endospeum plastids, the site of geranyl geranyl diphosphate formation. A third gene, lycopene β-cyclase, from *Narcissus pseudonarcissus,* controlled by the rice glutelin promoter, which carried a functional transit peptide allowing plastid import and selectable marker *aphIV*, were introduced. The resulting transgenic rice plants produce sufficient β-carotene converted to vitamin A and rice has been named as "Golden rice".

Shewmaker *et al.* (1999) have introduced the phytoene synthase gene into *Brassica napus,* which resulted in increased content of carotenoides. This work has been conducted by Monsanto to enhance the carotenoid levels of oilseed crops, with a focus on the accumulation of β-carotene in the seed of canola.

The major micronutrient deficiencies worldwide are iron deficiency, with 24 percent of the world's population (up to 60% in developing countries) or 1.4 billion women suffering from iron-deficiency anemia, and vitamin A deficiency, affecting approximately 400 million children, or 7% of the world population. Iron deficiency is the consequence of (a) a very low amount of iron in the endosperm, (b) a high concentration of phytate (the major cause for inhibition of iron resorption in the intestine) and, (c) lack of high sulfur-containing proteins, which enhance iron resorption.

Consequently, Potrykus and his coworkers at Switzerland aimed to (a) increase iron content with a ferritin transgene from *Phaseolus vulgaris*, (b) reduce phytase in cooked rice with a transgene for heat-stable phytase from *Aspergillus fumigatus*, and (c) increase the resorption-enhancing effect from a transgenic sulfur-rich metallothionin-like protein from *Oryza sativa*. All transgenes are under endosperm-specific regulation. Goto *et al.* (1999) analyzed a series of transgenic rice plants for all the genes mentioned and have so far achieved a twofold increase in iron content and a high activity of the *A. fumigatus* phytase, reducing phytate completely after one hour of cooking. Expression of the metallothionin-like protein led to an increase in cystein of approximately 25% above control.

Biodegradable plastic

Commercial plastic producing plants are also several years away, although researchers have been making progress. Polyhydroxy butyrate (PHB), an aliphatic polyester with thermoplastic properties, was initially synthesized in plants by introducing the relevant genes acetoacetyl-CoA reductase and PHA synthase from the bacterium *Alcaligenes euthophus* into *Arabidopsis thaliana*. The original gene transfer worked, although the resulting plants were sickly and made very little PHB. The problem was solved by adding a sequence to the gene that causes the enzymes they make to be targeted to storage vesicles called plastids. This both provides the enzymes with high levels of a key starting compound for PHB synthesis and protects the rest of the plant cell from the possible harmful effects of PHB accumulation. As a result, PHB synthesis showed a 100-fold increase with no significant ill effects on plant growth or seed yield (Poirier *et al.*, 1995).

Proteins, peptides and vaccines

New developments in plant biotechnology have suggested that it may be possible to use genetically engineered plants and plant viruses to produce vaccines against human diseases, ranging from tooth decay to life threatening infections such as bacterial diarrhea, cholera, and AIDS. One of the earliest examples of the production of pharmaceutical polypeptides in plants exploited the natural high level expression of seed storage protein. The neuropeptide [leu] enkephalin was produced in *B. napus* as part of the seed storage protein 2S albumin (Krebbers and Kerckhove, 1990). Viral vectors have been used to produce chimaeric coat proteins in transgenic plants. Turpen *et al.* (1995) have described a method for engineering the capsid protein of TMV as either internal or C-terminal fusions with peptides carrying epitopes derived from malarial sporozites. Both internal and C-terminal fusion constructs yielded high titers of genetically stable recombinant virus which produced appropriate monoclonal anti-malarial antibodies in infected tobacco plants. The zona pellucida ZP3 protein has been fused with TMV capsid protein. ZP3 protein of mammalian oocyte has been a target for immune contraception and an epitope of 13 amino acids from murine ZP3 has now been expressed in plants as a fusion with TMV capsid protein. Mice immunized with the recombinant TMV particles developed antibodies against ZP3 (Beachy *et al.*, 1996). Theoretically, vaccination with the modified viral coat protein might work as a form of birth control, because antibodies to zona pellucida could prevent fertilization of egg. The same is true for epitopes derived from humal immunodeficiency virus which are expressed as alfalfa mosaic virus coat protein fusion products (Yusibov *et al.*, 1997). Cowpea mosaic virus (CMV) has also been engineered as an expression system for the production of foreign peptides such as epitope derived from human rhinovirus 14 and human immunodeficiency virus (HIV-1). The modified plant viruses elicit the production of mouse antibodies that neutralize virus in test tube experiments (McLain *et al.*, 1995). It indicates the possibility of producing a preventive HIV vaccine. Fusion coat proteins may represent a cost-effective way for generating vaccines (Table 24.7).

Many groups have been working on the expression of antibodies in plants. Immunoglobins, such as IgG, IgM and IgG-IgA chimeric antibodies, fab fragments, single chain antibodies and VH chains have been expressed at levels of over 2% of extractable protein. Holmberg *et al.* (1997) introduced the gene for *Vitreoscilla* hemoglobin (VHb) in tobacco. *Vitreoscilla* synthesizes elevated quantities of a novel homodimeric hemoglobin (VHb) in oxygen limiting environment. Expression of VHb in *E. coli* increases growth rate and final cell density in oxygen limited cultures. When VHb was expressed in transgenic tobacco it exhibited enhanced growth, on average 80–100% more dry weight. Transgenic plants contained on an average 30–40% more chlorophyll and 34% more nicotine. Thus, the introduction of this gene could lead to enhanced plant growth and yield. Hood *et al.* (1998) purified avidin from high expressing lines of transgenic maize and it made commercially available. Avidin is primarily localized in the embryo and represents 72% of aqueous soluble extracted protein from dry seeds. Potato tubers have also been used as a biofactory for high level production of a recombinant single chain Fv (SCFv) antibody. Recombinant antibodies accumulated up to 2% of the total soluble tuber protein (Artsaenko *et al.*, 1998).

Production of edible vaccines

Transgenic plants show the promise for use as low-cost vaccine production system. Researchers are taking several tacks to making plant-based vaccines, but the edible vaccines now under development by a team led by Charles Arntzen of Texas A & M University in USA are likely to be the cheapest and easiest to administer. The idea behind the edible vaccines is to have people take their dose by eating, as part of their diet, the plant that produces the vaccine. Arntzen's team started work aimed at developing oral vaccines to prevent enteric diseases, including cholera and diarrhea caused by bacteria such as *E. coli*, *Shigella* and *Salmonella*. Bacterial diarrheas are a leading cause of infant deaths in the developing world. The first report of the concept of using a plant expression system for production of an edible vaccine appeared in a publication under International Patent Cooperation Treaty, which described a means to express a surface protein (spa A) from *Streptomyces mutans* in tobacco to 0.02% of total leaf proteins. The hepatitis B surface antigen (HBsAg) has been expressed in tobacco (Mason *et al.*, 1992) though with a low level of expression (0.01% of soluble protein). The immunity to HbsAg has been demonstrated in mice.

Transgenic plants of tobacco and potato expressing the heat-labile enterotoxin B (LT-B) subunit from *E. coli* with a microsomal retention sequence have been produced (Haq *et al.*, 1995). Plants expressed the foreign peptides and when mice were fed the transgenic potato tubers it caused oral immunization. This demonstrated the feasibility of using transgenic plants as expression and delivery systems for oral edible vaccines. This edible plant vaccine against diarrhea expressed in potato is already in human trials (Tacket *et al.*, 1998). A problem with potatoes is that they must be cooked before they are eaten. The heat used in cooking will cause the vaccine protein to denature, reducing or eliminating its ability to elicit immunity.

The potato problem, however, may be nearing a solution–through a switch to a carrier that is eaten raw: bananas. Arntzen and his colleagues have introduced a foreign gene into banana plants and shown its expression. They did not use a gene for a potential vaccine protein in these experiments, but plan to put the *E. coli* enterotoxin gene in bananas. If that succeeds, Arntzen wants to add genes for additional vaccine proteins. Another antigen Norwolk virus capsid protein (NVCP) has also been expressed. The first step was to show that proteins made in plants could elicit immune response in animals. They achieved the goal by introducing the gene encoding a surface protein from the hepatitis B virus (HBV) into tobacco plants. Not only did the plants make the viral protein, but also

when injected into mice, it triggered production of antibodies that recognize the hepatitis B protein. Cholera toxin, which is very similar to *E. coli* LT has also been expressed in plants. Hein *et al.* (1996) generated tobacco plants expressing CT-A or CT-B subunits of the toxin. Cholera toxin-B subunit when expressed in potato was processed in a natural way. Antigenically it was found to be similar to the bacterial protein (Arakawa *et al.*, 1998a; 1998b). Similarly, a rabies virus coat glycoprotein gene has been expressed in tomato plants. Carrillo *et al.* (1998) expressed structural protein, VP1, of foot and mouth disease virus in *Arabidopsis*. The mouse that was immunized intraperitoneally with a leaf extract elicited immune response to synthetic peptides carrying various epitopes of VP1, or to complete VP1. Furthermore, all the mice immunized with leaf extract were protected against challenge with virulent foot and mouth disease virus.

Julian Ma and his group at Guy's Hospital in London reported progress in attempts to use plants to make an antibody that might be added to toothpaste to protect against bacteria that cause tooth decay. A hybrid monoclonal antibody (IgA/G), having constant regions of IgA and IgG fused, has been used successfully against human dental caries caused by the bacterium *Streptococcus mutans*, which causes tooth decay. This antibody against dental caries, expressed in tobacco is already in preclinical human trials (Ma *et al.*, 1995). The researchers hope that their plant antibodies can be formulated into toothpaste.

Therefore, transgenic plants show promise for use as low cost vaccine production systems. Attempts are being made to express many proteins of immunotherapeutic use at high levels in plants and to use them as bioreactors of the modern era (rev. Sharma *et al.*, 1999).

Commercial transgenic crops

Transgenic plant products have now been released as commercial varieties in a range of crops (Table 24.8). By all accounts, 1996 was a critical year in the history of plant biotechnology.

For the first time, farmers planted large areas under transgenic plants. It was in 1995 that potato varieties expressing Bt crystal protein toxic to Colorado potato beetle were made commercially available. In the following year, cotton varieties expressing the Bt crystal protein toxic to a number of lepidopteran pests, including the tobacco budworm and cotton bollworm, were introduced. At the same time, Bt corn hybrids with improved resistance to European cornborer became available to the farmers. All these commercial introductions have taken place in USA. *Bt* cotton varieties carrying the same technology as used in USA are being introduced in Australia. Transgenics for herbicide tolerance was the dominant trait in 1997. Transgenics have now been released for various traits from virus resistance to quality and for industrial use.

Impact of recombinant DNA technology

Plant improvement, whether as a result of natural selection or the efforts of plant breeder, has always relied on evolving, evaluating, and selecting the right combination of alleles. The techniques involved gene transfer through sexual methods, and gene flow was limited by the sexual compatibility. The application of these conventional or classical techniques has produced significant achievements in the yield improvement of major food crops. Useful traits such as resistance to diseases, insects, and pests have been transferred to crop varieties from noncultivated plants by the conventional method of breeding. The era of biotechnology started in 1973, when Herb Boyer and Stan Cohen developed recombinant DNA molecules or gene splicing methods, in which fragments of DNA were joined together to create a new genetic combination. Recombinant DNA techniques coupled with the development of methodologies for introducing foreign genes in plants have dramatically changed the scenario of plant breeding. It facilitates introduction of only specifically desirable genes without cotransfer of any undesirable genes from donor species, which normally occurs by conventional breeding meth-

ods. Recombinant DNA techniques allow mobilization of genes between completely unrelated genomes, or, in other words, all barriers of gene transfer between species, taxa, and kingdom have been broken.

Biotechnology exploded onto the scene of crop-based agriculture in 1984 with the first report of inheritance of a foreign gene engineered into tobacco plants. Prior to this time, many scientists speculated on the use of recombinant DNA techniques for improvement of organisms. However, the list of plant species that can be transformed has been extended to more than 120 species in at least 35 families. Success includes most major economic crops, vegetables, ornamental, medicinal, fruit, tree, and pasture plants. The initial publication of *Agrobacterium*-mediated transformation, and those that followed shortly thereafter, sparked a new industry in plant biotechnology. Many companies were quick to jump on the "biotech wagon" prompted by claims for the potential to modify crop plants. In the 1990s, there was an upsurge of private sector investment in agricultural biotechnology. Some of the first products were insect- and herbicide-resistant crop plants and "Flavr Savr" tomato with increased shelf life. The development of transgenic crops to these agrochemicals was the work of big multinational companies who were producing these chemicals. The companies were looking for their profits, and they invested heavily on seed industry to develop transgenic crops that were resistant to their trademark chemicals (herbicides and pesticides). In other words, multinational companies were able to sell their improved genetically modified seeds and also their herbicide and pesticide chemicals. The area under transgenic crops has been increasing steadily in developed countries and the production level has also gone up, and as per reports the benefits have been accrued by farmers.

There has been dramatic and continuing increase in the area under transgenic crops. From 2.8 million hectares in 1996, the area increased nearly 10-fold to 11 million hectares in 1997 and rose to 27.8 million hectares in 1998. The total

Table 24.8 Commercial releases of transgenic plant varieties

Trait	Crop	Product name	Organization
Insect resistance			
Bt	Cotton	Bollguard	Monsanto Co.
Bt	Cotton	–	Calgene
Bt	Potato	Newleaf	Monsanto Co.
Bt	Maize	Yield Guard	Monsanto Co.
Bt	Maize	Maximizer	Ciba Seeds
Bt	Maize	–	Novartis, Mycogen, Pioneer Hi-Bred
Virus resistance	Squash	Freedom II	Asgrow Seeds, Inc.
Virus resistance	Tobacco, tomato, capsicum	–	(China)
Virus resistance	Cantaloupe	–	Asgrow Seeds, Inc.
Virus resistance	Papaya	Rainbow	–
Herbicide resistance			
Glufosinate	Maize	Liberty Link	AgrEvo Canada Inc.
Glufosinate		Liberty Link	AgrEvo USA Co.
Sethoxydim		–	BASF
Glufosinate		–	DeKalb Genetics Corp.
Glyphosate		Roundup Ready	Monsanto Co.
Glufosinate		–	Plant Genetic Systems
Glufosinate	Oilseed rape, Canola	Innovator	AgrEvo Canada Inc.
Glufosinate		,,	AgrEvo USA Co.
Glyphosate		Roundup Ready	Monsanto Co.
Glufosinate		–	Plant Genetic Systems
Bromoxynil		–	Rhone-Poulenc
Bromoxynil	Cotton	BXN	Calgene Inc.
Sulfonylurea		–	DuPont
Glyphosate		Roundup Ready	Monsanto Co.
Glyphosate	Soybean	Roundup Ready	Monsanto Co.
Sulfonyl urea	Flax	Triffid	Univ. of Saskatchewan
Glufosinate	Chicory	–	Bejo-Zaden
Quality			
Vine ripened, shelf life	Tomato	Flavr Savr	Calgene
Vine ripened, shelf life	Tomato	Endless Summer	DNA Plant Technology
Delayed softening, paste consistency	Tomato	–	Zeneca
Modified starch	Potato		AVEBE
Increased vase life	Carnation		Florigene
Modified flower color	Carnation	–	Florigene
Oil characteristics	Canola	Laurical	Calgene
Male sterility hybrid system	Canola	–	Plant Genetics Systems

global area tilled with genetically modified crops hit 52.6 million hectares in 2001. For the year 2001, four countries grew 99% (52.1 million hectares) of the global transgenic crops. USA grew 35.7 million hectares of transgenic crops; Argentina, 11.8 million hectares; Canada, 3.2 million hectares; and China, 1.5 million hectares. Australia, Mexico, and South Africa also have plantations devoted to transgenic cotton and corn. The five principal transgenic crops are soybean, maize, cotton, rapeseed/canola, and potato. Globally, the principal GM crops were soybean occupying 33.3 million hectares in 2001 (63%), followed by GM corn at

9.8 million hectares (19%), transgenic cotton at 6.8 million hectares (13%) and GM canola at 2.7 million hectares (5%).

In 2001 herbicide tolerant soybean occupied 33.3 million hectres or (63%) of the total global area grown in seven countries. Bt maize occupied 5.9 million hectares or (11%) of the transgenic area in six countries. The other six GM crops occupied 5% or less of global transgenic crop area.

Since 1996, herbicide tolerance has consistently been the dominant trait with insect resistance second. In 2001, herbicide tolerance deployed in soybean, corn and cotton occupied 77% or 40.6% million hectares and 7.8 million hectares (15%) planted to Bt crops. Both herbicide and insect resistance deployed in cotton and corn occupied 8% or 4.2 million hectares of the global transgenic area in 2001.

Farmers decided to plant more genetically modified rapeseed, corn, soybean, and cotton because of more convenient and flexible crop management, higher productivity, and safer environment through decreased use of conventional pesticides.

The initial assessments of the net benefits from Bt insect-resistant transgenic crops of cotton, corn, and potato are estimated to be $33, $27.25 and $19 per acre, respectively. The herbicide tolerant gene(s) introduced into different crops has no effect on yield *per se*, but significantly better weed control was achieved, which in turn increased yield dependability. Farmers in USA using herbicide-tolerant soybean required only one application of herbicide. This decreased requirement of herbicides on herbicide-tolerant varieties translated to herbicide savings of 10 to 40%. It also resulted in better control of weeds and soil moisture conservation, improved yield dependability, no carryover of herbicide residues, and greater flexibility in agronomic management of the crop. Herbicide-tolerant canola in Canada lowered herbicide requirement from 570 g to 160 g active ingredient per acre, increased yield up to 20%, with an average increased yield of 9%.

In general, transgenic crops have been well received in North America and Canada, with a very high percentage of farmers who planted in one year electing to plant again in the next year. Total transgenic crop sales grew more than sixfold, from U.S. $235 million in 1996 to $1.2 to $1.5 billion in 1998. The market is projected to increase to $6 billion in 2005 and to $20 billion in 2010.

During this early phase of plant revolution, the benefits of plant engineering have been largely confined to farmers. Apparently most of the agrochemical companies envision a far more lucrative future when the plant revolution matures further. There is increasing evidence that transgenics can produce healthier food. Another opportunity is to use plants as chemical factories. The next major phase of the plant revolution will emphasize the engineering of desirable traits in plants, traits that are readily apparent to the consumer, such as golden rice and golden mustard with increased vitamin A content in rice and mustard respectively. Intensive efforts have also been made to channelize plant metabolism into producing industrial feedstocks, pharmaceuticals, and nutraceutical products. The selection of superior strains and the conventional breeding of soybeans have led to lines with reduced levels of antinutritional oligosaccharides, stachyose, raffinose, and galactose. Modification of the genome of soybean to desirable fats is leading to healthier foods and to useful industrial chemicals. Another effort that succeeded was to use soybean plants to produce vernolic acid and ricinoleic acid, derivatives of oleic acid that are used as hardeners in paints and plastics.

Another thrust of GMO bacteria is in agricultural sector. Leguminous plants such as soybean form symbiotic associations with *Rhizobium*, *Bradyrhizobium,* and *Frankia* bacteria, which fix atmospheric nitrogen to the soil by *nif* gene. Nowadays genetically modified *Rhizobia* are added to the soil as legume inoculum to reduce the need of nitrogenous fertilizer.

Nobel laureate Norman Borlaug estimates

that to meet projected food demands by 2025, average cereal yield must increase by 80% over the 1990 average. Biotechnology, one of the many tools of agricultural research and development, can contribute to food security by helping to promote sustainable agriculture centered on small holder farmers in developing countries. This biotechnology revolution is very relevant to the problems of food security, poverty reduction, and environmental conservation in the developing world. Biotechnology will help us do things that we could not do before, and do it in a more precise and safe way. Biotechnology will allow us to cross genetic barriers that we were never able to cross with conventional genetics and plant breeding. In the past, conventional breeders were forced to bring along many other genes with the genes of interest, such as insect or disease resistance, that one wished to incorporate in a new variety. These extra genes often had a negative effect, and it took years of breeding to remove them. But with use of biotechnological tools, one has the more precise ways to introduce only the desired gene.

For many, biotechnology raises important questions relating to ethics, intellectual property rights, and biosafety. For some, biotechnology is a lightening rod for visceral debate, with opposite factions making strong claims of peril. There have been widespread protests against the spread of agro-biotechnology. Some of the concerns come from scientists who fear that "novel" products will destroy agricultural diversity, thus changing agricultural patterns into unrecognizable and uncontrollable forms. The dominance of a highly concentrated private sector has raised fears of a new phase of comparative disadvantage and increased dependency in the developing world.

Various arguments against herbicide-tolerant transgenics in crop plants have come up:

1. Use of herbicide-tolerant transgenic crops can lead to transfer of herbicide tolerance genes to sexually compatible wild relatives or weeds, which can be a major potential threat to environment.

2. Transgenic crops can create "super weeds".
3. It would actually increase the dependence on a few herbicides rather than reducing herbicide usage.
4. It may increase the problem of weed control if weeds develop resistance to such herbicides through gene flow from transgenic crops.
5. Herbicide tolerance is being sought not only for environmentally comparatively acceptable herbicides but also for older, more toxic and persistent products.
6. Nonchemical means of weed control, such as crop rotation, dense plantings, cover cropping, ridge tillage, and others, however labour-intensive for farmers, are preferable than the use of any herbicide at all. Gene flow is the primary risk in releasing transgenic plants.

However, according to a 10-year study by a team of scientists in Britain, genetically improved crops show no signs of turning into superweeds. Researchers planted genetically improved varieties of oilseed rape, potato, maize, and sugar beet alongside conventional crops in 12 different habitats in the UK to see whether the plants could invade natural habitats. But the research team found that native wild plants displaced both genetically improved plants and ordinary crops and that conventional crops actually outlived the biotech ones. Scientists said most of the crops died out after 4 years, and after 10 years the only survivor was one type of nonbiotech potato. When the study began, the kind of biotech plants available were varieties engineered to make them better able to withstand spraying with herbicides or insect attack, not crops engineered to withstand drought or natural pests. It was not expected that herbicide tolerance would give plants an ecological advantage. The results do not mean that other genetic modifications could not increase weediness or invasiveness of crop plants, but they do indicate that arable crops are unlikely to survive for long outside cultivation.

Issue of Bt corn and monarch butterfly: Researchers at Cornell University in 1999 published a report that when Bt corn pollen was fed to monarch butterflies, it killed the larvae of these flies. This raised concerns in some quarters because it was widely and incorrectly interpreted to mean that GM crops were threatening nonpest insects, such as the monarch butterfly. But this report was contradicted when follow-up studies showed the effect of GM pollen on nontarget insects, including the monarch butterfly, to be negligible under real life conditions. In the first place, corn pollen is pretty heavy. It does not fly long distances. Also, most monarchs are moving at different times of the season when there is no corn pollen. It is argued that some of them might get killed by *Bt* corn pollen, but how many get killed when they are sprayed with insecticides is not known. Further, the scientific community discounted the original report because (i) it was conducted under artificial lab conditions; (ii) the larvae were allowed to eat only corn pollen (which they do not often encounter in the open environment), and (iii) there was no comparison group of larvae fed on ordinary corn pollen sprayed with regular Bt insecticide.

It is said that Bt insect-resistant genes in crops put stress on the insect pests, and they mutate, and there is always a risk of the emergence of Bt-resistant insects. Therefore, it is recommended that a refuge crop be grown along with Bt crops so that susceptible insects may multiply on this refuge crop. To contradict this, it is argued that this process has been going on with conventional insecticides also. It is a problem that can be managed. Besides, so many other alternative strategies have been proposed.

GM crops can cause allergenic reaction to human beings. It was discovered that GM soybean carrying the Brazil nut gene was allergenic to people who were allergic to Brazil nut. The project was abandoned before any customer was exposed to the soybean modified with the Brazil nut gene. The product never appeared in the market, and the public was not exposed to any risk. However, the allergenic results showed conclusively that the undesirable

properties of the Brazil nut protein were transferred when the gene was moved into a different species. If a gene produces an allergenic protein in one species, it will likely do so in a new species; similarly, if a gene is nonallergenic in one species, it is not likely to become allergenic when transferred to another.

GM potatoes cause suppression of immune systems and damage to internal organs of rats when fed on these potatoes. After considerable deliberations and review of all available data, the evidence did not support Dr Pustzai's conclusions on rat experiments. However, an important point emerged that genetic modification introduced in potato was a gene to make lectins, a toxin. The public was not told that the rats were being forced to eat a toxic lectin, and the toxin, not the method used to introduce it, might have been responsible for their illness.

Researchers often introduce a marker gene (antibiotic resistance) that encodes an easily detectable substance along with a desirable gene as a signal or marker to help determine which plant cells and tissues have taken up new genetic material. The mere presence of a marker gene (or any other foreign DNA) is not a food safety concern because scientific evidence shows DNA itself is not hazardous. The pertinent issue is whether the gene product, the protein made from the gene recipe, is safe. As a case in point, the kanamycin (an antibiotic) resistance gene has been used as a selectable marker in the production of Calgene's Flavr Savr® tomato. That same gene is found in harmless bacteria that are normally found on fresh fruits and vegetables, which are eaten everyday without undue harm. The kanamycin resistance gene is also found in bacteria that populate the human gastrointestinal tract. In considering safety, it is no more or less relevant to know that the gene is used as a selectable marker in genetically engineered plants than it is to know that it is a normal component of bacteria that live on many of our foods or in our intestines. Safety is not determined by how or why the gene (and its product) was introduced into food – it is determined solely by the characteristics of the gene product and our experience with it in the food supply. However, lately researchers are trying not to use such antibiotic resistance markers.

The use of viral-derived, virus-resistance genes has raised concerns about possible viral recombination and the development of novel viruses with altered host range or symptom severity. The use of engineered *Bt*-derived insect resistance has raised concerns about rapid evolution of insect resistance.

Future: There is little doubt that biotechnology is fashionable, but it is clearly not a passing fad. Some expectations have been met with the commercial development of herbicide and insect resistance in crops, clearly pointing to a strong future for biotechnology in agriculture. The future prospects in this area are also exciting, with preliminary indications that novel genes can be introduced to generate plant lines useful for production of materials ranging from pharmaceuticals to biodegradable plastics, vaccines, minerals, vitamins, fatty acids, carbohydrates, etc. The path from inception of the agricultural biotechnology industry to field production of commercial products has been crisscrossed by obstacles ranging from scientific and technological challenges to legal and regulatory hurdles, economic factors, and social concerns. The conservative assumption underlying regulations in many countries is that all transgenic plants are potentially hazardous. The hazards relate to the genes transferred or the phenotype produced, not to the gene transfer method used. There have been no reports to date of any harmful environmental effects or other hazardous unforeseen behavior of transgenic plants in the thousands of field trials conducted internationally. However, various concerns with direct relevance to our agricultural system have been raised. In particular, the use of terminator seed technology has become an important and very valid issue. Toxicology of foods must be addressed, not only for GM but also for conventional bred foods. Issues of superweeds or supervirulent strains of pathogens or insects may arise, but must be tested and analyzed scientifically.

Further reading

Pattanayak, D. and Kumar, P.A. 2000. Plant biotechnology: Current advances and future perspectives. *Proc. Ind. Acad. Sci. B.*, **6**: 266-310.

25

Genomics

Genomics is a rapidly emerging area of research, and in the coming years it will revolutionize our understanding of biology. Genomics is the study of how genes and genetic information are organized within the genome, and how this organization determines their function. Genomics has been described as either structural or functional. *Structural genomics* involves sequencing and mapping of genomes and a study of the structure of all gene sequences encoded in a fully sequenced genome. *Functional genomics* involves a study of the functions of all specific gene sequences and their expression in time and space in an organism. Plant genomics is a part of this larger field and embraces the study of whole genome of plants, their physical and molecular organization, evolution and functions of the myriad of constituent genes. What the genome contains and how the genome functions are addressed in this chapter. Genomes of eukaryotes comprise thousands of genes and millions of base pairs, which store massive data and information not only about genes themselves but also relating to their expression. The advent of genomics is a consequence of the discovery of elegant and simple procedures of sequencing of nucleic acids. This ability to sequence the genome has aroused tremendous importance because deciphering the sequence of bases would unlock the whole blueprint of the development of an organism. Thus if we understand the genomes of crops such as rice,

wheat, maize, *Arabidopsis*, and potato, we can ensure a better future with the capability of more perfect and precised genetic manipulation for enhanced yield and survival under adverse conditions. This chapter is devoted to understand the basic steps involved in sequencing of genome, location of genes in DNA sequences, and understanding how the genome functions. In the past, attention was directed at the expression pathways for individual genes, with groups of genes being considered only when the expression of one gene is linked to that of another. Now the issues have become more general and relate to the expression of the genome as a whole rather than an individual gene function.

Mapping of prokaryotic genomes

A number of microbial genomes have already been sequenced, followed closely by simple eukaryotic genomes such as yeast and the nematode *Caenorhabditis elegans* (Table 25.1). The first organism to have its genome sequenced was *Haemophilus influenzae* (Fleischmann *et al.*, 1995). It has a genome size of 1.8 Mbp encoding 1743 recognized genes. *Bacillus subtilis* with a genome of 4.2 Mbp has 4100 genes (Saier, 1998). The prototypical bacterium *E. coli* has a genome size of 4.6 Mbp with 4288 recognized genes (Blattner *et al.*, 1997). The genome of one eukaryote (the brewers yeast *Saccharomyces cerevisiae*) has been sequenced.

Table 25.1 List of organisms whose genome has been sequenced

Organism	Genome size (Mb)	No. of genes
Haemophilus influenzae	1.83	1,743
Helicobacter pylori	1.66	1,590
Methanococcus janaschii	1.66	1,680
E. coli	4.6	4,288
Bacillus subtilis	4.2	4,100
Yeast (*Saccharomyces cerevisae*)	13	6,000
Drosophila	139	13,601
Caenorhabditis elegans	97	19,099
Arabidopsis thaliana	140	27,000
Human	3,200	33,000–1,50,000 (40,000)

E. coli genome

In 1997, Fred Blattner and his group at the University of Wisconsin reported the complete nucleotide sequence of nonpathogenic laboratory strain of *E. coli* K-12 called MG1655. Their 15-year effort led to the assembly of one continuous DNA strand containing all the 4,639,321 bp of the circular *E. coli* genome. A similar feat was accomplished shortly thereafter by a consortium of Japanese research laboratories using the W3110 strain of *E. coli* K-12. The *E. coli* genome is very compact with a homogeneous distribution of genes throughout the chromosome. Numerous clusters of genes encoding functionally related proteins were identified as physically linked regulatory operons. This compact arrangement of genes is also observed in the genomes of three other bacteria that have been recently sequenced: (1) the *Bacillus subtilis* genome was found to contain 4100 protein coding sequences distributed across 4,214,810 bp, (2) *Mycobacterium tuberculosis*, a pathogenic microorganism with 4000 genes distributed across a 4,411,529-bp genome, and (3) the gastric pathogen *Helicobacter pylori*, which is responsible for peptic ulcer disease, has a minimum of 1590 proteins in a genome of 1,667,867 bp.

The *E. coli* DNA sequence identified 4288 actual and proposed gene coding sequences, of which 38% have no attributed function based on homology searches with all available databases. It was found that approximately 88% of the genome encodes proteins or RNAs, ~11% appears to be utilized for gene regulatory functions, and ~1% consists of repetitive DNA sequences. The average distance between *E. coli* genes is only 120 bp. The coding sequences are continuous and lack the noncoding introns that are commonly found in most eukaryotic genes. *E. coli* K-12 is the best characterized free-living organism and research over the past 50 years has provided an abundance of genetic, biochemical, and physiological data. Using this information, Blattner's group was able to assign function to ~3200 genes, as summarized in Table 25.2.

Mapping of eukaryotic genomes

Sequencing of the entire genomes in humans, *Arabidopsis thaliana*, fruit fly, rice and many other organisms is in progress. The major revolution in the study of genomes of eukaryotes was brought about due to the availability of recombinant DNA and PCR technologies that became available during 1970s and 1980s respectively. These technologies helped in the preparation of molecular maps for a variety of animal and plant genomes. Ideally the objective of genomics research in any species should be to sequence its entire genome and to decipher functions of all the different coding and noncoding sequences. In crop sciences, genomics has been promoted using *Arabidopsis* and rice as model plants. *Arabidopsis* is a weed plant found

Table 25.2 Gene functional groups identified in the DNA sequence of *E. coli* K-12

Functional class	Number of genes	Percent of total
Transport proteins	427	9.9
Energy metabolism	373	8.7
Replication, transcription, translation	352	8.2
Putative enzymes	251	5.9
Cell structure and membrane proteins	237	5.5
Nucleotide, amino acid, fatty acid metabolism	237	5.5
Regulatory functions	178	4.2
Miscellaneous gene functions	601	14.0
Hypothetical, unclassified, unknown	1632	38.1
Total	4288	100

Source: Miesfeld, R.L. (1999). *Applied Molecular Genetics,* Wiley-Liss, New York, 289 pp.

all over the world and is easy to grow and to use for genetic experiments as it has one of the smallest plant genomes with 145 million base pairs. *Arabidopsis* is the model plant for dicot plant families whereas rice is the model plant for monocot families. The rice genome is also very simple, only three times larger than that of *Arabidopsis*. Also the genomic organization in cereals is highly conserved irrespective of the plant species; genes are lined up in the same order on the chromosomes of different cereal species. This phenomenon is called *synteny.*

Base sequencing is the fundamental to any genome-sequencing project. Sanger's dideoxy chain termination method is now preferred over the Maxam and Gilbert technique. High throughput techniques using fluorescent dideoxynucleotide terminators, laser excitation of bands of newly synthesized DNA and sensing of their colors by photomultipliers attached to computers are employed. It is rather easy and simple to sequence prokaryotic genomes (see Chapter 17 for DNA sequencing techniques). But the actual execution of any eukaryote multicellular organism is rather complex because of the great mass of DNA (see Table 13.2). Humans have 3.2 Mbp of DNA. Many plants have comparable or larger genomes. *Arabidopsis*, known to possess the smallest genome among flowering plants, has 145 Mbp of DNA and 20,000–50,000 genes. Rice has about 420 Mbp whereas wheat has ~16

billion base pairs of DNA. If sequencing is done at one time, in a single lane, and in a single gel in the electrophoresis apparatus, one can handle DNA fragments that are only 1–2 kb pairs long while actual sequences may be read for only 200–500 bp. In eukaryotic genomes, the problem becomes acute by the fact that if DNA of such eukaryotic organism is digested to such a small size, millions of fragments would be formed, which would not only make the task unmanageable but would leave no exact location in the genome. Thus for any genome sequencing program, the following steps are undertaken. In this chapter the relevant points have been mentioned because the techniques and procedures have already been described in the earlier chapters in details.

1. *Construction of linkage maps with molecular markers*: Generally, classical genetic linkage maps is a prerequisite for such program. These maps are based on recombination frequencies, and the distance between the genes is measured in centimorgans (cM). From a genetic linkage map, one has to proceed for the construction of a "physical map" with landmarks along the whole length of the DNA molecules, and with distance measured in base pairs. In this regard, the concept of molecular marker maps was developed by Botstein and coworkers, which relied on restriction fragment length

polymorphism (RFLP). Other molecular markers such as random amplified polymorphic DNA (RAPD), microsatellites, amplified fragment length polymorphism (AFLP), etc. and maps have also been prepared (see Chapter 22). RFLP markers are also positioned initially by linkage analysis and thus only approximately, they represent real molecular markers that can be actually hybridized to cloned DNA fragments and thus allow the work to begin on construction of an actual physical map. The RFLP markers are then hybridized to the vectors containing the DNA fragments that are to be sequenced, enabling the positioning of various cloned DNA fragments along the chromosomes in the cytogenetic map.

2. *Gene libraries*: Construction of gene libraries is an essential step in genome analysis. DNA is digested into smaller fragments, whether by restriction endonucleases or by shearing is cloned into a vector. Cloning enables multiplication of the DNA fragments for further handling or distribution to different laboratories. Significant advances have taken place in the development of vectors that can accommodate a big fragment of DNA. Cosmids allow 40- to 55-kb-long fragments as compared to 10 kb-long fragments to be inserted in a plasmid vector. However, even cosmid vectors were not quite appropriate for the task of sequencing the large genomes of eukaryotes. The development of yeast artificial chromosomes (YACs) allowed up to 1000-kb fragments to be cloned and gave a real impetus to genome studies. Cloning of such large fragments reduces the number of potential gaps in the assembly of contigs. It has been reported that 7500 YAC clones with an average insert size of ~400 kb DNA can contain the entire human genome, but actually 60,000–70,000 clones have to be screened by a thousand or more probes. This is because of redundancy in preparing libraries and is essential to ensure that overlapping clones are available for the entire genome. To prepare libraries, after the colonies are individually picked from petri dishes, they are first transferred to 96-well microtiter plates for replication and distribution. Sometimes colonies are gridded on special nylon filters by a robot with a 96-prong spotting device, which enables repetitive stamping on large nylon filters for preparation of replicates. The nylon membranes can then be processed for screening of libraries.

3. *Screening of libraries and constructing contigs*: The nylon membrane with colonies can be made to rest on an agar culture medium from which cells can absorb nutrients and further multiply. But it can also be dried, which greatly facilitates the screening process since the filters can be bathed by a DNA probe such as by a RFLP marker (the cells are lysed *in situ* and yeast DNA denatured before hand). Any spot that hybridizes can be detected by autoradiography and the corresponding yeast clone located in the microtitre plate. However, RFLP markers are few and far between and often there is only one marker per YAC clone. So when the PCR techniques became available, to order the clones precisely, the concept of marking them by sequence tagged sites (STS) was developed. STS represents a landmark of a unique short ~300-bp DNA sequence that can be had from any region of the cloned DNA, the critical feature being that it should be possible for probes to be generated by PCR by defining a pair of sites for annealing primers. To generate STS, only single-pass partial sequencing of the clones is required and it is through such probes that an investigator can determine the overlap relationship between clones. It will indicate whether a clone under test is identical, shorter, or longer than a previously tested clone or represents a new clone altogether.

STS made it possible to mark YAC clones. Analogous to STSs are the ESTs (expressed sequence tags), which greatly helped in the sequencing of genomes. Like STSs, ESTs represent single-pass sequences, but unlike STSs, which are derived from genomic clones, ESTs represent short sequences of cDNAs from the 3' or 5' ends. ESTs are obtained far more easily and they are very useful because the cDNAs from which they are derived represent genes that are actually expressed and ESTs, therefore, greatly facilitate the task of annotation of the final sequence.

When ordered overlapping clones are available, contigs are constructed, which represent stretches of contiguous sequence ready clones. Generally some gaps remain and one has to resort to gap closure by chromosome walking for which special probes have to be generated from clones at the ends of contigs by either plasmid rescue or by inverse polymerase chain reaction. These probes are then employed to identify new YAC clones to bridge the gaps.

4. *Sequencing:* After a minimal set of overlapping YAC clones have been identified, the DNA in each clone is further fragmented either randomly or by restriction digestion and then subcloned in cosmids as per standard practice. Overlapping cosmid clones are identified by fingerprinting which is done by restriction digesting the clones, filtering out common bands of vectors with the aid of computers, and then comparing the digestion patterns to find contiguous clones. The cosmid clones are then ready for further breakage, which is done randomly either by shearing or by sonication into 1–2 kb fragments and subcloned into pUC plasmids or M13 phages. Finally the DNA from each subclone is extracted, reaction set up for *in vitro* DNA synthesis as necessary in the Sanger method employing the M13 phage, and sequenced.

5. *New vectors BACs and PACs and the shotgun approach:* Since the storage of YAC colonies requires special procedures and YAC clones were often found to be unstable or chimeric (incorporating more than one DNA fragment), efforts were made to develop alternative vectors such as BACs or PACs (Chapter 16). Satisfactory BAC vectors have been developed in which ~100-kb-long DNA fragments are being directly cloned. Overlapping BACs can be found by fingerprinting; one can then proceed directly to subcloning and sequencing. Since YACs can be multiplied only in yeast cells, BACs are maintained in bacteria, simplifying all presequencing steps. Thus one can do away with the initial ordering of BACs by RFLP or other genetic markers. With the aid of computers and special software, major sequencing effort can largely be restricted to BAC ends alone, which provide adequate information for determining overlaps. In this strategy, while the ends of every BAC are sequenced, the number of BACs that need to be sequenced fully is much smaller and is determined later. To begin construction of a contig, one starts with a fully sequenced "seed" BAC. The database search for BAC ends may reveal overlaps with about 30 other BACs. When two most important extreme-end BACs are located, they should be sequenced fully, continuously expanding search for new BACs for full sequencing at either end, until one tackles an entire chromosome. This random shot-gun method requires more extensive sequencing and powerful computers, but it has been claimed that it more than compensates for time and money that go in a conventional sequencing program, where considerable effort goes in for construction of linkage maps of molecular markers.

Gene location in DNA sequences

Genomics involves the study of what a genome contains and how the genome functions. A genome sequence is not an end to itself. Gene location in a DNA sequence is the prime aspect, followed by its function. Once a DNA sequence has been obtained, various methods can be employed to locate the genes, which can be simple sequence inspection by the computer and the methods that involve experimental analysis.

Sequence inspection

Genes are not a random series of nucleotides but a specific sequence with specific features, which determines whether the sequence is a gene or not. Although sequence inspection is not a powerful tool yet, but it gives an idea of gene location.

Genes are open reading frames: Genes that code for proteins comprise open reading frames (ORFs), which consist of a series of codons that specify the amino acid sequence of protein. It begins with an initiation codon (usually ATG) and ends with a termination codon (TAA, TAG, TGA). Searching a DNA sequence with ORFs that begin with an initiation codon and a termination triplet is one way of locating genes. The success to ORF scanning is the frequency with which termination triplets appear in the DNA sequence. If the DNA has random sequence and a GC content of 50%, each of the termination triplets (TAA, TAG, TGA) will appear, on an average, once every $4^3 = 64$ bp. If the GC content is >50%, the termination triplets being AT-rich will occur less frequently, perhaps every 100–200 bp. On the basis of start of ORF with an initiation codon and number of termination triplets, random DNA should not show ORF longer than 50 triplets. Most genes, on the other hand, are longer than 50 codons (the average lengths are 317 codons for *E. coli* and 483 codons for *S. cerevisiae*). ORF scanning in its simplest form therefore takes a figure of 100 codons as the shortest length of a putative gene and records positive hits for all ORFs longer than this. Although simple bacterial genome ORF scanning is an effective procedure for location of genes, it is less effective in higher eukaryotes because there is more space between real genes. But the main problem is that genes are often split and presence of introns, and so do not appear as continuous ORFs in the DNA sequence. In other words, genes of eukaryotes

do not appear in their DNA sequences as long ORFs and simple scanning cannot locate them. Three modifications to the simple procedure have been adopted:

a. *Codon bias:* Codon bias means that all the codons are not used equally frequently in the genes of a particular organism. The codon bias of the organism being studied is therefore written into the ORF scanning software.

b. *Exon–intron boundaries:* Exon–intron boundaries can be searched as these have distinctive sequence features. The consensus sequence of the upstream exon–intron boundary is usually described as

5′–AG↓GTAAGT—3′

(arrow indicates the precise boundary point)

The downstream intron–exon boundary is defined as

5′—Py Py Py Py Py PyNCAG↓—3′

(Py means pyrimidine nucleotide T or C; N is any nucleotide)

c. *Upstream control sequences:* Upstream control sequences can be used, although these are variable, for location of genes.

Besides these three simple extensions of simple ORF scanning, additional features or strategies are also possible with individual organisms. For example, vertebrate genomes contain CpG islands upstream of many genes, these being sequences of approximately 1 kb in which the GC content is greater than the average for the genome as a whole.

In eukaryotes, homology searches are conducted to test whether a series of triplets is a real exon or a chance sequence. In this analysis, DNA databases are searched to determine if the test sequence is identical or similar to any genes that have already been sequenced. The main use of homology searching is to assign functions to new discovered genes. However, at this point, it enables tentative exon sequences location by ORF scanning. If the tentative exon sequence gives one or more positive matches after a homology search, it is probably an exon. Some of the softwares used for prediction of coding regions are mentioned in Table 25.3.

Experimental techniques

These procedures locate genes by examining the RNA molecules that are transcribed from a DNA fragment. Techniques that map the positions of transcribed or expressed sequences in a DNA fragment can therefore be used to locate exons and entire genes. It should be kept in mind that the transcript is usually longer than the coding part of the gene because it begins several nucleotides upstream of the initiation codon and several nucleotides downstream of the termination codon. This analysis indicates that a gene is present in a particular region and it can locate the exon–intron boundaries.

Hybridization tests: This is a procedure for studying expressed sequences. RNA molecules can be separated by agarose gel electrophoresis and transferred to a nitrocellulose or nylon membrane by northern blotting (see Chapter 14). If a northern blot is probed with a labeled DNA fragment, RNAs expressed from the fragment are detected. Northern hybridization is a means of determining the number of genes present in a DNA fragment and the size of each coding region. There are two problems in such hybridization tests: (1) Some individual genes give rise to two or more transcripts of different lengths because some of their exons are optional and may or may not be retained in the mature RNA and (2) with many species it is not possible to make a preparation from an entire organism; so the RNA extract is obtained from a single organ or tissue. Therefore, any gene not expressed in that tissue or organ will not be present in the RNA population and will not be detected when the RNA is probed with DNA fragment being studied.

A second type of analysis uses DNAs rather than RNAs to overcome the problems with poorly expressed and tissue-specific genes. If a DNA fragment from one species is used to probe a Southern blot of DNAs from related species and one or more hybridization signals are obtained, it is likely that the probe contains one or more genes. This analysis is called *zoo-blotting*.

cDNA sequencing: Northern hybridization and zoo-blotting enable the presence or absence

Table 25.3. Commonly used software tools for sequence assembly, prediction of coding regions, annotation, detecting repeats and splice sites

Software	Function
TRANSCAN	Detects for RNA coding regions
FGENEH	A dynamic programing algorithm that uses linear discriminant functions. Used to identify genes
GENSCAN	Prediction program for coding regions and exon/intron splice site
Genquest	Program for sequence assembly, analysis and comparison
MUMmer	A whole genome alignment tool
REPEATMASKER	A repetitive element filter that screen a sequence against a library of repetitive sequences and searches for low complexity regions
Annotator	An interactive genome annotation tool
PSI-BLAST and variants	The most commonly used program for homology searches
Genefinder	Popular tool to detect coding regions and splice sites
Grail XGRAIL/ Grailant	Predicts exon, putative genes, detect promoter regions, poly-As, CpG islands, similarities in ESTs and repetitive segments
Net Plant Gene	Predicts exon/intron splice site
AAT	Analysis and annotation tool

Source: Singh (2000).

of genes in a DNA fragment to be determined, but give no positional information relating to the location of those genes in the DNA sequence. The easiest way is to sequence the relevant cDNAs. The success of cDNA sequencing depends on the frequency of appropriate cDNAs in the cDNA clone library that has been prepared. For this purpose, various methods of cDNA capture or cDNA selection have been devised, which are based around repeated hybridization between the DNA fragment being studied and the pool of cDNAs. Because the cDNA pool contains so many different sequences, it is generally not possible to discard all relevant clones by these repeated hybridizations, but it is possible to enrich the pool for those clones that specifically hybridize to the DNA fragment. This reduces the size of the library, which must then be screened under stringent conditions to identify the desired clones.

Completeness of individual cDNA molecules also determines the success or failure of this approach. Traditionally, cDNAs are made by copying RNA molecules into single-stranded DNA with reverse transcriptase and then converting single-stranded DNA into double-stranded DNA with a DNA polymerase (see Chapter 17 for cDNA cloning). There is always a chance that one of the strand synthesis reactions will not proceed to completion, resulting in a truncated cDNA. The presence of intramolecular base pairs in the RNA can also lead to incomplete copying. Truncated cDNAs may lack some of the information needed to locate the start and end points of a gene and its exon–intron boundaries. For location of the precise start and end points of gene transcripts, RACE (rapid amplification of cDNA ends) is performed (see Chapter 18). In this type of PCR, RNA rather than DNA is the starting material. The first step is to convert the RNA into cDNA with reverse transcriptase, after which the cDNA is amplified with *Taq* polymerase then followed by RACE for 5′ or 3′ end, depending on the desired information.

Other method for transcript mapping is heteroduplex analysis. The DNA region being studied is cloned as a restriction fragment in an M13 vector from which the DNA obtained is single stranded. When mixed with an appropriate RNA preparation, the transcribed sequence in the cloned DNA hybridizes with the equivalent mRNA, forming a double-stranded heteroduplex (Fig. 25.1). The start of this mRNA lies within the cloned restriction fragment; therefore, a part of the cloned fragment participates in the heteroduplex but the rest does not. The single-stranded regions can be digested by treatment with a single strand-specific nuclease S1. The size of the heteroduplex is determined by degrading the RNA component with alkali and electrophoresing the single-stranded DNA in an agarose gel. This size measurement is then used to position the start of the transcript relative to the restriction site at the end of cloned fragment.

Yeast (*S. cerevisiae*) genome

The yeast genome consists of 16 linear chromosomes, each containing a centromeric region required for chromosome segregation and special repeat sequences at the two ends called telomeres, which provide important functions during DNA replication and in chromosome maintenance. Unlike the *E. coli* chromosome, which contains a single origin of replication, eukaryotic chromosomes contain multiple initiation sites for bidirectional DNA replication. The nucleotide sequence of the entire *S. cerevisiae* genome has been determined and found to contain 12,068 kb of DNA. Sequence analysis has identified 5885 potential protein coding genes and another 455 RNA coding genes (tRNA, snRNA, and rRNA genes). Almost 70% of the genome is devoted to protein coding sequences, with the average distance between genes being ~2 kb. Interestingly, unlike most other eukaryotic genes, only 4% of nearly 6000 yeast genes have introns, and even then most of these genes contain only a single intron near the 5′ end of the coding sequence.

Human genome

On 26 June 2000, successful completion of the

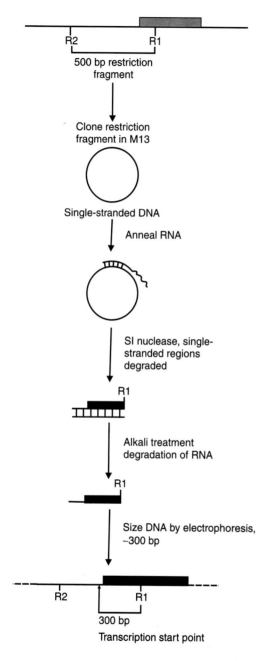

R2   R1

500 bp restriction
fragment

Clone restriction
fragment in M13

Single-stranded DNA

Anneal RNA

SI nuclease, single-
stranded regions
degraded

R1

Alkali treatment
degradation of RNA

R1

Size DNA by electrophoresis,
~300 bp

R2   R1

300 bp

Transcription start point

Fig. 25.1 Transcript mapping by S1 nuclease. Concept of the figure taken from *Genomes,* T.A. Brown, by kind permission of BIOS Scientific Publishers, Oxford, U.K.

sequencing of the Human Genome Project was announced jointly by two groups—The Human Genome Consortium and Celera Genomics. This was the first draft of human genome sequence, which is over 3 billion nucleotide long, consisting of 24 chapters, viz. 22 autosomes and X and Y sex chromosomes. Although the euchromatic regions have been sequenced, the Y chromosome and many gaps in other chromosomes are unlikely to be completed soon due to large regions of repeats.

Historical facts

1984–86   The idea for a large-scale effort to learn the sequence of the entire human genome was put forward. The practicality of sequencing was discussed at two meetings, one at Santa Cruz, USA, and another at Santa Fe, USA.

1988   Two government agencies of USA— Department of Energy (DOE) and the National Institute of Health (NIH) — drafted the first proposal to coordinate research and technical activities related to human genome.

1990   The project began in earnest when the projected goals for the first five years of a 15-year plan were published. Dr. James Watson pledges to decode the genome by 2005 at a cost of $3 billion. The project evolved over the 1990s and in its final stages had over 16 international centres: USA (8), Germany (3), Japan (2), France (1), China (1), and UK (1).

Two major technical advances that made the project feasible were the (1) development of yeast artificial chromosomes (YAC) vectors and (2) fluorescence-based detection of dideoxy-terminated fragments, which enabled near total automation of sequencing.

1990   A shortcut method to find fragments of human genes based on which whole genes can be identified, developed by Craig Venter, an NIH scientist.

1995	Sequencing of the genome of a bacterium (*Haemophilus influenzae*) by Venter's shot-gun method established.
1995–1996	Maps based on sequence tagged sites (Hudson *et al.*, 1995), human genetic map (Dib *et al.*, 1996), transcript map (Schuler *et al.*, 1996) published.
1997	STS-based radiation hybrid map published.
1998	Physical map of the genome presented (Dekoulas *et al.*, 1998).
May 1998	Craig Venter joins a new company that plans to complete the human genome in three years well ahead of the US government's target. The company was later named as Celera Genomics, Inc.
December 1998	The first full genome of the eukaryote roundworm animal *Caenorhabditis elegans* sequenced.
1999	Completion of the Sequencing of Chromosome 22 completed (Little, 1999; Dunham *et al.*, 1999).
March 2000	Two groups led by Craig Venter and Gerald M. Rubin sequenced the Genome of *Drosophila* using Venter's shot gun strategy.
May 2000	Sequencing of Chromosome 21 completed (Hattori *et al.*, 2000).
26 June 2000	The publicly funded Human Genome Project and Celera Genomics, Inc., USA, jointly announce the completion of a "rough draft" of the human genome. Eighty-six percent of the whole genome has been evaluated and classified to be highly accurate. The final draft was compiled from a total of 22 billion bases covering the total human DNA sequence almost sevenfold. The genome was sequenced in the past 6 months at the rate of 1000 bases per second.

Methodology: The substrate for sequencing was obtained by an elaborate process involving collecting DNA samples (blood from females and sperms from males) from multiple anonymous donors in strict accordance to the International Review Board Protocol. After screening, a few of the samples (~20–50 individuals) are combined and used for library construction followed by sequencing. Two approaches were followed for sequencing:

1. *Linear sequencing or "top down approach"*: The genome is segregated into smaller fragments in a stepwise manner, and when the pieces are small enough, they are cloned into vectors and sequenced. These cloned sections, which contained about 150,000 bp, are blown apart by a computer. After sequencing, these individual fragments are pieced together by backtracking until their point of origin in a given chromosome is reached. The advantage of this method is that any given fragment of a chromosome can be sequenced separately and the whole assembly can be done independent of other regions. This was the approach adopted by the Human Genome Sequencing Consortium in its initial stages. The first step in this method is generating genetic maps by plotting crossover frequencies among different gene loci.

2. *Shot-gun sequencing or "bottom up approach"*: Chromosomes are divided into sections and then the section of genomic DNA is broken into millions of small fragments and all the fragments are sequenced in an unbiased manner. Areas of overlapping DNA are matched which formed larger and larger segments of genomic DNA. Because of the degeneracy of this process, the total fragments generated typically cover the genome 10 times its original size. The overlapping of fragments is exploited in the assembly phase by computational exercise. Although prone to artefacts, especially in repeat regions, this process has been successfully applied for the sequencing of all the microbial genome sequences published and by Celera Genomics, Inc., for sequenc-

ing of human genome. This has been possible mainly due to the increase in accuracy of the algorithms performing the assembly.

In either approach the actual sequencing is performed in a manner of dideoxy chain termination method proposed by Sanger in 1977. Presently most of the sequencing is done by specially designed high-speed sequencers, which require little human intervention. Each of the four dideoxy terminators is tagged with different fluorescence markers, which are read by automated instruments (explained in Chapter 17). DNA sequencing machines developed by Perkin Elmer (PE Biosystem) can sequence up to 0.5 million bases per day.

As the sequencing of a given fragment that was inserted into a plasmid, cosmid, YAC, BAC, or PAC is completed, it is subjected to a set of computational procedures, which align the fragments, perform error checking, and detect potential protein coding regions. Once a region is predicted to be a coding segment, the features of the gene are analyzed (e.g. presence of specific signatures, CpG islands, etc.) and similarity searches run to establish their identity. Some of the programs that are commonly used are listed in Table 25.3. As can be easily made out, it is an exercise that is highly computation-intensive. Celera Genomics, Inc., announced that the assembly involved more than 500 million trillion calculations, making it the biggest exercise in the history of computation biology (Singh, 2000).

It is now known that 99.8% of the 3.2 billion base pairs between any two humans is the same and 0.2% different. For every 500 nucleotides, one nucleotide varies between two individuals. This means that any two individuals differ only in 6 million locations out of 3.2 billion locations. Now, if these locations have very little effect, then two individuals will be identical. More strikingly—some might not consider it so—humans are more than 98% similar to chimpanzees. So one has to understand that a variation in even a few locations in the DNA

sequence can lead to a severe disease and variation in humans. The study of variation will tell us not only what we are but also where we came from. It will also tell us what types of diseases we are susceptible to, because it will be possible to find out those variations that are associated with disease and those that are not. This exercise will involve functional analysis of the genome. The aim will be to find out how various genes are expressed in various tissues. This will be done in various tissues using DNA chips, where thousands of genes will be put on a single glass slide or silicon chip against which the messenger RNA isolated from various tissues will be hybridized. From this study it will be possible to pinpoint the expression of which gene has gone up and which gene has gone down in a tumor cell.

Arabidopsis genome

Three American groups, namely Meyerowitz, Somerville, and Goodman, at three different universities took the first steps leading to the organization of a multinational project on sequencing of the *Arabidopsis* genome, which has later come to be known as the AGI (*Arabidopsis* Genome Initiative). The first RFLP map was prepared by the Meyerowitz group and then a second by the Goodman group, which were later integrated (Hauge *et al.*, 1993). Early RFLP maps were prepared from F3 individuals whereas later maps were prepared by using recombinant inbreds. By 1996, the distance between RFLP markers had been reduced to >1.0 cM, which represents a physical spacing of about ~200 kb suitable for chromosome walking. Subsequently, other molecular markers such as RAPDs, CAPs, AFLPs, and microsatellite markers have been added. For sequencing of all the chromosomes, an international collaborative group comprising 25 laboratories worldwide under AGI was formed with various chromosomes assigned to different groups. Initially the conventional "clone-by-clone" was the dominant approach but later Venter's shot-gun approach for sequencing was followed.

cDNA libraries and ESTs were extensively used. By 1999, sequencing of Chromosome 2 and Chromosome 4 was made by American and European groups respectively. About 8000 genes coding for proteins have been found in these two chromosomes. At present, BAC and P1 phage based libraries are being prepared for sequencing. Progress has been so rapid that now the whole genome has been sequenced. Many questions are yet unanswered but some observations made are as follows:

- The genome of *Arabidopsis* is very compact; on an average, there is a gene every 4–5 kb and the approximate length of a coding region is 2–2.5 kb.
- The telomere and centromere regions are full of repeated DNA families.
- Several hundred transposons and pseudo-genes are present in the centromeric regions.
- There is an extensive duplication of genes, with many genes occurring as multigene families, such as those coding for rubisco, cab, and Ca++ binding proteins.
- Sometimes entire stretches of DNA and genes are duplicated between chromosomes (e.g. Chromosome 4 has a 37 gene sequence duplicated in Chromosome 5).
- Sometimes there are clusters of genes that are commonly regulated (e.g. by abscisic in Chromosome 2).
- Approximately 20% of the genes have signal sequences that target products into or-ganelles such as chloroplasts or mitochon-dria. It has been observed that many chloroplast and mitochondrial genes coded normally by organellar DNA also exist in the nuclear genome. Indeed, almost the entire mitochondrial genome of *Arabidopsis* is represented also on Chromosome 4 at the centromeric region.
- It has become possible to have a rough estimate of the number of genes in various functional groups. Now 50–60% of the ORFs and genes can be assigned some role on the basis of homology.

Rice genome

Scientists from Japan started the Rice Genome Programme (RGP) in 1991. However, in 1998, a second phase of RGP was launched with the objective of complete sequencing of the genome and also of functional analysis of genes. Presently, 10 countries are participating in this International Rice Genome Sequencing Programme (IRGSP). A new database known as the INE (Integrated Rice Genome Explorer) has also been established (pronounced I-ne, and refers to the rice plant in the Japanese language). The rice genome is estimated to comprise ~430 Mbp of DNA, which is about three times the amount of DNA in *Arabidopsis*.

As with *Arabidopsis*, the initial phase of rice project was taken up with the development of molecular marker maps. The first linkage map of 135 molecular markers was published in 1988. However, lately the densest of their molecular map derived from a cross between Nipponbare (a *japonica* variety) and Kaslath (an *indica* variety) comprised 2275 markers with an average distance of ~0.5 cM between markers, corresponding to a physical distance of ~150 kb. Many other types of markers have also been added. Similar to *Arabidopsis*, the early work began with random genomic clones, but in recent years cDNAs and ESTs have supplied a large number of markers. IRGSP has made considerable progress in preparing contig maps employing YACs and BACs.

Recently, Monsanto, a multinational company, in 2000 produced a draft of a rice (*Oryza sativa*) genome. The rice genome sequence was produced from rice cv. Nipponbare, a *japonica*, the same cultivar used by the IRGSP.

The Monsanto rice genome sequencing project produced approximately 399 Mb of sequences from 3391 BACs. Small amounts of contaminating sequences were removed from this initial dataset and redundant sequences were assembled into longer contigs. This process produced a draft dataset totaling 259 Mb of assembled sequence data.

From the sequence data available, it has been reported that the total number of genes in rice is not much larger than in *Arabidopsis*, and most of the genes in the two plants should show homology.

The results of functional genomics of model plants will increase the understanding of basic plant biology as well as the exploitation of genomic information for crop improvement. A series of important traits in valuable crops such as cereals, sugarcane, soybean, cotton, potato, and horticultural and ornamental crops can therefore be seen from a general perspective using gene function analysis of model plants. Recently, the isolation of genes determining plant height in *Arabidopsis* led to the identification of orthologous (similar and same function) dwarf genes in rice and other cereals. These were found to be the dwarf genes that were introduced into modern varieties by conventional breeding in the green revolution. This is the example of conservation of important traits in the plant kingdom and illustrates the potential of gene function discovery in model plants for application in a wide variety of crop plants.

Sequencing an entire genome is not possible in most of the cases. A rapid way to establish an inventory of expressed genes is to determine partial sequences of cDNA called expressed sequence tags (ESTs). Thus genomics technology is being used to develop ESTs, which provide a cost-effective and rapid approach for describing all the genes of an organism. The sequences are sufficiently accurate to unambiguously identify the corresponding genes in most cases. Thousands of sequences can thus be determined with a limited investment. EST information is present in public databases for a variety of species. There are more than 130,000 plant ESTs available in public databases (Pereira, 1999) from 19 plant species, including *Arabidopsis*, rice, tomato, maize, soybean, and cotton. Powerful programs for rapid sequence comparisons with databases are accessible via the Internet. EST databases have proven to be a tremendous resource for finding genes and for interspecies sequence comparison, and have provided markers for genetic and physical mapping and clones for expression analyses. The relative abundance of ESTs in libraries prepared from different organs and plants in different physiological conditions also provides preliminary information on expression patterns for the more abundant transcripts.

Functional genomics

Functional genomics involves the study of functions of all specific gene sequences and their expression in time and space in an organism. Once a new gene has been located and sequenced, the next question is what function it controls. It was thought from conventional studies on *E. coli* and yeast that most of the genes and their functions have been assigned. But genomic sequences revealed that there are large gaps in our knowledge. Of the 4288 protein coding genes in the *E. coli* genome sequence, only 1853 (43% of the total) had been previously identified (Blattner *et al.*, 1997). For *S. cerevisiae*, this figure was only 30%. Thus, there is a great challenge even if every gene in a genome is sequenced. We need to understand how the genome as a whole operates within the cell. Attempts are being made to find out the pattern of gene expression—which genes are switched on and which are switched off— in different tissues during different developmental stages.

The following approaches are used to determine the function of unknown genes.

Computer analysis

Computer analysis plays an important role in locating genes in the DNA sequences. For this, the most powerful tool available is *homology searching*. It locates genes by comparing the DNA sequence under study with all the other DNA sequences in the databases. The basis of homology searching is that the related genes have similar sequences and so a new gene is discovered by virtue of its similarity to an equivalent, already sequenced gene from a different organism. Homologous genes are the ones that share a common evolutionary ancestor, revealed by sequence similarities between the genes. A pair of homologous genes does not usually have identical sequences because the two genes undergo different random changes by mutation, but they have similar sequences because these random changes have operated on the same starting sequence, the common ancestral gene. Homology searching makes use of these sequence similarities. The basis of the analysis is that if a newly sequenced gene turns out to be similar to a second, previously sequenced gene, an evolutionary relationship can be inferred and the function of new gene is likely to be same or similar to the function of the second gene. In terms of homology and similarity, if the nucleotide sequences are identical by, for example, 80%, then percentage value is given to similarity and referred as the genes show 80% similarity rather than 80% homology.

It has been seen that homology analysis can provide information on the function of an entire gene or of segments within it. A homology search can be conducted either with a nucleotide or an amino acid sequence. It is best to conduct homology search using an amino acid sequence. Because there are 20 different amino acids in proteins but only four nucleotides in DNA, genes that are unrelated usually appear to be more different from one another when their amino acid sequences are compared. Thus results are likely to be less spurious if the amino acid sequence is used. Several software programs such as BLAST exist for carrying out this type of analysis. A positive match may give a clear indication of the function of a new gene or the implications of the match might be more subtle. In particular, genes that have no obvious evolutionary relatedness might have short segments that are similar to one another. The explanation of this is often that although the genes are unrelated, their proteins have similar functions and the shared sequence encodes a domain within each protein that is central to the shared function. Although the genes themselves have no common ancestor, the domains do, but

with their common ancestor occurring at a very ancient time, the homologous domains having subsequently evolved not only by single nucleotide changes, but also by more complex rearrangements that have created new genes within which the domains are found. An example that is quite often cited is that of the *tudor* domain, a 120 amino acid motif, first identified in *Drosophila melanogaster*. A homology search using the *tudor* domain as the test revealed that several known proteins contain this domain. These proteins are the RNA transport proteins during *Drosophila* oogenesis, a human protein with a role in RNA metabolism, and others whose activity appears to involve RNA in one way or the other. This suggests that the *tudor* domain sequence plays some part in RNA binding, RNA metabolism, or some other functions involving RNA. Information from computer analysis thus points the way to the types of experiment that should be done to obtain more clear-cut data on the function of the *tudor* domain.

Homology analysis has been conducted on *S. cerevisiae* genome for assigning functions to new genes. Of the 6000 genes in yeast, 30% had already been identified by conventional genetic analysis before the sequencing project started and 70% of the remaining genes were studied by homology analysis and were put in following categories.

By the homology analysis, 30% of the genes in the genome could be assigned functions. About half of these were clear homologs of genes whose functions had been previously established and about half had less striking similarities, restricted to domains only.

About 10% of all the yeast genes had homologs in databases, but the functions of these homologs were unknown. The homology analysis is thus unable to help in assigning functions to these genes. These yeast genes and their homologs are called *orphan families*.

The remaining yeast genes, about 30% of the total, had no homologs in the databases. A small portion (about 7%) was questionable ORFs, which might not be real genes because of unusual codon bias. The remainder 23% look like genes and are unique, called *single orphans*.

Experimental analysis

In conventional genetic analysis for study of gene function, the starting point is a phenotype and the objective is to identify the underlying gene or genes. The genetic basis of phenotype is usually studied by searching for mutant organisms in which the phenotype has become altered. The mutants might be obtained experimentally or present in a natural population. The gene or genes that have been altered in the mutant organism are then studied by genetic crosses, which can locate the position of gene(s) in the genome and also determine if the gene is the same as the one that has already been characterized. The gene can then be studied by molecular biology techniques and sequencing is done. In functional genomics, the problem addressed is in the opposite direction: it starts from the new gene or gene sequence and hopefully leads to the identification of its function, i.e. the associated phenotype.

The general principle of conventional analysis is that the gene responsible for a phenotype can be identified by determining which genes are inactivated in organisms that display a mutant version of the phenotype. If the starting point is the gene rather than the phenotype, the equivalent strategy would be to mutate the gene and identify the phenotypic change that results. This is the basis for most of the techniques used to assign functions to unknown genes. The following methods are employed to inactivate specific genes or overexpress genes for determining the function.

Gene inactivation: A specific gene can be inactivated if it is disrupted with an unrelated segment of DNA. This can be achieved by homologous recombination between the chromosomal copy of the gene and a second piece of DNA that shares some sequence identity with the target gene (Fig. 25.2). For present purposes, it is enough to know that if two DNA molecules have similar sequences, recombination can result in segments of the molecules being exchanged.

The gene inactivation procedure by homologous recombination has been explained with an example from *S. cerevisiae*. The gene inactivation procedure employs a deletion cassette, which carries a gene for antibiotic resistance preceded by the promoter sequences needed for expression in yeast and flanked by two restriction sites (Fig. 25.3). Before using the deletion cassette, new segments of DNA are attached as tails to either ends. These segments have sequences identical to parts of the yeast gene (start and end segments) that is going to be inactivated. This modified deletion cassette is put in a vector. This modified deletion cassette is then introduced into a yeast cell. Recombination occurs between the gene segments in the vector (DNA tails) and the chromosomal copy of the target gene in the yeast, replacing the yeast target gene with the antibiotic resistant gene. Cells in which the disruption has occurred are identified because they now express the antibiotic resistance gene and will therefore grow on an agar medium containing the antibiotic geneticin. The colonies lack the target gene, and their phenotypes can be examined to gain insight into the function

Gene inactivation procedure has been used in mice, which is analogous to that in yeast. On mice, experimentation is done for genes equivalent to those in humans, so that functions identified from this organism can be utilized on human genes. In such experiments, not one mutated cell but the whole mutant mouse is needed so that a complete assessment of the effect of gene inactivation on the phenotype can be seen. To achieve this, embryonic stem (ES) cells are used which are totipotent and can give rise to all types of differentiated cells leading to complete organism formation. The ES cell is engineered and then injected into a mouse embryo, which continues to develop and gives rise to a chimera. These chimeric mice are allowed to mate with one another. Some of the offsprings result from fusion of two engineered gametes and are therefore

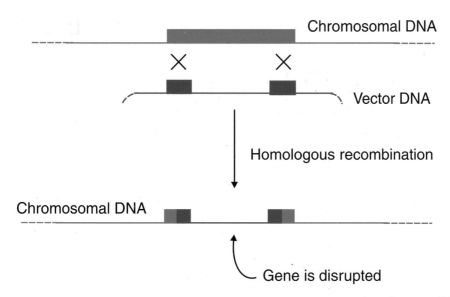

Fig. 25.2 Gene inactivation by homologous recombination. Concept of the figure taken from *Genomes,* T.A. Brown, by kind permission of BIOS Scientific Publishers, Oxford, U.K.

nonchimeric as every one of their cells carry the inactivated gene. These are called knockout mice, and it is hoped that examination of their phenotypes provides the desired information on the function of gene being studied.

Gene overexpression: In earlier techniques, gene inactivation was used to determine the function of a gene, but a complementary approach has been devised in which the test gene is much more active than the normal gene and the changes, if any, on the phenotype are determined. Thus there is a need to distinguish between a phenotype change that is due to specific function of an overexpressed gene and a less specific phenotypic change that reflects the abnormality of the situation. To overexpress a gene, a special type of cloning vector, multicopy, is used, which directs the synthesis of as much protein as possible. The vector must also contain a highly active promoter so that each copy of the test gene is converted into large quantities of mRNA, leading to high amounts of protein. Some genes functions have been identified in mouse by using this approach.

Inactivation and overexpression can determine the overall expression of a gene, but they cannot provide detailed information on the activity of a protein coded by a gene. For example, it might be that part of a gene codes for an amino acid sequence that directs its protein product to a particular compartment in the cell or is responsible for its action in response to a particular signal. To test these hypotheses, it would be necessary to delete, mutate, or alter a relevant part of the gene and then to determine its effect. This can be achieved by various procedures of site-directed mutagenesis (explained in Chapter 19). After mutagenesis, the gene sequence must be introduced into the host cell so that homologous recombination can replace the existing copy of the gene with the modified version.

One can also abolish expression by antisense DNA or by activating transposons. Considerable amount of work has been done on the insertion and mobilization of the well-known transposons of maize, namely the *Ac/Ds*, *Spm/dspm* and *Mu* systems in model plants such as tobacco and *Arabidopsis*. Special strategies to identify transposon-tagged lines in large populations have been made (see Chapter 20).

Fig. 25.3 Gene inactivation by the use of a yeast deletion cassette. The start and end segments of the target gene are inserted into the restriction sites and the vector introduced into yeast cells. Recombination between the gene segments in the vector and the chromosomal copy of the target gene results in the disruption of target gene. Cells in which the disruption has occurred are identified because they now express the antibiotic resistance gene and will grow on an agar medium containing geneticin, Kan^R – Gene for geneticin resistance; –■– : yeast promoter sequences; R$_1$, R$_2$ – Restriction sites. Concept of the figure taken from *Genomes,* T.A. Brown, by kind permission of BIOS Scientific Publishers, Oxford, U.K.

Clues to the function of a gene can be obtained by determining where and when the gene is expressed. If gene expression is restricted to a particular organ or tissue of a multicellular organism, or to a single set of cells within an organ or tissue, then reporter genes and immunochemistry can be used to locate where and when genes are expressed. The pattern of gene expression within an organism can be studied with a reporter gene (*uid*A, *lux,* GFP, etc.), explained in Chapter 21. For the reporter gene to give a reliable indication of where and when a test gene is expressed, the reporter is subjected to same regulatory signals as the test gene. This is achieved by replacing the ORF of the test gene with the ORF of the reporter gene, the regulatory signals being contained in the region of DNA upstream of an ORF (Fig. 25.4). The reporter gene will now display the same expression pattern as the test gene, which can be determined by examining the organism for the reporter signal.

Immunocytochemistry is used to locate the position of a protein within the cell. It makes use of an antibody that is specific for the protein of interest and thereby binds to this protein and no other. The antibody is labeled so that its position in the cell can be visualized. For low-resolution studies, fluorescent labeling and light microscopy are used.

Promoter

Target gene

Homologous recombination replaces the target gene with the reporter

Reporter gene

Fig. 25.4 A reporter gene. Concept of the figure taken from *Genomes*, T.A. Brown, by kind permission of BIOS Scientific Publishers, Oxford, U.K.

Gene function can often be determined if a protein coded by the gene interacts with other known proteins. If an interaction to a well-characterized protein is identified, it might be possible to infer the function of unknown protein. There are several methods for studying protein–protein interactions, but two most useful approaches are discussed here.

Phage display: In this approach, a special type of cloning vector based on λ bacteriophage or M13 filamentous bacteriophage is used (Clackson and Wells, 1994). The vector is so designed that a new gene that is cloned into it is expressed in such a way that its protein

product is synthesized as a fusion with one of its coat proteins. The phage protein therefore carries the foreign protein into the phage coat, where it is displayed and hence its name "phage display." In such an approach, the cloning vector used for phage display is a bacteriophage genome with a unique restriction site located within a gene for a coat protein (Fig. 25.5). To create a display phage, the DNA sequence coding for the test protein is ligated into the restriction site so that a fused reading frame is produced in which the series of codons continues unbroken from the coat protein gene into the test gene. This is transformed into *E. coli*, where the recombinant molecule directs synthesis of a hybrid protein, which is made up of the test protein fused to coat protein. Phage particles produced by these transformed bacteria therefore display the test protein in their coats. To make this approach a powerful strategy, a phage display library is produced by ligating many different DNA fragments into the cloning vector. The library contains a collection of clones displaying a range of proteins. After transformation, the bacteria are plated on to an agar medium so that plaques are produced. Each plaque contains a phage displaying a different protein. The lawn of plaques is blotted on to a nitrocellulose membrane and the test protein applied. If the test protein interacts with any of the displayed proteins, it will bind to the membrane. Binding is directed either because the test protein is labeled or by treating the membrane with a labeled antibody specific for the test protein.

Yeast two-hybrid system: The two-hybrid system makes use of a *S. cerevisiae* strain that lacks a transcription factor for a reporter gene. Transcription factors are responsible for controlling the expression of genes. There are two segments of a transcription factor coded by separate genes: a DNA-binding domain for binding to a DNA sequence upstream of gene and an activation domain that activates RNA polymerase to copy the gene into RNA. One of these genes, usually the one for the DNA-binding domain, is ligated to the gene for the protein whose interaction needs

(A) Production of a display phage

(B) Use of phage display library

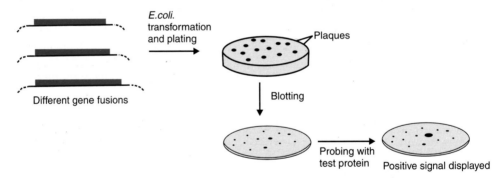

Fig. 25.5 Phage display approach for studying protein-protein interaction: **A.** The cloning vector is a bacteriophage genome with unique restriction sites located within a gene for coat protein. To create display phage, the DNA sequence coding for the test protein is ligated into the restriction site so that a fused reading frame is produced in which the series of codons continues from the coat protein gene into the test gene. After *E. coli* transformation, this recombinant molecule directs synthesis of a hybrid protein made up of the test protein fused to the coat protein. Phage particles therefore display the test protein in their coats. the technique was originally carried out with the gene III coat protein of the filamentous phage f1, but has now been extended to other phages including λ. **B.** A phage display library is produced by ligating many different fragments of DNA into cloning vector. After transformation, the bacteria are plated on to an agar medium so that plaques are produced. Each plaque contains a phage displaying a different protein. It is blotted on a nitrocellulose membrane and is probed with a test protein. If the test protein interacts with any of the displayed proteins then it will bind to the membrane. Concept of the figure taken from *Genomes,* T.A. Brown, by kind permission of BIOS Scientific Publishers, Oxford, U.K.

to be studied (e.g. protein X). This protein can be from any organism (Fig. 25.6). After introduction into yeast, the engineered gene will specify synthesis of a fusion protein made up of the DNA-binding domain of the transcription factor attached to protein X. The second gene, the one

(A) The two-hybrid system

(B) Screening for protein interactions using the two-hybrid system

HYBRID 1
(DNA-binding
domain)

HYBRID 2
(Activation
domain)

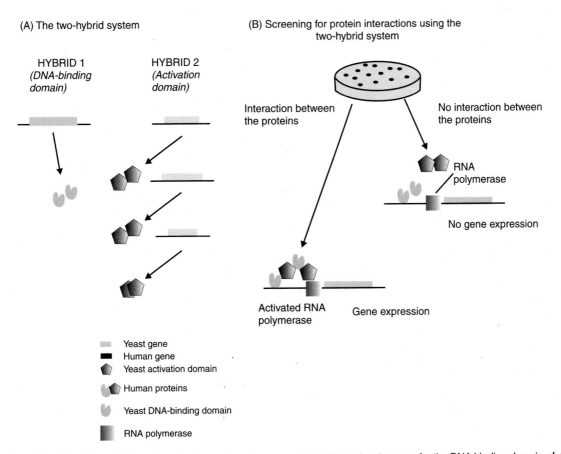

Interaction between
the proteins

No interaction between
the proteins

RNA
polymerase

No gene expression

Activated RNA
polymerase

Gene expression

Yeast gene
Human gene
Yeast activation domain
Human proteins
Yeast DNA-binding domain
RNA polymerase

Fig. 25.6   Yeast two hybrid system: A.- A gene for a human protein is ligated to the gene for the DNA binding domain of a yeast transcription factor. After transformation of yeast, this construct specifies a fusion protein, part human protein and part transcription factor. On the right, various human DNA fragments are ligated to the gene for the activation domain of the transcription factor, which specifies a variety of fusion proteins. B. – The two sets of constructs are mixed and cotransformed into yeast. A colony in which the reporter gene is expressed contains fusion proteins whose human segments interact, thereby bringing the DNA binding and activation domains of the transcription factor into proximity and enabling the RNA polymerase to be activated. Concept of the figure taken from *Genomes,* T.A. Brown, by kind permission of BIOS Scientific Publishers, Oxford, U.K.

for activation domain of the transcription factor, is ligated with a mixture of DNA fragments so that different constructs are made. The two sets of constructs are mixed and cotransformed into yeast. A colony in which the reporter gene is expressed contains fusion proteins whose segments interact, thereby bringing the DNA-binding and activation domains of the transcription factor into proximity and enabling the RNA polymerase to be activated. If this happens, the protein attached to the activation domain of the transcription factor must be one that can interact with protein X.

Patterns of gene expression

It is well known that all organisms regulate expression of their genes so that only those genes are switched on whose product is needed. To understand which genes are active in particular tissues at particular times and to assess the relative degrees of activity of the genes, the following approaches are employed.

Gene expression assay by measuring levels of RNA transcripts

The first stage of gene expression involves copying the gene into an RNA transcript. Identification of a gene transcript is therefore the most direct way of determining whether the gene is switched on. This can be done by immobilizing the DNA copy of the gene on a solid support and hybridizing it with a sample of RNA extracted from the cells being studied. In fact, the RNA is usually converted into cDNA prior to being used as the hybridization probe, as cDNA synthesis is the most convenient way of obtaining labeled sequences representing RNA molecules. If transcripts of the gene are present in the RNA extract, a hybridization signal is seen. To assay the entire mRNA content of a cell, referred to as *transcriptome,* this basic procedure has to be repeated for every coding gene in the genome. In such situations, the DNA chip strategy should be employed.

SAGE (serial analysis of gene expression)

SAGE is a PCR method developed by Bert Vogelstein and Ken Kinzler to identify differences in steady-state mRNA levels between two RNA samples. SAGE is based on the idea that the relative proportion of gene-specific ESTs in a cDNA pool reflects the relative abundance of the corresponding mRNA transcripts in the original RNA preparation. By determining the sequence of a statistically significant number of ESTs in one cDNA pool, relative to another, it is possible to identify gene transcripts that are differentially represented.

SAGE is a logical extension of EST sequencing. The basic idea of SAGE is to generate very short 3′ cDNA sequence tags using various mRNA samples. A flow scheme is shown for the generation of cDNA tags from just one RNA population (Fig. 25.7). In this scheme, cDNA synthesis is initiated with a biotin-labeled oligo-dT primer that permits physical separation of 3′ ESTs using magnetic beads coated with streptavidin. The two enzymes used for tag delineation are *Nla*III, which recognizes the sequence CTAG, and *Bsm*FI, a type II restriction endonuclease that cleaves 10 bp downstream of the sequence GGGAC. Restriction enyme digested products are adaptor mediated ligated and PCR amplified to link large number of short ESTs (10–15 bp) into a single array for automated DNA sequencing.

These short tags have proven sufficiently long to unambiguously identify corresponding genes in databases (Velculescu et al., 1997). Expression patterns for different genes are reflected by the relative abundance of individual tags. SAGE patterns have been studied in humans and yeast, but the technique has not been applied to plants. A prerequisite for the identification of tags is the availability of large sequence databases for the species under study. The technique is powerful but not very convenient for the comparison of many different samples and for the study of rare transcripts.

A comparison of EST databases from different plants reveals similarity in genes for specific functions and diversity in coding sequences. To supplement the rDNA and PCR technologies, a new technology described as DNA chip technology has emerged and will play a great role in the field of genomics (Gupta et al., 1999).

DNA chip technology

DNA chips are the microscopic array of single-stranded DNA molecules immobilized on a solid surface by biochemical analysis. DNA chips are also called as *gene chips* or *biochips* or *microarrays.* DNA chips contain a high density of DNA microarrays, which are derived from either cDNA (from identified expressed gene sequences) of an organism (also called as cDNA microarrays) or synthesized short oligonucleotide microarrays prepared by photolithography.

Principle: Hybridization of an unknown sample to an ordered array of immobilized DNA molecules of known sequences produces a specific hybridization pattern that can be analyzed or compared with a given sample. The basic idea is to attach known probe DNA sequences covalently on to a solid support surface and the

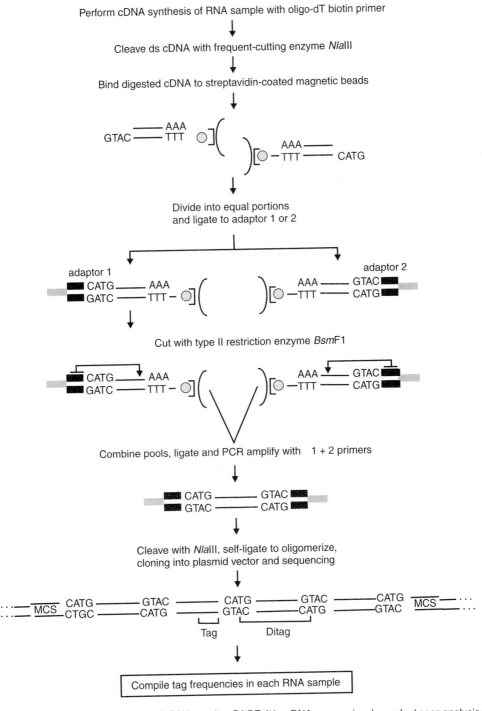

Fig. 25.7 A flow diagram for the generation of cDNA tags in a SAGE. It is a DNA sequencing dependent gene analysis method to identify differentially expressed gene transcripts. Concept of the figure taken from Miesfeld: *Applied Molecular Genetics*. Copyright © 1999 John Wiley and Sons. This material is used by permission of John Wiley and Sons, Inc.

sample RNA is poured over the probe area for hybridization. Before doing so, however, the sample RNA or cDNA is tagged with a fluorescent dye. Because each DNA microchip contains a standardized set of DNA sequences, it is referred to as a probe whereas the labeled experimental DNA or RNA is called the target or sample. By using autoradiography, laser scanning, fluorescence detection devices, enzyme detection system, etc. one can easily read the chip surface, and the hybridization pattern can be both qualitatively and quantitatively analyzed. In ordinary northern analysis, various types of RNA molecules remain immobilized in the gel and the DNA molecules constituting the probe are initially free in solution.

Types of DNA chips

1. Oligonucleotide-based chips
2. cDNA-based chips

Oligonucleotide-based chips

Oligonucleotide chips are based on a method to synthesize large amounts of different oligonucleotides *in situ* on a glass support using light-directed, solid phase, combinatorial chemistry developed by Affymetrix (Santa Clara, CA, USA). This type of DNA chip contains a high density of short oligonucleotides (10–25 mers) microarray, which is prepared by a photolithographic technique, a process in which light is used to direct the simultaneous synthesis of many different chemical compounds, and synthesis occurs on a solid support.

Integration of two key technologies forms the cornerstone of the method. The first technology, light-directed combinatorial chemistry, enables the synthesis of hundreds and thousands of discrete compounds at a high resolution in precise locations on a substrate. The second laser confocal fluorescence scanning permits measurement of molecular interactions on the array. Such arrays may contain 100,000 to 400,000 oligonucleotides immobilized within an area of 1.6 cm^2.

It is not always possible to produce cDNA from each type of mRNA, this being a difficult and time-consuming process. So rather than using cDNA, ESTs are prepared from an EST library. As mentioned earlier, an EST is a short oligonucleotide from either the start or end of the cDNA. For example,

5′ – TACGGGCAATACGATGGGCC – 3′
3′ – ATGCCCGTTA – 5′ (EST).

Therefore, the EST oligonucleotide is immobilized on a solid surface, which facilitates binding of sample DNA, which in turn is detected by different detection techniques.

A key advantage of this approach over nonsynthetic methods is that the photoprotected version of the four DNA building blocks allows chips to be manufactured directly from sequence databases, thereby removing the uncertain and burdensome aspects of sample handling and tracking.

Production of oligonucleotide microarrays/chips: Oligonucleotide microarrays are produced by *in situ* synthesis techniques using phosphoramidites, which utilizes photolithographic deprotection of 5′ OH groups at specific sites on the chips, so that individual specific bases may be added at these deprotected sites during chemical synthesis.

Procedure: A glass wafer can be taken as a support. This surface can be either a nonderivatized glass microscopic slide or an aminoalkylsilane derivatized one. The basic purpose of derivatization is to clean the surface of solid support, because hybridization is highly sensitive to any foreign bodies present on the surface. The procedure is shown in Fig. 25.8. The solid surface is derivatized with bis(2-hydroxyethyl) aminopropylthiethoxy silane. The 5′-*O*(α-methyl-6-nitropiperonyloxy carbonyl)-*N*-2′-deoxynucleoside phosphoramidites are then attached to the derivatized substrate through a synthetic linker such as 4,4′-dimethoxytrityl (DMT)-hexaethyloxy-O-cyanoethyl phosphoramidite [X]. The surface is then selectively activated for DNA synthesis by shining light through a photomask [MI]. Photolithographic masks having apertures are used to cover those regions where a particular

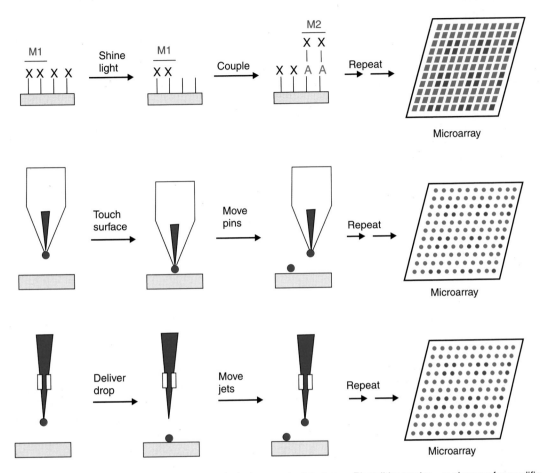

Fig. 25.8 Three approaches to microarray manufacturing are depicted: **a.** – Photolithography – a glass wafer modified with photolabile protecting group (X) is selectively activated for DNA synthesis by shining light through a photomask (M1). The wafer is then flooded with a photoprotected DNA base (A-X), resulting in spatially defined coupling on the chip surface. A second photomask (M2) is used to deprotect defined regions of the wafer. Repeated deprotection and coupling cycles enable the preparation of high density oligonucleotide microarrays. **b** – Mechanical microspotting – a biochemical sample is loaded into a spotting pin by capillary action, and a small volume is transferred to a solid surface by physical contact between the pin and the solid substrate. After the first spotting cycle, the pin is washed and a second sample is loaded and deposited to an adjacent address. Robotic control systems and multiplexed pinheads allow automated microarray fabrication. **c** – Ink jetting – a biochemical sample is loaded into a miniature nozzle equipped with a piezoelectric fitting and an electric current is used to expel a precise amount of liquid from the jet onto the substrate. After the first jetting step, the jet is washed and a second sample is loaded and deposited to an adjacent address. A repeated series of cycles with multiple jets enables rapid microarray production.

nucleotide is not desired to be added. Light with a wavelength of more than 280 nm through microapertures (~800 × 12,800 µm) of photolithograpghic masks is used. The glass wafer is then flooded with a photoprotected DNA base [A–X], resulting in defined coupling on the chip surface.

Coupling reaction: 3′-O-phosphoramidite activated deoxynucleoside (its 5′-hydroxyl is protected with a photolabile group) is presented on the surface that was illuminated, and coupling occurs at sites that were exposed to light. The substrate is then rinsed. A second photomask is used to deprotect defined regions of the wafer for next coupling reaction. Repeated deprotection and coupling cycles enable the preparation of high-density oligonucleotide microarrays.

Highly efficient strategies can be used to synthesize any arbitrary probe at any discrete, specified location on the array in a minimum number of chemical steps. For example, the complete set of $4N$ polydeoxynucleotides of length N, or any subset of this set, can be synthesized in only $4 \times N$ chemical cycles. Thus giving a reference sequence, a DNA chip can be designed that consists of a highly dense array of complementary probes with no restriction on design parameters. The amount of nucleic acid information encoded on the chip in the form of different probes is limited only by the physical size of the array and the achievable lithographic resolution (Fodor, 1997).

Advantages

1. The photoprotected versions of the four DNA building blocks allow chips to be manufactured directly from sequence databases.
2. The use of synthetic agents minimizes chip-to-chip variation by ensuring a high degree of precision in each coupling phase.
3. Photolithography has enabled the synthesiography has enabled the synthesices at individual grid positions and allowed miniaturization of the arrays.

Disadvantage

It needs photolithographic masks, which are expensive and time-consuming to design and build.

Ink jets: It is a "drop on demand" delivery approach to manufacture microarrays. A biochemical sample of phosphoramidite solution is loaded into a miniature nozzle equipped with a piezoelectric fitting and an electric current is used to expel a precise amount of liquid from the jet on to the precise location on a surface (Fig. 25.8). This drop on demand technology allows high-density gridding of virtually any biomolecule of interest including cDNAs, genomic DNAs, antibodies, and small molecules. This approach has been used to prepare microarrays of single cDNAs at a density of 10,000 spots/cm². Because ink jetting does not require direct surface contact, piezoelectric delivery is theoretically amenable to very high throughput. Piezoelectric-based delivery of phosphoramidite reagents has recently been used for the manufacture of high-density oligonucleotide microarrays (Blanchard, 1998).

cDNA-based chips

These are often derived from cDNAs, and are therefore also called cDNA microarrays. These are produced by taking cDNAs or ORFs that are PCR amplified and immobilized on a solid support of a microscopic slide or nylon filter. This procedure, originally developed by Davis, Brown, and their colleagues at Stanford, has been commercialized by Synteni/Incyte Companies, USA. The attachment of cDNAs or ORFs on a solid surface is done by delivery/deposition of DNA probe aliquots at specific sites on the surface of chip. The delivery can be performed manually or robotically. The following two delivery technologies are extensively used.

Mechanical microspotting: Microspotting is a miniaturized version of earlier DNA spotting techniques and encompasses a family of related deposition technologies that enable automated microarray production by printing small quantities of premade substances onto solid surfaces (Fig. 25.8). This was commercialized by Synteni (Freemont, CA, USA). Printing is accomplished by direct surface contact between the printing substrate and a delivery mechanism that contains an array of tweezers, pins, or capillaries that serve to transfer biochemical samples to the surface. The microarrays currently manufactured by Synteni contain as many as 10,000 groups of cDNAs in an area of ~3.6 cm². Microspotting is unlikely to produce the densities available by photolithography, but improvements are being made in microspotting technologies for automated production of chips containing 100,000 groups of cDNAs in an area of ~6.5 cm².

The following steps are involved in microspotting for production of cDNA microarrays:

1. Cloned cDNA fragments are amplified by PCR, which are referred to as probe DNA samples.
2. A 1 × 3 inch microslide with chemically modified surface is prepared for microspotting.
3. Probe DNA samples prepared as above are loaded each into a spotting pin by capillary action and a small volume is transferred to chemically modified solid surface at a specified position. This is achieved by a physical contact between the pin and the solid surface.
4. After the first spotting cycle, the pin is washed and the second sample is similarly loaded and deposited at an adjacent position.

These steps are speeded up by robotic control systems and multiplexed pinheads.

Advantages

The procedure is easy to use, low cost, versatile, and affordable.

Disadvantage

One disadvantage of microspotting is that each sample must be synthesized, purified, and stored prior to microarray fabrication.

Ink jets: Ink jets provide another approach to the manufacture of microarrays, as explained earlier.

Hybridization and detection methods

Hybridization of the target DNA to a microarray yields sequence information. The target DNA is labeled and incubated with microarray. If the target DNA has regions complementary to the probes on the array, the target DNA will hybridize with these probes. For the detection of hybridization patterns on DNA chips, the technique of reverse dot blot is utilized. In dot blots where the target DNA is dot blotted on the membrane and the probes are labeled opposed to this for DNA chips, the probes are anchored in the form of microarrays and the target DNA is labeled. Once hybridization is completed, the hybridization is detected by various methods, some rely on the use of an enzyme system whereas others use radiolabeling, and/or fluorescence.

Under a fixed set of hybridization conditions, e.g. target concentration, temperature, buffer and salt concentration, etc., the fraction of probes bound to targets will vary with the base composition of the probe and the extent of the target–probe match. In general, for a given length, probes with high GC content will hybridize more strongly than those with high AT content. Similarly, probes matching the target will hybridize more strongly than will probes with mismatches, insertions, and deletions.

Radioactive and nonradioactive detection methods have been discussed earlier in a chapter on gene cloning. Nonradioactive methods involve biotin or digoxigenin labeling whereas radioactive labeling requires direct detection through autoradiography, gas phase ionization, and phosphorimagers. However, there are drawbacks with the detection methods involving radioactivity. To avoid these problems, fluorochromes may be used, which allow direct detection due to fluorescence. In this procedure, one target DNA may be labeled with more than one fluorochromes for hybridization of microarray on the DNA chips, which also allows multiplexing.

A target sequence is fluorescently tagged and then injected into the chamber, where the target hybridizes to its complementary sequences on the array. Laser excitation enters through the back of the array, focused at the interface of the array surface and the target solution. Fluorescence emission is collected by a lens and passes through a series of optical filters to a sensitive detector. By simply scanning the laser beam or translating the array, or a combination of both, a quantitative two-dimensional fluorescent image of hybridization intensity is rapidly obtained.

Hybridization patterns can be screened using automatic scanners (e.g. Scan Array 3000).

These detection systems are based either on lens-based systems (epifluorescent and confocal microscopes) or on charge coupled device (CCD) based systems. Lens-based systems are not well suited to detect small quantities; therefore, CCD-based detection systems have been developed to detect small quantities of array bound molecules. In this method, labeled target DNA is hybridized to an immobilized probe on a silicon wafer, which is then placed on the CCD surface and a signal is generated. A fluorescence microscope fitted with a CCD camera and computer is used. This allows sequencing by hybridization with an oligonucleotide matrix.

Two versions of commercial readers are available: a first-generation system from Molecular Dynamics as well as a recently released high-performance system from Hewlett-Packard. Chip production is now in a scalable format, with a production capacity of ~5000 to 10,000 chips per month (Fodor, 1997).

Double-stranded DNA chips: Presently, DNA chip technology has been extended to fabrication of double-stranded DNA chips on a solid surface. In one of the approaches, single-stranded oligonucleotides (up to 40 mers) are immobilized onto a surface via the 3′ end. To these, a 16-mer primer is hybridized, which can be extended with a Klenow fragment, thereby converting the immobilized DNA into a double-stranded form (Fig. 25.9a). By incorporating a fluorescent label into the DNA as a marker, it was shown that double-stranded DNA is accessible to protein. An alternative approach is to couple 12-mer oligonucleotides to a gold support and then hybridize these oligos to double-stranded DNA with a complementary single-stranded end, followed by treatment with DNA ligase, which results in surface-coupled double-stranded DNA (Fig. 25.9b).

Double-stranded DNA chips can be used to for the following:
1. Systematically study DNA–protein interaction;
2. Study how eukaryotes regulates gene expression; and
3. Pinpoint nucleotide–amino acid interaction.

Fig. 25.9 Approaches for making double stranded DNA arrays. **a** – Single stranded oligonucleotides are synthesized on a glass surface in the presence of primer, polymerase and dNTPs. Since all the strands are not of unit length and thus not accessible to the primer and polymerase will not be converted into double stranded DNA of desired length and sequence. **b** – 12-mer oligonucleotides are coupled to a solid support. Hybridization to double stranded DNA with a complementary single stranded end, followed by treatment with DNA ligase, results in surface coupled double stranded DNA.

Characteristic features of DNA chips/microarrays

1. *Parallelism*: Allows parallel acquisition and analysis of massive data and a meaningful comparison between genes or gene products represented in microarrays.
2. *Miniaturization*: Involves miniaturization of DNA probes and reaction volumes, thus reducing time and reagent consumption.
3. *Speed*: Highly sensitive and allows rapid data acquisition.
4. *Multiplexing*: Multiple samples can be analyzed in a single assay. Labeling and detection methods that involve multicolor fluorescence allow comparisons of multiple samples in a single DNA chip. Multiplexing also increases the accuracy of comparative analysis by eliminating complicating factors such as chip-to-chip variation, discrepancies in reaction conditions, and other shortcomings inherent in comparing separate experiments.
5. *Automation*: Manufacturing technologies permit the mass production of DNA chips and automation leads to proliferation of microarray assays by ensuring their quality, availability, and affordability.
6. *Combinatorial synthesis*: Using the combinatorial synthesis strategy, a set of all 4^k oligonucleotides of the length k nucleotides (k-mers) can be generated in $4k$ synthesis cycles. For example, the set of all 4-mers (256) can be synthesized in 4 rounds, each round having 4 cycles, thus making a total of 16 cycles.

Applications of DNA chips

1. *Functional genomics:* Microarrays for gene expression analysis provided an integrated platform for functional genomics.
2. *Single nucleotide polymorphisms and point mutations:* Recently, single nucleotide polymorphisms (SNPs) as biallelic genetic markers have been extensively used as the markers of choice. Markers such as RFLPs and simple sequence length polymorphisms (SSLPs) are being utilized, but these markers have some drawbacks. They need gel-based assays and are therefore time consuming and expensive. SNPs have also the disadvantage of being biallelic whereas SSLPs are polyallelic, but they are abundant (more than 1 per 1000 bp). Genotyping individuals using SNPs needs only plus/minus assay, permitting easier automation. The approach used for this purpose relies on the capacity to distinguish a perfect match from a single base mismatch.

3. *DNA sequencing/resequencing:* In the late 1980s, a new approach toward DNA sequencing by hybridization (SBH) was simultaneously suggested by four groups (Southern, 1988; Bains and Smith, 1988; Lysov *et al.*, 1988; Drmanac *et al.*, 1988). The method involves manufacturing the sequencing DNA chips that contain a complete set of immobilized oligonucleotides of a particular size (e.g. 8-mers) and hybridizing the target DNA of unknown sequence (whose sequence is to be determined) on to these DNA chips. Hybridization patterns are then recorded by using one of the several suitable devices as discussed earlier. Identification and analysis of the overlapping oligomers that form perfect duplexes with the DNA of interest permits reconstruction of the target DNA sequence.

4. *Characterization of mutant populations exposed to various selection pressures:* This is done to collect information about the fitness value of a variety of alleles for each of the large number of genes in a species. This is particularly useful in those organisms where complete genome sequence is known. In such cases, deletions, insertions, or substitutions can be introduced and then the fitness value analyzed. Such an approach is called reverse genetics, where study is started from the DNA sequence and concluded with the phenotype.

5. *Diagnostics and genetic mapping*: DNA chips have already been prepared to detect mutant alleles. Cronin *et al.* (1996) designed a chip-based assay to detect multiple mutations in CFTR (cystic fibrosis) gene. Hacia *et al.* (1996) examined the *BRCA*-1 (cancer susceptible gene). Microarrays containing 96,000 20-mers were used to identify mutations over the entire 3.45-kb DNA belonging to exon 11. Oligonucleotide DNA chips have been used to detect β-thalassemia mutations in patients by hybridizing PCR-amplified DNA with the DNA chips (Yershov *et al.*, 1996). DNA chip technology has also been successfully applied to genotyping of hepatitis C virus in blood samples (Livache *et al.*, 1998).

6. *Proteomics (study of protein–protein interactions)*: Protein linkage maps can also be created using genomic sequence information.

7. *Agricultural biotechnology:* DNA chips with ESTs can be used to collect data on expression in an agricultural crop under different conditions. Transgenic plants can also be rapidly analyzed using microarray on DNA chips, and expression patterns under different environmental conditions can be predicted at the gene level. DNA microarray is also extensively used for the study of DNA polymorphism [e.g. single nucleotide polymorphisms (SNPs)] to develop molecular markers tagged to specific economic traits for molecula-marker-assisted selection in a breeding program.

Proteomics

Proteomics involves the study of the structure of all proteins encoded in a fully sequenced genome. Thus, it includes investigation of biological processes by the systematic analysis of a large number of expressed proteins for specific properties such as their identity, quantity, activity, and molecular interactions.

The term *proteome* was coined to refer to the total protein complement expressed by the genome.

The genome-wide analysis of mRNA expression by the microarray approach provides important clues about expression patterns and thus the function of gene products (discussed in the preceding section). However, for a substantial number of proteins there may be only a loose correlation between mRNA and protein levels. In addition, functions of proteins depend considerably on post-translational modifications and interactions with other proteins. Analysis of proteins is the most direct approach to define the gene function. Thus, whether the new sequence obtained corresponds to any functionality in terms of transcription and translation is to identify the proteins which the gene sequence codes. Therefore, information on mRNA levels is not sufficient to obtain a complete picture of the way gene expression is regulated within the cell. Protein expression data are more informative, but are much more difficult to obtain in a parallel fashion.

To understand the function of genes, efficient approaches for identification of proteins, for determining profiles of protein expression in different tissues and under different conditions, for identification of post-translational modifications of proteins in response to different stimuli, and for characterization of protein interactions are critical. Therefore protein analysis or proteomics is now a major subject of research. There are three major steps in proteome analysis:

1. Two-dimensional protein gel analysis,
2. Mass spectrometry (MS) for analysis of proteins, and
3. Database searching.

Two-dimensional protein gel electrophoresis

Two-dimensional PAGE is the method of choice to study the abundance and post-translational modifications of several hundred proteins in parallel. O'Farell made a path-breaking advancement in electrophorectic protein separation methods by combining isoelectric focusing (IEF) and sodium dodecyl sulfate (SDS) gel electrophoresis, resulting in the powerful two-dimensional gel electrophoresis (2D) technique (see Chapter 14). Several innovations in the basic 2D technique have made this technique suitable for a range of different applications. This method has been optimized for the separation of both soluble and membrane protein fractions as well as glycoproteins. 2D-separated proteins can be subjected to analysis of amino acid composition, immunological characterization, and peptide mapping. The amino acid sequence of 2D-separated proteins can be directly determined.

Mass spectrometry (MS) for analysis of proteins

Typically, the MS technique consists of a source to generate ions from the sample and an analyzer to separate and detect these ions according to their mass. The important development in proteomics research is the combination of 2D with mass spectrometry for the analysis of separated proteins. MS has essentially replaced the classical technique of Edman degradation because it is very sensitive, can also deal with protein mixtures, and offers much higher throughput. Simple MS such as MALD-TOF (matrix-assisted laser desorption ionization-time of flight) measures only the mass. However, tandem mass spectrometry also allows amino acid sequence determination. In this approach, proteins are separated by 2D and stained spots are excised and subjected to in-gel digestion with trypsin. Resulting peptides are separated by on-line high performance liquid chromatography (HPLC). The eluting peptides are subjected to tandem MS in which two stages of mass analysis are linked in series. The peptides are ionized by electrospray ionization and then delivered to first MS in which peptides are identified based on their mass to charge (m/z) ratio. The selected peptide is fragmented by collision with an inert gas (e.g. argon) and finally the second MS analysis separates the resulting peptide fragments. The MS spectrum of the peptide gives valuable information on the amino acid sequence. The amino acid sequence information helps in homology searching and cloning or database identification of the corresponding gene. New techniques allow automation of in-gel tryptic digestion of all the proteins in the 2D gel, followed by their transfer

to a membrane that can be scanned by MS to obtain diagnostic peptide masses for peptide mass fingerprinting. However, until the complete genome sequence is not available for peptide mass fingerprinting, the amino acid sequence approach of tandem MS is the method of choice. The synthesis of 2D + MS is the backbone of present-day proteomics research.

Database searching

Further support to proteomics research has been lent by emerging computer technologies. Identification of proteins by mass requires access to protein sequence database. The most commonly used databases are

- SWISS-2DPAGE for protein identification
- NCBI/BLAST – sequence database
- SWISS-PROT – sequence database
- SWISS-MODEL – for three-dimensional structure
- PROSITE – for domain structure
- GenBank and EMBL – DNA data banks

The science of bioinformatics is a cardinal part of present-day proteomics science, as the development of sophisticated software for an efficient analysis and storage of data with partially automated comparison of multiple 2D gels is needed in scaling proteomics to meet challenges put forth by genomics research.

Large two-dimensional gel databases exist for *E. coli*, yeast, and humans. In plants, more limited databases exist for maize, rice, and *Arabidopsis*. A database for plasma membrane proteins in *Arabidopsis* is under development, which not only provides information on abundance and post-translational modification, but also on the intracellular localization of a subset of proteins. This approach can be extended to other organelles, and even multisubunit protein complexes, provided that efficient separation procedures are available.

Other techniques based on MS of complex mixtures of protein fragments, may be developed in the future for the study of total protein complements. Protein interactions can be analyzed directly through proteomics science by performing coprecipitation studies with a bait protein followed by mass spectrometric read-out of the proteins. Post-translational modification of proteins is a key regulatory event in many cellular processes, including signaling, targeting, and metabolism. Current developments in proteomics enable the global analysis of post-translationally modified proteins. MS has been employed to characterize function-critical post-translational modifications, including phosphorylation and glycosylation.

Several newer methods of protein analysis such as affinity purification, antibody usage, yeast two-hybrid system, phage display (discussed earlier), etc. have been combined with 2D for proteomics research. DNA chips can also be used for the study of proteomics. DNA chip arrays can be used to identify the genes involved in protein–protein interactions. Protein–protein interactions can be studied using the yeast two-hybrid system. In this system, two fusion proteins are used for the activation of transcription of a reporter gene in yeast. The first fusion protein contains a DNA-binding domain fused to a protein of interest and the second fusion protein carries an acidic transcription activation domain fused to a second protein of interest. Specific interaction between two chimeric proteins leads to transcriptional activation of the reporter gene, which can be easily scored with color-based assays. The identity of the two proteins of interest is confirmed by sequence analysis of each clone thus identified. Phage representation libraries can also be used for DNA-chip-based detection systems. This involves the use of fusion proteins encoded by chimeric sequences of phage viral coat proteins and genes of interest (discussed earlier).

The proteome approach has been used in yeast to study gene function through the generation of knockout or overexpression mutants and for the analysis of changes in protein profiles on two-dimensional gels. In maize, comparison of proteomes of lines nearly isogenic for gene *opaque2* allowed the identification of new targets for the encoded transcription factors. In *Arabidopsis*, two-dimensional gel profiles were used to characterize developmental mutants and

allowed the hypothesis of the overproduction of cytokinins in one of the mutants, which was confirmed subsequently. Thus, proteomics research is proving important for characterization of individuals or lines, estimation of genetic variability within and between populations, establishment of genetic distances to be used in phylogenetic studies and characterization of mutants with localization of genes encoding several proteins. Selective applications of proteomics in animal and microbial systems include discovery of target molecules, designing/discovery of novel biomolecules and proteins (pharmaceuticals, industrial, and environmental applications) finding high value peptides/proteins, antibodies, vaccines, enzymes, therapeutic peptides, drug discovery, and biomolecular engineering.

Table 25.4. Examples of the use of 2D gel electrophoresis technique in plant systems

Application
Identification of somatic embryogenesis related proteins
Construction of wheat seed protein map
Scoring of polymorphism in *Saccharum* species
Identification of marker proteins for distinguishing *indica-japonica* rice
Preparation of data file of rice seed proteins
Tagging of plasma membrane proteins in *Arabidopsis*
Identification of lumenal and peripheral proteins in pea thylakoids

Proteomics is becoming a necessity in plant biology for deciphering the function and the role of genes in the plant genome sequencing projects. Some examples where 2D analysis has proved to be of great use in plants are shown in Table 25.4.

Further reading

Brown, T.A. 2000. *Genomes*. Bios Scientific Publishers Ltd., Oxford, UK.

Maheshwari, S.K., Maheshwari, N. and Sopory, S.K. 2001. Genomics, DNA chips and a revolution in plant biology. *Curr. Sci.*, **80**: 252-261.

Gupta, P.K., Roy, J.K. and Prasad, M. 1999. DNA chips, microarrays and genomics. *Curr. Sci.*, **77**: 875-884.

Dubey, H. and Grover, A. 2001. Current initiatives in proteomics research: The plant perspective. *Curr. Sci.*, **80**: 262-269.

26

Bioinformatics

In the last few decades, advances in molecular biology and the equipment available for research in this field of molecular biology have led to rapid sequencing of large portions of the genomes of several species. In fact, already, several bacterial genomes, simple eukaryotes (e.g. *Caenorhabditis elegans*, *Saccharomyces cerevisiae*), and *Drosophila*, *Arabidopsis* and rice have been sequenced. The Human Genome Project, designed to sequence all 24 of the human chromosomes has also been sequenced. Popular sequence databases, such as GenBank and EMBL, have been growing at exponential rates. This deluge of information has necessitated the careful storage, organization and indexing of sequence information. Bioinformatics is a science associated with database like activities involving persistent sets of data that are maintained in a consistent state over essentially indefinite periods of time. It is a computer-assisted interface discipline dealing with acquisition, storage, management, access and processing of molecular biology data. Thus, information technology has been applied to biology to produce a new field known as bioinformatics. It is an interdisciplinary scientific tool without barriers between various disciplines of science such as biology, mathematics, computer science and information technology.

The simplest tasks performed in bioinformatics concern the creation and maintenance of databases of biological information. Nucleic acid sequences (and the protein sequences derived from them) comprise the majority of such databases. While the storage and organization of millions of nucleotides is far from trivial, designing a database and developing an interface whereby researchers can both access existing information and submit new entries is only the beginning. This discipline helps us to collect, compile, analyze, process and represent the information in order to understand processes of life in healthy and diseased states and find new processes or better drugs and delivery systems for genes, drugs, and other genetic components for life improvement and for knowledge enhancement.

Biotechnology is instrumental in developing databases on genomics and proteomics pertaining to such developments in agriculture as well as other forms of life in general. Scientists are now working to segregate specific databases of different life forms from the information available in different databases. These databases are very essential for current and future biotechnology research.

The rapid advances in computer technology are responsible for the revolution in bioinformatics. Biological data that had once been dispersed in file drawers now wend their way into vast databases which by virtue of their comprehensive nature and instant cross accessibility, turns the information into a commodity more valuable than the sum of its parts. This has brought sweeping changes in the character of bioinformatics. Bioinformatics differs from computational biology and bioinformation structure. The most pressing tasks in bioinformatics involve the analysis of sequence information. Computational biology refers to algorithmic tools to facilitate biological analysis, whereas the bioinformation infrastructure comprises the entire array of information management systems, andytical tools and communication networks which support biology. Computational biology involves the following:

1. Finding the genes in the DNA sequences of various organisms.
2. Developing methods to predict the structure and/or function of newly discovered proteins and structural RNA sequences.
3. Clustering protein sequences into families of related sequences and the development of protein models.
4. Aligning similar proteins and generating phylogenetic trees to examine evolutionary relationships.

Development of bioinformatics

The development of bioinformatics started with the networking of computers and accumulation of data on genes and proteins. The 1970s witnessed several developments in protein and gene sequencing methods with the result that automated sequencing programs became possible which in turn generated a large Volume of data on various parameters of genes and proteins. The discipline developed further during the mid-1980s when computers became very popular for data storage and accessible to users. It was this period when various tools were developed to store, manage, and access the information through these of computers. Researchers started relying on computers for experimental data recording, analyzing, storing, browsing technical literature and modeling of proteins and genes. During the 1980s, for example, a software package "PC/GENE" was introduced by Intelli-Genetics, a US company (Moore *et al.*, 1988). This package could translate a given sequence

of gene to its protein end product with prediction of its secondary structure. This paved the way for the development of a number of new software packages for biotechnologists. The examples of early packages are the Staden Package for DNA sequences and PROPHET for data management and analysis in life sciences.

Internet gave a new dimension to computers and made a major impact on information technology and bioinformatics. During the 1990s, with the introduction of WWW (world wide web) developmental work in bioinformatics was given a boost with vast connectivity and accessibility to databases in a user friendly way. The success of internet came about with the advent of various networking tools and servers in the early and mid 1990s.

Bairoch (1991) established a moderate database pertaining to protein sequence and structural correlations on the "Net". This database was known as PROSITE, which was further strengthened with a database on sequence analysis and comparison of protein sequences known as SEQANALREE. The internet, which is an information superhighway, has practically compressed the world into a cyber colony through local area networks (LAN), wide area networks (WAN), metropolitan area networks (MAN) and other intranets. The development of internet and the emergence of WWW as a common vehicle for communication and access to databases provide exciting possiblities for bioinformatics. Tools for bioinformatics and computational biology include servers and computation boxes, cross-platform software, web-based interfaces to tools and databases, as well as, proprietary and public databases. At present, many bioinformatics software and data access tools are available. Strategies for searching the Net can be categorized, broadly, as useful for cataloguing the WWW directories, and search engines. Search engines are non-specific and produce voluminous results while directories are like 'yellow pages' which are browsed by the search engines.

The early 1990s also experienced the development of proteomics related databases on two dimensional polyacrylamide gel electrophoresis, known as SWISS-2DPAGE. Thereafter a fully organized and detailed protein sequence database, now famous as SWISS-PROT, was introduced. This database helps in providing a high level of annotation, a minimal level of redundancy, and a high degree of integration with other databases. Recent developments in this database include cross references to additional databases. A computer annotated supplement to SWISS-PROT has improved it significantly. This was possible with a variety of new documentation files and improvements to TrEMBL. The TrEMBL consists of entries in a SWISS-PROT like format derived from the translation of all coding sequences in the EMBL nucleotide sequence database. Some important databases which are commonly used by biotechnologists are those from Incyte, Pangea systems, PE-informatics, GCG, NCBI, PDB, SRS, UDB, SWISS-PROT, EMBLnet, ICBnet, Medline, Flybase, Mendelian inheritance in man (MIM), US patent and trademark office (USPTO) and SeqWeb. USDA maintains a crop genome database server at Cornell university at http://ars-genome.cornel.edu/. A number of universal resource locators for specific databases can be easily found on the web. There are various programs and tools available for making searches, algorithms analysis, modelling and the computer graphics of the databases on genomics and proteomics, which have been mentioned earlier also in this chapter.

Databases

Owing to high rate of databases and the need for researchers to have rapid access to new data, public databases have become the major medium through which macromolecular structural data, as well as genome mapping and sequencing data are published. A number of regularly updated lists of molecular biology databases and query servers are available via the internet shown in Table 26.1.

Successful public data services suffer from continually escalating demands from their user communities. It is probably important to realize that these databases will never completely satisfy a very large percentage of the user community. There is range of interests in biology and there is virtually no end to the depth and breadth of desirable information. In all cases, however, the principal requirements of the biomedical and biotechnology research communities on the public data service include:

1. Timeliness: The basic data should be available on an internet accessible server within days (or hours) of publication or submission.
2. Annotation: Deep, consistent annotation comprising supporting and ancillary information should be attached to each basic data object in the database.

Table 26.1 Selected World Wide Web (WWW) sources of information

Integrated bioinformation resources	
Integrated Genome Database	http://genome.dkfz-heidelberg.de/igd/>
Entrez Browser	http://www.ncbi.nim.nih.gov>
ExPASy	http://expasy.hcuge.ch>
SRS (Sequence Retrieval System)	http://www.ebi.ac.uk/srs/srsc>
Online directories of WWW-accessible bioinformation and analysis service	
Bio-wURLD (searchable collection of biology URLs)	http://www.ebi.ac.uk/htbi/bwurld.pl>
BRASS	http://mbisg2.sbc.man.ac.uk/gradschool/bioinf/brass95.html>
Harvard Biological Laboratories	http://golgi.harvard.edu/>
Biology Servers (Lawrence Berkeley Laboratory)	http://genome.lbl.gov/BioSrver.html>
The WWW Virtual Library: Biomolecules	http://golgi.harvard.edu/sequences.html>
Web Resources for Protein scientists	http://www.prosci.uci.edu/ProSciDos/>
Motif Bioinformatics Server (Stanford University)	http://stanford.edu:80/>

3. Supporting data: In some instances database users will need to examine the primary experimental data, either in the database itself, or by following cross-references back to network accessible laboratory databases.

4. Data quality: Database users and advisors have repeatedly asserted that data quality is of the highest priority. However, because the data services, in most cases lack access to supporting data, the quality of the data must remain the primary responsibility of the submitter.

5. Integration: Each data object in the database should be cross-referenced to representations of the same or related biological entities in other databases.

A search of recent literature has located more than 200 databases of molecular, structural, genetic or phenotypic data. Table 26.2 shows some of the specialized databases covering a diverse array of areas. They, therefore, provide valuable collections of organized, and usually validated data, and can be of broad benefit to researchers.

Online education in this field is also helpful.

Online Education

Online education is becoming a lifelong necessity and challenge for professionals who want to maintain their competitive advantage. With ever increasing pace of knowledge generation the skills and the information a student acquires through formal education soon becomes obsolete. With the internet and associated information

Table 26.2 A list of databases

Database subject content	Database
Molecular biology	
Protein sequences	Protein identification Resource (PIR), SWISS-PROT
Protein sequence motif	PROSITE, PRINTS, SBASE, BLOCKS
Protein sequence alignments	PIR-AL
Protein secondary structure assignments	DSSP
Protein structural alignments	FSSP
Predicted protein structures	HSSP, NRL_3D
Proteins classified according to structural similarities	SCOP
Enzymes	ENZYME
Restriction enzymes	REBASE
Two dimensional electrophoresis of proteins	SWISS-2D, QUEST
Databases of DNA and RNA sequences	GenBank, DNA databank of Japan (DDB), European
Molecular Biology laboratory (EMBL), Genome sequence database (GSDB), NCBI/BLAST	
CpG islands	CPGISLE
Expressed sequence tags (ESTs)	DbEST
Clustered, non-redundant ESTs	UniGene
Sequence tagged sites	DbSTS
Genetic regulatory elements; promoters	EPD
Transcription factors and sites	TFD, TRANSFAC
Genes of immunological importance	IMGT, Kabat
Genomic maps	RHDB, GDB, MGD
Cross species homologies	LISTA-HOP, LISTA-HON, XREFdb,
Metabolic pathways	EcoCyc, EMP, PUMA
Single organism data (integrated genetics, genomics, cell biology information	
Escherichia coli	ECDC, GenProtEc
Saccharomyces cerevisiae	YPD, SGD
Caenorhabditis elegans	AceDB
Drosophila	FlyBase
Human	IGD

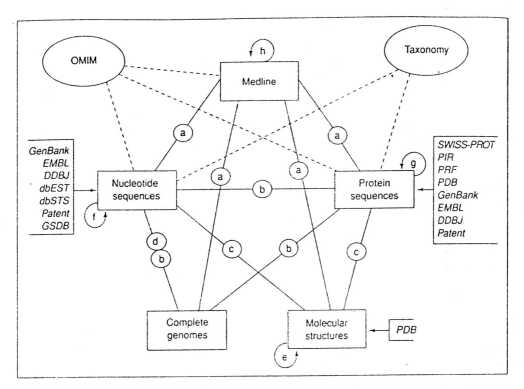

Fig. 26.1 The Entrez browser is an integrated set of databases, indexes, and access tools produced by the National Center for Biotechnology Information (NCBI; Bethesda, MD, USA). The original sources of data in the Entrez databases are shown in italics. Connections between the databases are based on: (a) literature citations in sequence or structure; (b) coding region features; (c) primary structure identity with sequence; and (d) alignments. Connections within databases (neighbors) are based on: (e) three-dimensional structural similarity; (f) nucleotide sequence similarity; (g) amino acid sequence similarity; and (h) term frequency statistics. OMIM (the on-line version of Mendelian Inheritance in Man (MIM) is not formally integrated within Entrez, but OMIM entries are linked and offer entry points into the Entrez system.

and communication technologies one can get up-to-date information. The online course consists of five components:

1. Web resources: It is an informational material on the course site itself or on external web pages linked to the course.
2. Tests and quizzes: It allows the students to assess their own understanding of the materials or have their understanding evaluated by others.
3. Learning activities offline: It is an individual study by a student from textbooks, labs, CD-ROMs, etc.
4. Home work: It is like assignments, collaborative projects, etc. which a student has to take.
5. Discussion forum: It begins with all the above

components by discussion, email conferencing, etc. to interact with one another.

Access to databases

Existing autonomous biological databases contain related data that are more valuable when interconnected with other related databases. However, the expertise to build and populate biological databases does not reside in any single institution. Therefore, biological databases are built by different teams, in different locations, for different purposes and using different data models and supporting database management systems. As a consequence, accessing the related data they contain is not straightforward. Experience with existing biological databases and their interfaces (e.g. the WWW) indicates that it is not possible to form usual queries across these databases, but to do so usually requires expertise in the semantic structure of each source database. There are two approaches:

One approach to database integration is construction of either a data warehouse or a monolithic database containing a combination of datasets from a variety of primary databases. Value added annotations and connections may be added either automatically or manually by domain experts. Examples of this approach are: Entrez (Fig. 26.1), Integrated Genomic database (IGD), the ExPASyWWW server and the sequence retrieval system. The popularity of these services with researchers is one indication of the perceived power of querying interrelated databases, rather than isolated databases.

The second approach depends on building most of the integrated intelligence into database query tools. Ideally, such tools should be capable of using knowledge of the schemata of remote databases to construct properly formed queries to each of the relevant remote databases, and capable of integrating the retrieved data into a coherent report for the user. The software modules that perform these functions have been termed 'mediators'. The principal advantage of an integration system based on mediators is that while it allows the individual databases to operate autonomously, it also enables them to function collectively as a federation. A federated database is a collection of essentially autonomous databases in which each constituent database may be implemented using a different schema and using a different database engine. The federated database provides defined access methods to the distributed databases and allows the user to view them collectively as a single database. While the development of mediator based query systems is a subject of continuing computer science research, recent implementations (e.g. the CPL/Kleisli system) have shown that it is possible to provide access both to relational databases (the Genome Database) and to unconventional sources of biological data [e.g.

the *Caenorhabditis elegans* genetic database (ACeDB), ASN.1 and BLAST] without building a monolithic database or writing a very application specific code for each query.

Some of the tools for bioinformatics and computational biology have been mentioned below as reported by Horton (1998).

A network computation server GeneMatcher system by Paracel has been designed for wide variety of supported algorithms that allows researchers to rapidly compare similarities and differences between DNA and protein sequence data. Gene Explorer 1.4 from Molecular Simulations (MSI), a computational environment for molecular biology and bioinformatics is compatible with several major genomic database systems, including those from Incyte, Pangea Systems, PF informatics, GCG, NCBI, PDB SRS, companies involved in bioinformatics.

Gene Explorer 1.4 automatically tells scientists when new database hits for previous experiments are found in database updates. In addition, the program helps researchers identify protein homologs with more sensitivity, using its protein threading technology.

SeqWeb from Genetics Computer Group (GCG) a part of Oxford Molecular Group has developed a web based interface that provides easy access to a core set of programs in the Wisconsin package, which are now accessible through Netscape Navigator or Microsoft Internet Explorer. With SeqWeb, users can use sequence files from their desktop, access group project data on the server and search a site's standard and proprietary databases through a web browser. SeqWeb links to local databases, as well as those on the internet, such as Medline, PDB, Flybase, Mendelian Inheritance in Man (MIM) and US Patent and Trademark Office patent database. Certain dynamic databases: Synergy from NetGenics, DiscoveryBase from Molecular Application Group; BioMerge3.0 from PE Informatics have been developed. These have been designed for providing centralized access to both proprietary and public genomic data

StackDB, a program from Pangea systems, provides an integrated dataset of the publicly available human ESTs and alignments. The sequence tag alignment and consensus knowledgebase or STACK is another public database of gene sequences expressed in the human genome. It features ESTs organized according to tissues and provides a comprehensive representation of each gene with alignments of its expressed fragments.

Applications

1. Data acquisition: High throughput biology requires automatic capture and conversion of biological data to symbolic digital representation without human intervention. Increasing the efficiency of current DNA sequencing systems will require base calling (the process of interpreting the signals from a fluorescent DNA sequencing instrument as a stream of nucleotides passing the detector) algorithms that provide increased accuracy, increased read-lengths and confidence estimations. These improvements will permit improvements in the automation of the assembly and finishing of DNA sequences.

2. Assembly of sequence contigs: Large scale sequencing projects require sequence assembly algorithms. Bioinformatics will help in generating contigs because softwares have been developed.

3. Predicting functional domains in genomic sequences: Advances in sequencing technology are expected to make the acquisition of large amount of anonymous sequences, which can be converted to its functions. It is likely that experimental determination of coding regions will be necessary to use computational methods to find all the exons in a genomic sequence. Computational approaches to recognize functional sites based on certain assumptions have been generated.

6. Sequence alignment: Once the sequence of nucleotides is obtained, it is compared to all the known sequences to see if any instruc-

tive similarities can be detected. Finding a similar sequence of known function can often be helpful in elucidating the function of a new cloned gene. For example, a quarter of positionally cloned human genes were found to have matches among *Saccharomyces* sequences of known function.

7. Searching structure databases: Newly determined protein structures are increasingly showing structural similarity to previously determined ones, even when no sequence similarity is detectable. New algorithms allow protein structures to be compared with databases of all known structures.

8. Determination of macromolecular structure: X-ray crystallography and nuclear magnetic resonance (NMR) are the major experimental methods used for deducing the macromolecular structure. Both methods produce extremely large amounts of data and are entirely dependent upon the availability of powerful computers and sophisticated processing algorithms.

9. Molecular evolution: Bioinformatics helps in constructing phylogenetic trees. Comparative genomics has provided a path of evolution to researchers through accumulated data on genomics and proteomics.

10. Bioinfomatics in medicine: The amount of data available through sequencing is bringing about a radical change in research methodology from a purely experimental approach to one in which experiments can be planned in front of a computer. The analysis of data will allow the identification of genetic markers of diseases (such as cancer), and the discovery of currently unknown physiological mechanisms which are at the basis of currently incurable diseases, offering clues for therapies including genetic therapy. Biocomputing techniques offer opportunities to biotech companies which want to give a stronger rational basis to the process of drug discovery. Experiments can be designed much more intelligently and the understanding of molecular interactions can be en-

hanced dramatically. Trial-and-error experiments can be replaced by predictions that allow for the design of custom-made drugs. Drugs based on rational design are 'smarter' in the sense that they are optimized for the target molecule in question and therefore tend to elicit far fewer side effects, which is a major problem for many rapies (e.g. cancer and AIDS therapies). Smarter design also calls for lower doses of drugs to achieve the desired effect. Furthermore, access to public databanks/resources improves information exchange on drug design research, and promotes cooperation and waste reduction resulting from unnecessary competition, duplication and precludes re-inventing the wheel type of situation.

11. Bioinformatics in agriculture: Agriculture can exploit the same computational tools, databases and strategies for investigating disease resistance markers, pest resistance genes and so on. As a result, crop harvests can increase and special plants can be engineered to alleviate food shortages in developing countries.

Challenges

1. Data management: The physical mapping of genomes and large scale sequencing projects require storage of information. In laboratory conditions data of a few 100 megabytes to nearly 100 gigabytes can be stored. Information technology talks in terms of terabytes (trillion bytes), petabytes (1000 terabytes) and exabytes (1000 petabytes). It is stated that all the words ever spoken by human beings amount to about five exabytes. Now a new, zettabyte term has been coined for such huge databases involving 1000 exabytes

and 1000 zettabytes as a yottabyte or 10^{24} bytes. High performance computers will be required to exploit the full potential of molecular and structural biology. High throughput biological research requires new and improved tools for managing information both at the laboratory level and at the national/international database levels.

2. Research is also needed for new and improved analytical methods. These methods are essential for turning molecular and structural data into biological knowledge.

3. The searching of biological databases via the WWW is becoming increasingly difficult. There are differences in database structure and nomenclature that hinder research efforts and standardizations have met with much resistance.

4. Audit and control of databases is becoming important because databases are becoming larger and larger. The complex computing environment and resource crunches would make it vital for information technology auditors to have practical guidance on conducting audits and also ensuring security in today's quickly changing computing environment.

5. The development of various strategies to surf WWW, and reaching a consensus on coining uniform definitions and adopting uniform platforms and technologies is another challenge in the area of bioinformatics.

Further reading

Benton, D. 1996. Bioinformatics – principles and potential of a new multidisciplinary tool. *Trends in Biotech.*, **14**: 261-279.

Tripathi, K.K. 2000. Bioinformatics: the foundation of present and future biotechnology. *Curr. Sci.*, **79**: 570-575.

27

Intellectual Property Rights

With the advent of biotechnology, one of the issues is the legal characterization of new inventions. Thus, an important part of the biotechnology industry concerns patenting or intellectual property rights. Loosely defined, intellectual property (IP) is a product of the mind. Intellectual property is intangible in contrast to real property (land) or physical property, which one can see, feel and use. With any type of property there are property rights. When IPs are expressed in a tangible form, they can also be protected. Intellectual property rights (IPRs) have been created to protect the right of individuals to enjoy their creations and discoveries. In fact, IPRs can be traced back to the fourteenth century, when European monarchs granted proprietary rights to writers for their literary works.

IPRs have been created to ensure protection against unfair trade practice. Owners of IP are granted protection by a state and/or country under varying conditions and periods of time. This protection includes the right to: (i) defend their rights to the property they have created; (ii) prevent others from taking advantage of their ingenuity; (iii) encourage their continuing innovativeness and creativity; and (iv) assure the world a flow of useful, informative and intellectual works.

The convention establishing the World Intellectual Property Organization (WIPO) in 1967, one of the specialized agencies of the United Nations System, provided that " Intellectual Property" shall include rights relating to
  i. Literary, artistic and scientific works;
 ii. Performance of performing artists, phonograms and broadcasts;
iii. Inventions in all fields of human endeavor;
 iv. Scientific discoveries;
  v. Industrial designs;
 vi. Trademarks, service marks, and commercial names and designations; and
vii. Protection against unfair competition and all other rights resulting from intellectual activity in the industrial, scientific, literary, or artistic fields.

PROTECTION OF INTELLECTUAL PROPERTY

Intellectual property is protected and governed by appropriate national legislations. The national legislation specifically describes the inventions that are the subject matter of protection and those which are excluded from protection. For example, methods of treatment of the humans or animals by surgery or therapy, inventions whose use would be contrary to law or morality, or inventions which are injurious to public health are excluded from patentability in the Indian legislation.

With the growing recognition of IPR, the importance of worldwide forums on IPs is realized. Companies, universities and industries want to protect their IPR internationally. In order to reach this goal, countries have signed numerous agreements and treaties and developed organizations.

World organizations

GATT   General Agreement on Tariffs and Trade
TRIPS Trade Related Aspects of Intellectual Property Rights
WIPO   World Intellectual Property Organization
UPOV   International Union for Protection of New Plant Varieties
WTO   World Trade Organization

GATT was framed in 1948 and was meant to be a temporary arrangement to settle amicably among countries disputes regarding who gets what share of the world trade. GATT serves as a code of rules for international trade and a forum to discuss and find solutions to trade problems of the member countries. Agriculture was included for the first time in the 8th round of GATT negotiations in 1986 at Uruguay. The Uruguay round discussions were started in 1986, but successfully concluded on December 15, 1993. The negotiations were signed in the form of an accord on April 15, 1994 at Marrakesh, Morocco, by about 124 countries and lead to the formation of World Trade Organization (WTO) on 1 January 1995. The Uruguay round covered 15 distinct areas and the Agreement on Agriculture (AoA) forms a part of the final act of the Uruguay negotiations. The long term objective of the agreement is to establish a fair and market oriented agricultural trading system and that a reform process should be initiated through the negotiation of commitments on support and protection and through the establishment of strengthened and more operationally effective GATT rules and disciplines. WTO is the rule-based body for all trade and trade-related issues. It seeks to reduce barriers to trade through mutually advantageous agreements. It differs from GATT and other bodies in that it (i) covers all aspects associated with trade with binding rules for all the members, (ii) has a dispute settlement board, and (iii) has a built in agenda for review.

WTO envisages a comprehensive scheme for protection of IPRs and the establishment of a legitimate reward system for the creative inputs of the inventors of intellectual property under the broad category of Trade Related Intellectual Property Rights (TRIPS). Various articles under TRIPS cover different forms of IP, *viz.* patents, trademarks, copyrights, designs, etc. All these forms of IP have a bearing on agricultural sector.

In 1961 at Paris, a "Union International Pour La Protection Des Obtentions Vegetables" UPOV (International Union for the Protection of New Plants Varieties) was signed for coordinating the intercountry implementation of PBR and it entered into force in 1968 with its headquarters based at Geneva.

Forms of protection

Usually IPRs are protected by the following legal theories:

   Patents
   Copyrights
   Trademarks
   Trade secrets
   Geographical indications
   Designs
   Know-hows

Of these, patents are the most important forms of protection for research and development organizations. One of the most important examples of IP is the processes and products that result from the development of genetic engineering techniques. Another example of IP is the development of crop varieties, which are protected through Plant Breeder Rights or PBRs. Thus, a better understanding of patents as the form of IP by research scientists and university/institute administrators will increase the pace of research for technological developments in biotechnology. Before describing the patents, the other forms of IP have been explained briefly.

Copyright

Copyright protects only the form of expression of ideas, not the ideas themselves. The creativity protected by copyright law is creativity in the choice and arrangement of words, musical notes, colors, shapes and so on. Copyright was created to provide protection to composers, writers, authors and artists to protect their original works against those who copy; those who take and use the form in which the original work was expressed by the author. The best examples of literary, scientific, and artistic works include productions such as books, pamphlets and other writings, dramatic or dramatico-musical works, choreographic works, musical compositions with or without words, cinematographic works, works of drawings, painting, architecture, photographic works, works of applied art, maps, plans, sketches, etc. Computer software/program is another mode of expression. A computer program is produced by one or more human authors but, in its final mode or form of expression, it can be understood directly only by a machine (the computer) and not by human readers (Gupta, 1999).

In biotechnology, the copyright may cover DNA sequence data that may be published. However, an alternative sequence coding for same protein may be prepared using wobble in the genetic code, so that the copyright is not infringed.

The rights bestowed by the law on the owner of copyright in a protected work are described as exclusive rights to authorize others to use the protected works. Copyright protection is generally valid in the country that grants the protection. Many countries respect the copyright of other countries, but if one wants protection in a certain country, it is best to apply for a copyright in that country. The agriculture industry uses copyright protection regularly. Directions on the use of a product and descriptions of products are few examples of copyright use.

In India, the Copyright Act, 1957 as amended in 1999 is in force. Literary work under the Copyright Act has been amended in 1994, which includes computer programs, tables, and compilation including computer databases. A computer program is separately defined in the act as "set of works expressed in words, codes, schemes or any other form including a machine readable medium capable of causing computer to perform a particular task or achieve a particular result." In India, the Copyright Act is under the charge of Ministry of Human Resource and Development.

The term of a copyright for published work is for the lifetime of the author plus 60 years.

Trademark

A trademark is a symbol that helps to distinguish one product or company from another. Symbols help the consumer identify products and/or a company and include designs, shapes, numbers, slogan, smell, sound or anything that helps the consumer to identify the products and/or companies.

In biotechnology research, laboratory equipments bear trademarks that are well known to workers in their field. Certain vectors useful in recombinant DNA technology are also known by their trademarks.

Commercial names and designations constitute another category of elements of intellectual property. Trade names are generally names, terms or designations that serve to identify and distinguish an enterprise and its business activities from those of other enterprises, whereas trademarks are used to differentiate between a company's product and all other products. For example slogan "Just Do It" identifies the Nike company.

Trademark law, unlike patent or copyright law, confers a perpetual right. So long as the trademark continues to identify a single source, anyone who uses a very similar mark may be liable for trademark infringement. The perpetual right of trademarks depends on the use. The basic idea of "use it or lose it" is essential to preserving trademark rights. A company cannot register a trademark and then not use it. The product for which the trademark was registered must be used commercially.

Trademark rights are so important that multinational companies spend large amount of money to maintain their respective trademarks around the world. Every country has different trademark laws. However, there are agreements to ensure that a company's trademark in one country is protected in another country. India has a Trade and Merchandise Act enacted in 1958 that has been amended in 1999.

Patent

A patent is a government granted exclusive right to an inventor to prevent others from practicing i.e. making, using or selling the invention. A patent is a personal property that can be licensed or sold like any other property. In USA, once a patent is issued, a patent gives the inventor the legal right for 17 years to create a monopoly by excluding others from creating, producing or selling the invention. In India, The Indian Patents Act of 1970 + 1999 amendment allows process patents, but no product patents for foods, chemicals, drugs and pharmaceuticals. In USA, a product patent is also allowed. If no product patent is available for a product, the same product may be manufactured by an alternative process without any infringement of the patent granted for the process. The duration of the patent in India is five years from the date of grant or 7 years from the date of filing the application, whichever is less.

There are three types of patents: (i) plant patent, (ii) design patent, and (iii) utility or regular patent.

Plant patents are granted for newly discovered asexually propagated plants. It provides protection for 20 years. A design patent protects ornamental characteristics. The life span of a design patent is only 14 years. Examples of design patents are the companies dealing with toys, souvenirs and industrial manufacturers. Utility patent is the most common patent used by universities and companies to protect their results of research and development. In USA, it has a life span of 20 years from the date of filing an application or 17 years from the date of grant (Erbisch and Velazquez, 1998).

There are three criteria to issue a utility patent.

i. *Novelty:* The inventor must establish that the invention is new or novel. The novelty requirement refers to the prior existence of an invention. If an invention is identical to an already patented invention, the novelty requirement is not met, so a patent cannot be issued.

ii. *Inventive:* It is invention and not merely discovery. It is nonobvious to one skilled in the field. The nonobvious requirement refers to the level of difficulty required to invent the technology. If the invention is so obvious that anyone having an ordinary skill would have thought of it, then it does not meet this requirement.

iii. *Industrial application:* It has a utility or is useful for the society. The useful requirement refers to the practical use of invention. If the invention provides a product that is required or needed in some manner, then it meets this requirement.

In the patent adequate disclosure should be made so that others can also work on it. It should (i) be a written description, (ii) enable other persons to follow, (iii) be adequate, and iv) have a deposit mechanism.

Most biotechnology inventions are filed as utility patents and not as plant patents. As a utility patent, it is possible to protect plant genes rather than just the plant, and to control the use of genetic material of a number of plants and for multiple uses such as disease resistance, herbicide resistance, or pharmaceutical or oil production.

The purpose of a patent is to promote the progress of science and useful arts. The patent law promotes this progress by giving the inventor the right of exclusion. In exchange for this right to exclude others, the inventor must disclose all details describing the invention, so that when the patent period expires, the public may have the opportunity to develop and profit from the use of invention.

Patent application

Filing of an application for a patent should be

completed at the earliest possible date and should not be delayed until the invention is fully developed for commercial working. Further, a published or disclosed invention cannot be patented. Hence inventors should not disclose their inventions in symposia or as research papers before filing a patent application. The preparation of a patent application is quite complex and generally a lawyer is required to draft the application. Writing of the claim is the most important component of the patent application. Claims are those portions of the patent that describe what can be accomplished with the invention and what is protected in the patent. This is neither to be too general nor to be too narrowly defined. Before filing a patent, search should be made on whether the patent has been filed earlier, and if a patent has been filed, what claim(s) is made in that patent. Certain patent search resources are given Table 27.1.

Patent is a document having the following components:

i. Title of invention, date of filing, date of issue, etc.
ii. Abstract containing a concise summary of disclosure in the invention.
iii. Description: The body of text with an introduction, specification, examples and the background information.
iv. Claims.
v. Any drawings, etc.
vi. Technical field to which the invention belongs.

A patent is enforced in the country that issues it. For each country a separate application is to be filed in that country where protection is sought.

International patenting and patent cooperation treaty

Patent cooperation treaty (PCT) is an agreement for international cooperation on patenting. India is a member of this treaty, which has 110 member countries. PCT does not provide an institutional authority for granting international patents. There is, in fact, nothing like an international patent, which may cover all countries or at least all those who are members of the PCT. PCT provides a platform from where an application filed in a member country can be used as a base to seek patent in any other PCT member countries on the choice of the applicant. PCT is helpful to an inventor for filing a patent application in one or more foreign countries. Under PCT, a patent application is first filed either in the home country or another country on the choice of the applicant. The date of this first filing is deemed as the priority date. Then one can file the application in other countries within 12 months of this date. But this multiple filing makes the task difficult, besides expenditure on filing, search, translation, etc. In contrast to this, PCT establishes an international system which requires filing of only one additional application which is called "International Application" (IA) at a designated patent office available in all PCT member countries and patent can be written in one of the languages approved in the PCT country where the IA is filed.

Use of technical information in patent documents

Patent literature has an important use as a source of useful scientific information:

Table 27.1 Patent search resources

Resource	Internet address
USPTO–Patent and Trademark Office	www.uspto.gov/web/menu/search.html
EPO–European Patent Office	http://www.european-patent-office.org/online/index.htm
Delphion Intellectual Property Network	http://www.delphion.com/
The French Office (INPI) patent database	http://www.inpi.fr/inpi/inbrevet.htm
Free Patent Search Site	http://www.Alliance-DC.org/
Biotechnology patents	http://www.nal.usda.gov/bic/Biotechpatents/

A) It is useful in searching for new technical solutions in the course of research and development and the creation and assimilation of new technology.

B) A scientist or technologist when beginning a new project should first study the patented literature and other information sources to find out what should not be invented, what technical solutions have already been found, when, by whom and where. The timely use of patent information makes it possible to avoid unnecessary expenditure and save time and resources. It also provides information on the trends of the development of knowledge in that field of research.

Revocation of patent

A patent is a monopoly granted by the statute in consideration of disclosure of invention. It is a limited monopoly for something that will pass to the public domain after expiry of the term of the patent. However, patent is a contract and it can always be revoked. Need for revocation of patent arises due to the fact that when patent application is filed, it is difficult to examine whether the innovation is new or it is dressed up. Also public deserves an opportunity to challenge the patent. Generally there are 4 months for opposing the patent and if it is not possible or any organization misses an opportunity for opposition, then it goes for revocation of patent. There are three modes of revocation of patents in India.

1. Central Government
2. Controller
3. Courts

There are various grounds of revocation of a patent:

i. in the public interest;
ii. nonworking of patent;
iii. patentee not entitled to the patent;
iv. patent wrongfully obtained;
v. failure to disclose the details;
vi. noncompliance with secrecy order;
vii. subject of claim is not an invention;
viii. not a patentable invention;
ix. subject matter lacks novelty;
x. lacks inventiveness;
xi. subject matter not useful;
xii. not sufficiently described;
xiii. claims not clearly defined

Patenting of biological material

Like in every other discipline, for biotechnology also, protection of intellectual property provides encouragement for creation of novel, innovative technologies, involving genetic engineering. The general prerequisites for patentability, namely novelty, inventiveness and utility apply to biotechnology inventions as well. As a rule, any new biological material is patentable, if obtained through nonbiological processes. Nonbiological processes are defined as those where the hand of man had a part to play.

In 1873 a US patent had been granted to Pasteur for "yeast free from germs of disease as an article of manufacture." But the US courts later held that the "discovery of some of the handiwork of nature" was unpatentable.

In 1980, Dr Chakrabarty got a patent on the discovery of an oil eating *Pseudomonas* bacterium, thus the microorganism life forms could be patented in USA.

In 1985, The US Patent and Trademark Office gave a ruling that plants previously under breeder's rights are patentable subject matter under industrial patent laws. Among higher plants, an important example is the patent for a "tryptophan-overproducing maize" obtained through tissue culture. The patent, known as the "Hibberd Patent," was issued to Molecular Genetics Research and Development, USA.

In 1987, in a case involving polyploid oysters, the patent office ruled that nonhuman multicellular organisms could be patented.

In 1988, a patent on the Harvard oncomouse was issued, which was another milestone in patenting of life forms. This was a transgenic mouse.

In 1993, Agracetus Company, USA, claimed broad patent protection over all genetically engineered cotton varieties regardless of how they are produced. The broad-based claim of patent protection over an entire species was unprecedented and was challenged.

In 1994, Agracetus Company, USA obtained a similar patent on all transgenic soybean plants in Europe.

In 1995, a patent was issued in USA to the National Institute of Health on all *ex-vivo* gene therapies.

In 1992, The US Government's National Institute of Health (NIH) applied for patents on thousands of human gene sequences. The patent application was denied in 1993 because NIH had no idea how the human gene sequences could be used or what role they will play in the human body. The main issue was whether these sequences are useful as required in the patent statute. This requirement was interpreted to mean that in the process or product to be patented there should be specific benefit in currently available form. Subsequently the word useful was interpreted to mean practical utility. Despite the above requirement, isolated genes, vectors and transformed cells expressing the hormone-angiogenesis factor (AGF) have been allowed a patent. Therefore, patent applications for isolated genes may in future either face a rigid attitude asking for specific practical utility in its currently available form or may receive a relaxation in view of its utility in immediate research with future potential of practical utility.

Microorganism patents: In the case of microorganisms, the first example was the classical judgment in the Diamond vs. Chakravarty case in 1980. In the Chakrabarty case, the US Supreme Court decided that a microorganism was not precluded from patentability solely because it was alive. Thus, a *Pseudomonas* bacterium manipulated to contain more than one plasmid (four plasmids were present) controlling the breakdown of hydrocarbons (therefore more useful in dispersing oil slicks than the natural organism containing only one such plasmid) was "a new bacterium with markedly different characteristics from any found in nature" and hence not nature's handiwork but that of inventor.

The "product of nature" objection therefore failed and the modified organism were held patentable. This precedent is being followed even today to define the patentability of microorganisms.

Plant patents: Plant patents are also obtainable in US, Europe, and Japan. The US Plant Patent Act of 1930 is restricted to asexually propagated plants and over 6500 such plant patents have been granted (mostly for rose and fruit trees). In the Hibberd case (1985), which involved a tryptophan-overproducing mutant, the patent office ruled that plants could be patented. Following the principle established in the Chakrabarty case, it was decided that normal US utility patents could be granted for other types of plant, for example, genetically modified plants. Among transgenic plants, herbicide-resistant cotton, canola, soybean, etc. and insect-resistant potato, cotton, maize, etc. have been patented. Plant patents have been granted by European Patent Office (EPO) from 1989. But in 1995, EPO severely restricted the scope of Plant Genetic Systems (Belgium) patent on herbicide resistant plants and allowed claims only on the herbicide resistant gene and the process used. In Japan, plant patents are allowed, but there are some disputes over territorial rights.

In Europe, patent law was originally considered unsuitable for protecting new plant varieties developed by traditional breeding methods. Special national laws of plant breeder's rights, which are also called Plant Variety Rights (PVRs), were therefore established in the 1960s in some countries as well as the International Union for the Protection of New Plant Varieties (UPOV, 1961). The UPOV convention was revised in 1991 and now does not prevent dual protection by PVR or patents. However, legal experts now generally recognize that patent law is better suited to protection of recombinant methods for producing transgenic plants and the resulting products.

Animal patents: Animal breeds produced by traditional methods have no legal system for their protection comparable to plant breeder's rights. Based on the microorganism and plant precedents, the US Commissioner of Patents declared in 1987 that the US patents would be granted for "non-naturally occurring non-human multicellular living organisms including animals." The first transgenic animal patent was issued in 1988 to Harvard University with claims covering the "oncomouse," one in which an oncogene has been introduced to make the animal more susceptible to cancer and therefore more sensitive in testing possible carcinogens. After initial reluctance by the EPO to grant corresponding European patent (and a successful appeal to the Appeals Board), the European patent was issued in 1992. More than 300 patent applications for transgenic animals have been filed, but so far few have been granted.

Gene patents: Patents on genes are available in all fields of biotechnology. For recombinant DNA inventions, the patent will claim the nucleotide sequence coding for the protein expression product, vectors, e.g. plasmids containing this sequence, microorganisms or higher organisms transformed with the sequence, and in appropriate cases the expression product itself (normally only if the product is new *per se*). Correspondingly process technology will also be claimed. The patentability of DNA sequences of unknown function is dubious and controversial. The Human Genome Organization accepts that patents should be granted for full-length genes but is against patenting fragmentary cDNA sequences having no established utility. Thus, a gene or sequence itself is not patentable. If the gene sequence is identified and it has a function and is proved to be novel, then it is patentable.

The genes that are synthesized artificially normally fulfill the requirements of patents. If the protein that the artificial gene makes and the organisms into which the gene is inserted are also novel and having desirable attributes, such a patent may extend to the protein and the organisms also.

In India, no life forms are patentable. In plant biotechnology, live plants (not the transgenic plants), naturally occurring microorganisms, micropropagation, tissue and organ culture techniques, biological control of pests or hybrid varieties cannot be protected using patents. Even techniques of cell fusion, protoplast fusion, and gene transfer cannot be patented.

Is cloning an inventive step: The laws prevailing in USA, Europe, and Japan, where biotechnology is most developed, have been more than generous on this issue and consider that cloning is indeed inventive. With the isolation of many disease-related genes, the number of sequences for which the patents will be filed will increase in coming years. However, as the gene cloning process becomes more standardized and more routine, patenting genes may become more difficult.

Geographical indications

The geographical indications (GI) for the purpose of the TRIPs agreement have been described as indications which could be used to recognize that a good has originated in a particular territory, region, or locality, where the given quality, reputation or other characteristic of the good are essentially attributable to its geographical origin. This is another way of protecting a country's biodiversity assets, which are uniquely endemic to certain geographical locations in the country and has been accepted in terms of quality and traits to that geographical region is through appropriate national legislations. Products such as Scotch whisky, Champagne, and California wines fall under this category. A recent instance of violations of possible geographical indications is the case of basmati rice of Indian and Pakistani origin.

GI can be used, when protected in legal terms, to prevent others from misleading the public or creating unfair competition in trade for the goods that have their origin from a particular territory, region, or locality. In the TRIPs, there is an important provision that if some one uses a GI as a trademark and it misleads the public as to the true place of origin, then registration of trademark is refused or invalidated *ex officio*.

India enacted The Geographical Indications Act in 1999. Prior to that there was no specific law governing GIs of any agricultural, natural, or manufactured goods, including food stuff. India, being party to the TRIPs agreement, was required to extend protection of GI for goods imported from other countries, which provide for such protection. A law to the effect in the country was necessary because other member countries would be obliged to give protection to the goods only if there are laws to protect the goods in the country of origin. The Indian GI act 1999 means an indication in relation to goods, which identifies such goods (agricultural goods, natural goods or manufactured goods or any goods of handicrafts or of industry, including food stuff) as originating, or manufactured in the territory of country, or a region or locality in that territory as the case may be, where a given quality, reputation or other characteristics of such goods is essentially attributable to its geographical origin. GI cannot be a trademark, the only exception is that trademark held prior to the Act or in good faith. According to GI, an indication includes any name (excluding the name of origin country, region or locality), geographical or figurative representation or any combination of these.

Trade secret

A trade secret is any information that gives a company a competitive edge over competitors and which the company maintains as secret and away from public knowledge. Trade secrets often include private proprietary information. Examples of trade secrets are the Coca Cola company brand syrup formula, Polaroid company instant film chemical formula, etc. The nature or the identity of a product is maintained secret for as long as the company can keep this information from becoming public knowledge.

Trade secret rights are mainly kept and enforced through agreements between employers and employees. These are nonpatented. Usually at the time employment begins, an employer makes an employee sign an agreement that grants the employer trade secret protection. The agreements protect the company by preventing its competitors from enticing key personnel since

these individuals cannot divulge the trade secret material without incurring severe penalties. Criminal prosecution of an employee who steals trade secrets from their employer is a recognized remedy.

Trade secrets have an unlimited term, which may be perpetual. The term is as long as it takes the public or a competitor to determine how to make the product and to ascertain the nature and identity of the trade secret. Disclosure of a trade secret and its unauthorized use can be punished by the court and the owner may be allowed compensation. However, if a trade secret becomes public knowledge by independent discovery or other means, then it is not punishable by court.

Trade secrets are much more common in industry where scholarly publication is not required. Trade secrets in the area of biotechnology may include (i) cell lines, (ii) gene transfer parameters, (iii) hybridization conditions, (iv) corporate merchandising plans, and (v) customer lists. However, since the biotechnological research being conducted in universities and institutes are expected to share their findings through publication and presentation in various conferences, it is almost impossible to maintain a trade secret.

Designs

The expression *designs* means only the feature of shape, configuration, pattern, or ornament applied to any article by any industrial process or means, whether manual, mechanical or chemical, separate or combined, which in the finished article appeal to end are judged solely by the eye. Design means the features of shape, pattern, etc. applied to an article or an object is protected and not the article or object itself. The features are conceived in the author's intellect. He gives those ideas conceived by him a material (visual) form as a pictorial illustration or as a specimen, prototype or model. In India, The Designs Act, 1911 as amended in 1999 is in force, in which the features are protected as design. The act confers exclusive right to apply a design to any article in any class in which the design is registered.

To qualify for registration, the design must
i. be a new or original design and
ii. not have been previously published in India.

Know-how

Know-how is another important form of intellectual property generated by R&D institutions that does not have the benefit of patent protection. This could be in the form of an aggregation of known processes/procedures and accumulation of data, a secret formulation, or a combination of any of these. Know-how is often transferred together with licensing of patents and under transfer of technology arrangements.

Plant breeders rights

Trade-related intellectual property rights are among the major areas of concern to India as the Indian economy is agriculture-based. Agriculture was included for the first time in the 8th round of Uruguay negotiations and an accord was signed in 1994. TRIPS are copyright, trademarks, trade secrets, designs, patents, etc. Since agriculture was included in the 1994 accord, there is a general feeling that enforcement of TRIPS will give rise to monopoly conditions and in particular our agriculture sector will be influenced. Section 3b, Article 27, on "Patentable Subject Matter" of TRIPS states that

Members may exclude from patentability
a. Diagnostic, therapeutic and surgical methods for the treatment of humans or animals.
b. Plant and animals other than microorganisms and essentially biological process production of plants or animals other than nonbiological and microbial process.

If the language of this provision is looked at critically, it would appear that all member countries are free to decide whether or not to treat plants and animals patentable. Depending upon its convenience, a member country can provide legal protection of plants and animals. In the USA plant varieties are protected under plant patent laws and also under the US Patents and Trademark Laws. The Harvard mouse or oncomouse, which is easily susceptible to cancer, is also patentable in US. Even the European Patent Office has approved the Harvard mouse as a patentable subject matter. Thus, this Article 27 Section 3b has the following implications: ,

a. It requires that microorganisms, nonbiological processes, and microbial processes must be patented.
b. Plant varieties must be protected by patents or by an effective *sui generis* system or by any combination thereof.

Life forms are not patentable in India. For protection of plant varieties, a *sui generis* system should be evolved. A *sui generis* system essentially means a system that is specially designed to protect plant varieties. The obvious model for this is the protection of plant breeders rights.

In The Indian Patent Act, 1970, all methods of agriculture and horticulture have been excluded from patentability. However, with the creation of WTO, it is incumbent upon member countries to have a legal system in place to provide protection of plant varieties. India is thus required to meet this obligation.

In order to understand the implication of IPRs on plant varieties and on agricultural production, one has to understand the plant breeders rights (PBRs) and how does it differ from patents. Plant varieties are generally protected in several countries (not in India) through PBR or plant variety rights (PVRs). PBR refers to the legal protection provided to a breeder, originator, or owner of a variety. The concept is based on the realization that if commercial plant breeding is to be encouraged for the benefit of agriculture and society, measures have to be taken to allow breeders to profit from their products (Mishra, 1999).

For over 60 years, different forms of protection of new plant varieties through system of PBRs have been in existence in industrialized countries. In 1961, at Paris, UPOV (International Union for the Protection of New Plants Varieties) was signed and it entered into force in 1968 with its headquarter based in Geneva. From 1961 to

1991 UPOV, there have been modifications in the plant breeders rights.

Organization – UPOV accomplishes its tasks through three committees: Consultative Committee, Administrative and Legal Committee, and Technical Committee. The technical committee is further assisted by five working parties, namely for agricultural crops, vegetables, fruit crops, ornamentals and forest trees, and for automation and computer programs.

Functions of UPOV

The convention has two main functions:
1. Prescribe minimum rights that must be granted to plant breeders by the member countries; in other words, it specifies a minimum scope of protection.
2. Establish standard criteria for grant of protection.

The purpose of the UPOV convention is to ensure that the member states of union acknowledge the achievement of breeders of new plant varieties by making available to them an exclusive property right on the basis of a set of clearly defined principles. To be eligible for protection, varieties have to be
   i. distinct from existing, commonly known varieties;
   ii. sufficiently homogeneous;
   iii. stable;
   iv. new in the sense that it must not have been commercialized prior to certain date established by reference to the date of application for protection.

The 1978 Act entered into force in 1981. The 1991 Act has not yet entered into force. Thus UPOV Acts of 1978 and 1991 have been summarized in Table 27.2. UPOV strengthened the position of PBR holder by eliminating the breeder's exemption from an essentially derived variety. An esentially derived variety is defined as a variety predominantly derived from another (initial) variety, which retains the expression of the essential characteristics from the genotype or combinations of genotypes of the initial variety. As a result of this modification, a breeder who inserts a single gene (e.g. disease resistant gene) in to a protected variety will now have to obtain the permission from the holder of the original right before marketing the new variety.

Plant breeders rights are related to the rights available to a patentee, where the model of protection is plant patent-based. Under the TRIPS agreement, a patent shall confer on its owner the exclusive right to prevent third parties not having owner's consent from the acts of making, using, offering for sale or importing. At times there are confusions between patent protection and plant variety protection by PBR. Hence a comparison of the two is presented in Table 27.3.

From the standpoint of developing countries, the 1991 UPOV Act is not desirable. For example, it has curtailed the rights of breeders to use the material of a protected variety for further research. It requires prior authorization of the breeder of the protected variety if it is to be used for further research by another breeder. This was not so in the 1978 Act. Moreover, the 1991 Act is against the farmers right to use, reuse their produce as seed and the right to dispose of their farm produce. These rights were available to farmers in the 1978 Act.

Advantages of PBR

1. It will encourage private research resulting in an increase in number of good varieties available to farmers.
2. It will increase competition.
3. Indian farmers and the seed industry will get an access to foreign plant varieties from the countries where IPR laws are already enforced.
4. More genetic variability will be available to Indian plant breeders, as a result of which they can develop better varieties.

Disadvantages of PBR

1. It will encourage monopolies in genetic material for specific traits.
2. The holder of PBR may produce less seed than the demand to increase prices for making more profit.

Table 27.2 A comparison between the two UPOV Acts on plant variety protection

Sr. No	Issue	1978 Act	1991 Act
1	Membership	Only a state can be the party	Intergovernmental organization competent for enacting and implementing legislation with binding upon on all its member states can also be the party
2	Discovery	Breeder is entitled to protection as discoverer irrespective of the origin, artificial or natural of the variation	A mere discovery is not sufficient. The breeder must also have developed the variety
3	National treatment	Member state may limit the right on a new variety to national of states which also apply that Act. A similar reciprocity rule may also be applied by a member state granting more extensive rights	Reciprocity rule does not apply. Operation of the principle of national treatment to one and all without qualification
4	Scope	Authorization of breeder is required for i. the production for purposes of commercial marketing of the propagating material; ii. the offering for sale of the propagating material; iii. the marketing of such material; iv. the repeated use of the new plant variety for the commercial production of another variety (e.g. hybrids); v. the commercial use of the ornamental plants or parts thereof as propagating material in the production of ornamental plants or cut flowers; vi. does not require authorization for use of the material for further research; vii. farmers can use/reuse his produce as seed and can dispose of his farm produce.	Authorization of breeder is essential for i. production or reproduction; ii. conditions for the purpose of propagation; iii. offering for sale; iv. selling or other marketing; v. exporting; vi. importing; vii. stocking. Furthermore, the 1991 Act specified four subject matters to which the breeder's right extends i. the protected variety itself ii. varieties which are not clearly distinguishable from the protected variety; iii. varieties which are essentially derived from the protected variety; iv. varieties whose production requires repeated use of protected variety.
5	Minimum number of species to be covered	Atleast 5 to start with Atleast 10 within 3 years Atleast 18 within 6 years Atleast 24 within 8 years	UPOV 1978 member states, all after 5 years transitional period. All after 10 years if only bound by 1991 Act. To start with while acceding 15 plant genera/species
6	Period of protection	18 years for grapevines and trees including rootstocks; 15 years for all other plants	25 years for grapevines and trees including rootstocks; 20 years for all other plants
7	Special title/patent protection	Each state is free for any of the two forms of protection except those where it was a practice before October 31, 1979 for providing protection to the same genera or species by both	No alternative forms of protection

Table 27.3 A comparison between patent protection and plant variety protection

Sr.No.	Issue	Patent protection	Plant variety protection
1	Object	Invention	Plant variety
2	Documentary examination	Required	Required
3	Field examination	Not required	Required
4	Plant material for testing	Not necessary	Required
5	Conditions for protection	a. Novelty b. Industrial applicability c. Unobviousness d. A disclosure	a. Commercial novelty b. Distinctiveness c. Uniformity d. Stability
6	Determination of scope of protection	Determined by claims of the patent	Fixed by national legislation
7	Use of protected variety for breeding further varieties	Require authorization of the patentee of the right holder	No authorization required
8	Use of propagating material of the protected variety grown by a farmer for subsequent planting on the same farm	Require authority of the patentee	Generally no authorization required
9	Term of protection	30 years	25 years for trees and vines; 20 years for others (1991 UPOV) while 18 years for trees and vines and 15 years for others (1978 UPOV)

3. Genetic diversity will be reduced because land races will go out of use.
4. It will inhibit free exchange of material and foster unhealthy practices.
5. Farmer's privilege to use his own seed will be threatened.
6. PBR may increase cost of the seed, which the poor farmer may not be able to afford. Thus, it will only help rich farmers.

Farmer's Rights

Farmer's rights is a concept developed and adopted in the Food and Agriculture Organization (FAO) as a resolution and endorsed by all member countries. It recognizes that farming communities have greatly contributed to the creation, selection, and preservation of genetic material and knowledge. The indigenous knowledge systems are similar to general scientific information in that they are part of public knowledge. The usual criteria for recognizing IPR, that is, novelty and nonobviousness, tend to ignore the knowledge systems of rural and tribal families. While the knowledge itself may not be patentable, the products of this knowledge

namely "folk" varieties and races and genetic diversity provide the basic raw material for modern plant breeding and biotechnology research (Swaminathan, 1994). The plant breeding and seed industry of the developed countries has used this knowledge for development of new varieties without compensating the farmers who had preserved the genetic stock. It is against this background that the concept of farmer's rights emerged in international fora.

Therefore, it is the obligation of world community to help these farmers. Thus, it has been strongly advocated that the farmers retain their traditional right to save, use, exchange, share and sell the propagated material and seed from harvest. But the right is restricted to noncommercial sale and branded seeds cannot be sold.

Convention on Biodiversity

In June 1992, 170 countries met in Rio-de-Janeiro to discuss the details of a proposed biodiversity treaty under the biodiversity convention. Although US has not signed this treaty, this Convention on Biodiversity (CBD) dealt with the sovereign rights of the nations over their genetic resources. This CBD has come into force in December 1993 as the first legal mechanism dealing with biological resources. In respect of biological resources, earlier status according to FAO International Undertaking on Plant Genetic Resources, a non-legally binding mechanism that had adhered to since inception in 1983 that "plant genetic resources are the heritage of mankind" and should be made available without restriction. But as per the CBD, which came into force in 1993 as the first legal mechanism dealing with biological resources, these plant genetic resources are deemed not to be the heritage of mankind, but are the properties of the countries and are therefore tradable commodities. This is in contrast to protection of IPRs, which decrees that natural products are the God's gift to mankind, and hence cannot be exploited exclusively by any one. The objective of the CBD, to be pursued in accordance with its relevant provisions, included (i) the conservation of biological diversity, (ii) the sustainable use of its components, and (iii) the fair and equitable sharing of the benefits arising out of the utilization of genetic resources. The latter may include appropriate (a) access to rightfully owned/ possessed genetic resources, (b) transfer of relevant rightfully held technologies, and (c) funding.

Countries need to legislate biodiversity bills (protecting nature's treasures), the basic principle being that no foreigner or foreign organization can take away any biological resources for research or commercial use without permission of the country of origin. No local organization will be allowed to transfer even research results on biological resources to any foreigner without permission. The legislation will also ensure that benefits between conservers of the resources and users are equitably shared. Even though all the signatories had confirmed such approaches for protection of their sovereign rights, most are yet to enact appropriate legislations in this regard. A weak point in the CBD is its ambiguous treatment for the equity. It may be seen that

whereas patenting of products of biotechnology is clearly recognized, there are no effective guidelines and conditions defined to recognize and reward the contributions of indigenous communities and the other informal innovators who have been responsible for nurturing, using, and developing biodiversity worldwide.

The key developments in India concerning legislative and regulatory provisions include the enactment of two legislations, one on the Protection of Plant Varieties and Farmer's Rights (PPVFR) and the other one on biological diversity. The Biological Diversity Bill, 2000 is under consideration.

Plant variety protection

TRIPS, one of the agreements bound to the WTO, mandates the countries signatory to it to put in place all legal and administrative changes required on IPR domains by January 2000 and 2005. Under this commitment, WTO member countries are required to grant patent to all inventions, whether processes or products, in all fields of technology (including agriculture). TRIPS further demands India, as also any other member country, to grant patent to microorganisms, microbiological, and nonbiological processes and plants and animals produced by nonbiological processes. Thus it implies that member countries should protect plant varieties either by patent or an effective system of *sui generis* protection, or a combination of these two. In this context, India initiated its legislative process to protect plant varieties. Although all the three options – patent, *sui generis*, and patent and *sui generis* – were available, India chose a *sui generis* system for protection of plant varieties. An act named as Protection of Plant variety and Farmer's Rights (PPVFR), 2001 has been passed.

The PPVFR Act, 2001 provides protection to three types of plant varieties:
  i. Newly-bred varieties
  ii. *Extant varieties*: The varieties that were released under the Indian Seed Act, 1966, and have not completed 15 years as on the date of application for their protection.

iii. *Farmer's varieties*: The varieties that have been traditionally cultivated, including landraces and their wild relatives which are in common knowledge, as well as those evolved by farmers.

Among the newly-bred varieties, essentially derived varieties (EDVs) are distinguished for the purpose of a different treatment for the grant of their protection. An EDV is defined as a variety that is predominantly derived from an initial variety and retains the essential characteristics of the latter while having limited difference(s) to qualify distinctiveness from the initial variety. The legislation stipulates novelty, distinctiveness, uniformity, and stability (acronym NDUS) as the essential requirements for gaining protection to all newly-bred varieties while novelty is not required for protection of extant and farmer's varieties.

Novel: A variety is deemed novel when it has not been placed for commercial use in India for a period not exceeding 12 months if bred in India. The variety bred outside India is deemed to have novelty if its protection is sought within a period not exceeding 4 years in the case of annuals and six years in case of vines and tress from the date of its registration in another country.

Distinctiveness: It means that the candidate variety should be different from all other varieties of common knowledge in India and outside for at least one essential character.

Uniformity: It means phenotypic similarity among plants subject to the variation that may be expected and allowed due to specific nature of the reproduction of the crop such as vegetative, self- and cross-pollinated. Hybrids are to be treated like self-pollinated plants.

Stability: It refers to the phenotypic similarity across generations.

The IPR accorded with a registration of a plant variety is called plant breeders right (PBR). The PBR grants right to the breeder, his successor, his agent, his licensee to exclude others from producing, offering for sale, selling, marketing, distributing, exporting, or importing the propagating material of the variety for a period of 15 years in the case of annuals and 18 years in the case of vines and tress. The PBR excludes researchers rights (RR) and Farmer's rights (FR). Researchers exemption allows the use of a protected variety by any person for conducting experiments or research or its use as an initial source for creating another variety without prior approval of the PBR holder, provided that such use does not involve repeated use of the protected variety as a parental line or reproduction or multiplication of its propagating material for commercial purpose. FR includes entitlement to save, use sow, re-sow, exchange, share or sell farm produce including seed of a protected variety in the same manner as the farmer was entitled before the coming into force of this Plant Variety Protection Act. However, this right to sell the seed does not cover the branded seed of a protected variety. Branded seed means any seed placed in a package or any other container and labeled to indicate that such seed is that of a protected variety.

The PPVFR Act is notable for farmer's rights and integration of the some of the important elements of CBD, such as prior informed consent for the use of Indian genetic resources, the concept of benefit sharing and creation of gene fund for strengthening agrobiodiversity conservation. The farmers right in this act provides right to register farmers varieties, right to claim compensation for under-performance of a protected variety from the promised level, benefit sharing for use of biodiversity conserved by farming community, protection from legal proceedings of any alleged infringement. According to the concept of benefit sharing, whenever a variety submitted for protection is bred with the possible use of a landrace, extant variety, or farmer's variety, a claim can be referred either on behalf of the local community or institution for a share of the royalty.

Some case studies on plant patents

Patenting of Basmati Rice in USA

A US patent No. 5663484 on "Basmati Rice lines

and grains" was granted to M/s RiceTec, Inc., Texas, USA on 2 September 1997. This patent application was filed on 8 July 1994. This patent allowed 20 claims on three new Basmati rice lines (Bas 867, RT1117, and RT1121) and grains, which was very broad, and hence apprehension that this patent covers all the Basmati rice. The Government of India challenged the patent. Only three claims – claims 15–17 – were challenged, which had defined rice grains in the patent without limitation of territory or growing seasons (photoperiod insensitivity). These claims were broadly worded and had wider implications in terms of posing a threat of infringement of patent even by traditional Basmati lines being exported to USA from India. The challenged claims were as follows:

15. A rice grain which has

   i.   a starch index of about 27 to 35

   ii.   a 2 acetyl-L-pyrroline content of about 150 ppb to about 2000 ppb.

   iii.   a length of about 6.2 mm to about 8.0 mm, a width of about 1.6 mm to about 1.9 mm and a length / width ratio of about 3.5 to about 4.5.

   iv.   a whole grain index of about 41 to 63.

   v.   a lengthwise increase of about 75% to about 150% when cooked and

   vi.   a chalk (starch?) index of less than about 20.

16. The rice grain of claim 15, which has a 2 acetyl-L-pyrroline content of about 350 ppb to about 600 ppb.

17. The rice grain of claim 15, which has a burst index of about 4 to about 1.

The challenge was filed on 28 April 2001, requesting for reexamination of patent. In response, M/S RiceTech, Inc. surrendered the three claims 15–17 along with another claim, which also concerned grains. On the basis of withdrawal of these claims by M/S RiceTec, India made a new representation for further reexamination of claims. USPTO has issued a notice on 14 August 2001 confirming claims

8, 9 and 11 and accepting amendments in claims 12 and 13, while all other claims stand cancelled. Eventually the patent is to be restricted to only three rice strains developed by M/S RiceTec, Inc. The title of the patent has also to be modified accordingly by excluding words basmati rice lines and grains.

Revocation of the turmeric patent

USPTO granted a patent US Patent No 5401504 on "Use of Turmeric in Wound Healing" on 28 March 1995 to University of Mississippi Medical Center, Jackson, USA. This patent generated great concern and public debate on creating private ownership to traditional knowledge.

The Council of Scientific and Industrial Research, Govt. of India, challenged this patent on the basis that prior knowledge on the wound healing property of turmeric existed and patent claim is not novel. Based on printed references predated to the date of patent application, the patent was revoked because it lacks novelty with respect to *prior art*. This patent was revoked on 21 April 1998.

Revocation of the neem patent

EPO granted European Patent No. 436257B1 to W.R. Grace and Co. and the US Department of Agriculture in 1995 on "Method of controlling fungi on plants by the aid of a hydrophobic extracted neem oil."

Legal opposition to the patent was that the invention lacked novelty and originality as there was an Indian testimony on prior knowledge of insecticidal and fungicidal properties of neem. Moral issues were also raised as grounds for this opposition.

On the basis of representation by filing affidavits and counter-affidavits, ultimately in May 2000, EPO concluded that there was adequate evidence in *prior art*, and therefore lack inventive step in the patent claim. The patent was therefore revoked.

The two patents cited for neem and turmeric demonstrate the importance of authentic documentation of traditional knowledge and database and the ability to provide convincing evidence of *prior art*. Thus, it is essential that for countries having tremendous biodiversity and where common practices as traditional knowledge are employed by tribal and local peoples, there must be proper documentation of information so that no one can misutilize that information for his/her own benefit.

Appendix I

Metric prefixes

Factor	Prefix	Symbol
10^{18}	exa	E
10^{15}	peta	P
10^{12}	tetra	T
10^{9}	giga	G
10^{6}	mega	M
10^{3}	kilo	k
10^{-3}	milli	m
10^{-6}	micro	μ
10^{-9}	nano	n
10^{-12}	pico	p
10^{-15}	femto	f
10^{-18}	atto	a

Examples: 1 ng (nanogram) $= 10^{-9}$ gram
1 pmol (picomole) $= 10^{-12}$ mol

Appendix II
Greek alphabets

Uper Cese type	Lower Cese type	Letter	Upper cese type	Lower Cese type	Letter
A	α	alpha	N	ν	nu
B	β	beta	X	χ	xi
Γ	γ	gamma	O	o	omicron
Δ	δ	delta	Π	π	pi
E	ε	epsilon	P	ρ	rho
Z	ξ	zeta	Σ	σ	sigma
H	η	eta	T	τ	tau
Θ	ϑ	theta	Y	υ	upsilon
I	ι	iota	Φ	φ	phi
K	κ	kappa	X	χ	chi
Λ	λ	lambda	Ψ	ψ	psi
M	μ	mu	Ω	ω	omega

Appendix III

Nucleic acid conversion factors
Average MW of a DNA base pair = 660 Da
1 pmol of 1000 bp DNA = 0.66 µg
1 µg of 1000 bp DNA fragment = 1.5 pmol = 3.0 pmol ends
To calculate picomoles of ends per microgram of linear double-strand DNA:

$$(2 \times 10^6)/660 \times \text{number of bases} = \text{pmol ends/µg DNA}$$

In solution:
1 A_{260} unit = ~ 50 µg/ml of double-strand DNA = 0.15 mM
1 A_{260} unit = ~ 40 µg/ml of double-strand RNA = 0.11 mM
1 A_{260} unit = ~ 33 µg/ml of single-strand DNA = 0.10 mM
To calculate the concentration of plasmid DNA in solution using absorbance at 260 nm:
(observed A_{260}) × (dilution factor) × (0.05) = DNA concentration in µg/µl
1000 bp DNA open reading frame = 330 amino acids = 37,000 Da protein

Appendix IV

Agarose gel percentages for resolution of linear DNA

Gel percentage	DNA size range (bp)
0.5	1,000–30,000
0.7	800–12,000
1.0	500–10,000
1.2	400–7,000
1.5	200–3,000
2.0	50–2,000

The agarose concentration does not significantly alter the migration of tracking dye, xylene cyanol and bromophenol blue, relative to the molecular weight of DNA fragments. Therefore, over the range of agarose gel percentages shown in this table, xylene cyanol will co-migrate with ~5 kb DNA fragments and bromophenol blue will co-migrate with ~0.5 kb DNA fragments.

Appendix V

Polyacrylamide gel percentages for resolution of DNA

Gel percentage	DNA size range (bp)
3.5	100–1,000
5.0	75–500
8.0	50–400
12.0	35–250
15.0	20–150
20.0	5–100

Appendix VI

Formulas for commonly used electrophoresis buffers

Resolution of nucleic acids		Resolution of proteins	
Buffer	Formula	Buffer	Formula
Tris-acetate (TAE)	40 mM Tris-acetate 1 mM EDTA	Separating gel buffer	1.5 M Tris-HCl, pH 8.8 0.4% SDS
Tris-phosphate (TPE)	90 mM Tris-phosphate 2 mM EDTA	Stacking gel buffer	0.5 M Tris-HCl, pH 6.8 0.4% SDS
Tris-borate (TBE)	45 mM Tris-borate 1 mM EDTA	Tris-glycine running buffer	25 mM Tris-base 192 mM glycine 0.1% SDS
Alkaline	50 mM NaOH 1 mM EDTA	Sample buffer	10% glycerol 2.3% SDS 62 mM Tris-HCl, pH 6.8

References

Adkins, S.W., Kunanuatchaidach, R. and Godwin, I.D. 1995. Somaclonal variation in rice-drought tolerance and other agronomic characters. *Aus. J. Bot.*, **43**: 201-209.

Altenbach, S.B., Pearson, K.W., Meeker, G., Staraci, L.C., and Sun, S.S.M. 1989. Enhancement of methionine content of seed proteins by the expression of a chimeric gene encoding a methionine rich protein in transgenic plants. *Plant Mol. Biol.*, **13**: 513-522.

Alwine, J.C., Kemp, D.J., Parker, B.A., Reiser, J., Renart, J., Stark, G.R., and Wahl, G.M. 1979. Detection of specific RNAs or specific fragments of DNA by fractionation in gels and transfer to diazobenzyloxymethyl paper. *Meth. Enzymol.*, **68**: 220-242.

Anzai, H., Yoneyama, K. and Yamaguchi, I. 1989. Transgenic tobacco resistant to a bacterial disease by the detoxification of a pathogenic toxin. *Mol. Gen. Genet.*, **219**: 492-494.

Aono, M., Saji, H., Sakamoto, A., Tanaka, K., Kondo, N. and Tanaka, K. 1995. Paraquat tolerance of transgenic *Nicotiana tabacum* with enhanced activities of glutathione reductase and superoxide dismutase. *Plant Cell Physiol.*, **36**: 1687-1691.

Arihara, A., Kumagai, R., Koyama, H., and Ojima, K. 1991. Aluminium tolerance of carrot (*Daucua carota* L.) plants regenerated from selected cell cultures. *Soil Sci. Plant Nutr.*, **37**: 699-705.

Arakawa, T., Chong, D.K.X., and Langridge, W.H.R. 1998a. *Nature Biotechnol.*, **16**: 292-297.

Arakawa, T., Yu, J., Chong, D.K.X., Hough, J., Engen, P.C., and Langridge, W.H.R. 1998b. *Nature Biotechnol.*, **16**: 934-938.

Arkin, A.P. and Youvan, D.C. 1992. Optimizing nucleotide mixtures to encode specific subsets of amino acids for semi-random mutagenesis. *Biotechnology*, **10**: 297-300.

Artsaenko, O., Kettig, R., Fiedler, U., Conrad, U. and During, K. 1998. Potato tubers as a biofactory for recombinant antibodies. *Mol. Breed.*, **4**: 313-319.

Arumuganathan, K. and Earle, E.D. 1991. Nuclear DNA content of some important plant species. *Plant Mol. Biol. Rep.*, **9**: 208-221.

Atkins, D., Young, M., Uzzell, S., Kelly, L., Fillatti, J. and Gerlach, W.L. 1995. The expression of antisense and

ribozyme genes targeting citrus exocortis viroid in transgenic plants. *J. Gen. Virol.*, **16**: 1781-1790.

Audy, P., Palukaitis, P., Slack, S.A. and Zaitlin, M. 1994. Replicase mediated resistance to potato virus Y in transgenic tobacco plants. *Mol. Plant Microbe Interact.*, **7**: 15-22.

Azpiroz-Leehan, R. and Feldman, K.A. 1997. T-DNA insertion mutagenesis in *Arabidopsis* going back and forth. *Trends Genet.*, **13**: 152-156.

Bains, W. and Smith, G.C. 1998. *J. Theor. Biol.*, **135**: 303-307.

Baird, E., Cooper-Bland, S., Waugh, R., Demaine, M., and Powell, W. 1992. Molecular characterization of inter- and intraspecific somatic hybrids of potato using randomly amplified polymorphic DNA (RAPD) markers. *Mol. Gen. Genet.*, **233**: 469-475.

Bairoch, A. 1991. *Nucleic Acids Res.*, **19**: 2241-2245.

Balcells, L., Swinburne, J., and Coupland, G. 1991. Transposons as tools for the isolation of gene. *Trends Biotech.*, **9**: 31-37.

Baltimore, D. 1970. RNA dependent DNA polymerase in virions of RNA tumor virus. *Nature*, **226**: 1209-1211.

Barcak, G.J. and Wolf, R.E. Jr. 1986. A method for unidirectional deletion mutagenesis with application to nucleotide sequencing and preparation of gene fusions. *Gene*, **49**: 119-128.

Barclay, I.R. 1975. High frequencies of haploid production in wheat (*Triticum aestivum*) by chromosome elimination. *Nature (London)*, **256**: 410-411.

Barun, A.C. and Stonier, R. 1958. Morphology and physiology of plant tumours. *Protoplasmatologia*, **10**: 1-93.

Beachy, R.N., Fitchen, J.H., and Hein, M.B. 1996. *Ann. N.Y. Acad. Sci.*, **792**: 43-49.

Behenke, M. 1979. Selection of potato callus for resistance to culture filtrates of *Phytophthora infestans* and regeneration of resistant plants. *Theor. Appl. Genet.*, **55**: 69-71.

Bergmann, L. 1960. Growth and division of single cells of higher plants *in vitro*. *J. Gen. Physiol.*, **43**: 841-851.

Bhojwani, S.S. and Razdan, M.K. 1990. *Plant Tissue Culture: Applications and Limitations*. Elsevier Science Publishers, Amsterdam

Blanchard, A. 1998. In: *Genetic Engineering, Principles and Methods*, Setlow, J. (Ed.). Vol. 20.: Plenum Press, New York, pp. 111-124.

Blattner, F.R., Plunkett, G., Bloch, C.A., Perna, N.T., Burland, V., Riley, M., Collado-Vides, J., Glasner, J.D., Rode, C.K. and Mayrew, G.F. 1997. The complete genome sequence of *Escherichia coli* K-12. *Science,* **227**: 1453-1474.

Blaydes, D.F. 1966. Interaction of kinetin and various inhibitors on the growth of soybean tissue. *Physiol. Plant.,* **19**: 748-753.

Bolivar, F., Rodriguez, R.L., Betlach, M.C. and Boyer, H.W. 1977. Construction and characterization of new cloning vehicles I. Ampicillin resistant derivatives of the plasmid pMBq. *Gene,* **2**: 75-93.

Bolivar, F., Rodriguez, R.L., Greene, P.J., Betlach, M.C., Heynecker, H.L., Boyer, H.W, Crosa, J.H., and Falkow, S. 1977. Construction and characterization of new cloning vehicles II. A multipurpose cloning system. *Gene,* **2**: 95-113.

Botstein, D., White, R.L., Skolnik, M. and Davis, R.W. 1980. Construction of a genetic linkage map in man using restriction fragment length polymorphism. *Am. J. Human Genet.,* **32**: 314-331.

Bourgin, J.P. and Nitsch, J.P. 1967. Obtentionde Nicotians haploides a partir d'etamines cultivees *in vitro. Ann. Physiol. Veg. (Paris),* **9**: 377-382.

Bowler, C., Slooten, L., Vandenbranden, S., De Rycke, R., Botterman, J., Sybesma, C., Van Montagu, M. and Inze, D. 1991. Manganese superoxide dismutase can reduce cellular damage mediated by oxygen radicals in transgenic plants. *EMBO J.,* **10**: 1723-1732.

Bozorgipour, R. and Snape, J.W. 1997. An assessment of somaclonal variation as a breeding tool for generating herbicide tolerant genotypes in wheat (*Triticum aestivum* L.). *Euphytica,* **94**: 335-340.

Bradley, J.M., Davies, K.M., Deroles, S.C., Bloor, S.J. and Lewis, D.H. 1998. The maize Lc regulatory gene upregulates the flavonoids biosynthetic pathway of petunia. *Plant J.,* **13**: 381-392.

Brettell, R.I.S., Goddard, B.V.D. and Ingram, D.S. 1979. Selection of Tms-cytoplasm maize tissue cultures resistant to *Drechslera maydis* T-toxin. *Maydica,* **24**: 203-213.

Brisson, N., Paszkowski. J., Penswick, J.R, Gronenborn, B., Potrykus, I., and Hohn, T. 1984. Expression of a bacterial gene in plants using a viral vector. *Nature,* **310**: 511-514.

Broglie, K., Chet, I., Holliday, M., Crossman, R., Biddle, P., Knowlton, S., Manain, I.J., and Brogile, R. 1991. Transgenic plants with enhanced resistance to fungal pathogen *Rhizoctonia solani. Science,* **254**: 1194-1197.

Broun, P., Boddupalli, S. and Somerville, C. 1998. A bifunctional oleate 12-hydroxylase desaturase from *Lesquerella fendleri. Plant J.,* **13**: 201-210.

Brown, P.T.H. 1989. DNA methylation in plants and its role in tissue culture. *Genome,* **31**: 717-729.

Brummer, E.C., Nickell, A.D., Wilcox, J.R. and Shoemaker, R.C. 1995. Mapping the fan locus controlling linolenic acid content in soybean oil. *J. Hered.,* **86**: 245-247.

Bulow, L., Holmberg, N., Lilius, G. and Bailey, J.E. 1999. The metabolic effects of native and transgenic hemoglobin on plants. *Trends Biotechnol.,* **17**: 21-24.

Burkee, D.T., Carle, G.F. and Olson, M.V. 1987. *Science,* **236**: 806-812.

Byeong Geun, O.H., Lim, S.J., Yang, S.J. and Hwang, H.G. 1997. *R. D. A. J. Crop Sci.,* **39**: 94-100.

Cargaff, E. 1951. Structure and function of nucleic acids as cell constituents. *Fed. Proc.,* **10**: 654-659.

Carle, G.F., Frank, M. and Olson, M.V. 1986. Electrophoretic separation of large DNA molecules by periodic inversion of electric field. *Science* **232**: 65.

Carlson, P. 1973. Methionine sulfoximine resistant mutants of tobacco. *Science,* **180**: 1366.

Carrillo, C., Wigdorovitz, A., Oliveros, J.C., Zamorano, P.I., Sadir, A.M., Gomez, N., Salinas, J., Escribano, J.M. and Borca, M.V. 1998. *J. Virol.,* **72**: 1688-1690.

Chakraborty, N., Chakraborty, S., Kesarwan, M., Azam, M. and Datta, A. 1998. Proceedings of IUBS Symposium, Taipei, Taiwan, 1997, pp. 125-131.

Chaleff, R.S. 1981. *Genetics of Higher Plants.* Cambridge Univ Press. Cambridge.

Chargaff, E. 1951. Structure and function of nucleic acids as cell constituents. *Fed. Proc.,* **10**: 654-659.

Charnay, P., Mellon, P. and Maniatis, T. 1985. Linker scanning mutagenesis of the 5' flanking region of the mouse β-major globin gene: sequence requirements for transcription in erythroid and nonerythroid cells. *Mol. Cell. Biol.,* **5**: 1498-1511.

Chawla, H.S. 1998. *Biotechnology in Crop Improvement.* International Book Distributing Co., Lucknow.

Chawla, H.S. and Wenzel, G. 1987a. *In vitro* selection of barley and wheat for resistance against *Helminthosporium sativum. Theor. Appl. Genet.,* **74**: 841-845.

Chawla, H.S. and Wenzel, G. 1987b. *In vitro* selection for fusaric acid resistant barley plants. *Plant Breed.,* **99**: 159-163.

Chawla, H.S., Cass, L.A. and Simmonds, J.A. (1999). Development and environmental regulation of anthocyanin pigmentation in wheat tissues transformed with anthocyanin regulatory genes. *In Vitro Cell Dev. Biol.-Plant,* **35**: 403-408.

Cheng, M., Fry, J.E., Pang, S., Zhou, H., Hironaka, C.M., Duncan, D.R., Conner, T.W. and Wan, Y. 1997. Genetic transformation of wheat mediated by *Agrobacterium tumefaciens. Plant Physiol.,* **115**: 971-980.

Chilton, M.D., Drummond, M.H., Merlo, D.J., Sciaky, D., Montoya, A.L., Gordon, M.P. and Nester, E.W. 1977. Stable incorporation of plasmid DNA into higher plant cells: the molecular basis of crown gall tumorigenesis. *Cell,* **11**: 263-271.

Chong, D.K., Roberts, W., Arakawa, T., Illes, K., Bagi, G., Slatter, C.W. and Langridge, W.H. 1997. Expression of human milk protein beta casein in transgenic potato plants. *Transgenic Res.,* **6**: 289-296.

Chu, G., Vollrath, D. and Davis, R.W. 1986. Separation of large DNA molecules by contour clamped homogeneous electric fields. *Science,* **234**: 1582-1585.

Chu, C.C. 1978. The N6 medium and its applications to anther culture of cereal crops. *In: Proc. Symp. Plant Tissue Culture,* Science Press, Peking, pp. 43-45.

Chuang, C.C., Ouang, T.W., Chia, H., Chou, S.M. and Ching, C.K. 1978. A set of potato media for wheat anther culture. *In: Proc. Symp. Plant Tissue Culture,* Science Press, Peking, pp. 51-56.

Clackson, T. and Wells, J.A. 1994. *In vitro* selection from protein and peptide libraries. *Trends Biotechnol.,* **12**: 173-184.

Cocking, E.C. 1960. A method for the isolation of plant protoplasts and vacuoles. *Nature,* **187**: 962-963.

Cohen, S.N., Chang, A.C.Y. and Hsu, L. 1972. Nonchromosomal antibiotic resistance in bacteria: genetic transformation of *Escherichia coli* by R-factor DNA. *Proc. Natl. Acad. Sci., USA,* **69**: 210-2114.

Collins, F.S. and Weissman, S.M. 1984. Directional cloning of DNA fragments at a large distance from an initial probe: a circularization method. *Proc. Natl. Acad. Sci. USA,* **81**: 6812-6816.

Cooper, B., Lapidot, M., Heike, J.A., Dodds, J.A. and Beachy, R.N. 1995. A defective movement protein of TMV in transgenic plants confers resistance to multiple viruses whereas functional analogs increases susceptibility. *Virology,* **206**: 307-313.

Cornish, E.C. 1995. Antisense ACC oxidase RNA delays carnation petal senescence. *Hort. Sci.,* **30**: 970-972.

Crick, F.H.C. 1958. On protein syntesis. *Symp. Soc., Expt. Biol.,* **12**: 138-163.

Cronin, M.T. *et al.* 1996. *Hum. Mutat.* **7**: 244.

Cronin, S., Verchot, J., Haldeman-Cachill, R. S. and Carrington, J.G. 1995. Long distance movement factor: A transport function of the potyvirus helper component proteinase. *Plant Cell,* **7**: 549-559.

Crow, J.F. and Dove, W.F. 1988. A diamond anniversary: the first chromosome maps. *Genetics,* **118**: 389-391.

De Block, M., Herrera-Estrella, L., van Montagu, M., Schell, J. and Zambryski, P. 1984. Expression of foreign genes in regenerated plants and their progeny. *EMBO. J.,* **3**: 1681-1689.

De Fossard, R.A., Baker, P.K. and Bourne, R.A. 1977. The organ culture of nodes of four species of *Eucalyptus. Acta Hortic.,* **78**: 157-165.

De la Feunte, H.M., Mosqueda-Cano, G., Alvarez-Morales, A. and Herrera-Estrella, L. 1992. Expression of a bacterial phaseolotoxin resistant ornithyl transcarbamylase in transgenic tobacco plants confers resistance to *Pseudomonas syningae* pv

phaseolicola. BioTechnology, **10**: 905.

De la Pena, Lorz, H. and Schell, J. 1987. Transgenic rye plants obtained by injecting DNA into young floral tillers. *Nature,* **325**: 274-276.

De Paepe, R., Prat, D., and Huguet, T. 1982. Heritable nuclear DNA changes in doubled haploid plants obtained from pollen cultures of *Nicotiana sylvestris. Plant Sci Lett.,* **28**: 11-28.

Dekoulas, P., Schuler, G.D., Gyapay, G., Beasley, E.M. *et al.* 1998. A physical map of 30,000 human genes. *Science,* **282**: 744-746.

Dib, C., Faure, S., Fizames, C., Samson, D., Drouot, N., Vignal, A., Millasseau, P., Marc, S., Hazan, J., Seboun, E., Lathrop, M., Gyapay, G., Morissette, J. and Weissenback, J. 1996. A comprehensive genetic map of the human genome-based on 5264 microsatellites. *Nature,* **380**: 152-154.

Dieterich, K. 1924. Uber Kultur von Embryonen ausserhalb des Samens. *Flora (Jena),* **117**: 379-417.

Donson, J., Kearney, C.M., Hilf, M.E. and Dawson, W.O. 1991. systemic expression of a bacterial gene by a tobacco mosaic virus based vector. *Proc. Natl. Acad. Sci. USA,* **88**: 7204-7208.

Dorffling, K., Dorffling, H., Lesselich, G., Luck, E., Zimmerman, C., Melz, G. and Jurgens, H.U. 1997. Heritable improvement of frost tolerance in winter wheat by *in vitro* selection of hydroxyproline resistant proline overproducing mutants. *Euphytica,* **93**: 1-10.

Drmanac, S., Kita, D., Labat, I., Hauser, B., Schmidt, C., Burczak, J.D. and Drmanac, R. 1998. *Nat. Biotechnol.,* **16**: 54-58.

Dudits, D., Hadlaczky, Gy., Levi, E., Fejer, O., Haydu, Z. and Lazar, G. 1977. Somatic hybridization of *Daucus carota* and *D. capillifolius* by protoplast fusion. *Theor. Appl. Genet.,* **51**: 127-132.

Dunham, I., Shimizu, N., Roe, B.A. and Chissoe, S. 1999. The DNA sequence of human chromosome 22. *Nature,* **402**: 489-495.

Durante, M., Geri, C., Grisvad, J., Guille, E., Parenti, R., and Buaitti, M. 1983. Variation in DNA complexity in *Nicotiana glauca* tissue cultures. I. Pith dedifferentiation *in vitro. Protoplasma,* **114**: 114-118.

During, K., Porsch, P., Flauddung, M. and Lorz, H. 1993. Transgenic potato plants resistant to the phytopathogenic bacterium *Erwinia carotovora. Plant J.,* **3**: 587-598.

Edallo, S., Zucchinali, C., Perenzin, M. and Salamini, F. 1981. Chromosomal variation and frequency of spontaneous mutation associated with *in vitro* culture and plant regeneration in maize. *Maydica,* **26**: 39-56.

Edington, B.V. and Nelson, R.S. 1992. Utilization of ribozymes in plants: plant viral resistance. In: *Gene Regulation–Biology of Antisense RNA and DNA,* Erickson, R.P. and Izant, J.G. (Eds.), Raven Press, New York.

Eichholtz, D., Robitaille, H.A. and Hasegawa, P.M. 1979. Adventive embryology in apple. *Hort. Sci.*, **14**: 699–700.

Erbisch, F.H. and Velazquez, C. 1998. Introduction to intellectual properties. In: *Intellectual Property Rights in Agricultural Biotechnology* (Erbisch, F.H. and Meredia, K.M., eds.), CAB International, Wallingford, U.K., pp. 3-18.

Evans, D.A., Sharp, W.R., Ammirato, P.V. and Yamada, Y. 1983. *Handbook of Plant Cell Culture*. Macmillan Publishing Company, New York

Fahleson, J., Lagercrantz, U., Eriksson, I. and Glimelius, K. 1993. Genetic and molecular analysis of sexual progenies from somatic hybrids between *Eruca sativa* and *Brassica napus*. Swedish Univ. of Agril. Sci., Dissertation, ISBN 91-567-4684-8.

Fedoroff, N., Wessler, S. and Shure, M. 1983. *Cell*, **35**: 235-242.

Feinberg, A.P. and Vogelstein, B. 1983. A technique for radiolabelling DNA restriction endonuclease fragments to high specific activity. *Anal. Biochem.*, **132**: 6.

Feinberg, A.P. and Vogelstein, B. 1984. Addendum: A technique for radiolabelling DNA restriction endonuclease fragments to high specific activity. *Anal. Biochem.*, **137**: 266.

Fischhoff, D.A., Bowdish, K.S., Perlak, F.J., Marrone, P.G., McCormick, S.M., Niedermayer, E.J., Rochester, E.J., Rogers, S.G. and Fray, R.T. 1987. Insect tolerant transgenic tomato plants, *Biotechnology*, **5**: 807-813.

Flavell, A.J., Smith, D.B. and Kumar, A. 1992. Extreme heterogeneity of Ty1-Copia group transposons in plants. *Mol. Gen. Genet.*, **231**: 233-242.

Fleischmann, R.D., Adams, M.D., White, O., Clayton, R.A., Kirkness, E.F., Kerlavage, A.R., Bult, C.J., Tomb, J.F., Dougherty, B.A. and Merrick, J.M. 1997. Whole genome random sequencing and assembly of *Haemophilus influenzae* Rd. *Science*, **269**: 496-512.

Fodor, S.P.A. 1997. Massively parallel genomics. *Science*, **270**: 393-394.

Forlani, G., Nielsen, E., and Racchi, M.L. 1992. A glyphosate resistant 5-enol pyruvyl shikimate 3 phosphate synthase confers tolerance to a maize cell line. *Plant Science Limerick*, **85**: 9-15.

Foroughi-Wehr, B. and Friedt, W. 1984. Rapid production of recombinant barley yellow mosaic virus resistant *Hordeum vulgare* lines by anther culture. *Theor. Appl. Genet.*, **67**: 377-382.

French, R., Janda, M., and Ahlquist, P. 1986. Bacterial gene inserted in an engineered RNA virus: Efficient expression in monocotyledonous plant cells. *Science*, **23**: 1294-1297.

Frohman, M.A., Dush, M.K. and Martin, G.R. 1988. Rapid production of full length cDNAs from rare transcripts: Amplification using a single gene specific oligonucleotide primer. *Proc. Natl. Acad. Sci. USA*, **85**: 8998-9002.

Galbraith, D.W. and Mauch, T.J. 1980. Identificatin of fusion of plant protoplasts. II. Conditions for the reproducible fluorescence labellig of protoplasts derived from mesophyll tissue. *Z. Pflanzenphysiol.*, **98**: 129-140.

Gamborg, O.L., Miller, R.A., and Ojima, K. 1968. Nutrition requirements of suspension cultures of soybean root cells. *Expt. Cell. Res.*, **50**: 151-158.

Gardiner, A., Lass, R.W., and Patterson, D. 1986. Fraction of large mammalian DNA restriction fragments using vertical pulse field gradient gel electrophoresis. *Somat. Cell Mol. Gen.*, **12**: 185.

Gautheret, R.J. 1939. Sur la possibilitie de realiser la culture indefinie des tissue de tubercules de carotte. *C.R. Acad. Sci. Paris*, **208**: 118-120.

Gautheret, R.J. 1959. La culture des tissus vegetauz: Techniques et realisations. Masson, Paris.

Gengenbach, B.G., Green, C.E. and Donovan, C.D. 1977. Inheritance of selected pathotoxin resistance in maize plants regenerated from cell cultures. *Proc. Natl. Acad. Sci. USA*, **74**: 5113-5117.

Gierl, A., Lutticke, S., and Saedler, H. 1988. TnpA product encoded by the transposable element En-1 of *Zea mays* is a DNA binding protein. *EMBO J.*, **7**: 4045-4053.

Gierl, A. and Saedler, H. 1992. Plant transposable elements and gene tagging. *Plant Mol. Biol.*, **19**: 39-49.

Gilliard, L.U., Mckinney, E.C., Asmussen, M.A. and Meagher, R.B. 1998. Detection of deleterious genotypes in multigenerational studies. I. Disruptions in individual *Arabidopsis* actin genes. *Genetics*, **149**: 717-725.

Glimelius, K., Eriksson, T., Grafe, R. and Muller, A. 1978. Somatic hybridization of nitrate reductase deficient mutants of *Nicotiana tabacum* by protoplast fusion. *Physiol. Plant.*, **44**: 273-277.

Golemboski, D.B., Lomonossoff, G.P. and Zaitlin, M. 1990. Plants transformed with a tobacco mosaic virus nonsturctural gene sequence are resistant to the virus. *Proc. Natl. Acid. Sci. USA*, **87**: 6311.

Goto, F., Yoshihara, T., Shigemoto, N., Toki, S. and Takaiwa, F. 1999. *Nature Biotechnol.*, **17**: 282-286.

Grandbastein, M.A., Spielman, A. and Caboche, M. 1989. Tnt 1, a mobile retroviral-like transposable element of tobacco isolated by plant cell genetics. *Nature*, **337**: 376-380.

Grayburn, W.S., Collins, G.B. and Hildebrand, D.F. 1992. *BioTechnology*, **10**: 675-678.

Grimsley, N., Hohn, T., Davis, J.W. and Hohn, B. 1987. *Agrobacterium*-mediated delivery of infectious maize streak virus into maize plants. *Nature*, **325**: 177-179.

Grunstein, M. and Hogness, D.S. 1975. colony hybridization– A method for the isolation of cloned DNAs that contain a specific gene. *Proc. Natl. Acad. Sci. USA*. **72**: 3961-3965.

Grunstein, M. and Hogness, D.S. 1975. Colony hybridization-A method for the isolation of cloned DNAs that contain a specific gene. *Proc. Natl. Acad. Sci. USA,* **72**: 3961-3965.

Guha, S. and Maheshwari, S.C. 1964. *In vitro* production of embryos from anthers of *Datura. Nature,* **204:** 497.

Gulick, P.J. and Dvorak, J. 1990. *Gene,* **95:** 173-177.

Gupta, P.K., Roy, J.K., and Prasad, M. (1999). DNA chips, microarrays and genomics. *Curr. Sci.,* **77:** 875-884.

Gupta, V.K. 1999. Basic concepts of IPRs and their management in R & D. In: *Lecture notes on Patents.* Technology Information, Forecasting and Assessment Council, New Delhi.

Haberlandt, G. 1902. Culturversuchemit isolierten Pflanzenzellen. *Sitzungsber. Math. Naturwiss. Kl. Kais. Akad. Wiss. Wien.,* **111:** 69-92.

Hacia, J.G., Brody, L.C., Chee, M.S., Fodor, S.P.A. and Collins, F.S. 1996. *Nature Genet.,* **14:** 441-447.

Hain, R., Biessler, B., Kindl, H., Schroeder, G. and Stocker, R. 1990. Expression of a stilbene synthase gene in *Nicotiana tabacum* results in synthesis of phytoalexin resveratrol. *Plant Mol. Biol.,* **15:** 325.

Hannig, E. 1904. Zur Physiologie pflanzlicher embryonen. I. Uber die cultur von Cruciferen-Embryonen ausserhalb des Embryosacks. *Bot. Ztg.,* **62:** 45-80.

Hansen, L.N. and Earle, E.D. 1995. Transfer of resistance to *Xanthomonas campestris* pv *campestris* into *Brassica oleracea* by protoplast fusion. *Theor. Appl. Genet.,* **91:** 1293-1300.

Haq, T.A., Mason, H.S., Clements, J.D. and Arntzen, C.J. 1995. *Science,* **268:** 714-716.

Harding, K. 1991. Molecular stability of the ribosomal RNA genes in *Solanum tuberosum* plants recovered from slow growth and cryopreservation. *Euphytica,* **55:** 141-146.

Harms, C.T., Potrykus, I. and Widholm, J.M. 1981. Complementation and dominant expression of amino acid analogue resistance markers in somatic hybrid clones from *Daucus carota* after protoplast fusion. *Z. Pflanzenphysiol.,* **101:** 377-385.

Harrison, B.D., Mayo, M.A. and Baulcombe, D.C. 1987. Virus resistance in transgenic plants that express cucumber mosaic virus satellite RNA. *Nature,* **328:** 799-801.

Hattori, M. *et al.* 2000. *Nature,* **405:** 311-319.

Hauge, P.M. *et al.* 1993. *Plant J.,* **3:** 745-754.

Hein, M.B., Yeo, T.-C., Wang, F. and Sturtevent, A. 1996. *Ann. N.Y. Acad. Sci.,* **792:** 51-56.

Hernould *et al.* 1992. *Bulletin di Information,* **34:** 17-18.

Herrera-Estrella, L., Depicker, A., van Montagu, M. and Schell, J. 1983. Expression of chimaeric genes transferred into plant cells using a Ti plasmid derived vector. *Nature,* **303:** 209-213.

Hiei, Y., Ohta, S., Komari, T. and Kumashiro, T. 1994. Efficient transformation of rice (*O. sativa* L.) mediated by

Agrobacterium and sequence analysis of the boundaries of the T-DNA. *The Plant J.,* **6:** 271-282.

Higuchi, R., Krummel, B., and Saiki, R.K. 1988. A general method of *in vitro* preparation and specific mutagenesis of DNA fragments: study of protein and DNA interactions. *Nucl. Acids Res.,* **16:** 7351-7367.

Hinnen, A., Hicks, J.B. and Fink, G.R. 1978. Transformation of yeast. *Proc. Natl. Acad. Sci. USA,* **75:** 1929-1933.

Hirsch, R.E., Lewis, B.D., Spalding, E.P. and Sussman, M.R. 1998. A role of AKT1 potassium channel in plant nutrition. *Science,* **280:** 918-921.

Holmberg, N., Lilius, G., Bailey, J.E., and Bulow, L. 1997. Transgenic tobacco expressing *Vitroscilla* hemoglobin exihibits enhanced growth and altered metabolite production. *Nat. Biotechnol.,* **15:** 244-247.

Holton, T.A., Brugliera, F., Lester, D.R., Tanaka, Y., Hyland, C.D., Menting, J.G.T., Lu, C.Y., Farcy, E., Stevenson, T.W. and Cornish, E.C. 1993. Cloning and expression of cytochrome p450 genes controlling flower color. *Nature,* **366:** 276-279.

Hood, E.E., Howard, J.A. and Tannotti, E.L. 1998. Transgenic corn: a new source of valuable industrial products. In: *Proc. Corn Utilization and Tech. Conf.,* Missouri, USA, pp. 101-104.

Horsch, R.B., Fraley, R.T., Rogers, S.G., Sanders, P.R., Lloyd, A. and Hoffman, N. 1984. Inheritance of functional foreign gene in plants. *Science,* **223:** 496-498.

Horton, B. 1998. Sequence cyberbiology. *Nature,* **393:** 603-606.

Hudson, D.J. *et al.* 1995. *Science,* **270:** 540-546.

Ignacimuthu, S. 1998. *Plant Biotechnology.* Oxford & IBH Publishing Co. Pvt. Ltd., Delhi.

Ioannou, P.A. *et al.* 1994. *Nature Genet.,* **6:** 84-89.

Ishida, Y., Saito, H., Ohta, S., Hiei, Y., Komari, T. and Kumashiro, T. 1996. High efficiency transformation of maize (*Zea mays* L.) mediated by *Agrobacterium tumefaciencs. Nature Biotech.,* **14:** 745-750.

Ishizaki-Nishizawa, O., Fuji, T., Azuma, M., Sekiguchi, K., Murata, N., Ohtani, T. and Toguri, T. 1996. Low temperature resistance of higher plants is significantly enhanced by a non-specific cyanobacterial desaturase. *Nature Biotech.,* **14:** 1003-1006.

Izawa, T., Tohru, O., and Nakano, T. 1997. Transposon tagging in rice. *Plant Mol. Biol.,* **35:** 219-229.

Jacard, P. 1908. *Bull. Soc. Vaud. Sci. Nat.,* **44:** 223-270.

Jackson, D.A., Symons, R.H. and Berg, P. 1972. Biochemical method for inserting new genetic information into DNA of Simian virus 40: circular SV 40 DNA molecules containing lambda phage genes and the galactose operon of *Escherichia coli. Proc. Natl. Acad. Sci., USA,* **69:** 2904-2909.

James, C. 1997. Global status of transgenic crops in 1997. ISAAA Briefs No. 5. ISAAA: Ithaca, N.Y. pp. 31.

James, C. 1999. Global review of commercialized transgenic crops; 1998, ISAAA Brief No. 8.

Jan, V.V.S., Macedo, C.C. de., Kinet, J.M. and Bouharmont, J. 1997. Selection of Al-resistant plants from a sensitive rice cutivar, using somaclonal variation, *in vitro* and hydroponic cultures. *Euphytica,* **97:** 303-310.

Jeffreys, A.J., Wilson, V. and Thein, S.L. 1985. *Nature,* **314:** 67-73.

Johal, G.S. and Briggs, S.P. 1992. Reductase activity encoded by the HM1 disease resistance gene in maize. *Science,* **258:** 985.

Jones, A.L., Johansen, I.E., Bean, S.J., Bach, I. and Maule, A.J. 1998. Specificity of resistance to pea seed borne mosaic potyvirus in transgenic peas expressing the viral replicase (*Nib*) gene. *J. Gen. Virol.* **79:** 3129-3137.

Jones, D.A., Thomas, C.M., Kim, E. and Jones, D.J. 1994. Isolation of the tomato cf-9 gene for resistance to *Cladosporium fulvum* by transposon tagging. *Science,* **266:** 789-793.

Jones, J.D.G., Dean, C., Gidoni, D., Bond-Nutter, D., Lee, R., Bedrock, J. and Dunsmuir, P. 1988. Expression of bacterial chitinase protein in tobacco leaves using pro photosynthetic promoters. *Mol. Gen. Genet.,* **212:** 536-542.

Kao, K. and Michayluk, C. 1975. Growth of cells and protoplasts at low population density. *Planta,* **126:** 105-110.

Kao, K.N. and Michayluk, M.R. 1974. A method for high frequency intergeneric fusion of plant protoplasts. *Planta,* **115:** 355-367.

Kasha, K.J. and Kao, K.N. 1970. High frequency haploid production in barley (*Hordeum vulgare* L.). *Nature, (London).* **225:** 874-876.

Kasuga, M., Liu, Q., Miura, S., Yamaguchi-Schinozaki, K., and Shinozaki, K. 1999. Improving plant drought, salt and freezing tolerance by gene transfer of a single stress inducible transcription factor. *Nat. Biotechnol.,* **17:** 287-291.

Kawchuk, L.M., Martin, R.R. and Pherson, J.MC. 1991. Sense and antisense RNA-mediated resistance to potato leaf roll virus in Russet Burbank potato plants. *Mol. Plant-Microbe Interact.,* **4:** 247-253.

Keller, H., Pamboukdjian, N., Ponchet, M., Poupet, A., Delon, R., Verrier, J.L., Roby, D. and Ricci, P. 1999. Pathogen-induced elicitin production in transgenic tobacco generates a hypersensitive response and nonspecific disease resistance. *Plant Cell,* **11:** 223-235.

Keller, W.A. and Melchers, G. 1973. The effect of high pH and calcium on tobacco leaf protoplast fusion. *Z. Naturforsch.,* **288:** 874-876.

Kemble, R.J. and Shepard, J.F. 1984. Cytoplasmic DNA variation in a protoclonal population. *Theor. Appl. Genet.,* **69:** 211-216.

Kishore, P.B.K., Hong, Z., Miao, G.H., Hu, C.A.A. and Verma, D.P.S. 1995. Overexpression of D1-pyrroline-5

carboxylate synthetase increases proline production and confers osmotolerance in transgenic plants. *Plant Physiol.,* **108:** 1387-1394.

Klee, H.J. 1993. Ripening physiology of fruit from transgenic tomato (*Lycopersicon esculentum*) plants with reduced ethylene synthesis. *Plant Physiol.,* **102:** 911-916.

Klein, T.M., Wolf, E.D., Wu R. and Sanford, J.C. 1987. High velocity microprojectiles for delivering nucleic acids into living cells. *Nature,* **327:** 70-73.

Klercker, J.A. 1892. Eine methods zur isoliering lebender Protoplastern. *Oefvers. Vetenskaps Adad., Stockholm,* **9:** 463-471.

Knudson, L. 1922. *Bot. Gaz.,* **73:** 1-25.

Knudson, L. 1946. A new method for the germination of orchid seeds. *Am. Orchid Soc. Bull.,* **15:** 214-217.

Knutzon, D.S., Thurmond, J.M., Huang, Y.S., Chaudhary, S., Bobik, E.G. Jr., Chan, G.M., Kirchner, S.J. and Mukherji, P. 1998. Identification of D5 desaturase from *Mortierlla alpina* by heterologous expression in bakers yeast and canola. *J. Biol. Chem.,* **273:** 29360-29366.

Kohayashi, R.S., Stomme, J.R. and Sinden, S.L. 1996. Somatic hybridization between *S. ochranthum* and *L. esculentum. Plant Cell Tiss. Org. Cult.,* **45:** 73-78.

Kole, P.C. and Chawla, H.S. 1993. Variation of *Heiminthosporium* resistance and biochemical and cytological characteristics in somaclonal generations of barley. *Biologia Plantarum,* **35:** 81-86.

Konieczny, A. and Asubell, F.M. 1993. *Plant J.,* **4:** 403-410.

Kotte, W. 1922a. Wurzelmeristem in Gewebekultur. *Ber. Deut. Ges.,* **40:** 269-272.

Kotte, W. 1922b. Kulturversuche mit isolierten Wurzelspitzen. *Beitr. Allg. Bot.,* **2:** 413-434.

Krebbers, E. and van de Kerckhove, J. 1990. *Trends Biotechnol.,* **8:** 1-3.

Kriete, G., Neuhaus, K., Perlick, A.M., Puhler, A. and Broer, I. 1996. Male sterility in transgenic tobacco plants induced by tapetum specific deacetylation of externally applied non-toxic compound N-acetyl L-phosphinothricin. *Plant J.,* **9:** 809-818.

Krumbiegel, G. and Schieder, O. 1979. Selection of somatic hybrids after fusion of protoplasts from *Datura innoxia* and *Atropa belladona. Planta,* **145:** 371-375.

Kruse, A. 1974. An *in vivo/ in vitro* embryo culture technique. *Hereditas,* **73:** 157–161.

Kunkel, T.A., Roberts, J.D. and Zakour, R.A. 1987. Rapid and efficient site-specific mutageneis without phenotypic selection. *Meth. Enzymol.,* **154:** 367-382.

Kunze, R. and Starlinger, P. 1989. The putative transposase of transposable element Ac from *Zea mays* interacts with subterminal sequences of Ac. *EMBO J.,* **8:** 3177-3185.

Kyozuka, J., Kaneda, T. and Shimamoto, K. 1989. Production of cytoplasmic male sterile rice (*Oryza sativa* L.) by cell fusion. *BioTechnology,* **7:** 1171-1174.

La Rue, C.D. 1947. Growth and regeneration of the endosperm of maize in culture. *Am. J. Bot.*, **34**: 585-586.

Laibach, F. 1925. Das Taubverden von Bastardsmen und die kunsliche Aufzucht fruhabsterbender Bastardembryonen. *Z. Bot.*, **17**: 417-459.

Laibach, F. 1929. Ectogenesis in plants. Methods and genetic possibilities of propagating embryos otherwise dying in the seed. *J. Hered.*, **20**: 201-208.

Lakshami Sita, G., Raghava Ram, N.V. and Vaidyanathan, C.S. 1980. Triploid plants from endosperm cultures of sandalwood by experimental embryogenesis. *Plant Sci. Lett.*, **20**: 63-69.

Lampe, L. and Mills, C.O. 1933. Growth and development of isolated endosperm and embryo of maize. *Abstr. Paper Bot. Soc.*, Boston.

Lander, E.S. and Botstein, D. 1989. *Genetics,* **121**: 185-199.

Landes, M., Walther, H., Gerber, M., Auxier, B. and Brown, H. 1996. New possibilities for post-emergence grass weed control with cycloxydim and sethoxydim in herbicide tolerant corn hybrids. In: *Proceedings of the Second International Weed Control Congress,* Denmark, pp. 869-874.

Laporte, M.M., Galagon, J.A., Shapiro, J.A., Boersig, M.R., Shewmaker, C.R. and Sharkey, T.D. 1997. Sucrose phosphate synthase activity and yield analysis of tomato plants transformed with maize sucrose phosphate synthase. *Planta,* **203**: 253-259.

Larkin, P.J. and Scowcroft, W.R. 1981. Somaclonal variation – a novel source of variability from cell cultures for plant improvement. *Theor. Appl. Genet.,* **60**: 197-214.

Lazar, M.D., Chen, T.H.H., Gusta, L.V., and Kartha, K.K. 1988. Somaclonal variation for freezing tolerance in a population derived from Norstar winter wheat. *Theor. Appl. Genet.,* **75**: 480-484.

Lee, J.H., Hubel, A. and Schoffl, F. 1995. *Plant J.,* **8**: 603-612.

Lee, Y.T., Lim, M.S., Kim, H.S., Shin, H.T., Kim, C.H., Baes, S.H. and Cho, C.I. 1989. An anther derived new high quality variety with disease and insect resistance "Hwacheong byeo". *Res. Rep. of Rural Development Administration,* Rice, **31**: 27-34.

Lelivelt, C.L.C. 1993. Introduction of beet cyst nematode resistance from *Sinapis alba* L. and *Raphanuss sativus* L. into *Brassica napus* L. (oil seed rape) through sexual and somatic hybridization. Ph.D. Thesis, Univ. of Wageningen, The Netherlands.

Liang, P. and Pardee, A.B. 1992. Differential display of eukaryotic messenger RNA by means of the polymerase chain reaction. *Science,* **257**: 967-971.

Linsmaier, E.M. and Skoog, F. 1965. Organic growth factor requirements of tobacco tissue cultures. *Physiol. Plant.,* **18**: 100-127.

Litt, M. and Lutty, J.A. 1989. *Am. J. Jum. Genet.,* **44**: 397-401.

Little, P. 1999. The book of genes. *Nature,* **402**: 467-468.

Litz, R.E. and Conover, R.A. 1982. *In vitro* somatic embryogenesis and plant regeneration from *Carica papaya* L. ovular callus. *Plant Sci. Lett.* **26**: 153-158.

Livache, T., Brigitte, F., Roget, A., Morchand, J., Bidan, G., Teoule, R. and Mathus, G. 1998. *Anal. Biochem.,* **255**: 188-194.

Lloyd, A.M., Walbot, V. and Davies, R.W. 1992. *Arabidopsis* and *Nicotiana* anthocyanin production activated by maize regulators R and C1. *Science,* **258**: 1773-1775.

Lodge, J.K., Kaniewski, W.K. and Tumer, N.E. 1993. Broad spectrum virus resistance in transgenic plants expressing pokeweed antiviral proteins. *Proc. Natl. Acad. Sci. USA,* **90**: 7089-7093.

Lonnig, W.E. and Huijser, P. 1994. Gene tagging by endogenous transposons. *Plant Mol. Biol. Manual* **K1**: 1-15.

Lorenz, M., Werhe, A., and Borner, T. 1997. Cloning and sequencing of RAPD fragments amplified from mitochondrial DNA of male sterile and male fertile cytoplasm of sugar beet (*Beta vulgaris* L.). *Theor. Appl. Genet.,* **94**: 273-278.

Lorz, H. and Scowcroft, W.R. 1983. Variability among plants and their progeny regenerated from protoplasts of Su/su heterozygotes of *Nicotiana tabacum.* *Theor. Appl. Genet.,* **66**: 67-75.

Lysov, Y.P., Florentiev, V.L., Khorlin, A.A., Karpko, K.V., Shik, V.V. and Mirzabekov, A.D. 1998. *Dokl. Acad. Sci. USSR,* **303**: 1508-1511.

Maliga, P., Lazar, G., Joo, F., Nagy, A.H. and Menczel, L. 1977. Restoration of morphogenic Potential in *Nicotiana* by somatic hybridization. *Mol. Gen. Genet.,* **157**: 291-296.

Malik, M.K., Slovin, J.P., Hwang, C.H. and Zimmerman, J.L. 1999. *Plant J.,* **20**: 89-99.

Mandal, A.B., Chowdhury, B. and Sheeja, T.E. 1999. Development and characterization of salt tolerant somaclones in rice cultivar pokkali. *Ind. J. Expt. Biol.,* **38**: 74-79.

Mantell, S.H., Mathew, J.A. and McKee, R.A. 1985. *Principles of Plant Biotechnology: An Introduction to Genetic Engineering in Plants.* Blackwell Scientific Publications, Oxford.

Mariani, C., de Beuckeleer, M., Truettner, J., Leemans, J., and Goldberg, R.B. 1990, *Nature,* **347**: 737-741.

Mariani, C., de Beuckeleer, M., Truettner, J., Leemans, J., and Goldberg, R.B. 1992, *Nature,* **357**: 384-387.

Martin, G.B., Brommonschenkel, S.H., Chunwongsee, J., Frary, A., Ganal, M.W., Spivey, R., Wu, T., Earle, E.D., and Tanksley, S.D. 1993. Map based cloning of a protein kinase gene conferring disease resistance in tomato. *Science,* **262**: 1432-1436.

Mason, H.S., Lam, D.M.K. and Arntzen. C.J. 1992. *Proc. Natl. Acad. Sci. USA,* **89**: 11745-11749.

Maxam, A.M. and Gilbert, W. 1977. A new method for sequencing DNA. *Proc. Natl. Acad. Sci. USA,* **74**: 560-564.

McClintock, B. 1948. Mutable loci in maize. *Carneigie Inst. Wash. Yearb.,* **47**: 155-169.

McClintock, B. 1951. Chromosome organization and genic expression. *Cold Spring Harbor Symp. Quant. Bio..* **16**: 13-47.

McCoy, T.J. and Philips, R.L. 1982. Cytogenetic variation in tissue culture regenerated plants of *Avena sativa*: High frequency of chromosome loss. *Can. J. Genet. Cytol.,* **24**: 37-50.

McGarvey, P.B., Hammond, J., Dienelt, M.M., Hooper, D.C., Fu, Z.F., Dietzscold, B., Koprowski, H. and Michaels, F.H. 1995. *Trends Biotechnol.,* **13**: 1448-1487.

McKersie, B.D., Bowley, S.R., Harjanto, E. and Leprince, O. 1996. Water deficit tolerance and field performance of transgenic alfalfa overexpressing superoxide dismutase. *Plant Physiol.,* **111**: 1177-1181.

McKnight, S.L. and Kingsbury, R. 1982. Transcriptional control signals of a eukaryotic protein coding gene. *Science,* **217**: 316-324.

McLain, L., Porta, C., Lomonossoff, G.P., Durrani, Z. and Dimmock, N.J. 1995, *AIDS Res. Hum. Reteroviruses,* **11**: 327-334.

Mehta, S.L. and Santha, I.M. 1996. Plant biotechnology for development of non-toxin strains of *Lathyrus sativus*. In: *Summer Short Course on Enhancement of Genetic Yield Potential of Pulse Crops*. Kanpur, India, pp. 65-73.

Melchers, G. and Labib, G. 1974. Somatic hybridization of plants by fusion of protoplasts. I. Selection of light resistant hybrids of haploid light sensitive varieties of tobacco. *Mol. Gen. Genet.,* **135**: 277-294.

Melchers, G., Sacristan, M.D. and Holder, A.A. 1978. Somatic hybrid plants of potato and tomato regenerated from fused protoplasts. *Carlsberg Res. Commun.,* **43**: 203-218.

Memelink, J., Swords, K.M.M., Staehlin, L.A. and Hoge, J.H.C. 1994. Southern, Northern and Western blot analysis. *Plant Mol. Biol. Manual* **F1**: 1-21.

Meyers, R.M., Lerman, L.S. and Maniatis, T. 1985. A general method for saturation mutagenesis of cloned DNA fragments. *Science,* **229**: 242-246.

Miah, M.A.A., Pathan, M.S. and Quayam, H.A. 1996. Production of salt tolerant rice breeding line via doubled haploid. *Euphytica,* **91**: 285-288.

Michelmore, R.W., Pran, I. and Kesseli, R.V. 1991. Identification of markers linked to disease resistance genes by bulked segregants analysis: a rapid method to detect markers in specific genomic regions by using segregating populations. *Proc. Natl. Acad. Sci. USA,* **88**: 9828-9832.

Mishra, J.P. 1999. Biotechnology and intellectual property rights. *Yojana,* May Issue, pp. 15-20.

Mittler, R., Shulaev, V. and Lam, E. 1995. Coordinated activation of programmed cell death and defense mechanism in transgenic tobacco plants expressing a bacterial proton pump. *Plant Cell,* **7**: 29-42.

Mohan, M., Nair, S., Bentur, J.S., Prasada Rao, U., and Bennet, J. 1994. RFLP and RAPD mapping of rice Gm2 gene that confers resistance to biotype 1 of gall midge (*Orseolia oryzae*). *Theor. Appl. Genet.,* **87**: 782-788.

Molina, A. and Garcia, O.F. 1997. Enhanced tolerance to bacterial pathogens caused by transgenic expression of barley lipid transfer protein LTP2. *Plant J.,* **12**: 669-675.

Molvig, L., Tabe, L.M., Eggum, B.O., Moore, A.E., Craig, S., Spencer, D. and Higgins, T.J.V. 1997. Enhanced methionine levels and increased nutritive value of seeds of transgenic lupin (*Lupinus angustifolius*) expressing a sunflower seed albumin gene. *Proc. Natl. Acad. Sci. USA,* **94**: 8393-8398.

Momma, K., Hashimoto, W., Ozawa, S., Kawai, S., Katsube, T., Takaiwa, F., Kito, M., Utsumi, S. and Murata, K. 1999. Quality and safety evaluation of genetically engineered rice with soybean glycinin: analyses of the grain composition and digestibility of glycinin in transgenic rice. *Biosci. Biotechnol. Biochem.,* **63**: 314-318.

Moore, J., Engelberg, A. and Bairoch, A. 1988. *Biotechnique,* **6**: 566-572.

Morel, G. 1960. Producing virus-free cymbidium. *Am. Orchid Soc. Bull.,* **29**: 495-497.

Morel, G. 1965. Clonal propagation of orchids by meristem culture. *Cymbidium Soc. News,* **20**: 3-11.

Morel, G. and Martin, C. 1952. *C.R. Acad. Sci.,* **235**: 1324-1325.

Morel, G. and Martin, C. 1952. Cuerison de pommes de terre atteintics de maladie a virus. *C.R. Acad. Agric. Fr.,* **41**: 471-474.

Mourgues, F., Brisset, M.N. and Chevreau, E. 1998. Strategies to improve plant resistance to bacterial diseases through genetic engineering. *Trends Biotechol.,* **16**: 203-210.

Mu, S., Liu, S., Zhou, Y., Qian, N., Zhang, P., Xie, H., Zhang, F. and Yan, Z. 1977. Induction of callus from apple endosperm and differentiation of the endosperm plantlets. *Sci. Sinica,* **20**: 370-377.

Muir, W.H. 1953. Cultural conditions favouring the isolation and growth of single cells from higher plants *in vitro*. Ph.D. Thesis, University of Winconsion, USA.

Mullins, M.G. and Srinivasan, C. 1976. Somatic embryos and plantlets from an ancient clone of the grapevine (cv. Cabernet-Sauvignon) by apomixis *in vitro*. *J. Exp. Bot.,* **27**: 1022-1030.

Mullis, K.B., Faloona, F., Scharf, S.J., Saiki, R.K., Horn, G.T. and Erlich, H.A. 1986. Specific enzymatic amplification of DNA *in vitro*: the polymerase chain reaction. *Cold Spring Harbor Symp. Quant. Biol.,* **51**: 263-273.

Murashige, T. 1974. *Ann. Rev. Plant Physiol.,* **25**: 135-166.

Murashige, T. and Skoog, F. 1962. A revised medium for rapid growth and bioassays with tobacco tissue cultures. *Physiol. Plant,* **15**: 473-497.

N. Guessan, F.K., Quisenberg, S.S. and Croughan, T.P. 1994. Evaluation of rice anther culture line for tolerance to the rice weevil (Coleoptera: Curculionidae). *Environ. Entomol.,* **23**: 331-336.

Nakano, H., Tashiro, T., and Maeda, E. 1975. Plant differentiation in callus tissue induced from immature endosperm of *Oryza sativa* L. *Z. Pflanzenphysiol.,* **76**: 444-449.

Nei, M. and Li, W.H. 1979. *Proc. Natl. Acad. Sci. USA.,* **76**: 5269-5273.

Neumann, E., Schaeffer-Ridder, M., Wang, Y. and Hofschneider, P.H. 1982. Gene transfer into mouse lymphoma cells by electroporation in high electric fields. *EMBO J.,* 841-845.

Nilsson, B., Uhlen, M., Josephson, S., Gatenbeck, S., and Philipson, L. 1983. An improved positive selection plasmid vector constructed by oligonucleotide mediated mutagenesis. *Nucl. Acids Res.,* **11**: 8019-8039.

Nitsch, J.P. 1951. Growth and development of *in vitro* excised ovaries. *Am. J. Bot.,* **38**: 566-577.

Nitsch, J.P. and Nitsch, C. 1969. Haploid plants from pollen grains. *Science,* **163**: 85-87.

Nobercourt, P. 1939. Cultures en serie de tissus vegetaux sur milieu artificiel. *C.R. Seanc. Soc. Biol.,* **205**: 521-523.

Nordlee, J.A., Taylor, S.L., Townsend, J.A., Thomas, L.A. and Bush, R.K. 1996. Identification of a Brazil nut allergen in transgenic soybeans. *New Eng. J. Med.,* **334**: 688-692.

Norelli, J.L., Aldwinckle, H.S., Destephano-Beltran, L. and Jatnes, L.M. 1994. Transgenic "mailing 26" apple expressing the attacin E gene has increased resistance to *Erwinia amylovora. Euphytica,* **77**: 123-128.

Norelli, J.L., Jensen, L.A., Momol, M.T., Milis, J.Z., Ko, K-S., Aldwinckle, H.S. and Cummis, J.M. 1996. Increasing the resistance of apple rootstocks to fire blight by genetic engineering: A progress report. *Acta Hortic.,* **411**: 409.

Norrander, J., Kempe, T., and Messings, J. 1983. Construction of improved M13 vectors using oligodeoxynucleotide directed mutagenesis. *Gene,* **26**: 101-106.

Ochmann, H., Gerber, A.S. and Hartl, D.L. 1988. Genetic application of an inverse polymerase chain reaction. *Genetics,* **120**: 621-623.

Ogawa, T., Hori, T. and Ishida, I. 1996. Virus-induced cell death in plants expressing the mammalian 2'5' oligoadenylate system. *Nat. Biotech.,* **14**: 1566-1569.

Olson, M., Hood, L., Cantor, C., and Dostein, D. 1989. A common language for physical mapping of the human genome. *Science,* **254**: 1434-1435.

Oono, K. 1978. Test tube breeding of rice by tissue culture. In: *Proceedings of the Symposium on Methods in Crop Breeding.* Tropical Agriculture Res Series No. 11, Ministry of Agriculture and Forestry, Ibaraki, Japan.

Ortiz, D.F. and Stromner, J.M. 1990. The *Mu 1* maize transposable element induces tissue-specific aberrant splicing and polyadenylation in two *Adh 1* mutants. *Mol. Cell Biol.,* **10**: 2090-2095.

Paran, I. and Michelmore, R.W. 1993. Development of reliable PCR-based markers linked to downy mild resistance gene in lettuce. *Theor. Apply Genet.,* **85**: 985-993.

Paszhowski, J., Shillito, R.D., Saul, M., Mandak, V., Hohn, T. *et al.* 1984. Direct gene transfer of plants. *EMBO J.,* **3**: 2712-2722.

Pelletier, G., Primard, C., Ferault, M, Vedel, Fe., Cherit, P., Renard, M. and Delourme, R. 1998. Use of protoplast in plant breeding: Cytoplasmic aspects. In: *Progress in Protoplast Research* (Puite, K.J., Dons, J.J.M., Huising, H.J., Kool, A.C., Koorneef, M, and Krens, F.A., eds.). Kluwer Academic Publishers, London, pp. 169-176.

Pereira, A. (1999).Plant genomics in revolutionizing agricultural research. *Biotechnology and Development Monitor,* **40**: 2-7.

Perl, A.R., Perl, T., Galili, S., Aviv, D., Shalgi, E., Malkin, S. and Galun, E. 1993. Enhanced oxidative stress defense in transgenic potato expressing tomato Cu, Zn superoxide dismutases. *Theor. Appl. Genet.* **85**: 568-576.

Peschke, V.M. and Phillips, R.L. 1992. Genetic variation of somaclonal variation in plants. *Adv. Genet.,* **30**: 41-75.

Pierik, R.L.M. 1987. *In vitro culture of Higher Plants.* Martinus Nijhoff Publishers, Dordrecht.

Pilons-Smits, E.A.H., Ebskamp, M.J.M., Paul, M.J., Jenken, M.J.W., Weisbeek, P.J. and Smeekens, S.C.M. 1995. Improved performance of transgenic fructan accumulating tobacco under drought stress. *Plant Physiol.,* **107**: 125-130.

Pinto, Y.M., Kok, R.A. and Baulcombe, D.C. 1999. Resistance to rice yellow mosaic virus (RYMV) in cultivated African rice varieties containing RYMV transgenes. *Nat. Biotechnol.,* **17**: 702-707.

Pitcher, L.H., Repeti, P. and Zilinskas, B.A. 1994. Overproduction of ascorbate peroxidase protects transgenic tobacco plant against oxidative stress (Abstract No. 623). *Plant Physiol.,* **105**: S-169.

Poirier, Y., Nawrath, C. and Somerville, C. 1995. *BioTechnology,* **13**: 142-150.

Poustka, A. and Lehrach, H. 1986. Jumping libraries and linking libraries: the next generation of molecular tools in mammalian genetics. *Trends Genet.,* **2**: 174-179.

Poustka, A., Pohl, T.M., Barlow, D.P., Frischauf, A.M. and Lehrach, H. 1987. Construction and use of human chromosome jumping libraries from NotI digested DNA. *Nature,* **325**: 353-355.

Powell-Abel, P.A., Nelson, R.S., De, B., Hoffman, N., Rogers, S.G., Fraley, R.T. and Beachy, R.N. 1986. Delay of disease development in transgenic plants that express tobacco mosaic virus coat protein gene. *Science,* **232**: 738-743.

Power, J.B., Cummins, S.E. and Cocking, E.C. 1970. Fusion of isolated plant protoplasts. *Nature (London),* **225**: 1016-1018.

Primard, C., Vedel, F., Mathieu, C., Pelletier, G. and Chevre, A.M. 1988. Interspecific somatic hybridization between *Brassica napus* and *Brassica hirta* (*Sinapis alba* L.). *Theor. Appl. Genet.,* **75**: 546-552.

Prober, J.M., Trainor, G.L., Dam, R.J., Hobbs, F.W., Robertson, C.W., Zagursky, R.J., Cocuzza, A.J., Jensen, M.A. and Baumeister, K. 1987. A system for rapid DNA sequencing with fluorescent chain terminating didexoynucleotides. *Science,* **238**: 336-341.

Raghavan, V. 1976. *Experimental Embryogenesis in Vascular Plants.* Academic Press, New York.

Raina, A. and Datta, A. 1992. Molecular cloning of a gene encoding a seed specific protein with nutritionally balanced amino acid composition from *Amaranthus. Proc. Natl. Acad. Sci. USA,* **89**: 11774-11778.

Rajasekaran, K., Grula, J.W. and Anderson, D.M. 1996. Selection and characterization of mutant cotton (*Gossypium hirsutum* L.) cell lines resistant to sulfonylurea and Imidazolinone herbicides. *Plant Sci. Limerick,* **119**: 115-124.

Rambaud, G., Bellamy, A. and Dubreuca, A. 1997. Molecular analysis of the fourth progeny plants derived from a Cytoplasmic male sterile chicory hybrid. *Plant Breed.,* **116**: 481-486.

Rangan, T.S., Murashige, T., and Bitters, W.P. 1968. *In vitro* initiation of nucellar embryos in monoembryonic *Citrus. Hort. Sci.,* **3**: 226-227.

Rangaswamy, N.S. 1958. Culture of nucellar tissue of *Citrus in vitro. Experientia,* **14**: 111–112.

Rathinasabapathi, B. and King, J. 1992. Physiological response of herbicide resistant cell variants of *Datura innoxia* to branched chain amino acids. *Plant Sci. Limerick,* **81**: 191-198.

Ratzkin, B. and Carbon, J. 1977. Functional expression of cloned yeast DNA in *Escherichia coli. Proc. Natl. Acad. Sci. USA,* **74**: 487-491.

Razdan, M.K. 1983. *An Introduction to Plant Tissue Culture.* Oxford and IBH Publishing Co. Pvt Ltd., New Delhi.

Reddy, A.S. and Thomas, T.L. 1996. Expression of a cyanobacterial D6 desaturase gene results in gamma linolenic acid production in transgenic plants. *Nat. Biotechnol.,* **14**: 639-642.

Reidhaar-Olson, J.F. and Sauer, R.T. 1988. Combinatorial cassette mutagenesis as a probe of the informational

content of protein sequences. *Science,* **241**: 53-57.

Reinert, J. 1958. Morphogenese und ihre knotrolle an gewebekulturen aus carotten. *Naturwissenschaften,* **45**: 344-345.

Reinert, J. 1959. Uber die knotrolle der morphogenese und die induktion von advientive embryonen an gewebekulturen aus Karotten. *Planta,* **58**: 318-333.

Robbins, W.J. 1922a. Cultivativation of excised root tips under sterile conditions. *Bot. Gaz.,* **73**: 376-390.

Robbins, W.J. 1922b. Effect of autolyzed yeast and peptone on growth of excised corn root tips in the dark. *Bot. Gaz.,* **74**: 57-79.

Rokka, V.M., Tauriamen, A., Pietila, L. and Pehu, E. 1998. Interspecific somatic hybrids between wild potato *Solanum acaule* and anther derived dihaploid potato (*Solanum tuberosum* L.). *Plant Cell Rep.,* **18**: 82-88.

Saier, M.H. 1998. Genome sequencing and informatics: New tools for biochemical discoveries. *Plant Physiol.,* **117**: 1129-1133.

San Noeum, L.H. 1976. Haploides d' *Hordeum vulgare* L. par culture in vitro d' ovaires non fecondes. *Ann. Amelior. Plantes,* **26**: 751-754.

Sanford, J.C., Devit, M.J., Russell, J.A., Smith, F.D., Harpending, P.R., Roy, M.K. and Johnston, S.A. 1991. An improved helium driven biolistic device. *Technique,* **3**: 3-16.

Sanger, F., Nicklen, S. and Coulson, A.R. 1977. DNA sequencing with chain termination inhibitors. *Proc. Natl. Acad. Sci. USA,* **74**: 5463-5467.

SanMiguel, P. and Bennetzen, J.L. 1998. Evidence that a recent increase in maize genome size was caused by the massive amplification of intergene retrotransposons. *Ann. Bot.,* **82**: 37-44.

Sano, T., Nagayama, A., Ogawa, T., Ishida, I. and Okada, Y. 1997. Transgenic potato expressing a double stranded RNA-specific ribonuclease is resistant to potato spindle tuber viroid. *Nat. Biotechnol.,* **15**: 1290-1294.

Sarkar, G. and Sommer, S.S. 1990. The megaprimer method of site directed mutagenesis. *Biotechniques,* **8**: 404-407.

Sathish, P., Gamborg, O.L., and Nabors, M.W. 1997. *Theor. Appl. Genet.,* **95**: 1203-1209.

Savin, K.W., Baudinette, S.C., Graham, M.W., Michael, M.Z., Nugent, G.D., Lu, C.Y., Chandler, S.F. and Cornish, E.C. 1996. Antisense ACC oxidase delays carnation petal senescence. *Hort. Sci.,* **30**: 970-972.

Sayanova, O., Smith, M.A., Lapinskas, P., Stobart, A.K., Dobson, G., Christie, W.W., Shewry, P.R. and Napier, J. 1997. Expression of a borage desaturase cDNA containing an N-terminal cytochrome b5 domain results in the accumulation of high levels of D6 desaturated fatty acids in transgenic tobacco. *Proc. Natl. Acad. Sci. USA,* **94**: 4211-4216.

Schake, S.A., Wong-Vega, L. and Allen, R.D. 1995.

Analysis of tobacco plants that overexpress pea chloroplastic manganese superoxide dismutase (abstract No. 260). *Plant Physiol.*, **108**: S-62.

Scharf, S.J., Horn, G.T. and Erlich, H.A. 1986. Direct cloning and sequence analysis of enzymatically amplified genomic sequences. *Science*, **233**: 1076-1078.

Schmulling, T., Schell, J. and Spena, A. 1988. Single gene from *Agrobacterium rhizogenes* influence plant development. *EMBO J.*, **7**: 2621-2629.

Schuler, G.D. *et al.* 1996. *Science*, **274**: 540-546.

Schwartz, D.C. and Cantor, C.R. 1984. Separation of yeast chromosome sized DNA by pulsed field gradient gel electrophoresis. *Cell*, **37**: 67-75.

Sehgal, C.B. 1974. Experimental studies on maize endosperm. *Beitr. Biol. Pflanzen.*, **46**: 233-238.

Sen Gupta, A., Heinen, J.L., Holady, A.S., Burke, J.J. and Allen, R.D. 1993. Increased resistance to oxidative stress in transgenic plants that overexpress chloroplastic Cu/Zn superoxide dismutase. *Proc. Natl. Acad. Sci. USA*, **90**: 1629-1633.

Sharma, A.K., Mohanty, A., Singh, Y. and Tyagi. A.K. 1999. Transgenic plants for the production of edible vaccines and antibodies for immunotherapy. *Curr. Sci.*, **77**: 524-529.

Sharp, W.R., Sondahl, M.R., Caldas, L.S., and Maraffa, S.B. 1980. In: *Horticultural Reviews*, Vol. 2. (Janick, J., ed.). AVI Publishing Co, Westport, Conn., USA, p. 268.

Shewmaker, C., Sheehy, J., Daley, M., Colburn, S. and Ke, D.Y. 1999. Seed-specific expresssion of phytoene synthase: increase in carotenoids and other metabolic effects. *Plant J.*, **20**: 401-412.

Siebert, P. and Larrick, J. 1992. Competitive PCR. *Nature*, **359**: 557-558.

Sigarava, M.A. and Earle, E.D. 1997. Transfer of a cold tolerant ogura male sterile cytoplasm into cabbage. *Theor. Appl. Genet.*, **99**: 213-220.

Skoog, F. and Miller, C.O. 1957. Chemical regulation of growth and organ formation in plant tissues cultivated *in vitro*. In: *Biological Action of Growth Substances. Symp. Soc. Exp. Biol.*, **11**: 118-131.

Slightom, J.L. 1991. Custom polymerase chain reaction engineering of a plant expression vector. *Gene*, **100**: 251-255.

Smirnoff, N. and Bryant, J.A. 1999. DREB takes the stress out of growing up. *Nat. Biotechnol.*, **17**: 229-230.

Smith, H.O. and Nathans, D. 1973. A suggested nomenclature for bacterial host modification and restriction systems and their enzymes. *J. Mol. Biol.*, **81**: 419-423.

Sommer, H., Beltran, J.P., Huijser, P., Pape, H., Lonnig, W.E., Saedler, H., and Schwartz-Sommer, Z.S. 1990. Deficiens, a homoeotic gene involved in the control of flower morphogenesis in *Antirrhinum majus*: The protein shows homology to transcription factors. *EMBO J.*, **9**: 605-613.

Song, W.Y., Wang, G.L., Chen, L.L., Kim, H.S. and Pi, L.Y. 1995. A receptor kinase like protein encoded by the rice disease resistane gene, *Xa21*. *Science*, **270**: 1804-1806.

Southern, E.M. 1975. Detection of specific sequences among DNA fragments separated by gel electrophoresis. *J. Mol. Biol.*, **98**: 503-517.

Southern, E.M. 1998. International patent application PCT GB 89/00460.

Sowa, A.W., Duff, S.M.G., Guy, P.A. and Hill R.D. 1998. Altering hemoglobin levels changes energy status in maize cells under hypoxia. *Proc. Natl. Acad. Sci. USA*, **95**: 10317-10321.

Sree Ramulu, K., Dijkhuis, P., Hanisch, C.H. and Groot de, B. 1985. Patterns of DNA and chromosome variation during *in vitro* growth in various genotypes of potato. *Plant Sci.*, **41**: 69-78.

Stanssens, P., Opsomer, C., Mc Keown, Y.M., Kramer, W., Zabeau, M., and Fritz, H.J. (1989). Efficient oligonucleotide directed construction of mutations in expression vectors by the gapped duplex DNA using alternating selectable markers. *Nucl. Acids Res.*, **17**: 4441-4454.

Stefanov, K.L., Djilianov, D.L., Vassileva, Z.Y., Batchvarova, R.B., Atanassov, A.I., Kuleva, L.V. and Popov, S.S. 1996. Changes in lipid composition after *in vitro* selection for glyphosate tolerance in tobacco. *Pesticide Sci.*, **46**: 369-374.

Stefenson, B.J., Jin, Y., Lossnagel, B.G., Rasmussen, J.B.K. and Kao, K. 1995. Genetics of multiple disease resistance in a doubled haploid population of barley. *Plant Breed.*, **114**: 50-54.

Steinecke, P., Herget, T. and Schreier, P.H. 1992. Expression of a chimeric ribozyme gene results in endonucleolytic cleavage of target mRNA and a concomitant reduction of gene expression *in vivo*. *EMBO J.*, **11**: 1525-1530.

Stevenson, T.W. and Cornish, E.C. 1993. Cloning and expression of cytochrome p450 genes controlling flower color. *Nature*, **366**: 276-279.

Steward, F.C., Caplin, S.M. and Miller, F.K. 1952. Investigations on growth and metabolism of plant cell. I. New techniques for the investigation, metabolism, nutrition and growth in undifferentiated cells. *Ann. Bot.*, **16**: 57-77.

Steward, F.C., Mapes, M.O., and Mears, K. 1958. Growth and organized development of cultured cells. II Organization in cultures grown from freely suspended cells. *Am J. Bot.*, **45**: 705-708.

Stirpe, F., Barbieri, L., Battelli,M.G., Soria, M. and Lappi, D.A. 1992. Ribosome inactivating proteins from plants: present status and future prospects. *BioTechnology*, **10**: 405-412.

Street, H.S. (ed.)1973. *Plant Tissue and Cell culture.* Blackwell Scientific Publications, Oxford.

Strittmatter, G., Janssens, J., Opsomer, C. and Botterman, J. 1995. Inhibition of fungal disease development in plants by engineering controlled cell death. *BioTechnol.,* **13**: 1085-1089.

Struhl, K., Cameron, J.R., and Davis, R.W. 1976. Functional genetic expression of eukaryotic DNA in *Escherichia coli. Proc. Natl. Acad. Sci. USA,* **73**: 1471-1475.

Sun, T., Goodman, H.M. and Asubell, F.H. 1992. *Plant Cell,* **4**: 119-128.

Suzuki, K., Zue, H., Tanaka, Y., Fukui, Y., Mizutani, M. and Kusumi, T. 1997. Molecular breeding of flower color of *Torenia fournieri. Plant Cell Physiol.,* **38**: 538-540.

Swaminathan, M.S. 1994. Draft Plant Varieties Recognition and Protection Act: Rationale and Structure. In: *GATT Accord: India's Strategic Response* (Remachandria, V., ed.)., Commonwealth Publishers, New Delhi, pp. 189-243.

Tabata, M., Migukami, H., Hirasoka, N. and Konoshima, M. 1974. Pigment formation in callus cultures of *Lithospermum erythrorhizon. Phytochemistry,* **13**: 927-932.

Tacke, E., Salamani, F. and Rohde, W. 1996. Genetic engineering of potato for broad spectrum protection against virus infection. *Nat. Biotechnol.,* **14**: 1597-1601.

Tacket, C.O., Mason, H.S., Losonsky, G., Clements, J.D., Levine, M.M. and Arntzen, C.J. 1998. *Nature Med.,* **4**: 607-609.

Takabe, I., Labib, G. and Melchers, G. 1971. Regeneration of whole plants from isolated mesophyll protoplasts of tobacco. *Naturwissen.,* **58**: 318-320.

Takabe, I., Otsuki, Y. and Aoki, S. 1968. Isolation of tobacco mesophyll cells in intact and active state. *Plant Cell Physiol. (Tokyo),* **9**: 115-124.

Tanaka, Y., Tsuda, S. and Takaaki, K. 1998. Application of recombinant DNA to floriculture. In: *Applied Plant Biotechnology.* Chopra, V.L., Malik, V.S. and Bhatt, S.R. (Eds.), Oxford & IBH, New Delhi, pp. 181-235.

Tanaka, Y., Yonekura, K., Fukuchi-Mizutani, M., Fukui, Y., Fujiwara, H., Ashikari, T. and Kusumi, T. 1996. Molecular and biochemical characterization of three anthocyanin synthetic enzymes from *Gentiana triflora. Plant Cell Physiol.,* **37**: 711-716.

Tanksley, S.D., Ganal, M.W. and Martin, G.W. 1995. Chromosomal landing: a paradigm for map based gene cloning in plants with large genomes. *Trends Genet.,* **11**: 63-68.

Tarczynscki, M.C., Jensen, R.G. and Bohnert, M.J. 1993. Stress protection of transgenic tobacco by production of the osmolyte mannitol. *Science,* **239**: 508-510.

Tareghyan, M.R., Collin, H.A., Putwain, P.D. and Mortimer, A.M. 1995. Characterization of somaclones of soy-bean resistant to imazethapyr. In: *Brighton Crop Protection Conference: Weeds. Proceedings of the International Conference,* Brighton, UK, pp. 375-380.

Taylor, J.W. and Eckstein, F. 1985. The rapid generation of oligonucleotide directed mutations at high frequency using phosphorothioate modified DNA. *Nucl. Acids Res.,* **13**: 8765-8785.

Temin, H.M. and Mizutani, S. 1970. RNA dependent DNA polymerase in virions of Rous Sarcoma virus. *Nature,* **226**: 1211-1213.

Thorpe, T.A. 1981. *Plant Tissue Culture: Methods and Applications in Agriculture.* Academic Press Inc., New York.

Tien, P. and Gussui, N. 1991. Satellite RNA for the biocontrol of plant diseases. *Adv. Virus Res.,* **39**: 321.

Tien, P. Zhang, X., Qiu, B. and Wu, G. 1987. Satelliete RNA for the control of plant diseases caused by cucumber mosaic virus. *Ann. Appl. Biol.,* **111**: 143.

Toubart, P., Desiderio, A., Salvi, G., Cervone, F., Daroda, L., De Lorenzo, G., Bergmann, C., Darvill, A.G. and Albersheim, P. 1992. Cloning and characterization of the gene encoding the endo polygalacturonase inhibiting protein (PGIP) of *Phaseolus vulgaris* L. *Plant J.,* **2**: 367-373.

Triglia, T., Peterson, M.G. and Kemp, K.J. 1988. A procedure for *in vitro* amplification of DNA segments that lie outside the boundaries of known sequences. *Nucl. Acids Res.,* **16**: 8186.

Truve, E., Kelve, M., Aaspollu, A., Kuuksalu, A., Seppannen, P. and Sarma, M. 1994. Principles and background for the construction of transgenic plants displaying multiple virus resistance. *Arch. Virol. (Suppl.),* **9**: 41-50.

Turpen, T.H., Reinl, S.J., Charroenvit, Y., Hoffman, S.L., Fallarme, V. and Grill, L.K. 1995. *BioTechnology,* **13**: 53-57.

Uchimiya, H., Ohgawara, T., Kato, H., Akiyama, T. and Harada, H. 1983. Detection of two different nuclear genomes in parasexual hybrid by ribosomal RNA gene analysis. *Theor. Appl. Genet.,* **64**: 117.

Vacin, C. and Went, O. 1949. Culture solution for orchid seedings *Bot. Gaz.,* **110**: 605-610.

Vaeck, N., Reynaerts, A., Hofte, H., Jansens, S., Beukeleer, M.D., Dean, C., Zabeau, M., Montagu, M.C. and Leemans, J. 1987. Transgenic plants protected from insect attack. *Nature,* **328**: 33-37.

Vardi, E., Sela, I., Edelbaum, O., Livneh, O., Kuznetsova, L. and Stram, Y. 1993. Plants transformed with a cistron of a potato virus Y protease (Nia) are resistant to virus infection. *Proc. Natl. Acad. Sci. USA,* **90**: 7513-7517.

Vasil, I.K. 1984. *Cell Culture and Somatic Cell Genetics of Plants: Laboratory Procedures and Their Applications.* Academic Press, Inc., New York.

Velculescu, V.E., Zhang, L., Vogelstein, B. and Kinzler, K.W. 1995. Serial analysis of gene expression. *Science,* **270**: 484-487.

Voelker, T.A., Worrell, A.C., Anderson, L., Bleibaum, J., Fane, C., Hawkins, D.J., Radke, S.E. and Davis, H.M. 1992. Fatty acid biosynthesis redirected to medium chains in transgenic oilseed plants. *Science,* **257**: 872-874.

Vogel, J.M. and Morgante, M. 1995. *Plant genome III.,* Abstract, pp. 84.

Vollrath, D. and Davis, R.W. 1987. Resolution of DNA molecules greater than 5 Mb by contour-clamped homogeneous electric fields. *Nucl. Acids Res.,* **15**: 7865.

Vos, P., Hogers, R., Bleeker, M., Reijans, M., van de Lee T., Hornes, M., Fritjers, A., Pot, J., Peleman, J., Kuiper, M., and Zabeau, M. 1995. AFLP: a new technique for DNA fingerprinting. *Nucl. Acids Res.,* **23**: 4407-4414.

Walbot, V. 1992. Strategies for mutageneis and gene cloning using transposon tagging and T-DNA insertional mutagenesis. *Ann. Rev. Plant Physiol. Plant Mol. Biol.,* **43**: 49-82.

Wallin, A., Glimelius, K. and Eriksson, T. 1974. The induction of aggregation and fusion of *Daucus carota* protoplasts by polyethylene glycol. *Z. Pflanzenphysiol.,* **74**: 64-80.

Wang, C.X., Wang, L.Q., Bai, Z., Wang, F., Wang, F., Zheng, L.P., Li, A.J. and Wang, C. 1997. Tissue culture selection of somaclonal variation line R111 of sorghum. *Acta Agriculturae Borealii-Sinica,* **12**: 49-53.

Wang, P., Zoubenko, O. and Tumer, N.E. 1998. Reduced toxicity and broad spectrum resistance to viral and fungal infection in transgenic plants expressing pokeweed antiviral protein II. *Plant Mol. Biol.,* **38**: 957-964.

Wang, T. and Chang, C. 1978. Triploid Citrus plantlets from endosperm culture. In: *Proc. Symp. Plant Tissue Culture,* Sci. Press, Peking, pp. 463-468.

Waston, J.D. and Crick. F.H.C. 1953. The structure of DNA. *Cold Spring Harbor Symp. Quant. Biol.,* **18**: 123-131.

Webb, R.P. and Allen, R.D. 1995. Overexpression of a pea cytosolic ascorbate peroxidase in *Nicotiana tabacum* confers protection against the effects of paraquat (abstract No. 272). *Plant Physiol.,* **108**: S-64.

Wegener, D., Steinecke, P., Herget, T., Petereit, I., Philipp, C. and Schreier, P.H. 1994. Expression of a reporter gene is reduced by ribozyme in transgenic plants. *Mol. Gen. Genet.,* **245**: 465-470.

Wells, J.A., Vasser, M. and Powers, D.B. 1985. Cassette mutagenesis: an efficient method for generation of multiple mutations at defined sites. *Gene,* **34**: 315-323.

Welsh, J. and McClelland, M. 1990 Fingerprinting genomes using PCR with arbitrary primers. *Nucl. Acids Res.,* **8**: 7213-7218.

Wenzel, G. and Uhrig, H. 1981. Breeding methods for nematode and virus resistance in potato via anther culture. *Theor. Appl. Genet.,* **138**: 293-297.

White, P.R. 1934. Potentially unlimited growth of excised tomato root tips in a liquid medium. *Plant Physiol.,* **9**: 585-600.

White, P.R. 1939. Potentially unlimited growth of excised plant callus in an artificial nutrient. *Am. J. Bot.,* **26**: 59-64.

White, P.R. 1943. *A Handbook of Plant Tissue Culture.* Jacquess Cattell Press, Tempe, Arizona.

White, P.R. 1954. *The Cultivation of Animal and Plant Cells.* Ronald Press Co., New York.

White, P.R. 1963. *The Cultivation of Animal and Plant Cells.* Ronald Press Co., New York, 239. pp.

Williams, J.G.K., Kubelik, A.R., Livak, K.J., Rafalski, J.A. and Tingery, S.V. 1990. DNA polymorphisms amplified by arbitrary primers are useful as genetic markers. *Nucl. Acids Res.,* **18**: 6531-6535.

Williams, M.N.V., Pande, N., Nair, S., Mohan, M. and Bennet, J. 1991. Restriction fragment length polymorphism analysis of polymerase chain reaction products amplified from mapped loci of rice (*Oryza sativa* L.) genomic DNA. *Theor. Appl. Genet.,* **82**: 489-498.

Witherspoon, W.D., Wornsman, E.A., Gooding, G.V. and Rufby, R.C. 1991. Characterization of gametoclonal variant controlling virus resistance in tobacco. *Theor. Apply Genet.,* **81**: 1-5.

Wright, T.R., Bascomb, N.F., Sturner, S.F., and Penner, D. 1998. Biochemical mechanism and molecular basis for ALS-inhibiting herbicide resistance in sugar beet (*Beta vulgaris*) somatic cell selections. *Weed Sci.,* **46**: 13-23.

Wu, K.S., Jones, R., Dannenberg, L., and Scolnik, P.A. 1994. *Nucleic Acids Res.,* **22**: 3257-3258.

Wu, G. Shott, B.J., Lawrence, E.B., Levine, E.B., Fitzsimmons, K.C. and Shah, D.M. 1995. Disease resistance conferred by expression of a gene encoding H_2O_2 generating glucose oxidase in transgenic potato plants. *Plant Cell,* **7**: 1357-1368.

Xu, D., Duan, X., Wang, B., Hong, B., Ho, T.D., and Wu, R. 1996. Expression of a late embryogenesis abundant protein gene HVA1 from barley confers tolerance to water deficit and salt stress in transgenic rice. *Plant Physiol.,* **110**: 249-257.

Yamada, Y. and Fujita, Y. 1983. Production of useful compounds in culture. In: *Handbook of Pland Tissue Culture.* Vol. 1 (Evans, D.A., Sharp, W.R., Ammirato, P.V. and Yamads, Y., eds.). Macmillan, New York, pp. 717-728.

Yanisch-Perron, C., Vieira, J., and Messing, J. 1985. Improved M13 phage cloning vectors and host strains: nucleotide sequences of the M13 mp18 and pUC19 vectors. *Gene,* **33**: 103-119.

Ye, X., Al-Babili, S., Kloeti, A., Zhang, J., Lucca, P., Beyer, P. and Potrykus, I. 2000. Engineering the Provitamin A (Beta carotene) biosynthetic pathway into (carotenoid free) rice endosperm. *Science* **287**: 303-305.

Yershov, G., Barsky, V., Belgovsky, A., Kirillov, E., Kreindlin, E., Ivanov, I., Parinov, S., Guschin, D., Drobishev, A., Dubiley, S. and Mirzabekov, A. 1996. DNA analysis and diagnostics on oligonucleotide microchips. *Proc. Natl. Acad. Sci. USA*, **93**: 4913-4918.

Yoshimura, S., Yoshimura, A., Iwata, N., McCouch, S., Abenes, M.L., Baraidon, M.R., Mew, T. and Nelson, R.J. 1995. Tagging and combining bacterial blight resistance genes in rice using RAPD and RFLP markers. *Mol. Breed.*, **1**: 375-387.

Young, N.D., Kumar, L., Menancio-Hautea, D., Danesh, D., Talekar, N.S., Shanmugasundaram, S. and Kim, D.H. 1992. RFLP mapping of major bruchid resistance gene in mungbean (*Vigna radiata* L. Wilczek). *Theor. Appl. Genet.*, **84**: 839-844.

Yu, K. and Pauls, K.P. 1993. Identification of a RAPD marker associated with somatic embryogenesis in alfalfa. *Plant Mol. Biol.*, **22**: 269-277.

Yuan, R. 1981. Structural and mechanism of multifunctional restriction endonucleases. *Ann. Rev. Biochem.*, **50**: 285-315.

Yusilbov, V., Modelska, A., Steplewski, K., Agadjanyan, M., Weiner, D., Hooper, D.C., Koprowski, H. and Yusibov, V. 1997. *Proc. Natl. Acad. Sci, USA*. **95**: 2481-2485.

Zaenen, I., Van Larebeke, N., Teuchy, H., van Montagu. M., and Schell, J. 1974. Supercoiled circular DNA in crown gall inducing *Agrobacterium* strains. *J. Mol. Biol.*, **86**: 107-127.

Zao, R.T., Gao, S.G., Qiao, Y.K., Zhu, H.M. and Bi, Y.J. 1994. A new approach of screening salt tolerance variants by anther culture to cultivate salt tolerant wheat variety. *Acta Agriculturae Boreali Sinica*, **9**: 34-38.

Zapata, F.J., Alejar, M.S., Torrizo, L.B., Novero, A.U., Singh, V.P. and Senadhira, D. 1991. Field performance of anther culture derived lines from F1 crosses of *indica* rice under saline and non-saline land. *Theor. Appl. Genet.*, **83**: 6-11.

Zelcer, A., Aviv, D. and Galun, E. 1978. Interspecific transfer of cytoplasmic male sterility by fusion between protoplasts of normal *Nicotiana sylvestris* and X-ray irradiated protoplasts of male sterile *N. tabacum*. *Z. Pflanzenphysiol.*, **90**: 397-407.

Zemetra, R.S., Schotzko, D.J., Smith, C.M. and Lauver, M. 1993. *In vitro* selection for Russian wheat aphid (*Diuraphis noxia*) resistance in wheat (*Triticum aestivum*). *Plant Cell Rep.*, **12**: 312-315.

Zhang, G., Bharaj, T.S., Lu, Y., Virmani, S.S. and Huang, H. 1997. Mapping of the RF-3 nuclear fertility restoring gene for WA cytoplasmic male sterility in rice using RAPD and RFLP markers. *Theor. Appl. Genet.*, **94**: 118-127.

Zhang, G., Angeles, E.R., Abenes, M.L.P., Khush, G.S. and Huang, N. 1994. Molecular mapping of a bacterial blight resistance gene on chromosome 8 in rice. *Rice Genet. Newsl.*, **11**: 142-144.

Zhang, Q., Shen, B.Z., Dai, K.K., Mei, M.H., Saghai Maroof, M.A. and Li, Z.B. 1994. Using bulked extremes and recessive class to map genes for photoperiod sensitive genic male sterility in rice. *Proc. Natl. Acad. Sci. USA.*, **91**: 8675-8679.

Zhang, L., Ku, J. and Birch, R.G. 1999. Engineered detoxification confers resistance against a pathogenic bacterium. *Nat. Biotechnol.*, **17**: 1021-1024.

Zietkiewicz, E., Rafalski, A. and Labuda, D. 1994. *Genomics.*, **20**: 176-183.

Glossary

Abiotic stress	The effect of non-living factors which can harm living organisms. These non-living factors include drought, extreme temperature, salt, minerals, etc.
Acclimatization	The adaptation of a living organism (plant, animal or microorganism) to a changed environment that subjects it to physiological stress.
Adaptor	A synthetic single-stranded oligonucleotide that, after self-hybridization, produces a molecule with cohesive ends and an internal restriction endonuclease site. When the adaptor is inserted into a cloning vector by means of cohesive ends, the internal sequence provides a new restriction endonuclease site.
A-DNA	A right-handed DNA double helix that has 11 base pairs per turn. DNA exists in this form when partially dehydrated.
Adventitious	Adjective used to describe organs or embryos developing from positions on the plant from which they would not normally be derived, e.g. shoots from leaves, roots, and embryos from any cell other than a zygote.
Agarose gel electrophoresis	A process in which a matrix composed of a highly purified form of agar is used to separate larger DNA and RNA molecules. *See* electrophoresis.
Allele	(Gr. *Allelon*, of one another, mutually each other); allelomorph (adj: allelic, allelomorphic). One of a pair, or series, of variant forms of a gene that occur at a given locus in a chromosome.
Alu sequences	A family of 300 bp sequences occurring nearly a million times in the human genome.
Amino acid	An organic compound containing an amino ($-NH_2$) and an acid carboxyl group ($-COOH$).
Amphidiploid	A species derived by doubling of chromosomes in the F_1 hybrid of two species.
Amplification	1. Replication of a gene library in bulk. 2. Duplication of gene(s) within a chromosomal segment. 3. Creation of many copies of a segment of FNA by polymerase chain reaction.
Amplified fragment length polymorphism (AFLP)	A type of DNA molecular marker generated by digestion of genomic DNA with two restriction enzymes to create many DNA fragments, which are amplified by PCR.
Analogous	Features of organisms or molecules that are superficially or functionally similar but have evolved in a different way or contain different compounds.
Androgenesis	Development of plants from male gametophytes.
Aneuploidy	The loss or gain of chromosomes from the normal euploid number by various processes.

Annealing	The process of heating (denaturing step) and slowly cooling (renaturing step) double stranded DNA to allow the formation of hybrid DNA or complementary strands of DNA or of DNA and RNA.
Antiauxin	A chemical that interferes with auxin response. Antiauxin may or may not involve prevention of auxin transport or movement in plants. Some antiauxins are said to promote morphogenesis *in vitro*, such as 2,3,5-tri-iodobenzoic acid (TIBA) or 2,4,5-trichlorophenoxyacetate (2,4,5-T), which stimulate the growth of some cultures.
Anticoding strand	The strand of DNA double helix that is actually transcribed. Also known as antisense or template strand.
Anticodon	Triplet of bases carried by the tRNA molecule which are complementary to the codon in the mRNA.
Antigen	A molecule that is capable of stimulating the production of neutralizing antibody proteins when injected into an organism.
Antioxidants	A group of chemicals which prevent oxidation. For example, ascorbic acid, citric acid. They retard senescence and browning of tissue.
Antiparallel strands	The two strands of the double helix are organized in opposite orientation, so that the 5' end of one strand is aligned with the 3' end of the other strand.
Apical dominance	The phenomenon of suppression of growth of an axillary bud in the presence of the terminal bud on the branch.
ARS (Autonomous replicating sequence)	Any eukaryotic DNA sequence that initiates and supports chromosomal replication. They have been isolated in yeast cells.
Aseptic	Free from all microorganisms.
Asexual embryogenesis	The sequence of events whereby embryos develop from somatic cells, also known as somatic embryogenesis.
Autoradiography	A technique for the detection of radioactively labeled molecules by overlaying the specimen with photographic film. When the film is developed an image is produced which corresponds to the location of the radioactivity.
Autotrophic	Characteristic of plants that are capable of manufacturing their own food, as in photosynthesis.
Auxins	A class of plant growth regulators that stimulates cell division, cell elongation, apical dominance and root initiation. 2,4-D, IAA, IBA, NAA are some of the auxins commonly used in plant cell and tissue culture.
Auxotroph	A mutant organism that will not grow on a minimal medium; it requires the addition of some growth regulators.
Avirulent	Lacking virulence; a microorganism lacking the properties that normally cause disease.
Axenic culture	Free of external contaminants and internal symbionts, which is generally not possible with surface sterilization alone and incorrectly used to indicate aseptic culture.
Axillary	An adjective describing the relative position of a bud, i.e. axillary bud, in the axils of leaves.
Axillary bud proliferation	A technique of micropropagation of plants in culture, which is achieved primarily through hormonal inhibition of apical dominance and stimulation of lateral branching.
BAC	Bacterial artificial chromosome; a cloning vector system for isolating

	genomic DNA based on the F-factor plasmid, the bacterial sex or fertility plasmid, which has a low copy number because of the strict control of replication. They accept an insert size of 200 to 500 kb.
Bacteriophage	A virus that infects and replicates in bacteria. Also called simply phage.
Base pair (bp)	It is a partnership of A with T or C with G in a DNA double helix; other pairs can be formed in RNA under certain circumstances.
Batch culture	A cell suspension culture in which cells are grown in a finite volume of nutrient medium and follow a sigmoid pattern of growth.
B-DNA	The normal form of DNA found in biological systems. It exists as a right-handed helix.
Binary vector system	A two-plasmid system in *Agrobacterium tumefaciens* for gene transfer in plant cells. One plasmid contains the virulence gene (responsible for transfer of the T-DNA), and another plasmid contains the T-DNA borders, the selectable marker and the DNA to be transferred.
Biodiversity	The variability among living organisms from all sources including, *inter alia*, terrestrial, marine and other aquatic ecosystems and the ecological complexes of which they are part.
Bioinformatics	It is a science associated to biomolecular database-like activities involving persistent sets of data that are maintained in a consistent state over essentially indefinite periods of time.
Biotic stress	Stress resulting from living organisms, which can harm plants, such as viruses, fungi, bacteria, parasitic weeds and harmful insects.
Biotransformation	The conversion of a small part of molecule in the structure and composition by means of biological systems.
Blunt end	The end of a DNA duplex molecule in which neither strand extends beyond the other, also known as flush end.
Browning	Discoloration due to phenolic oxidation of freshly cut surfaces of explant tissue. In later stages of culture, such discoloration may indicate a nutritional or pathogenic problem, generally leading to necrosis.
C value	It is the total amount of DNA in a haploid genome.
CAAT box	It is a part of conserved sequence located upstream of the startpoints of eukaryotic transcription units; it is recognized by a large group of transcription factors. It has the consensus sequence GGCCAATCT; it occurs around 75 bases prior to the transcription initiation site.
Callus	1. A tissue consisting of dedifferentiated cells generally produced as a result of wounding or of culturing tissues in the presence of an auxin in particular. 2. Actively dividing non-organized masses of undifferentiated and differentiated cells often developing from injury (wounding) or in tissue culture in the presence of growth regulators.
Cap	The structure found on the 5′-end of eukaryotic mRNA which consist of an inverted methylated guanosine residue.
Capsid	External protein shell or coat of a virus particle.
cDNA	Complementary DNA, a fragment of DNA which has been produced from an RNA sequence by reverse transcription. Messenger RNA is commonly used to synthesize cDNA.
cDNA clone	A double stranded DNA molecule that is carried in a vector and was

	synthesized *in vitro* from an mRNA sequence by using reverse transcriptase and DNA polymerase.
cDNA cloning	A method of cloning the coding sequence of a gene, starting with its mRNA transcript. It is normally used to clone a DNA copy of a eukaryotic mRNA.
cDNA library	A collection of cDNA clones that were generated *in vitro* from the mRNA sequences isolated from an organism or a specific tissue or cell type or population of an organism.
Caulogenesis	Stem organogenesis; induction of shoot development from callus.
Cell culture	The growing of cells *in vitro* derived from multicellular organisms.
Cell line	Cells (originating from a primary culture) successfully subcultured for the first, second, etc. time. A cell line arises from a primary culture.
Cell suspension	Cells in culture in moving or shaking liquid medium, often used to describe suspension cultures of single cells and cell aggregates.
centiMorgan	One percent recombination between two loci (cM).
Central dogma	The basic concept that in nature genetic information generally can flow only from DNA to RNA to protein. It is now known that information contained in RNA molecules of certain viruses (called retroviruses) can also flow back to DNA.
Chelate	Complex organic molecule that can combine with cations and does not ionize. Chelates can supply micronutrients to plants at slow, steady rates. Usually used to supply iron to plant cells.
Chemiluminescence	The emission of light from a chemical reaction.
Chemostat	An open continuous culture in which cell growth rate and cell density are held constant by a fixed rate of input of a growth limiting nutrient.
Chimeric DNA	A recombinant DNA molecule containing unrelated genes.
Chimeric gene	A semi-synthetic gene, consisting of the coding sequence from one organism, fused to promoter and other sequences derived from a different gene. Most genes used in transformation are chimeric.
Chromosome jumping	A technique that allows two segments of duplex DNA that are separated by thousands of bp (about 200 kb) to be cloned together.
Chromosome walking	Isolation of sequential pieces of genomic DNA in a set of clones to form a contiguous overlapping set (a contig). This is usually accomplished by isolating the end fragment of one clone, which is used to rescreen the library for the next overlapping clone i.e. sequential isolation of overlapping molecular clones so as to span large chromosomal intervals. This process (walking) is repeated along the chromosome until the gene of interest is reached.
Circularization	A DNA fragment generated by digestion with a single restriction endonuclease will have complementary 5' and 3' extensions (sticky ends). If these ends are annealed and ligated, the DNA fragment will have been converted to a covalently closed circle or circularized.
Cistron	It is the genetic unit defined by the *cis/trans* test. A DNA fragment or portion that specifies or codes for a particular polypeptide.
Clone	A group of cells, tissues or plants, which are in principle genetically identical. A clone is not necessarily homogeneous.
Cloning vector	A small, self-replicating DNA molecule – usually a plasmid or phage into

which foreign DNA is inserted in the process of cloning genes or other DNA sequences of interest. It can carry inserted DNA and be perpetuated in a host cell. Also called a cloning vehicle, vector, or vehicle.

Coding sequence	That portion of a gene which directly specifies the amino acid sequence of its protein product. Non-coding sequences of genes include control regions, such as promoters, operators and terminators as well as intron sequences of certain eukaryotic genes.
Coding strand	The strand of duplex DNA which contains the same base sequence (after substituting U for T) found in the mRNA molecule resulting from transcription of that segment of DNA. Also known as sense strand. The mRNA molecule is transcribed from the other strand known as template or antisense strand.
Codon	A group of three nucleotides coding for an amino acid.
Cohesive ends	DNA with single stranded ends, which are complementary to each other, enabling the different molecules to join each other. Also known as protruding ends, sticky ends, overhang. Bacteriophage lambda has cohesive ends (the 'cos' site) which allows the formation of concatamers.
Co-integrate vector	A two plasmid system for transferring cloned genes to plant cells. The cloning vector has a T-DNA segment that contained cloned genes. After introduction into *A. tumefaciens*, the cloning vector DNA undergoes homologous recombination with a resident disarmed Ti plasmid to form a single plasmid carrying the genetic information for transferring the genetically engineered T-DNA region to plant cells.
Colony hybridization	A technique for using *in situ* hybridization to identify bacteria carrying chimeric vectors whose inserted DNA is homologous with some particular sequence.
Complementary DNA	DNA synthesized by reverse transcriptase from RNA template. *See* cDNA.
Composite transposon	A transposable element formed when two identical or nearly identical transposons insert on either side of a transposable segment of DNA, such as the bacterial transposon Tn5.
Concatamer	DNA segment made up of repeated sequences linked end to end.
Consensus sequence	The nucleotide sequence that is present in the majority of genetic signals or elements that perform a specific function.
Constitutive promoter	An unregulated promoter that allows for continuous transcription of its associated gene.
Contig	A set of overlapping clones that provide a physical map of a portion of a chromosome. It refers to contiguous map.
Contig map or contiguous map	The alignment of sequence data from large, adjacent regions of the genome to produce a continuous nucleotide sequence across a chromosomal region.
Continuous culture	A suspension culture continuously supplied with nutrients by the inflow of fresh medium but the culture volume remains constant.
Copy number	The average number of molecules of a plasmid or gene per genome contained in a cell.
Cos ends	The 12-base, single strand, complementary extensions of bacteriophage lambda DNA. Also known as *cos* sites.
Cosmids	Plasmid vector into which phage lambda *cos* sites have been inserted, as a result the plasmid DNA can be packaged *in vitro* in the phage coat. This

	vector can accommodate around 40 kb fragment of DNA. Cosmids replicate as plasmids.
Cos sites	*See* cos ends.
Cot curve	When duplex DNA is heated, it dissociates into single strands. When temperature is lowered, complementary strands tend to anneal or re-nature. The extent of renaturation depends on the product of DNA concentration in moles of nucleotides per liter, and time in seconds. A graph showing the proportion of renatured DNA against cot is known as cot curve. The cot at which half the DNA has renatured is the half-cot, a parameter indicating the degree of complexity of the DNA.
Crown gall	A disease of plants in which tumor forms. The causal agent is the bacterium *Agrobacterium tumefaciens*. A tumor-inducing portion of the bacterial genome (Ti plasmid) may be used experimentally as a genetic tool to incorporate (vector) genetic information into plant cells.
Cryopreservation	Storage and preservation at ultra-low temperature of −196°C.
Cryoprotectant	An agent able to prevent freezing and thawing damage to cells.
Cybrid	A somatic hybrid in which nucleus is derived from one parent and cytoplasm is derived from both the parents.
Cytokinins	A class of plant growth regulators which cause cell division, cell differentiation, shoot differentiation, and the breaking of apical dominance. Some of the cytokinins are BAP, kinetin, zeatin, 2-iP.
Dedifferentiation	Reversion of differentiated to non-differentiated cells.
Denaturation	Conversion of DNA or RNA from double stranded to single stranded form; Mostly accomplished by heating at high temperature of 94°C.
de novo	It means from the beginning, arising, anew, afresh.
Differentiation	A process in which unspecialized cells develop structures and functions characteristic of a particular type of cell. The development of cells or tissues with a specific function and/or the regeneration of organs or organ-like structures (roots, shoots, etc.) or proembryos.
Dihaploid	An individual (denoted by 2n = 2x) which arises from a tetraploid (2n = 4x).
Diploidization	Process of doubling the chromosome complement (genome) of a cell.
Direct repeat	Two or more stretches of DNA within a single molecule which have the same nucleotide sequence in the same orientation. For example:
	TATTA...TATTA
	ATAAT...ATAAT
Disarm	To delete from a plasmid or virus genes that are cytotoxic or tumor-inducing
Disease free	This should be interpreted to mean free from any known disease.
Disinfection	Full elimination of internal microorganisms from a culture.
DNA amplification	Multiplication of a piece of DNA in a test tube into many thousands of millions of copies. The most commonly used process is PCR.
DNA chip	*See* DNA microarray
DNA fingerprinting	*See* genetic fingerprinting
DNA microarray	A small glass surface to which an array of DNA fragments, each with a defined location, is fixed. DNA fragments can be cDNA or short synthesized oligonucleotides. A solution of fluorescently labeled DNA fragments is hybridized to the chip, spots to which hybridization occurs are visible as fluorescence.

DNA polymerase	Enzyme that catalyzes the phosphodiester bonds in the formation of DNA.
Doubling time	Term used in tissue culture and shoot propagation for the time necessary to double the number of cells/shoots *in vitro*.
Downstream	A term used to describe the relative position of sequences in a nucleic acid pertaining to the direction in which the nucleic acid is synthesized i.e. towards the 3' end. For example, the coding region is downstream of the initiation codon. In molecular biology, site of initiation of transcription, is designated as +1. Thus downstream nucleotides are marked with plus signs, e.g. +2, +20.
Electrophoresis	A technique that separates charged molecules (DNA, RNA, proteins) on the basis of relative migration in a appropriate matrix (such as agarose, polyacrylamide) subjected to an electric field.
Embryogenesis	Process by which an embryo develops from a fertilized egg cell or asexually from a group of cell(s).
Embryo culture	The culture of embryos on nutrient medium.
Embryoids	Embryo-like structures produced as a consequence of differentiation processes such as embryogenesis and androgenesis.
Endomitosis	Duplication of chromosomes without division of the nucleus, resulting in increased chromosome number within a cell. Chromosome strands separate but the cell does not divide.
Endonuclease	An enzyme that catalyzes the cleavage of DNA at internal positions, normally cutting DNA at specific sites.
Enhancer sequence	A sequence found in eukaryotes and certain eukaryotic viruses, which can increase transcription of a gene when located (in either orientation) up to several kilobases from the gene concerned. These sequences usually act as enhancers when on the 5' side (upstream) of the gene in question. However, some enhancers are active when placed on the 3' side (downsteam) of the gene.
Epicotyl	The upper portion of the axis of a plant embryo or seedling above the cotyledons.
Epigenesis	Describes the developmental process whereby each successive stage of normal development is built on the foundations created by the preceding stages of development. For example, an embryo is built up from a zygote, a seedling from an embryo, and so on.
Epigenetic	A term referring to the non-genetic causes of a phenotype.
Epigenetic variation	Non-hereditary variations which is at the same time reversible; often the result of a changed gene expression.
Episome	It is a plasmid, a genetic extrachromosomal element which replicates within a cell independently of the chromosome and can integrate into the host bacterial DNA (for example F-fertility factor in *E. coli*). Plasmids and F factors are episomes.
Eukaryotes	Cellular organisms having a membrane bound nucleus within which the genome of the cell is stored as chromosome composed of DNA and proteins. For example, algae, fungi, plants and animals.
Excise	Cut out (with knife, scalpel, etc.) and prepare a tissue, organ, etc. for culture.

Exon	The coding regions of a gene that are presented in the final mRNA.
Exonuclease	An enzyme that digests DNA or RNA from the ends of strands. 5' exonucleases requires a free 5' end and degrades the molecule in 5' → 3' direction. 3' exonucleases require a free 3' end and degrade the molecule in opposite direction.
Exonuclease III	An *E. coli* enzyme that removes nucleotides from the 3' hydroxyl ends of double stranded DNA. Also known as exodeoxyribonuclease III.
Explant	A piece of tissue used to initiate tissue culture.
Exponential phase	A phase in culture in which cells undergo maximum rate of cell division. It follows the lag phase and precedes the linear growth phase in most batch propagated suspension cultures.
Expressed sequence tag (EST)	Short cDNA sequence. So called because it represents part of the sequence of an expressed gene.
Expression vector	A vector that has been constructed in such a way that inserted DNA molecule is put under appropriate promoter and terminator sequences for proper transcription and translation.
F factor	*See* episome
Filter sterilization	Process of sterilizing a liquid by passage through a filter, with pores so small that they are impervious to micro-organisms.
Flaming	A technique for sterilizing instruments by heating on a flame after dipping in alcohol.
Flow cytometry	A technique used to sort cells, nuclei or other biological materials by means of flow through apertures of defined size.
Friable	A term commonly used to describe a crumbling or fragmenting callus. A friable callus is easily dissected and readily dispersed into single cells or clumps of cells in solution.
Functional genomics	It is a study of the functions of all specific gene sequences and their expression in time and space in an organism.
Fusogen	A fusion inducing agent used for agglutination of protoplasts in somatic hybridization.
Gametophyte	Cells and tissues of the haploid stage of the life cycle of plants.
Gapped DNA	A duplex DNA molecule with one or more internal single stranded regions.
Gel electrophoresis	*See* electrophoresis.
Gene	A segment of genome with a specific sequence of nucleotides which has a specific biological function. A unit of inheritance.
Gene cloning	The process of synthesizing multiple copies of a particular DNA sequence using a bacterial cell or another organism as a host. Used in genetic engineering as cloning. *See* cloning.
Gene library	Random collection of cloned fragments in a vector that ideally includes all the genetic information of that species; For example, wheat, rice, *Arabidopsis*; sometimes called shot gun collection.
Genetically modified organism (GMO)	An organism that has been modified by the application of recombinant DNA technology.
Genetic engineering	Changes in the genetic constitution of cells by introduction or elimination of specific genes using molecular biology techniques. This is a non-sexual

	method of gene transfer. *See* gene cloning.
Genetic fingerprinting	A technique in which an individual DNA is analyzed to reveal the pattern of repetition of particular nucleotide sequences throughout the genome. The unique pattern of DNA fragments are identified by Southern hybridization or polymerase chain reaction.
Genetic map	The linear array of genes on a chromosome based on recombination frequencies (linkage map).
Genetic marker	A DNA sequence used to mark or track a particular location (locus) on a particular chromosome.
Genetics	The science of heredity and variation.
Genetic transformation	Transfer of extracellular genetic information (DNA) among and between species with the use of vectors (e.g. bacterial or viral).
Genome	1. A complete set of chromosomes (hence of genes) inherited as a (haploid) unit from one parent.
	2. The entire complement of genetic material (genes + non-coding sequences) present in each cell of an organism or in a virus or organelle.
Genomic library	A collection of clones containing the genomic DNA sequences of an organism. Typically these molecules are propagated in bacteria or virus.
Genomics	Genomics is the study of how genes and genetic information are organized within the genome, and how this organization determines their function.
Gibberellins	Group of plant growth regulators, which induce among other things, cell elongation and cell division.
Habituation	The acquired ability of cells to grow and divide independently of growth regulators.
Hairpin loop	A region of double helix formed by base-pairing within a single strand of DNA or RNA which has folded back on itself.
Haploid	Plant with half the number of chromosomes (denoted by 'n') due to reduction division of the diploid (= 2n).
Hardening off	Gradual acclimatization of *in vitro* grown plants to *in vivo* conditions. Adapting plants to outdoor conditions by gradually withholding water, lowering the temperature, increasing light intensity, or reducing the nutrient supply.
Heredity	Transmission of traits from parents to offspring.
Heteroduplex	A double stranded DNA molecule or DNA-RNA hybrid formed by base-pairing between two strands that do not have completely complementary nucleotide sequences.
Heterogeneous nuclear RNA (hnRNA)	Large RNA molecules, which are unedited mRNA transcripts, found in the nucleus of an eukaryotic cell.
Heterokaryon	A cell in which two or more nuclei of unlike genetic make-up are present.
Heterologous probe	A DNA probe that is derived from one species and used to screen for a similar DNA sequence from another species.
Homokaryon	Cell with two or more identical nuclei as a result of fusion.
Homopolymer	A nucleic acid strand that is composed of one kind of nucleotides, for example, CCCCCCC.
Homopolymeric tailing	*See* tailing.
Hydrogen bond	A relatively weak bond formed between a hydrogen atom and a nitrogen

	or oxygen atom with an unshared electron pair.
Hypocotyl	Portion of an embryo or seedling below the cotyledons, which is a transitional area between stem and root.
Immobilized cells	Cells entrapped in matrices such as alginate, agarose, polyacrylamide designed for use in bioreactors for production of metabolites.
Inoculation	Placing of an explant in or on a nutrient medium.
Insertion element or insertion sequence (IS)	Generic term for DNA sequences found in bacteria capable of genome insertion. IS carries only the genetic functions involved in transposition.
Insertion site	A unique restriction site in a vector DNA molecule into which foreign DNA can be inserted. It is also used to describe the position of integration of a transposon or insertion site element.
In situ	In Latin means 'in place'. Meaning in the natural place or in the original place.
Integrating vector	A vector that is designed to integrate cloned DNA into host cell chromosomal DNA.
Intercalating agent	A chemical capable of inserting between adjacent base pairs in a DNA molecule.
Intergeneric	A cross between two different genera.
Intergenic regions	DNA sequences located between genes. They comprise a large percentage of a genome with no known function.
Intervening sequence	*See* intron.
Interspecific	A cross between two different species of a genus.
Intron or intervening sequence	Sequence of DNA interrupting the coding sequence of a gene. This is the portion of a gene that is transcribed but does not appear in the final mRNA transcript.
Inverted repeat	Two regions of a nucleic acid molecule which have the same nucleotide sequence but in an inverted orientation. Inverted repeats are known as palindromic sequences, such as

$$5' \; GCACTTG... \qquad\qquad ...CAAGTGC \quad 3'$$
$$3' \; CGTGAAC... \qquad\qquad ...GTTCACG \quad 5'$$

In vitro	Literally means "in glass". Living in test tubes, outside the organism or in an artificial environment. Now applied for the growth of a tissue in any type of culture container.
In vivo	Literally means "in living". The natural condition in which organisms reside. Refers to biological processes that take place within a living organism or cell under normal conditions.
Isoelectric focusing	Variant of gel electrophoresis, which separates macromolecules on the basis of their isoelectric point rather than their size.
Isoenzymes	Multiple molecular forms of an enzyme exhibiting similar or identical catalytic properties.
Jumping genes	*See* transposon.
Jumping library	*See* chromosome jumping.
Juvenile phase	Early phase of development in which an organism is juvenile and incapable of sexual reproduction. The period in the life of a plant during which no flowering can be induced. During the juvenile phase a plant has often very special characteristics (morphology, physiology etc.) which are different from the adult phase.

kb	Kilobase used interchangeably with kilobase pairs. A length unit equal to 1000 bases (in RNA) or base pairs (in DNA). DNA sizes are often expressed in kilobases.
Klenow fragment	Piece obtained from DNA polymerase I of *E. coli* by proteolytic cleavage; it has both polymerase and $3' \rightarrow 5'$ exonuclease activities but lacks the $5' \rightarrow 3'$ exonuclease activity.
Lag phase	It describes the first of the five growth phases of most batch propagated cell suspension/callus cultures in which inoculated cells in fresh medium adapt to the new environment and prepare to divide.
Lagging strand	The strand of DNA that is synthesized discontinuously during replication, because DNA synthesis can proceed only in the $5' \rightarrow 3'$ direction.
Laminar air flow cabinet	Cabinet for inoculation, which is kept sterile by a continuous non-turbulent flow of sterilized air.
Leader sequence	It is the non-translated sequence at the 5' end of mRNA that precedes the initiation codon.
Leading strand	The strand of DNA that is synthesized continuously during replication.
Library	A collection of cloned DNA fragments. The library may consist of cDNAs or genomic clones (fragments cloned directly from cellular DNA).
Ligation	Joining of DNA fragments to produce a single DNA molecule by the formation of phosphodiester bonds. Ligases are enzymes which perform this function.
LINEs (Long interspersed nuclear elements)	Families of long (~6500 bp), moderately repetitive (about 10,000 copies). LINEs are cDNA copies of functional genes present in the same genome, also known as processed pseudogenes.
Linker	A synthetic double stranded oligonucleotide that carries the sequence for one or more restriction endonuclease sites useful for gene splicing.
Liposomes	Artificial phospholipid vesicles.
Log phase	Also known as logarithmic phase or exponential growth phase. The steepest slope of the growth curve, the phase of vigorous growth during which cell number doubles every 20–30 minutes.
Long terminal repeat (LTR)	A string of bases that occurs at each end of the genome of a retrovirus that has become integrated into the host genome. Involved in the integration process.
Lux	The unit of measurement for illuminance.
M13	A single stranded DNA bacteriophage used as a vector for DNA sequencing.
Macronutrient	An essential element required by plants in relatively large quantities.
Map distance	The standard measure of distance between loci, expressed in centiMorgans (cM).
Marker assisted selection (MAS)	The use of DNA markers to increase the response to selection in a population.
Megabase (Mb)	A length of DNA consisting of 10^6 base pairs (if double stranded) or 10^6 bases (if single stranded). 1 Mb = 10^3 kb = 10^6 bp.
Megabase cloning	The cloning of very large DNA fragments. *See* cloning.
Melting temperature (*Tm*)	The temperature at which a double stranded DNA or RNA molecule denatures into separate single strands. The *Tm* is characteristic of each DNA species and gives an indication of its base composition. DNAs rich

in G:C base pairs are more resistant to thermal denaturation than A:T rich DNA.

Mericloning
A propagation method using meristem shoot tips in culture to proliferate multiple buds, which can then be separated, rooted and planted out.

Meristem
Undifferentiated but determined tissue, the cells of which are capable of active cell division and differentiation into specialized and permanent tissue such as shoots and roots. The principal meristems in the flowering plants occur at the tips of the stems and roots (apical meristems), between xylem and phloem of vascular bundles (cambium) in the cortex (cork cambium), in young leaves and, in many grasses, at the bases of internodes (intercalary meristems).

Meristemoid
A localized group of cells in callus tissue, characterized by an accumulation of starch, RNA and proteins, and giving rise to adventitious shoots or roots.

Messenger RNA
The RNA that specifies the amino acid sequence for a particular polypeptide chain.

Microarray
See DNA microarray.

Micronutrient
An essential element required by plants in relatively small quantity.

Micropropagation
In vitro clonal propagation of plants.

Microsatellite
A form of variable number of tandem repeat in which the repeat units typically range from 1 to 10 bases.

Microspore
Immature pollen grain.

Minisatellite
A form of variable number of tandem repeat in which the repeat units typically range from 10 to 100 bases.

Monoploid
A cell or individual which has one genome denoted by x and is the lowest number of chromosomes of a polyploid series. For example, barley has n = x = 7, *Triticum aestivum* has x = 7 and n = 21.

Morphogenesis
Developmental pathways in differentiation which results in the formation of recognizable tissues.

Multicopy plasmids
These plasmids are present in bacteria at amounts greater than one per chromosome.

Mutant
A cell which has undergone a heritable change which has resulted due to a change in its genes or chromosomes.

Nanometer
One millionth of a milliliter or 10^{-9} m. A millimicron which is equal to ten angstroms.

Nick translation
A procedure for labeling DNA *in vitro* using DNA polymerase I.

Non-template strand
Also known as sense or coding strand of DNA. In transcription, the non-transcribed strand of DNA.

Northern blotting
A technique for the transfer of RNA from an agarose gel to a nitrocellulose filter on which it can be hybridized to a complementary DNA.

Northern hybridization
Hybridization of a labeled DNA probe to RNA fragments that have been transferred from an agarose gel to a nitrocellulose filter. *See* northern blotting.

Nuclease
A class of enzymes that degrades DNA or RNA molecules by cleaving phosphodiester bonds that link adjacent molecules.

Okazaki fragments
The short stretches of 1000–2000 bases produced during discontinuous replication which are later joined into a covalently intact strand.

Oligomer	A molecule formed from a small number of monomers.
Oncogene	A gene that causes cells to grow in an uncontrolled manner, and that causes cancer.
Oncomouse	A mouse that has been genetically modified to incorporate an oncogene; oncogenes causes cells to undergo cancerous transformation.
Ontogeny	Developmental life history of an organism.
Open continuous culture	A continuous culture in which inflow of fresh medium is balanced by outflow of a corresponding volume of culture. Cells are constantly washed out with the outflowing liquid.
Open reading frame	The part of a protein coding gene that is translated in to protein. Contains a series of triplets coding for amino acids with no termination codons. Sequence is (potentially) translatable into protein. This term is often used when, after the sequence of a DNA fragment has been determined, the function of the encoded product is not known.
Operator	A site on DNA at which a repressor or activator protein binds to control the transcription from initiating at an adjacent promoter.
Operon	A number of contiguous genes under coordinate control (usually in prokaryotes). It consists of one or more genes that encode one or more polypeptide(s) and the adjacent site (promoter and operator) that controls their expression by regulating the transcription of the structural genes.
Organ culture	Culture of an organ *in vitro* in a way that allows development and/or preservation of the originally isolated organ.
Organogenesis	Type of morphogenesis which results in the formation of organs viz. shoots, roots either adventitious or *de novo* origin from callus or cell suspension cultures.
Origin of replication	The nucleotide sequence at which initiation of DNA synthesis (replication) take place.
Osmosis	Diffusion from areas of high concentration to areas of lower concentration of a solvent through a differentially permeable membrane.
Osmotic potential (value)	Potential brought about by dissolving a substance, especially in water.
Osmoticum	Agents such as PEG, mannitol, glucose, sucrose, etc. which maintain the osmotic potential of a nutrient medium equivalent to that of cultured cells. Because of this osmotic equilibrium, cells are not damaged *in vitro*.
PAC	P1 derived artificial chromosome; a cloning system for isolating genomic DNA based on the F factor plasmid, as in BACs, but also containing some of the elements of the bacteriophage P1 cloning system.
P1 clones	A cloning system for the isolating genomic DNA that uses elements from bacteriophage P1 (i.e. recombination sites, loxP and a packaging site, pac, etc.).
Packed cell volume (PCV)	A test for determining the viability of cells. It is the percentage volume of cells in a set volume of culture after sedimentation (packing) by means of low speed centrifugation.
Palindrome	It is a sequence of DNA that is the same when one strand is read left to right or the other is read right to left; consists of adjacent inverted repeats.
Parthenogenesis	Production of an embryo from a female gamete without the participation of a male gamete.

pBR322	A standard plasmid cloning vector. One of the first plasmid vectors widely used for cloning DNA in *E. coli*.
pH	The negative logarithm of the hydrogen ion concentration of any solution for measuring acidity or alkalinity.
Phagemid	Cloning vectors that contain components derived from both phage chromosome and plasmids.
Photoperiod	The length of day or period of daily illumination provided or required by plants for reaching the reproductive stage.
Physical map	A map showing physical locations on a DNA sequence, such as restriction sites and sequence tagged sites.
Phytohormone	The generic name for all classes of substances that stimulate growth or other processes in plants.
Picogram	10^{-12} grams, a unit commonly used to express the DNA content per cell or per nuclear genome in a plant.
Plasmid	An extrachromosomal, autonomous circular DNA molecule found in certain bacteria, capable of autonomous replication. Plasmids can transfer genes between bacteria and are important tools of transformation in genetic engineering. They may be present in one to many copies per cell and are transferred independently of host chromosomes.
Plating efficiency	The percentage of cells plated which give rise to cell colonies.
Polyacrylamide gel electrophoresis (PAGE)	A method for separating nucleic acid or protein molecules according to their size. The molecules migrate through the inert gel matrix under the influence of an electric field. *See* electrophoresis.
Polyadenylation	Post-transcriptional addition of polyadenylic acid tail to the 3' end of eukaryotic mRNA.
Polycistronic mRNA	A mRNA that includes coding regions of more than one gene.
Polyembryony	When two or more embryos are formed after fertilization.
Polylinker	A segment of DNA that contains a number of different restriction endonuclease sites. Also known as multiple cloning site (MCS).
Polymerase chain reaction (PCR)	An *in vitro* procedure that amplifies enzymatically a particular DNA sequence which is flanked by two oligonucleotide (deoxy) primers that share identity to the opposite DNA strands. It involves multiple cycles of denaturation, annealing of oligo primers and synthesis of DNA using a thermostable DNA polymerase.
Polysomes	Complexes of ribosomes bound together by a single mRNA molecule; also known as polyribosomes.
Positional cloning	The process that commences with searching for markers linked to a particular inherited trait, then using those markers to identify the approximate location of the gene responsible for the trait, and then using various cloning strategies to identify, isolate and characterize the gene.
Pribnow box	Consensus sequence near the mRNA start point of prokaryotic genes. *See* TATA box.
Primary culture	A culture started from cells, tissues, or organs taken directly from an organism. A primary culture may be regarded as such until it is subcultured for the first time. It then becomes a cell line.
Primer	It is a short sequence (often of RNA) that is paired with one strand of DNA

	and provides a free 3' OH end at which a DNA polymerase starts synthesis of a deoxyribonucleotide chain.
Primordium	The earliest detectable stage of an organ.
Probe	A labeled nucleic acid molecule used to detect the presence of its complementary strand by hybridization.
Processed pseudogene	A copy of functional gene, which has no promoter, no introns and is not itself transcribed. Pseudogenes are thought to originate from the integration into the genome of cDNA copies synthesized from mRNA molecules by reverse transcriptase.
Proembryo	Early embryonic stage. A group of cells arising from the division of the fertilized egg cell or somatic embryo before those cells which are to become the embryo are recognizable.
Promoter	Region of DNA which is recognized by RNA polymerase in order to initiate transcription.
Promoter –10 sequence	It is the consensus sequence TATAATG centered about 10 bp before the startpoint of a bacterial gene. It is involved in the initial melting of DNA by RNA polymerase.
Promoter –35 sequence	It is the consensus sequence centered about 35 bp before the startpoint of a bacterial gene. It is involved in initial recognition by RNA polymerase.
Proteome	It refers to the total protein complement expressed by the genome.
Proteomics	It is a study of the structure of all proteins encoded in a fully sequenced genome.
Protoplast	A plasmalemma bound vesicle consisting of a naked cell formed as a consequence of the removal of cell wall by mechanical or enzymatic means.
Pulsed field gel electrophoresis (PFGE)	A procedure used to separate very large DNA molecules by alternating the direction of electric current in a pulsed manner across a gel.
Quantitative trait locus (QTL)	A locus that affects a quantitative trait. The plural form (quantitative trait loci) is also abbreviated as QTL.
Random amplified polymorphic DNA (RAPD)	A PCR based technique for detecting polymorphism at DNA level. In RAPD uses single short (usually 10-mer) synthetic oligonucleotide primers for PCR.
Recombinant DNA	DNA molecules in which sequences that are not naturally contiguous to each other are placed next to each other by *in vitro* manipulation. The result of combining DNA fragments from different sources.
Regeneration	In tissue culture, it is used for development of new organs or plantlets from a tissue, callus culture or from a bud.
Regulatory gene	A gene that codes for an RNA or protein product whose function is to control the expression of other genes.
Relaxed replication control	It refers to the ability of some plasmids to continue replicating after bacteria cease dividing.
Renaturation	The reassociation of two nucleic acid strands after denaturation.
Repeated DNA sequence	A sequence of nucleotides which occurs more than once in a genome. Repeated sequences may be present in a few to many millions of copies. The individual repeated sequence may be only a few nucleotides in length up to several kb.
Repetitive DNA	DNA sequences that are present in a genome in multiple copies, sometimes a million times or more.

Replicon	The segment of the genome in which DNA is replicated and by, definition, contains an origin of replication.
Reporter gene	It is the coding unit of a test gene whose product is easily assayed during genetic transformation (e.g. GUS, chloramphenicol transacetylase, etc.). It may be connected to any promoter of interest so that expression of the gene can be used to assay promoter function.
Restriction enzyme	An endonuclease with the ability to cleave DNA at the point where a certain base sequence occurs.
Restriction site	A DNA base sequence recognized by a restriction enzyme.
RFLP	Restriction fragment length polymorphism. The occurrence of variation in the length of DNA fragments that are produced after cleavage with a type II restriction endonuclease. The differences in DNA lengths are due to the presence or absence of recognition site(s) for that particular restriction enzyme.
Restriction map	It is a linear array of sites on DNA cleaved by various restriction enzymes.
Retroelement	Any of the integrated retroviruses or the transposable elements that resemble them.
Retroposon; retrotransposon	A transposable element that moves via reverse transcription (i.e. from DNA to RNA to DNA) but lacks the long terminal repeat sequences.
Retroviruses	A class of eukaryotic RNA viruses that can form double stranded DNA copies of their genome by using reverse transcription; the double stranded form integrates into chromosomes of an infected cell.
Reverse transcriptase	An enzyme that uses RNA molecule as a template for the synthesis of a complementary DNA strand. Also known as RNA dependent DNA polymerase.
Ribonuclease	Any enzyme that hydrolyzes RNA.
Ribosomal binding site	A conserved sequence of nucleotides near the 5' end of a bacterial mRNA molecule that facilitates the binding of the mRNA to the small ribosomal subunit. Also known as Shine-Delgarno sequence.
Ribozyme	RNA molecule that can catalyze chemical reactions, often cutting other RNAs.
RNase (Ribonuclease)	A group of enzymes that catalyze the cleavage of nucleotides in RNA.
S1 nuclease	An enzyme that specifically degrades RNA or single stranded DNA to 5' mononucleotides.
Satellite DNA	It consists of highly repetitious sequences (million of copies) typically in the range of 5 to 500 bases (identical or related).
Scarification	The chemical or physical treatment given to some seeds (where the seed coats are very hard or contain germination inhibitors) in order to break or weaken the seed coat sufficiently to permit germination.
Secondary metabolites	Metabolites synthesized by plants besides primary metabolites, which are not essential for its survival (e.g. alkaloids, steroids).
Selectable marker	A gene whose expression allows the identification of: 1) a specific trait or gene in an organism. 2) Cells that have been transformed or transfected with a vector containing the marker gene.

Sense RNA	A primary transcript (RNA) that contains a coding region (contiguous sequence of codons) that is translated to produce a polypeptide.
Sepsis	Destruction of tissue by pathogenic microorganisms or their toxins, especially through infection of a wound.
Sequence tagged site (STS)	Short, unique DNA sequence (usually 200 to 500 bp) that by being able to be amplified by PCR is uniquely tagged to the site on the chromosome from which it is amplified.
Shine-Delgarno sequence	A conserved sequence in prokaryotic mRNAs that is complementary to a sequence near the 5' terminus of the 16S ribosomal RNA and is involved in the initiation of translation. *See* ribosomal binding site.
Shoot tip	Terminal (0.5–2.0 mm) portion of a shoot comprising the meristem dome and subjacent leaf and stem tissues.
Shot gun collection	Cloning of an entire genome in the form of randomly generated fragments.
Shuttle vector	A plasmid constructed to have origins for replication for two hosts (e.g. *E. coli* and *S. cerevisiae*) so that it can be used to carry a foreign sequence in either prokaryotes or eukaryotes. It can therefore be used to 'shuttle' gene(s) from one to another. For example, the Yep, pJDB219 a shuttle vector with *E. coli* and *S. cerevisiae* origin.
Simple sequence repeat	*See* microsatellite.
SINEs	Short interspersed nuclear elements. Families of short (150 to 300 bp), moderately repetitive elements of eukaryotes, occurring about 100,000 times in a genome.
Single copy sequence	A sequence of nucleotides that occurs only once in a genome.
Single node culture	Culture of separate lateral buds, each carrying a piece of stem tissue.
Site directed mutagenesis	The introduction of base changes – mutations – into a piece of DNA at a specific site, using recombinant DNA methods.
Site specific mutagenesis	A technique to change one or more specific nucleotides within a cloned gene in order to create an altered form of a protein with one or more specific amino acid changes. Also known as oligonucleotide directed mutagenesis.
Small nuclear DNA (snRNA)	Short RNA transcripts of 100–300 bp that associate with proteins to form small nuclear ribonucleoprotein particles (snRNPs). Most snRNAs are components of spliceosomes that excise introns from pre mRNAs in RNA processing.
Somaclonal variation	The variability generated by the use of a tissue culture cycle.
Somatic	It refers to the vegetative or non-sexual stages of a life-cycle.
Somatic embryogenesis	It is the formation of a bipolar structure containing both shoot and root meristems either directly from the explant (adventitive origin) or *de novo* origin from callus and cell culture induced from the explant.
Somatic hybridization	A technique of fusing protoplasts from two contrasting genotypes for production of hybrids or cybrids which contain various mixtures of nuclear and/or cytoplasmic genomes, respectively.
Southern transfer (blot)	Transfer of DNA from an electrophoresis gel into a nitrocellulose or nylon membrane. The membrane is then used in hybridization experiments.
Southern hybridization	A procedure in which a cloned, labeled DNA is hybridized to DNA fragments on a Southern blot.

Splicing	Removal of introns from the primary transcript and covalently joining of exons during the maturation of eukaryotic mRNA.
Split gene	In eukaryotes, structural genes are typically divided up (split) by a number of non-coding regions called introns.
Sterile	Medium or object with no perceptible or viable microorganisms.
Sterilization	Procedure for the elimination of microorganisms.
Sticky ends	Single stranded ends left on a restriction fragment by many type II restriction endonucleases. Also known as cohesive ends, protruding ends, overhang.
Structural gene	A gene codes for any RNA or polypeptide product other than a regulator.
Structural genomics	It involves a study of the structure of all gene sequences in a fully sequenced genome.
Subculture	Subdivision of a culture and its transfer to fresh medium.
Surface sterilization	Destruction of microorganisms from the surface of the plant parts with the help of chemical agent.
Suspension culture	A culture consisting of cells or cell aggregates initiated by placing callus tissues or sometimes seedlings in an agitated liquid medium.
Synchronous culture	A culture in which the cell cycles (or a specific phase of the cycle) of the majority of cells are synchronous.
TATA box	Eukaryotic promoter region analogous to the Pribnow box. It is a conserved A*T rich septamer found about 25–30 bp upstream of a gene, which is the target site of eukaryotic RNA polymerase II.
Tailing	The *in vitro* addition of same nucleotide by the enzyme terminal transferase, to the 3'-hydroxyl ends of a duplex DNA molecule. Also known as homopolymeric tailing.
Tandem array	The existence of two or more identical DNA sequences in series, i.e. end to end.
Taq polymerase	A heat stable DNA polymerase isolated from thermophilic bacterium *Thermus aquaticus*, and used in PCR. *See* Polymerase chain reaction.
Target site duplication	A sequence of DNA that is duplicated when a transposable element inserts; usually found at each end of the insertion.
Ti plasmid	Tumor-inducing plasmid often responsible for crown gall (tumor) induction in plants. Ti plasmids are used as vectors to introduce foreign DNA into plant cells. *See* vector.
Tautomeric shift	The transfer of a hydrogen atom from one position in an organic molecule to another position.
T-DNA	The segment of DNA in Ti plasmid of *Agrobacterium tumefaciens* that is transferred to plant cells and inserted into the chromosomes of the plant.
T4 DNA ligase	An enzyme from bacteriophage T4 infected cells and that catalyses the joining of duplex DNA molecules and repairs nicks in DNA molecules. The enzyme requires that one of the DNA molecules has a 5' phosphate group and the other DNA molecule has a 3' hydroxyl group.
Template	An RNA or single stranded DNA molecule on which a complementary nucleotide strand is synthesized. Also known as antisense strand.
Template strand	The polynucleotide strand that a polymerase uses for determining the sequence of nucleotides during the synthesis of a new nucleic acid strand.

Terminal transferase	An enzyme that adds nucleotides to the 3' terminus of DNA molecules.
Thermolabile	A substance which disintegrates or is unstable upon heating.
Tissue	A group of cells that performs a collective function.
Tissue culture	The *in vitro* culture of protoplasts, cells, tissues, organs, embryos or seeds on a nutrient medium under aseptic conditions.
Totipotency	The ability of individual cells to express the phenotype of the whole plant from which it is derived.
Trailer	Non-translated sequence at the 3' end of a messenger RNA following the termination codon.
Trans acting	A term describing substances that are diffusable and that can affect spatially separated entities within cells.
Transcription	Process through which an RNA is formed along a DNA template. The enzyme DNA dependent RNA polymerase catalyses the formation of RNA.
Transfection	The transfer of DNA to an eukaryotic cell.
Transferred DNA	*See* T-DNA.
Transformant	A cell that has been genetically altered through the uptake of foreign DNA.
Transformation	Introduction of DNA into the cells of an organism by a method other than conventional sexual crossing.
Transgene	A gene from one genome that has been incorporated into the genome of another organism. Often refers to a gene that has been introduced into a multicellular organism.
Transgenesis	The introduction of a gene or genes into animal or plant cells, which leads to the transmission of the input gene (transgene) to successive generations.
Transgenic	An organism in which a foreign gene (a transgene) is incorporated into its genome. The transgene is present in both somatic and germ cells and is inherited by offspring in a Mendelian fashion.
Transient	Of short duration.
Translation	Copying of mRNA into polypeptide.
Transposase	An enzyme coded by a transposon gene and that catalyses the movement of a DNA sequence to a different site in a DNA molecule, i.e. enzyme activity involved in insertion of transposon at a new site.
Transposition	The process whereby a transposon or insertion sequence inserts itself into a new site on the same or another DNA molecule.
Transposon	A movable genetic element. A DNA element which has the ability to move from one chromosomal position to another. It can insert at random into plasmids or the bacterial chromosome independently of the host cell recombination system.
Transposon tagging	The insertion of a transposable element into or nearby a gene, thereby marking that gene with a known DNA sequence.
Turbidostat	An open continuous culture into which fresh medium flows in response to an increase in the turbidity of culture.
Upstream	It identifies sequences proceeding in the opposite direction from expression. In molecular biology, the stretch of DNA base pairs that lie in the 5' direction from the site of initiation of transcription. Usually the first transcribed base is designated as +1 and the upstream nucleotides are marked with minus

signs. For example, the promoter is upstream from the transcription unit, the initiation codon is upstream of the coding region.

Vector
A plasmid, phage, virus or bacterium used to deliver selected foreign DNA for cloning and in gene transfer.

Vegetative propagation
The asexual propagation of plants by detachment of some parts of the plant, e.g. a cutting, and its subsequent development into a complete plant.

Vermiculite
Coarse aggregate material made from expanded mica having a high cation exchange capacity and high water holding capacity. It is used as a rooting medium and a soil additive.

Viability
The capacity to live and develop normally.

Viable
Capable of germinating, living, growing and developing.

Virion
An infectious virus particle. A plant pathogen that consists of a naked RNA molecule of approximately 250–350 nucleotides, whose extensive base pairing results in a nearly correct double helix.

Viroid
An infectious entity similar to a virus but smaller, consisting only of a strand of nucleic acid without the protein coat characteristic of a virus.

Virulence
The degree of ability of an organism to cause disease.

Virus
An infectious particle composed of a protein capsule and a nucleic acid core (DNA or RNA), which is dependent on a host organism for replication.

Virus-free
Plant, animal, cell, tissue or meristem which exhibits no viral symptoms or contains no identifiable virus particles.

Vitrified, water-soaked
Cultured tissue having leaves and sometimes stems with a glassy, transparent or wet and often swollen appearance.

Vitrification
An undesirable condition of *in vitro* tissues characterized by succulence, brittleness and glassy appearance. The process of vitrification is a general term for a variety of physiological disorders that lead to shoot tip and leaf necrosis.

Western blot
A technique in which protein is transferred from an electrophoretic gel to a cellulose or nylon support membrane following electrophoresis. A particular protein molecule can then be identified by probing the blot with a radiolabeled antibody which binds only the specific protein to which the antibody was prepared.

Wetting agent
A substance that improves surface contact by reducing the surface tension of a liquid. For example, Triton X-10™ added to disinfecting solutions, promotes the disinfection process.

Wide hybridization
When individuals from two different species of the same genus or from two different genera are crossed.

Wobble hypothesis
An explanation of how one tRNA may recognize more than one codon. The first two bases of the mRNA codon and anticodon pair properly, but the third base in the anticodon has some wobble that permits it to pair with more than one base.

x-number
The basic number of chromosomes in a species. Monoploid = x

YAC
Yeast artificial chromosome; a cloning system in yeast comprising an autonomously replicating linear vector containing an exogenous DNA insert, flanked by a yeast centromere and two telomere seeding sequences. They can accommodate foreign DNA inserts up to 1000 kb.

Yeast episomal vector (Yep)	A cloning vector for the yeast that uses 2 μm plasmid as origin of replication and is maintained as an extrachromosomal nuclear DNA molecule.
Z-DNA; zig-zag DNA	A form of DNA duplex in which the double helix is wound in a left hand, instead of a right hand, manner. DNA adopts the Z configuration when purines and pyrimidines alternate on a single strand. For example

5' GCGCGCGC 3'   5' CACACACACA 3'
3' CGCGCGCG 5'   3' GTGTGTGTGT 5'

Zoo blot	Hybridization of cloned DNA from one species to DNA from other organisms to determine the extent to which the cloned DNA is evolutionary conserved.
Zygote	A product of the fusion of male and female gametes.

Author Index

Subject Index

Italic page nos. indicates Figures.
Bold page nos. indicates Glossory.